AF856240

Florida's

B.E.S.T.

Standards for

MATH

with CalcChat® and CalcView®

Grade 6

Accelerated
Volume 2

Ron Larson
Laurie Boswell

Big Ideas Learning™

Erie, Pennsylvania
BigIdeasLearning.com

Big Ideas Learning, LLC
1762 Norcross Road
Erie, PA 16510-3838
USA

For product information and customer support, contact Big Ideas Learning at **1-877-552-7766** or visit us at ***BigIdeasLearning.com***.

Cover Design:
Betsi Santos and Mary Rose

Front Matter:
bgblue/DigitalVision Vectors/Getty Images

Printed in the U.S.A.

ISBN 13: 978-1-63896-311-0

5 6 7 8 9—25

A Single Authorship Team

Written by renowned authors, Dr. Ron Larson and Dr. Laurie Boswell, *Florida's B.E.S.T. Standards for MATH* offers a seamless math pedagogy from Kindergarten through Algebra 2. Together, Ron and Laurie provide a consistent voice that encourages students to make connections through cohesive progressions and clear instruction. Since 1992, Ron and Laurie have authored over 50 mathematics programs.

"Each time Laurie and I start working on a new program, we spend time putting ourselves in the position of the reader. How old is the reader? What is the reader's experience with mathematics? The answers to these questions become our writing guides. Our goal is to make the learning targets understandable and to develop these targets in a clear path that leads to student success."

Ron Larson, Ph.D., is well known as lead author of a comprehensive and widely used mathematics program that ranges from elementary school through college. He holds the distinction of Professor Emeritus from Penn State Erie, The Behrend College, where he taught for nearly 40 years. He received his Ph.D. in mathematics from the University of Colorado. Dr. Larson engages in the latest research and advancements in mathematics education and consistently incorporates key pedagogical elements to ensure focus, coherence, rigor, and student self-reflection.

"My passion and goal in writing is to provide an essential resource for exploring and making sense of mathematics. Our program is guided by research around the learning and teaching of mathematics in the hopes of improving the achievement of all students. May this be a successful year for you!"

Laurie Boswell, Ed.D., is the former Head of School at Riverside School in Lyndonville, Vermont. In addition to authoring textbooks, she provides mathematics consulting and embedded coaching sessions. Dr. Boswell received her Ed.D. from the University of Vermont in 2010. She is a recipient of the Presidential Award for Excellence in Mathematics Teaching and later served as president of CPAM. Laurie has taught math to students at all levels, elementary through college. In addition, Laurie has served on the NCTM Board of Directors and as a Regional Director for NCSM. Along with Ron, Laurie has co-authored numerous math programs and has become a popular national speaker.

1 Numerical Expressions and Factors

2 Fractions and Decimals

3 Integers, Number Lines, and the Coordinate Plane

4 Operations with Integers

5 Operations with Rational Numbers

6 Algebraic Expressions and Properties

7 Equations and Inequalities

8 Ratios and Rates

9 Percents

10 Area, Surface Area, and Volume

11 Statistical Measures and Data Displays

12 Probability and Populations

7 Equations and Inequalities

Chapter Learning Target:
Understand equations.

Chapter Success Criteria:
- ◆ I can identify key words and phrases that indicate equality.
- ◆ I can write word sentences as equations or inequalities.
- ■ I can solve equations and inequalities.
- ■ I can use equations or inequalities to model real-life problems.

◆ Surface ■ Deep

Rock Climbing

GO DIGITAL

7 MTR *Mathematicians who apply mathematics to real-world contexts connect mathematical concepts to everyday experiences.*

STEAM Video — Rock Climbing

1. How can you use pitches to estimate the amount of time it will take to climb a rock wall?

In the video, Alex and Tory visit a rock-climbing site and learn about the mathematics of rock climbing.

3 MTR 2. **CHOOSE A METHOD** Are there any other methods you could use to estimate the amount of time it will take to climb a rock wall? Explain.

3. You know two of the three pieces of information below. Explain how you can find the missing piece of information.

    Average climbing speed

    Height of rock wall

    Time to complete climb

STEAM Performance Task — Planning the Climb

After completing this chapter, you will be able to use the concepts you learned to answer the questions in the *STEAM Performance Task*. You will be given information about two rock-climbing routes.

Route 1: 500 feet, 125 feet per pitch

Route 2: 1200 feet, 8 pitches

You will find the average speed of the climbers on Route 1 and the amount of time it takes to complete Route 2.

GO DIGITAL

Getting Ready for Chapter 7 WITH CalcChat®

Chapter Exploration

1 MTR **Actively Participate in Effortful Learning Collectively**
Work with a partner to prepare for concepts in this chapter.

Every equation that has an unknown variable can be written as a question. Write a question that represents the equation. Then answer the question.

	Equation	Question	Answer to Question
1.	$x + 3 = 7$		
2.	$5 - x = 2$		
3.	$3x = 12$		
4.	$x \div 5 = 3$		
5.	$20 \div x = 4$		

Write an equation that represents the question. Then answer the question.

	Question	Equation	Answer to Question
6.	What number can be added to 7 to get 12?		
7.	What number can be subtracted from 11 to get 3?		
8.	What number can be multiplied by 10 to get 30?		
9.	What number can be divided by 7 to get 3?		

7 MTR **10.** **MODELING REAL LIFE** Your friend says that he will be 21 years old in 7 years. How old is he now?

Vocabulary

The following terms are defined in this chapter. Think about what each might mean and record your thoughts.

equation

inverse operations

solution

inequality

GO DIGITAL

7.1 Writing Equations in One Variable

Learning Target: Write equations in one variable and write equations that represent real-life problems.

Success Criteria:
- I can identify key words and phrases that indicate equality.
- I can write word sentences as equations.
- I can create equations to represent real-life problems.

Algebraic Reasoning

MA.6.AR.2.2 Write and solve one-step equations in one variable within a mathematical or real-world context using addition and subtraction, where all terms and solutions are integers.

MA.6.AR.2.3 Write and solve one-step equations in one variable within a mathematical or real-world context using multiplication and division, where all terms and solutions are integers.

Exploration 1 Writing Equations

5 MTR MAKE A PLAN

What information do you need to solve part (b)?

Work with a partner. Customers order drinks at a coffee shop from the menu board shown.

a. The equation $4x = 16$ represents the purchase of one customer from the menu board. What does the equation tell you about the purchase? What cannot be determined from the equation?

b. The four customers in the table buy multiple drinks of the same type. For each customer, find the total cost and write an equation that represents the situation. Then determine how many drinks each customer buys. Explain your reasoning.

	Drink	Amount Used for Payment	Change Received
Customer A	12 oz Latte	$15	$3
Customer B	Americano	$10	$4
Customer C	Coffee	$20	$12
Customer D	32 oz Cold Brew	$20	$10

GO DIGITAL

7.1 Lesson

Key Vocabulary
equation, *p. 494*

An **equation** is a mathematical sentence that uses an equal sign, =, to show that two expressions are equal.

Expressions	***Equations***
$4 + 8$	$4 + 8 = 12$
$x + 8$	$x + 8 = 12$

To write a word sentence as an equation, look for key words or phrases such as *is*, *the same as*, or *equals* to determine where to place the equal sign.

Example 1 Writing Equations

Write each word sentence as an equation.

a. The sum of a number n and 7 is 15.

The sum of a number n and 7 is 15.

$n + 7 \quad = 15$ *Sum of* means *addition*.

▶ An equation is $n + 7 = 15$.

b. A number y decreased by 4 is -3.

A number y decreased by 4 is -3.

$y - 4 \quad = -3$ *Decreased by* means *subtraction*.

▶ An equation is $y - 4 = -3$.

c. 48 equals 12 times a number p.

48 equals 12 times a number p.

$48 \quad = \quad 12p$ *Times* means *multiplication*.

▶ An equation is $48 = 12p$.

GO DIGITAL

Try It

Write the word sentence as an equation.

1. 9 less than a number b equals 2.

2. The product of a number g and -5 is 30.

3. A number k increased by 10 is the same as -24.

4. The quotient of a number q and 4 is 12.

5. -2 is the same as the sum of a number w and 9.

Example 2 B.E.S.T. Test Prep: Writing an Equation

Ten workers decorate 25 tables for a wedding. Each table is decorated as shown. Let c be the total number of yellow and purple candles. Which equation can you use to find c?

Ⓐ $c = 25 + (4 \times 6)$

Ⓑ $c = 25(4 + 6)$

Ⓒ $c = 10(25 + 4 + 6)$

Ⓓ $c = 10(4 + 6)$

Use a verbal model to write an equation.

Verbal Model Total number of candles = Number of tables • Number of candles on each table

Variable Let c be the total number of candles.

Equation $c = 25 \cdot (4 + 6)$

▶ The correct answer is Ⓑ.

Try It

6. **WHAT IF?** Each worker decorates one table. Which equation can you use to find c?

GO DIGITAL

In-Class Practice

1 I don't understand yet. | 2 I can do it with help. | 3 I can do it on my own. | 4 I can teach someone else.

7. VOCABULARY How are expressions and equations different?

4 MTR

8. DIFFERENT WORDS, SAME QUESTION Which is different? Write "both" equations.

4 less than a number n is 8.

A number n is 4 less than 8.

A number n minus 4 equals 8.

4 subtracted from a number n is 8.

9. OPEN-ENDED Write a word sentence for the equation $28 - n = -5$.

2 MTR

10. MAKE A CONNECTION You purchase x items for \$4 each. Explain how the variable in the expression $4x$ and the variable in the equation $4x = 20$ are similar. Explain how they are different.

Example 3 Modeling Real Life

7 MTR

At Fairchild Tropical Botanic Garden, exhibits are in the *Uplands* and the *Lowlands*. The area of the Uplands is 23 acres, which is 37 acres less than the area of the Lowlands. Find the area of the Lowlands.

Located near Miami, Florida, Fairchild is one of the premier conservation and education-based gardens in the world.

Understand the problem.

You know that exhibits are in two regions. You are given the area of one, and the difference of the two areas. You need to find the area of the other.

Make a plan.

Use a verbal model to write an equation relating the area of the Lowlands, the area of the Uplands, and the difference of the areas. Then use the equation to solve the problem.

Solve and check.

Verbal Model

Area of Lowlands (acres)	−	Area of Uplands (acres)	=	Difference of the areas (acres)

Variable Let a be the area of the Lowlands.

Equation $a \quad - \quad 23 \quad = \quad 37$

Reading

One *acre* is equal to 43,560 square feet.

An equation is $a - 23 = 37$. You can think of the equation as asking the question *"37 is 23 units to the left of what number on a number line?"*

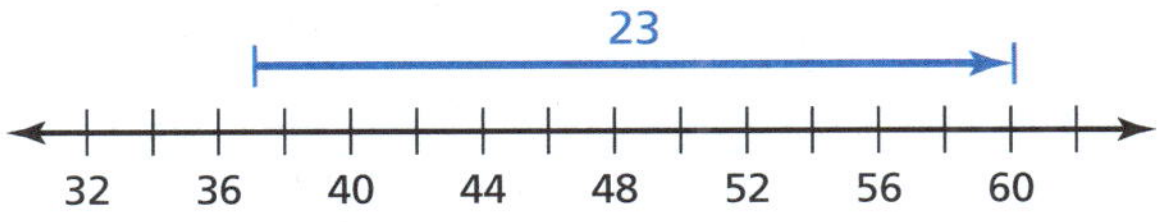

► So, the area of the Lowlands is 60 acres.

Check Verify that $60 - 23 = 37$.

$$\begin{aligned} 60 - 23 &= 60 - 20 - 3 \\ &= 40 - 3 \\ &= 37 \checkmark \end{aligned}$$

GO DIGITAL

In-Class Practice

1 I don't understand yet. | 2 I can do it with help. | 3 I can do it on my own. | 4 I can teach someone else.

11. After four rounds, 74 teams are eliminated from a robotics competition. There are 18 teams remaining. Write and solve an equation to find the number of teams at the start of the competition.

12. The red fluorite crystal has a mass of 30 grams. The mass of the blue copper sulfate crystal is two-thirds the mass of the red fluorite crystal. Write an equation you can use to find the mass (in grams) of the blue copper sulfate crystal.

13. **Dig Deeper** A customer receives $20 in change after using $200 to purchase several sessions at Rock Crawlers Climbing Gym. Each session cost is $15. There is a special for $3 off per session if you purchase 10 or more sessions. How many sessions did the customer purchase?

7 MTR **25. MODELING REAL LIFE** You hit a golf ball 40 yards. It travels one-third of the distance to the hole. Write an equation you can use to find the distance d (in yards) from the tee to the hole.

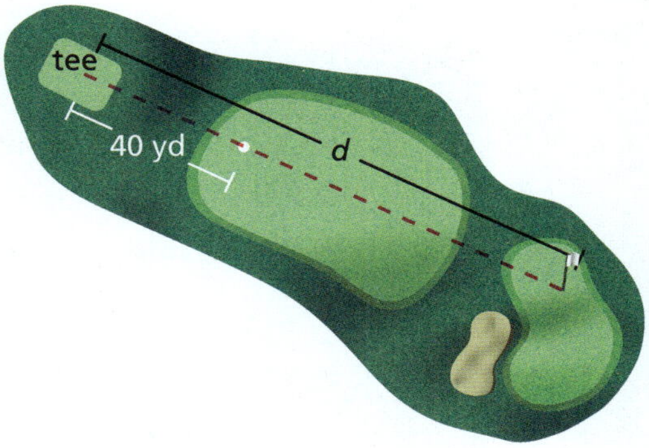

GEOMETRY Write an equation you can use to find the value of *x*.

26. Perimeter of triangle: 16 in.

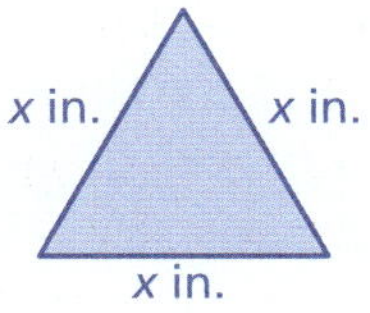

27. Perimeter of square: 30 mm

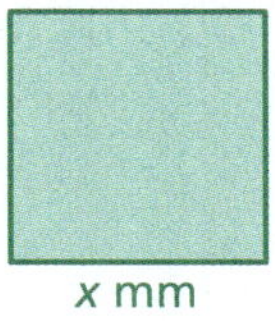

7 MTR **28. MODELING REAL LIFE** You sell instruments at a Caribbean music festival. You earn \$326 by selling 12 sets of maracas, 6 sets of claves, and x djembe drums. Find the number of djembe drums you sold. (See Example 3.)

29. PROBLEM SOLVING Apollo 11 departed from Kennedy Space Center, performed a lunar landing, and then returned to Earth. The total time Apollo 11 traveled to and from lunar orbit was about 108 hours. Apollo 11 spent about 44 more hours traveling to lunar orbit than from lunar orbit.

a. How long did it take Apollo 11 to reach lunar orbit?

7 MTR **b. INVESTIGATE** Research why the journey to lunar orbit and the journey from lunar orbit took different amounts of time.

5 MTR **30. DECOMPOSE A PROBLEM** You eat a basket of 24 strawberries while walking to the beach. You walk each block at the same pace. When you are halfway there, half of the berries are gone. After walking 3 more blocks, you still have 5 blocks to go. You reach the beach 28 minutes after you began. One-sixth of your strawberries are left.

a. Is there enough information to find the time it takes to walk each block? Explain.

b. Is there enough information to find how many strawberries you ate while walking the last block? Explain.

31. GEOMETRY A square is cut from a rectangle. The side length of the square is half of the unknown width w. The area of the shaded region is 75 square inches. Write an equation you can use to find the width (in inches).

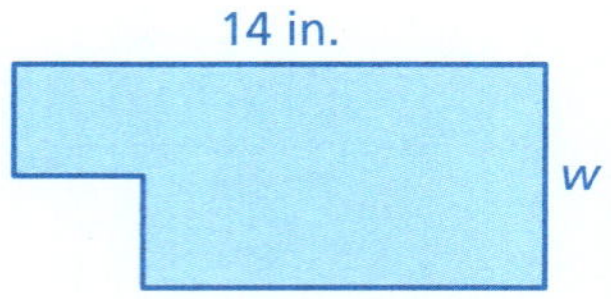

7.2 Finding Unknown Values in Equations

Learning Target: Find unknown values in equations.

Success Criteria:

- I can determine whether a value is a solution of an equation.
- I can use mental math to solve equations.
- I can use models to solve equations.

Algebraic Reasoning

MA.6.AR.2.1 Given an equation or inequality and a specified set of integer values, determine which values make the equation or inequality true or false.

MA.6.AR.2.4 Determine the unknown decimal or fraction in an equation involving any of the four operations, relating three numbers, with the unknown in any position.

Exploration 1 Using Mental Math to Find Unknown Values

Work with a partner. Write a question that represents the equation. Then use mental math to answer the question. Compare your answers with another pair of students.

Equation	Question	Answer
a. $x + 1 = 5$		
b. $4a = 28$		
c. $g - 0.3 = 0.2$		
d. $\frac{1}{4}k = \frac{1}{5}$		
e. $\frac{w}{2} = 2.5$		

GO DIGITAL

Exploration 2 Using Models to Find Unknown Values

Work with a partner.

a. Write an equation represented by each model. Then find the unknown values. Compare your answers with another pair.

i.

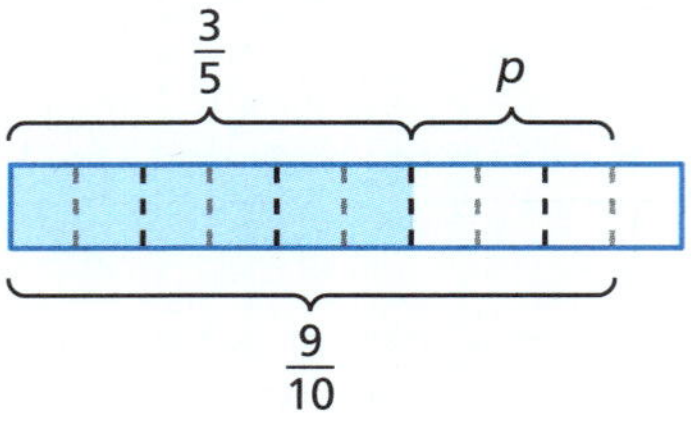

ii.

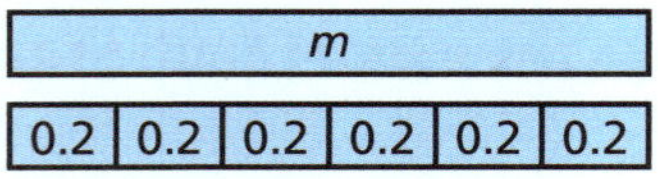

b. Write a question that represents the equation $\frac{x}{3} = \frac{3}{4}$. Then use a model to find x. Compare your model and your answer with another pair.

4 MTR COMPARE METHODS

When is it convenient to use mental math to find an unknown value in an equation? When is it helpful to use a model?

7.2 Lesson

Key Vocabulary
solution, *p. 505*

Remember

The symbol $\neq$ means *is not equal to*.

Equations may be true for some values and false for others. A **solution** of an equation is a value that makes the equation true.

Value of x	$x + 3 = 7$	Are both sides equal?
3	$3 + 3 \stackrel{?}{=} 7$ $6 \neq 7$ ✗	no
4	$4 + 3 \stackrel{?}{=} 7$ $7 = 7$ ✓	yes
5	$5 + 3 \stackrel{?}{=} 7$ $8 \neq 7$ ✗	no

So, the value $x = 4$ is a solution of the equation $x + 3 = 7$.

Example 1 Checking Solutions

Tell whether the given value is a solution of the equation.

a. $p + 10 = 38;\ p = 18$

$18 + 10 \stackrel{?}{=} 38$ Substitute 18 for p.

$28 \neq 38$ ✗ Sides are *not* equal.

▶ So, $p = 18$ is *not* a solution.

b. $4y = 56;\ y = 14$

$4(14) \stackrel{?}{=} 56$ Substitute 14 for y.

$56 = 56$ ✓ Sides are equal.

▶ So, $y = 14$ is a solution.

c. $12 + k = -2k;\ k = -4$

$12 + (-4) \stackrel{?}{=} -2(-4)$ Substitute -4 for k.

$8 = 8$ ✓ Sides are equal.

▶ So, $k = -4$ is a solution.

Try It Tell whether the given value is a solution of the equation.

1. $a + 6 = 17;\ a = 9$

no

2. $12 = 9 - g;\ g = -3$

yes

3. $35 = 5n + 2n;\ n = 5$

yes

4. $\frac{q}{2} = q - 10;\ q = 14$

no

GO DIGITAL

You can use algebraic reasoning, drawings, or mental math to solve equations.

Example 2 Solving Equations Involving Addition or Subtraction

Another Method You can think of the equation as asking the question "*8 is how many units to the right of 3 on a number line?*"

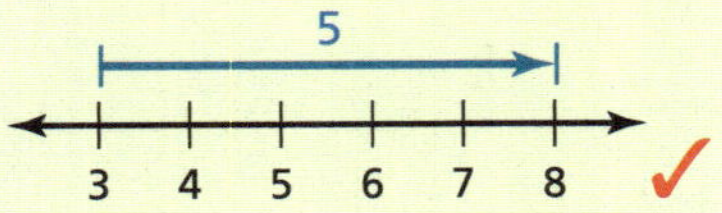

a. Solve $3 + k = 8$.

Use mental math.

$$3 + \square = 8$$
$$3 + 5 = 8$$

Think, "*How many more than 3 is 8?*"

▶ 8 is 5 more than 3. So, $k = 5$.

b. Solve $12 = x - 4$.

Use mental math.

$$12 = \square - 4$$
$$12 = 16 - 4$$

Think, "*12 is 4 less than what number?*"

▶ 12 is 4 less than 16. So, $x = 16$.

c. Solve $m - 0.2 = 0.6$.

Use mental math.

$$\square - 0.2 = 0.6$$
$$0.8 - 0.2 = 0.6$$

Think, "*What number is 0.2 greater than 0.6?*"

▶ 0.8 is 0.2 greater than 0.6. So, $m = 0.8$.

d. Solve $\frac{3}{4} = \frac{5}{8} + g$.

Draw a model. You can think of the equation as asking the question "*$\frac{3}{4}$ is how much greater than $\frac{5}{8}$?*"

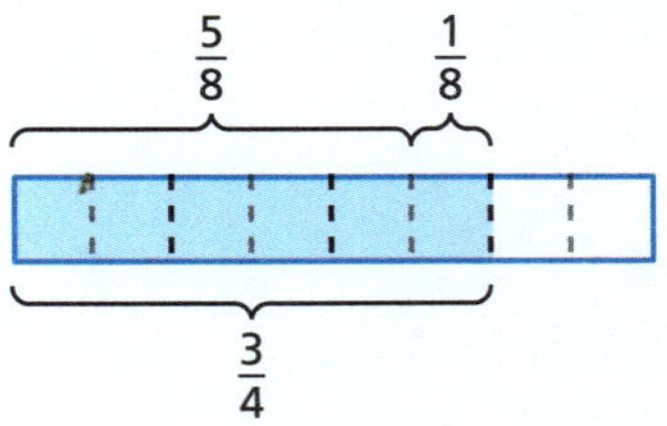

▶ $\frac{3}{4}$ is $\frac{1}{8}$ greater than $\frac{5}{8}$. So, $g = \frac{1}{8}$.

3 MTR REFLECT ON YOUR METHOD

In part (d), why is it more helpful to draw a model than to use mental math?

____

____

____

GO DIGITAL

Try It

Solve the equation. Explain your method.

5. $9 - h = 2$

h = 7

6. $b + 4.5 = 4.8$

b = 0.3

7. $3.7 = 5.1 - w$

W = 1.4

8. $\frac{3}{10} = \frac{4}{10} - a$

$a = \frac{1}{10}$

9. $x + \frac{1}{6} = \frac{2}{3}$

$x = \frac{3}{6} = \frac{1}{2}$

10. $1\frac{1}{4} = 2\frac{1}{2} - k$

$K = \frac{5}{4}$

Example 3 Solving Equations Involving Multiplication or Division

Another Method Draw a tape diagram.

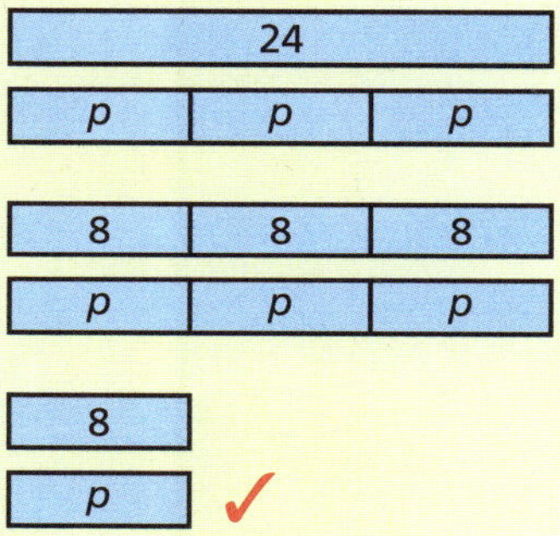

a. Solve $3p = 24$.

Use mental math.

$3 \times \square = 24$ Think, "3 times what number is 24?"

$3 \times 8 = 24$

▶ 3 times 8 equals 24. So, $p = 8$.

b. Solve $\frac{1}{6} = \frac{1}{4}x$.

Draw a model. Because you can think of the equation as asking the question "*$\frac{1}{6}$ is $\frac{1}{4}$ of what number?*," you know that x is 4 times $\frac{1}{6}$.

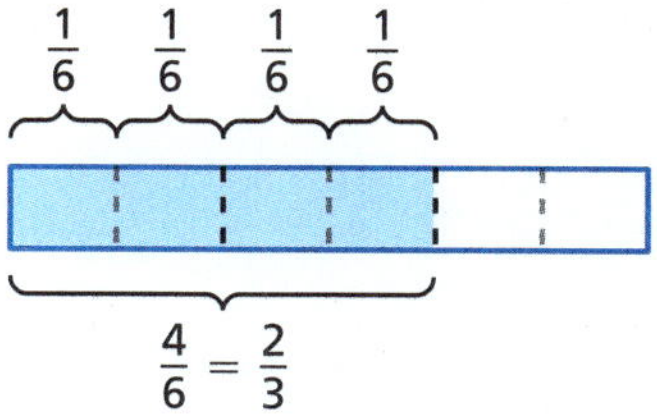

▶ 4 times $\frac{1}{6}$ is $\frac{4}{6}$, or $\frac{2}{3}$. So, $x = \frac{2}{3}$.

c. Solve $\frac{y}{3} = 0.6$.

Draw a tape diagram. You can think of the equation as asking the question "*What number divided by 3 is 0.6?*"

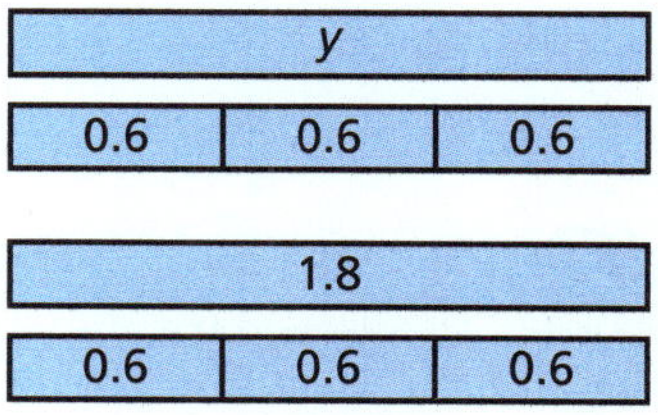

▶ 1.8 divided by 3 equals 0.6. So, $y = 1.8$.

2 MTR MODEL A PROBLEM

Show how you can use a number line to model the equation in part (c).

GO DIGITAL

Try It **Solve the equation. Explain your method.**

11. $7 = \frac{z}{6}$

z = 42

12. $0.2h = 1.8$

h = 9

13. $n \div 5 = \frac{1}{4}$

n = $\frac{5}{4}$

In-Class Practice

1 I don't understand yet. **2** I can do it with help. **3** I can do it on my own. **4** I can teach someone else.

CHECKING SOLUTIONS **Tell whether the given value is a solution of the equation.**

14. $n + 8 = 42;\ n = 36$

15. $-45 = -9g;\ g = -5$

SOLVING EQUATIONS **Solve the equation. Explain your method.**

16. $5 + k = 19$

17. $1.25 = 2.75 - x$

18. $\frac{2}{3}n = \frac{1}{2}$

19. $z \div 1.3 = 4$

2 MTR **20.** **MAKE A CONNECTION** Write and solve an equation represented by the model.

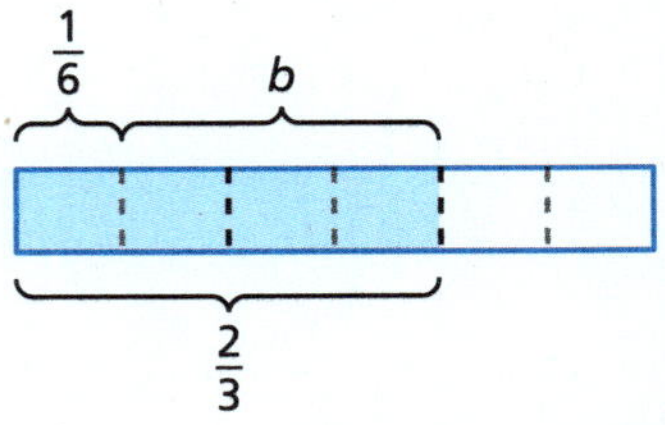

Example 4 Modeling Real Life

7 MTR

You have $\frac{1}{3}$ cup of lime juice to make *mojo*, a traditional Cuban sauce. A recipe calls for $\frac{1}{2}$ cup. The equation $\frac{1}{2} = \frac{1}{3} + n$ represents the additional amount n (in cups) that you need. Do you have enough lime juice when you borrow an additional $\frac{1}{4}$ cup of lime juice from a neighbor?

Draw a model to find the additional amount n of lime juice that you need. You can think of the equation as asking the question *"$\frac{1}{2}$ is how much greater than $\frac{1}{3}$?"*

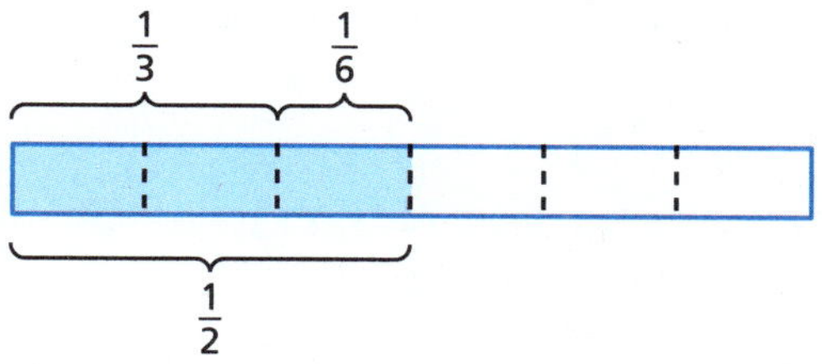

$\frac{1}{2}$ is $\frac{1}{6}$ greater than $\frac{1}{3}$. So, $n = \frac{1}{6}$.

You need an additional $\frac{1}{6}$ cup of lime juice. Compare the amount that you borrow with the additional amount that you need.

Amount from friend (cups): $\frac{1}{4} = \frac{1 \times 3}{4 \times 3} = \frac{3}{12}$

Amount needed (cups): $\frac{1}{6} = \frac{1 \times 2}{6 \times 2} = \frac{2}{12}$

Because $\frac{3}{12} > \frac{2}{12}$, $\frac{1}{4}$ cup $> \frac{1}{6}$ cup.

▶ Because $\frac{1}{4}$ cup is greater than $\frac{1}{6}$ cup, you have enough lime juice.

GO DIGITAL

In-Class Practice

1 I don't understand yet. **2** I can do it with help. **3** I can do it on my own. **4** I can teach someone else.

21. A veterinarian wants your dog to gain weight. This situation is represented by the equation $16.9 + p = 18.5$, where the units are in pounds.

a. Explain the meaning of each term in the equation.

b. Your dog gains 1.4 pounds. Use two different methods to determine whether your dog gained enough weight. Explain each method.

22. You grow herbs in your kitchen using the planter shown. The base of the planter is rectangular with an area of $\frac{7}{8}$ square foot. Write and solve an equation to find the length of the base.

$\frac{1}{2}$ ft

SOLVING EQUATIONS INVOLVING ADDITION OR SUBTRACTION **Solve the equation. Explain your method.** (See Example 2.)

▶**23.** $a + 7 = 13$

24. $34 = m + 18$

25. $k + 19 = 5$

26. $58 = 47 - g$

27. $d - 4.3 = 0.4$

28. $5.6 - b = 3.3$

29. $\frac{5}{7} = m + \frac{2}{7}$

30. $\frac{5}{8} = \frac{1}{4} + n$

31. $1\frac{1}{3} = 2\frac{2}{9} - w$

SOLVING EQUATIONS INVOLVING MULTIPLICATION OR DIVISION **Solve the equation. Explain your method.** (See Example 3.)

32. $3f = 33$

▶**33.** $6 = \frac{18}{x}$

34. $48 = 8g$

35. $4b = 36$

36. $4.2 = 6r$

37. $3 = \frac{1}{5}p$

38. $\frac{1}{6}y = \frac{1}{2}$

39. $\frac{7.2}{w} = 0.9$

40. $3 = q \div 1.8$

4 MTR **YOU BE THE TEACHER** **Your friend uses mental math to solve the equation. Is your friend correct? Explain your reasoning.**

41.

$3.4 = x - 0.6$

$3.4 = 4 - 0.6$

3.4 is 0.6 less than 4.

So, $x = 4$.

42.

$\frac{1}{3}y = \frac{1}{9}$

$\frac{1}{3} \cdot 3 = \frac{1}{9}$

$\frac{1}{3}$ times 3 equals $\frac{1}{9}$.

So, $y = 3$.

7 MTR **43.** **MODELING REAL LIFE** You are training for a road race. You need to run a total of 6.8 miles in the next two days to reach your weekly mileage goal. Today you run 2.7 miles. Solve the equation $2.7 + d = 6.8$ to find the distance d (in miles) that you need to run tomorrow to reach your weekly mileage goal. (See Example 4.)

7 MTR **44.** **MODELING REAL LIFE** Your piano teacher tells you to practice piano $1\frac{1}{4}$ hours each day. You practice piano in the morning and have only $\frac{1}{2}$ hour left to practice after school. Solve the equation $1\frac{1}{4} - t = \frac{1}{2}$ to find how much time you practiced piano in the morning.

7 MTR **45.** **MODELING REAL LIFE** Your class goes on a field trip to Gatorland in Orlando. A total of 56 students are on the trip. The students are divided into groups of 7. Solve the equation $56 \div x = 7$ to find the total number of groups.

7 MTR **46.** **MODELING REAL LIFE** A parking meter near a local park charges a fee of \$2.25 per hour. You spend \$6.75 at the meter. Solve the equation $2.25h = 6.75$ to find the number of hours of parking you pay for. How many hours of parking do you pay for if you spend \$13.50?

7 MTR **47.** **MODELING REAL LIFE** Your goal is to swim the 50-meter freestyle in 30.25 seconds by the end of the season. You swim the 50-meter freestyle in 33.65 seconds at the first meet. The equation $33.65 - t = 30.25$ represents the time t (in seconds) you need to reduce your time by to achieve your goal. Do you reach your goal if you reduce your time by 3.35 seconds?

48. **B.E.S.T. Test Prep** What value of x makes the equation $14.4 \div x = 12$ true?

Ⓐ 0.12

Ⓑ 1.2

Ⓒ 12.2

Ⓓ 0.012

1 MTR **49.** **HELP A CLASSMATE** Your classmate wants to solve the equation $\frac{1}{2} = \frac{1}{8} + x$ using a tape diagram. Explain how your classmate can complete the work.

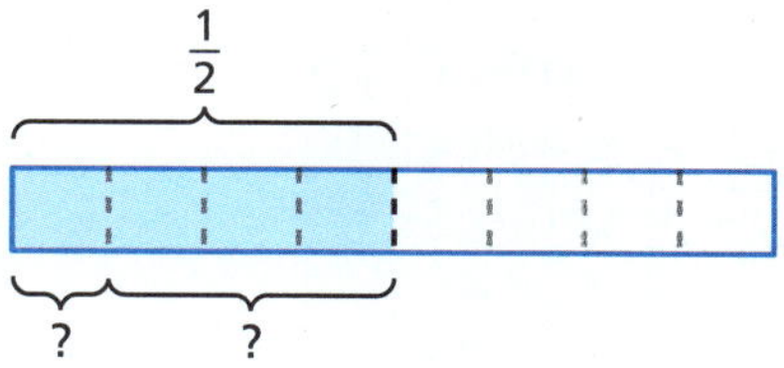

7 MTR **50. MODELING REAL LIFE** You are making bracelets for yourself and your friends. It takes 1.8 yards of string to make one bracelet. The equation $s \div 4 = 1.8$ represents the amount of string s you need to make the bracelets. You have 22 feet of string. How much string do you have left over when you make four bracelets? Justify your answer.

51. OPEN-ENDED Write a real-life situation that involves solving an equation using addition or subtraction. Then write a real-life situation that involves solving an equation using multiplication or division. Solve each problem.

6 MTR **52. ASSESS REASONABLENESS** You are doing a circuit of different exercises. You spend between 25 and 30 seconds on each exercise and rest when needed between exercises. You complete the circuit in 174.2 seconds. Your friend hears your time and says that you must have completed 7 exercises. Is your friend's guess reasonable? Explain.

53. NUMBER SENSE Find a value of x that makes the statement true. Explain how you found your answer.

$$\frac{18}{x} = x - 3$$

7.3 Solving Equations Using Addition or Subtraction

Learning Target: Write and solve equations using addition or subtraction.

Success Criteria:
- I can apply the Addition and Subtraction Properties of Equality to generate equivalent equations.
- I can solve equations using addition or subtraction.
- I can create equations involving addition or subtraction to solve real-life problems.

Algebraic Reasoning
MA.6.AR.2.2 Write and solve one-step equations in one variable within a mathematical or real-world context using addition and subtraction, where all terms and solutions are integers.

Exploration 1 Solving an Equation Using a Tape Diagram

Work with a partner. A student solves an equation using the tape diagrams shown.

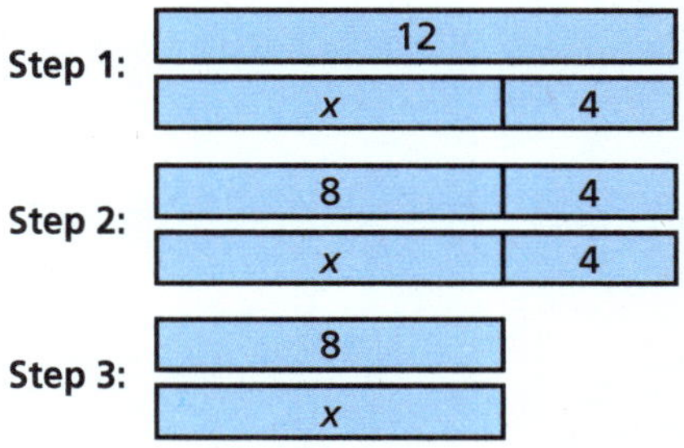

a. What equation did the student solve? What is the solution?

b. Explain how the tape diagrams in Steps 2 and 3 relate to the equation and its solution.

GO DIGITAL

Exploration 2 Solving Equations Using Algebra Tiles

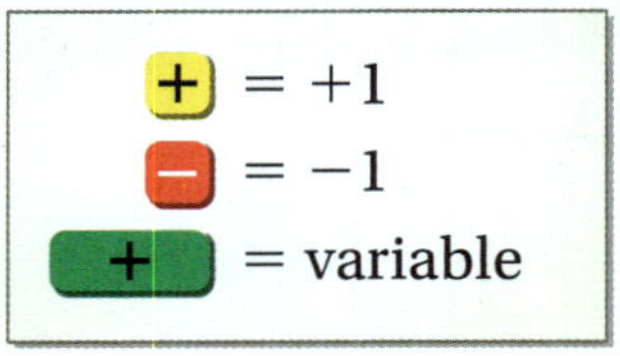

Work with a partner. You can use the algebra tiles shown at the left to model and solve equations.

a. How can you use algebra tiles to model a sum of terms that equals 0? Explain your reasoning.

b. Write the equations modeled by the algebra tiles. Explain how you can use algebra tiles to solve each equation. Then find the solutions.

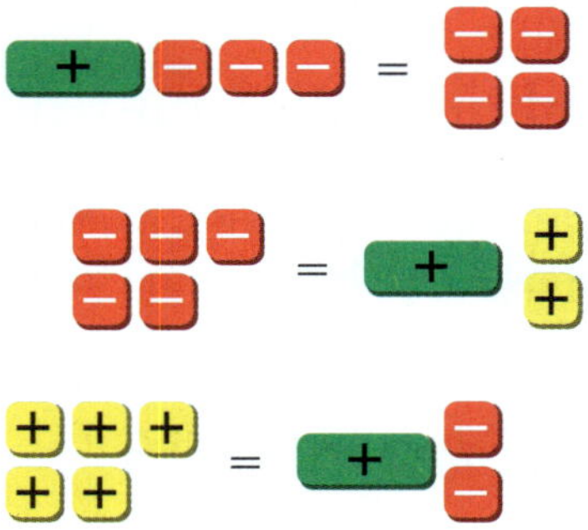

c. How can you solve each equation in part (b) without using algebra tiles?

5 MTR **RELATE CONCEPTS**

How can you use the relationship between addition and subtraction to solve $x + 3 = -5$?

7.3 Lesson

Key Vocabulary
inverse operations, *p. 519*

You can use *inverse operations* to solve equations. **Inverse operations** "undo" each other. Addition and subtraction are inverse operations.

You can add or subtract any number on each side of an equation to generate an equivalent equation. For example, $x + 4 = 5$ and $x + 2 = 3$ are equivalent equations.

$$\begin{aligned} x + 4 &= 5 \\ -2 &= -2 \\ x + 2 &= 3 \end{aligned}$$

Key Ideas

Addition Property of Equality

Words When you add the same number to each side of an equation, the two sides remain equal.

Numbers

$$\begin{aligned} 8 &= 8 \\ +5 &\quad +5 \\ 13 &= 13 \end{aligned}$$

Algebra

$$\begin{aligned} x - 4 &= 5 \\ +4 &\quad +4 \\ x &= 9 \end{aligned}$$

Subtraction Property of Equality

Words When you subtract the same number from each side of an equation, the two sides remain equal.

Numbers

$$\begin{aligned} 8 &= 8 \\ -5 &\quad -5 \\ 3 &= 3 \end{aligned}$$

Algebra

$$\begin{aligned} x + 4 &= 5 \\ -4 &\quad -4 \\ x &= 1 \end{aligned}$$

Example 1 Solving Equations Using Addition

a. Solve $x - 2 = 6$.

	$x - 2 = 6$	Write the equation.
Undo the subtraction. →	$+2 \quad +2$	Addition Property of Equality
	$x = 8$	Simplify.

▶ The solution is $x = 8$.

Check

$x - 2 = 6$

$8 - 2 \stackrel{?}{=} 6$

$6 = 6$ ✓

b. Solve $-11 + x = -8$.

$-11 + x = -8$	Write the equation.
$+11 \quad +11$	Addition Property of Equality
$x = 3$	Simplify.

▶ The solution is $x = 3$.

2 MTR USE ANOTHER METHOD
Show how you can use a number line or algebra tiles to solve the equations in Example 1.

____

____

____

____

____

GO DIGITAL

Try It **Solve the equation. Check your solution.**

1. $k - 3 = 1$

$k = 4$

2. $n - 10 = 4$

$n = 14$

3. $15 = r - 6$

$r = 21$

4. $-14 = -3 + p$

$p = -11$

Example 2 Solving Equations Using Subtraction

a. Solve $x + 2 = -9$.

	$x + 2 = -9$	Write the equation.
Undo the addition. →	$-2 \quad -2$	Subtraction Property of Equality
	$x = -11$	Simplify.

▶ The solution is $x = -11$.

Check

$x + 2 = -9$

$-11 + 2 \stackrel{?}{=} -9$

$-9 = -9$ ✓

b. Solve $26 = 11 + x$.

$26 = 11 + x$	Write the equation.
$-11 \quad -11$	Subtraction Property of Equality
$15 = x$	Simplify.

▶ The solution is $x = 15$.

Another Method

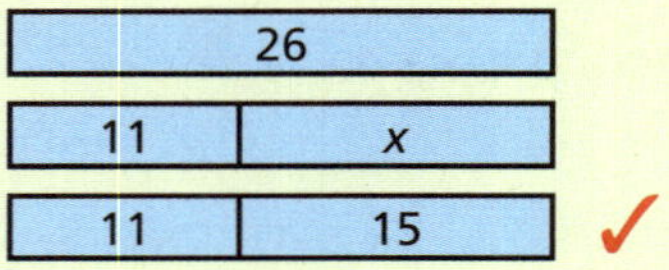

Try It **Solve the equation. Check your solution.**

5. $s + 8 = 17$

$s = 9$

6. $9 = y + 6$

$y = 3$

7. $13 + m = -20$

$m = 33$

8. $-14 = 18 + d$

$d = -32$

In-Class Practice

1 I don't understand yet. **2** I can do it with help. **3** I can do it on my own. **4** I can teach someone else.

SOLVING EQUATIONS **Solve the equation. Check your solution.**

9. $x - 8 = -12$

10. $b + 14 = 33$

11. $17 = y + 23$

12. $-6 = 31 + h$

GO DIGITAL

In-Class Practice

1 I don't understand yet. **2** I can do it with help. **3** I can do it on my own. **4** I can teach someone else.

13. WRITING When solving $x + 5 = 16$, why do you subtract 5 from the left side of the equation? Why do you subtract 5 from the right side of the equation?

14. REASONING Do the equations have the same solution? Explain your reasoning.

$$x - 8 = 6$$
$$x - 6 = 8$$

5 MTR **15. STRUCTURE** Just by looking at the equation below, find the value of x. Explain your reasoning.

$$x + 6 + 2x = 2x + 6 + 4$$

Example 3 Modeling Real Life

7 MTR

Your aunt gives you $50 to help buy the reflector telescope shown. After you buy the telescope, you have $25 left. How much money did you have before your aunt gave you $50?

Use a verbal model to write an equation that represents the situation. Then solve the equation.

Verbal Model Starting amount (dollars) + Aunt's money (dollars) − Telescope price (dollars) = Amount left (dollars)

Variable Let s be the starting amount.

Equation

$s + 50 - 275 = 25$	Write the equation.
$s + 50 - 275 + 275 = 25 + 275$	Addition Property of Equality
$s + 50 = 300$	Simplify.
$s + 50 - 50 = 300 - 50$	Subtraction Property of Equality
$s = 250$	Simplify.

Addition and subtraction preserve units. For example, when you add dollars to dollars, you get dollars.

▶ You had $250 before your aunt gave you money.

Another Method Solve the problem by working backwards from $25.

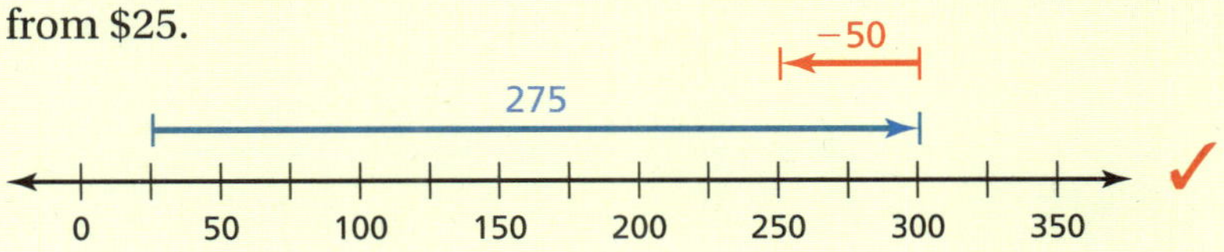

GO DIGITAL

In-Class Practice

1 I don't understand yet. 2 I can do it with help. 3 I can do it on my own. 4 I can teach someone else.

16. On September 22, 2019, an 18-foot, 4-inch-long Burmese python was captured at Big Cypress National Preserve, breaking the previous record at the preserve by 16 inches. Write and solve an equation to find the length (in inches) of the second-largest snake ever captured at the preserve at that time.

17. **Dig Deeper** You get in an elevator and go up 2 floors and down 8 floors before exiting. Then you get back in the elevator and go up 4 floors before exiting on the 12th floor. On what floors did you get in the elevator?

7.3 Practice WITH CalcChat® AND CalcView®

Review & Refresh

Tell whether the given value is a solution to the equation.

1. $d - 9 = 13;\ d = 22$
2. $3n + 2 = -14;\ n = -4$
3. $16 = 4h - 2h;\ h = 8$

Divide. Write the answer in simplest form.

4. $\frac{1}{2} \div \frac{1}{4}$
5. $12 \div \frac{3}{8}$
6. $8 \div \frac{4}{5}$
7. $\frac{7}{9} \div \frac{3}{2}$

8. Find the GCF of 42 and 70.

Evaluate the expression.

9. $(2 + 5^2) \div 3$
10. $6 + 2^3 \cdot 3 - 5$
11. $4 \cdot [3 + 3(20 - 4^2 - 2)]$

12. You and your friend run a 400-meter race. Your time is 82.6 seconds. Your friend's time is 22.5 seconds more than half of your time. Find your friend's time.

GO DIGITAL

Concepts, Skills, & Problem Solving

2 MTR **MODEL A PROBLEM** **Write and solve the equations represented by the algebra tiles.** (See Exploration 2.)

13. 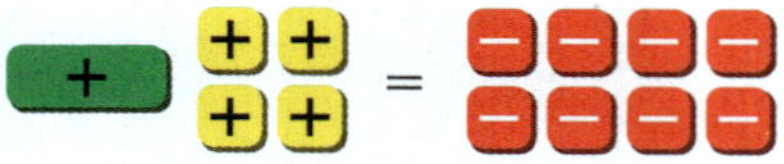

14. 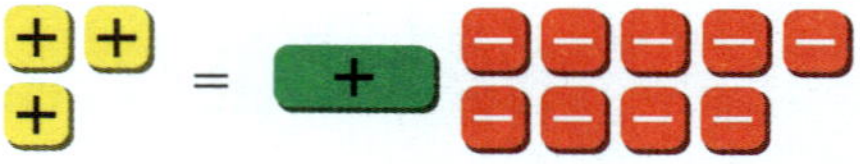

SOLVING EQUATIONS **Solve the equation. Check your solution.** (See Examples 1 and 2.)

15. $y - 7 = 3$

16. $z - 3 = 13$

▶ **17.** $8 = r - 14$

18. $p + 5 = 8$

▶ **19.** $k + 6 = 18$

20. $64 = h + 30$

21. $f - 27 = 19$

22. $25 = q + 14$

23. $-12 + x = -32$

24. $-10 = t + 15$

25. $13 = 43 + w$

26. $35 + j = -51$

4 MTR **YOU BE THE TEACHER** **Your friend solves the equation. Is your friend correct? Explain your reasoning.**

27.

$$\begin{aligned} x + 7 &= 13 \\ \underline{+7} \quad & \underline{+7} \\ x &= 20 \end{aligned}$$

28.

$$\begin{aligned} 34 &= y - 12 \\ \underline{-12} \quad & \underline{\quad +12} \\ 22 &= y \end{aligned}$$

7 MTR

29. MODELING REAL LIFE The main span of the Sunshine Skyway Bridge is 366 meters long. The bridge's main span is 30 meters shorter than the main span of the Dames Point Bridge. Write and solve an equation to find the length (in meters) of the main span of the Dames Point Bridge. (See Example 3.)

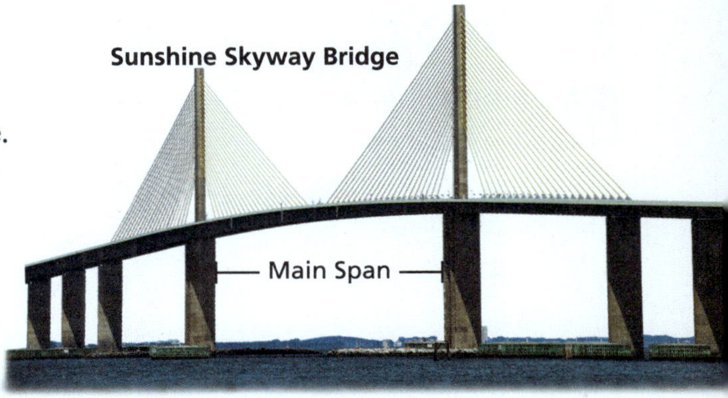

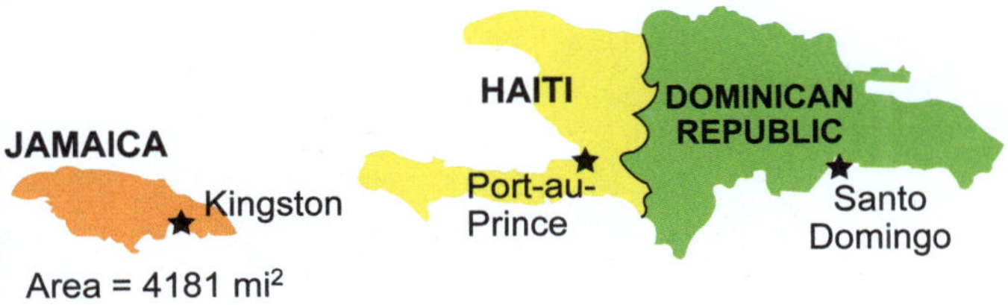

30. PROBLEM SOLVING The area of Jamaica is 6460 square miles less than the area of Haiti. Find the area (in square miles) of Haiti.

WRITING EQUATIONS **Write the word sentence as an equation. Then solve the equation.**

31. 13 subtracted from a number w is 15.

32. How many units to the left of 34 is 7 on a number line?

33. 93 is the sum of a number g and 58.

34. How many units to the right of -18 is 46 on a number line?

5 MTR

35. **PATTERNS** In a *magic square,* the sum of the numbers in each row, column, and diagonal is the same. Find the values of a, b, and c. Justify your answers.

a	37	16
19	25	b
34	c	28

36. **REASONING** On Saturday, you spend \$33, give \$15 to a friend, and receive \$20 for mowing your neighbor's lawn. You have \$21 left. Use two methods to find how much money you started with that day.

37. **Dig Deeper** You have \$15.

Bumper Cars:	\$2
Super Pendulum:	\$3 more than Ferris wheel
Giant Slide:	\$1 less than bumper cars
Ferris Wheel:	\$1 more than giant slide

a. How much money do you have left if you ride each ride once?

b. Do you have enough money to ride each ride twice? Explain.

7.4 Solving Equations Using Multiplication or Division

Learning Target: Write and solve equations using multiplication or division.

Success Criteria:
- I can apply the Multiplication and Division Properties of Equality to generate equivalent equations.
- I can solve equations using multiplication or division.
- I can create equations involving multiplication or division to solve real-life problems.

Algebraic Reasoning
MA.6.AR.2.3 Write and solve one-step equations in one variable within a mathematical or real-world context using multiplication and division, where all terms and solutions are integers.

Exploration 1 Solving an Equation Using a Tape Diagram

Work with a partner. A student solves an equation using the tape diagrams shown.

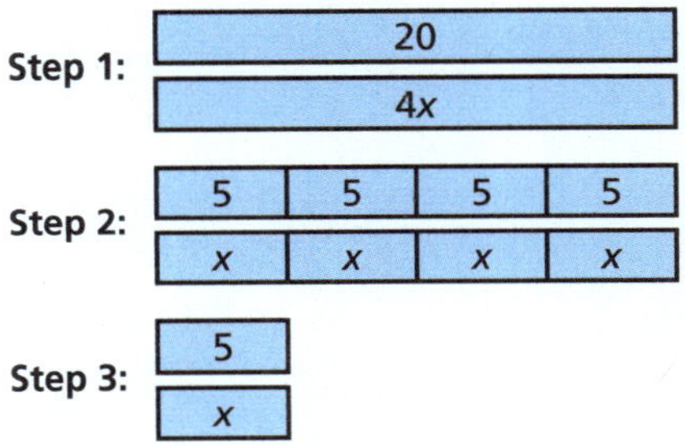

a. What equation did the student solve? What is the solution?

b. Explain how the tape diagrams in Steps 2 and 3 relate to the equation and its solution.

GO DIGITAL

Exploration 2 Solving Equations Using Algebra Tiles

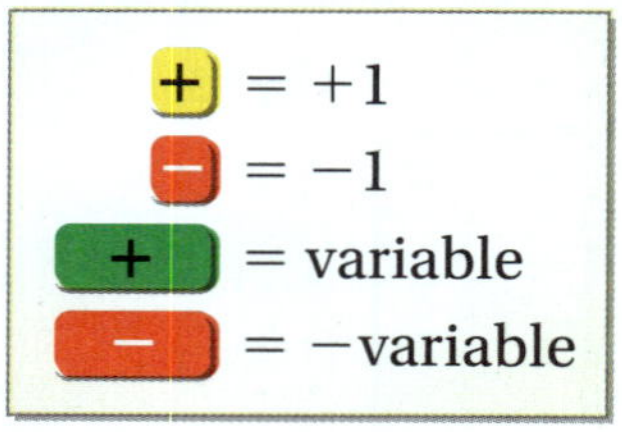

Work with a partner.

a. Write the equations modeled by the algebra tiles. Explain how you can use algebra tiles to solve each equation. Then find the solutions.

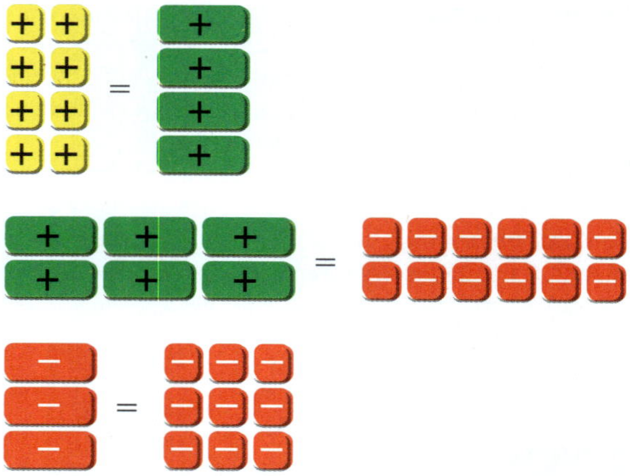

b. How can you solve each equation in part (a) without using algebra tiles?

4 MTR **CONSTRUCT AN ARGUMENT**

How can you use properties of equality to solve equations of the form $ax = b$?

7.4 Lesson

Key Ideas

Multiplication Property of Equality

Words When you multiply each side of an equation by the same nonzero number, the two sides remain equal.

Numbers

$$\frac{8}{4} = 2$$
$$\frac{8}{4} \cdot 4 = 2 \cdot 4$$
$$8 = 8$$

Algebra

$$\frac{x}{4} = 2$$
$$\frac{x}{4} \cdot 4 = 2 \cdot 4$$
$$x = 8$$

Multiplicative Inverse Property

Words The product of a nonzero number n and its reciprocal, $\frac{1}{n}$, is 1.

Numbers $5 \cdot \frac{1}{5} = 1$

Algebra $n \cdot \frac{1}{n} = \frac{1}{n} \cdot n = 1, n \neq 0$

Remember

Inverse operations "undo" each other. Multiplication and division are inverse operations.

Example 1 Solving Equations Using Multiplication

a. Solve $\frac{w}{4} = 12$.

$\frac{w}{4} = 12$ Write the equation.

Undo the division. → $\frac{w}{4} \cdot 4 = 12 \cdot 4$ Multiplication Property of Equality

$w = 48$ Simplify.

▶ The solution is $w = 48$.

Check

$$\frac{w}{4} = 12$$
$$\frac{48}{4} \stackrel{?}{=} 12$$
$$12 = 12 \checkmark$$

b. Solve $5x = 35$.

$5x = 35$ Write the equation.

Use the Multiplicative Inverse Property. → $\frac{1}{5} \cdot 5x = \frac{1}{5} \cdot 35$ Multiplication Property of Equality

$x = 7$ Simplify.

▶ The solution is $x = 7$.

Another Method

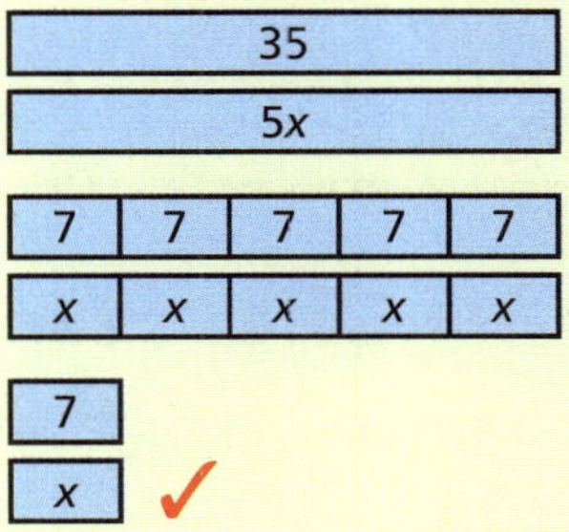

GO DIGITAL

Try It

Solve the equation. Check your solution.

1. $\frac{a}{8} = 6$

2. $14 = 2y$

3. $\frac{z}{-2} = -9$

You can multiply or divide each side of an equation by any number to generate an equivalent equation. For example, $4x = 32$ and $2x = 16$ are equivalent equations.

$$4x = 32$$
$$\frac{4x}{2} = \frac{32}{2}$$
$$2x = 16$$

Key Idea

Division Property of Equality

Words When you divide each side of an equation by the same nonzero number, the two sides remain equal.

Numbers

$$8 \cdot 4 = 32$$
$$8 \cdot 4 \div 4 = 32 \div 4$$
$$8 = 8$$

Algebra

$$4x = 32$$
$$\frac{4x}{4} = \frac{32}{4}$$
$$x = 8$$

Example 2 Solving an Equation Using Division

Solve $-15 = -5b$.

$-15 = -5b$ Write the equation.

Undo the multiplication. → $\frac{-15}{-5} = \frac{-5b}{-5}$ Division Property of Equality

$3 = b$ Simplify.

▶ The solution is $b = 3$.

Another Method You can interpret the equation as asking the question "*How many groups of −5 are in −15?*"

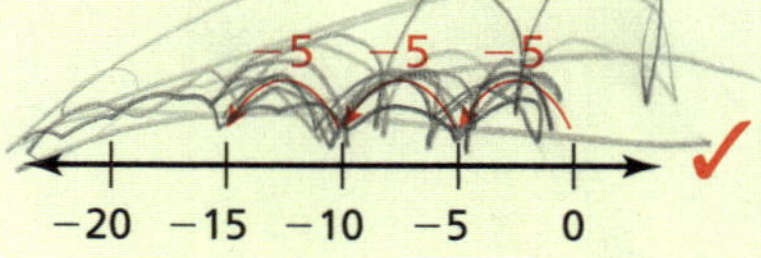

Try It

Solve the equation. Check your solution.

4. $p \cdot 3 = 18$

5. $12q = 60$

6. $-81 = 9r$

In-Class Practice

1 I don't understand yet. 2 I can do it with help. 3 I can do it on my own. 4 I can teach someone else.

SOLVING EQUATIONS **Solve the equation. Check your solution.**

7. $-36 = \frac{y}{3}$

8. $8s = 56$

GO DIGITAL

In-Class Practice

1 I don't understand yet. 2 I can do it with help. 3 I can do it on my own. 4 I can teach someone else.

4 MTR

9. WHICH ONE DOESN'T BELONG? Which equation does *not* belong with the other three? Explain your reasoning.

$\frac{x}{3} = 36$ $\quad 3x = 36$ $\quad \frac{x}{6} = 2$ $\quad 4x = 48$

5 MTR

STRUCTURE Just by looking at the equation, find the value of x. Explain your reasoning.

10. $5x + 3x = 5x + 18$

11. $8x + \frac{x}{2} = 8x + 6$

Example 3 Modeling Real Life

7 MTR

The area of a rectangular LED sky screen is 7500 square meters. The width of the sky screen is 30 meters. What is the length of the sky screen?

Use the given information to draw a diagram. Then substitute for the area A and the width w in the formula for the area of a rectangle, $A = \ell w$. Solve for the length ℓ.

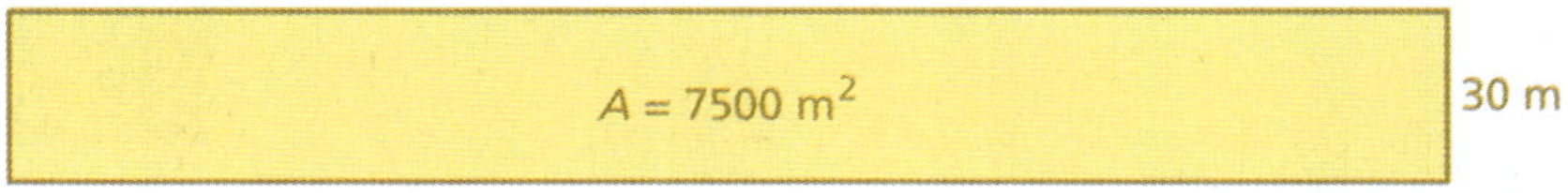

$A = \ell w$	Use the formula for area of a rectangle.
$7500 = 30\ell$	Substitute 7500 for A and 30 for w.
$\frac{7500}{30} = \frac{30\ell}{30}$	Division Property of Equality
$250 = \ell$	Simplify.

▶ So, the sky screen is 250 meters long.

GO DIGITAL

In-Class Practice

1 I don't understand yet. | 2 I can do it with help. | 3 I can do it on my own. | 4 I can teach someone else.

Area = 1625 mm^2

12. The area of the screen of the smart watch is shown. What are possible dimensions for the length and the width of the screen? Justify your answer.

13. A rock climber climbs 720 feet per hour. Write and solve an equation to find the number of minutes it takes for the rock climber to climb 288 feet.

14. **Dig Deeper** A gift card stores data using a black, magnetic stripe on the back of the card. Find the width w of the stripe.

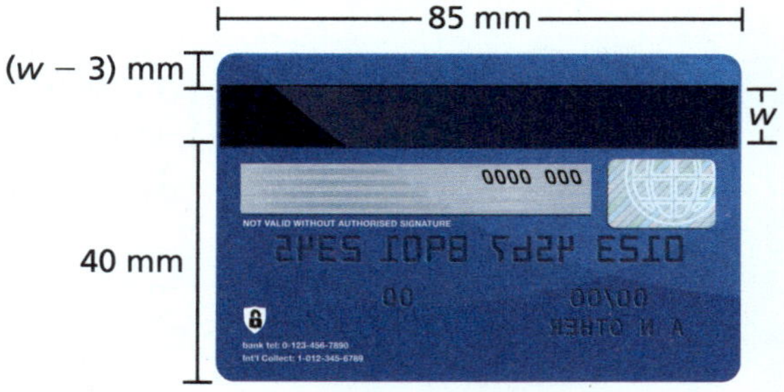

Area = 4675 mm^2

7.4 Practice WITH CalcChat® AND CalcView®

Review & Refresh

Solve the equation. Check your solution.

1. $y - 5 = 6$

2. $m + 7 = 8$

3. $-2 = 1 + p$

4. What is the value of a^3 when $a = 4$?

Multiply. Write the answer in simplest form.

5. $\frac{1}{5} \bullet \frac{2}{9}$

6. $\frac{5}{12} \times \frac{4}{7}$

7. $2\frac{1}{3} \bullet \frac{3}{10}$

8. $1\frac{3}{4} \times 2\frac{2}{3}$

Multiply.

9. $\begin{array}{r} 0.4 \\ \times\ 0.9 \\ \hline \end{array}$

10. $\begin{array}{r} 0.78 \\ \times\ 0.5 \\ \hline \end{array}$

11. 2.63×4.31

12. 1.115×3.28

Concepts, Skills, & Problem Solving

USING TOOLS **Use the model to solve the equation.** (See Exploration 1.)

13.

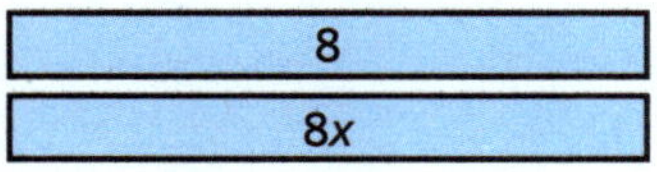

14.

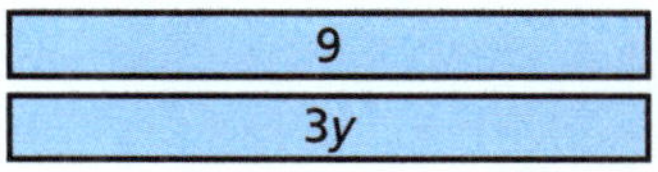

GO DIGITAL

SOLVING EQUATIONS **Solve the equation. Check your solution.** (See Examples 1 and 2.)

▶**15.** $\frac{s}{10} = 7$

16. $6 = \frac{t}{5}$

▶**17.** $3a = 12$

18. $5 \cdot z = 35$

19. $40 = 4y$

20. $42 = 7k$

21. $7x = 105$

22. $13 = d \div 6$

23. $-9 = v \div 5$

24. $\frac{x}{2} = -7$

25. $\frac{s}{-4} = 10$

26. $y \div (-6) = 1$

4 MTR

27. YOU BE THE TEACHER Your friend solves the equation $x \div 4 = 28$. Is your friend correct? Explain your reasoning.

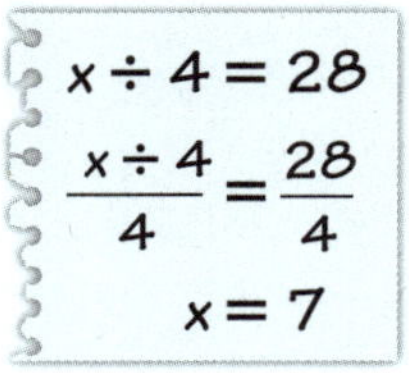

3 MTR

28. CHOOSE A METHOD Show how you can solve the equation $3x = 9$ by multiplying each side by the reciprocal of 3 and by using algebra tiles. Which method do you prefer? Explain.

2 MTR

29. MAKE A CONNECTION What equation does the tape diagram represent? Explain how to find the value of $6x$ using the tape diagram.

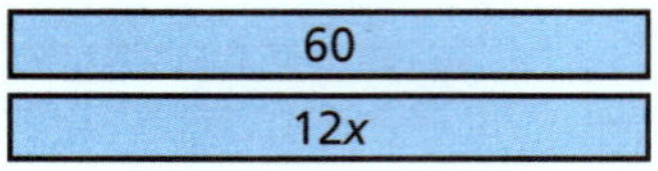

30. **MAKE A CONNECTION** Write an equation that you can solve using the number line. Then solve the equation.

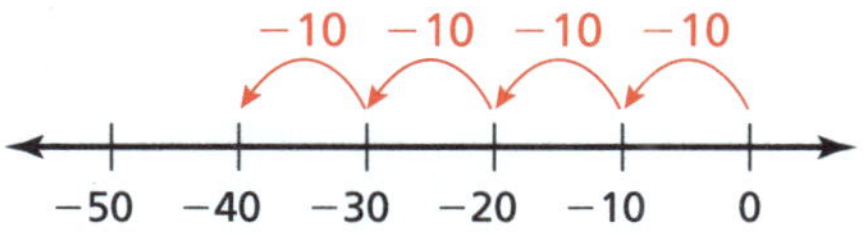

GEOMETRY Solve for *x*. Check your answer.

31. Area = 45 square units

32. Area = 176 square units

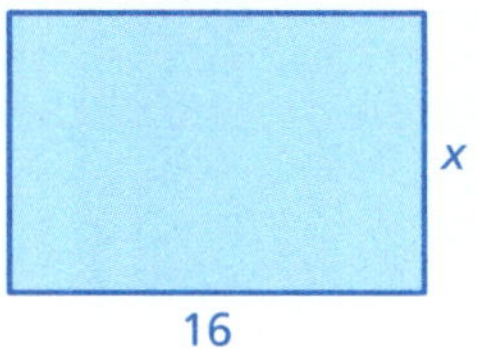

33. **MODELING REAL LIFE** Forty-five basketball players participate in a three-on-three tournament. Write and solve an equation to find the number of three-person teams in the tournament. (See Example 3.)

34. **STRUCTURE** A series of hourly weather reports tracks changes in air pressure caused by a hurricane. In the last hour, the change in pressure was 4 times the change during the previous hour. Write and solve an equation that represents the change in pressure during the previous hour when the change during the last hour was −24 millibars.

4 MTR

35. DISCUSS MATHEMATICAL THINKING A florist uses f flowers to make floral arrangements. Compare the equations $2 = f \div 30$ and $f \div 20 = 3$. What does each equation represent in this situation? Why are the solutions of the equations the same?

36. Dig Deeper A slush drink machine fills 1440 cups in 24 hours.

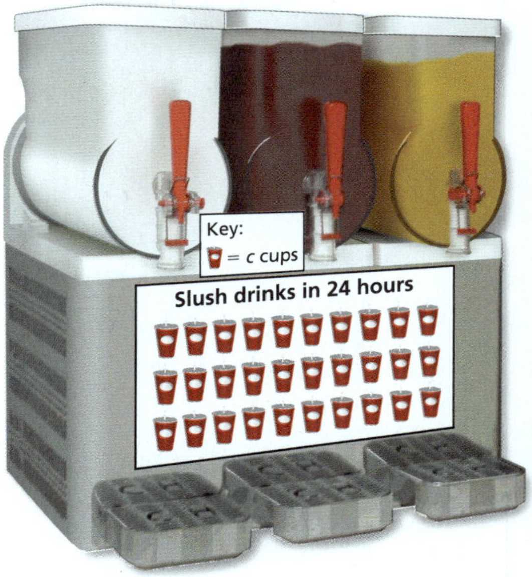

a. Find the number c of cups each symbol represents.

b. To lower costs, you replace the cups with paper cones that hold $\frac{1}{5}$ less. Find the number n of paper cones that the machine can fill in 24 hours.

37. NUMBER SENSE The area of a picture is 100 square inches. The length is 4 times the width. Find the length and width of the picture.

7.5 Writing and Graphing Inequalities

Learning Target: Write inequalities and represent solutions of inequalities on number lines.

Success Criteria:
- I can write word sentences as inequalities.
- I can determine whether a value is a solution of an inequality.
- I can graph the solutions of inequalities.

Algebraic Reasoning

MA.6.AR.1.2 Translate a real-world written description into an algebraic inequality in the form of $x > a$, $x < a$, $x \geq a$ or $x \leq a$. Represent the inequality on a number line.

MA.6.AR.2.1 Given an equation or inequality and a specified set of integer values, determine which values make the equation or inequality true or false.

Exploration 1 Understanding Inequality Statements

2 MTR MAKE A CONNECTION

You know the inequality symbols < and >. What do the symbols ≤ and ≥ mean?

____

____

____

____

____

Work with a partner.

a. Your class started more than 3 minutes late. Shade the part of the number line that represents when your class started.

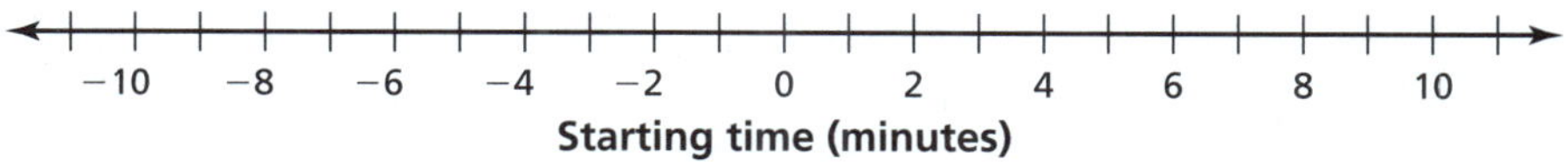

b. You need at least 3 tablespoons of lime juice for a recipe. Shade the part of the number line that represents how much lime juice you need.

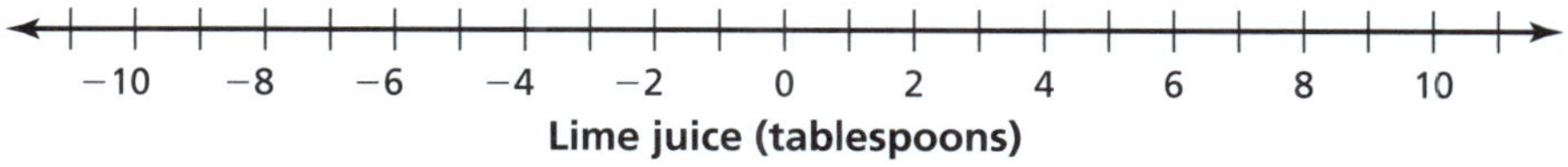

c. The temperature is at most 3 degrees Celsius. Shade the part of the number line that represents what the temperature could be.

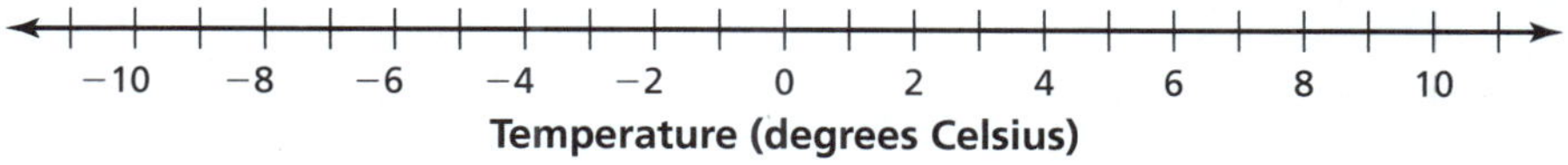

d. You are fewer than 3 feet above sea level. Shade the part of the number line that represents what your elevation might be.

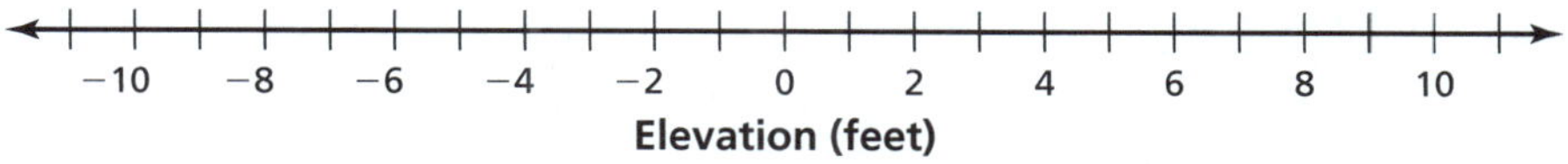

GO DIGITAL

7.5 Lesson

Key Vocabulary
inequality, *p. 538*
solution of an inequality, *p. 539*
solution set, *p. 539*
graph of an inequality, *p. 540*

An **inequality** is a mathematical sentence that compares expressions. It contains the symbols $<$, $>$, $\leq$, or $\geq$. To write a word sentence as an inequality, look for the following phrases to determine where to place the inequality symbol.

Inequality Symbols

Symbol	$<$	$>$	$\leq$	$\geq$
Key Phrases	• is less than • is fewer than	• is greater than • is more than	• is less than or equal to • is at most • is no more than • maximum	• is greater than or equal to • is at least • is no less than • minimum

Example 1 Writing Inequalities

Write each word sentence as an inequality.

a. A number c is less than -4.

A number c | is less than | -4.

c $<$ -4

▶ An inequality is $c < -4$.

b. A number k plus 5 is greater than or equal to 8.

A number k plus 5 | is greater than or equal to | 8.

$k + 5$ $\geq$ 8

▶ An inequality is $k + 5 \geq 8$.

c. 16 is no more than 4 times a number q.

16 | is no more than | 4 times a number q.

16 $\leq$ $4q$

▶ An inequality is $16 \leq 4q$.

Try It **Write the word sentence as an inequality.**

1. A number n is greater than 1.

$n > 1$

2. Twice a number p is fewer than 7.

$2p < 7$

3. A number w minus 3 is less than or equal to 10.

$w - 3 \leq 10$

4. -6 is at least a number z divided by 2.

$-6 \geq \frac{z}{2}$

GO DIGITAL

A **solution of an inequality** is a value that makes the inequality true. An inequality can have more than one solution. The set of all solutions of an inequality is called the **solution set**.

Value of x	$x + 3 \le 7$	Is the inequality true?
3	$3 + 3 \overset{?}{\le} 7$ $6 \le 7$ ✓	yes
4	$4 + 3 \overset{?}{\le} 7$ $7 \le 7$ ✓	yes
5	$5 + 3 \overset{?}{\le} 7$ $8 \not\le 7$ ✗	no

Reading

The symbol $\not\le$ means *is not less than or equal to.*

Example 2 Checking Solutions

Tell whether the given value is a solution of the inequality.

a. $x + 1 > 7;\ x = 8$

$x + 1 > 7$ — Write the inequality.

$8 + 1 \overset{?}{>} 7$ — Substitute 8 for x.

$9 > 7$ ✓ — Add. 9 is greater than 7.

▶ So, 8 is a solution of the inequality.

b. $7y < -29;\ y = -4$

$7y < -29$ — Write the inequality.

$7(-4) \overset{?}{<} -29$ — Substitute -4 for y.

$-28 \not< -29$ ✗ — Multiply. -28 is *not* less than -29.

▶ So, -4 is *not* a solution of the inequality.

c. $5z \ge z + 12;\ z = 3$

$5z \ge z + 12$ — Write the inequality.

$5(3) \overset{?}{\ge} 3 + 12$ — Substitute 3 for z.

$15 \ge 15$ ✓ — Simplify. 15 is greater than or equal to 15.

▶ So, 3 is a solution of the inequality.

GO DIGITAL

Try It **Tell whether 3 is a solution of the inequality.**

5. $-b + 10 < 6$

No

6. $3n - n \geq 6$

yes $6 \geq 6$

7. $10 \leq 18 \div x$

no

The **graph of an inequality** shows all the solutions of the inequality on a number line. An open circle, ○, is used when a number is *not* a solution. A closed circle, ●, is used when a number is a solution. An arrow to the left or right shows that the graph continues in that direction.

Example 3 Graphing an Inequality

Reading

The inequality $g > 2$ is the same as $2 < g$.

Graph $g > 2$.

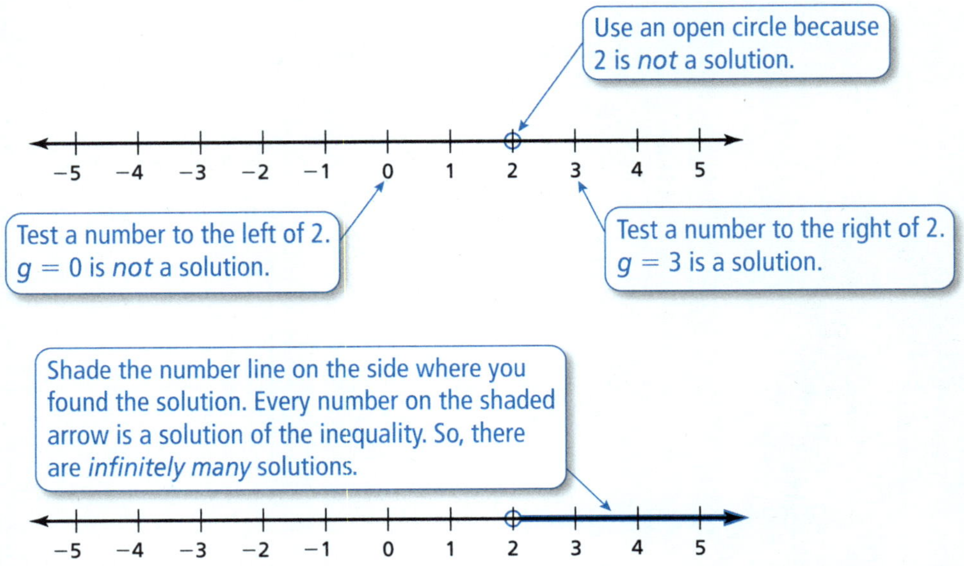

Try It **Graph the inequality on a number line.**

8. $a < 4$

9. $f \leq 7$

10. $n > 0$

11. $-3 \leq p$

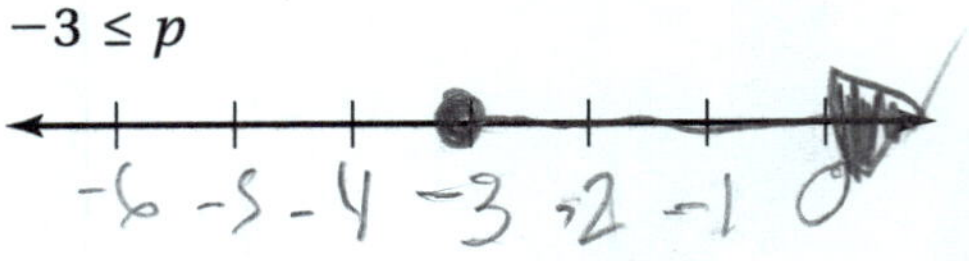

GO DIGITAL

In-Class Practice

1 I don't understand yet. 2 I can do it with help. 3 I can do it on my own. 4 I can teach someone else.

4 MTR **12. DIFFERENT WORDS, SAME QUESTION** Which is different? Write "both" inequalities.

A number n is at most 3.	A number n is no more than 3.
A number n is less than or equal to 3.	A number n is at least 3.

CHECKING SOLUTIONS **Tell whether the given value is a solution of the inequality.**

13. $p + 5 \leq 12; p = 6$

14. $w + 12 < -4; w = -16$

GRAPHING AN INEQUALITY **Graph the inequality on a number line.**

15. $n > 8$

16. $q \leq -4$

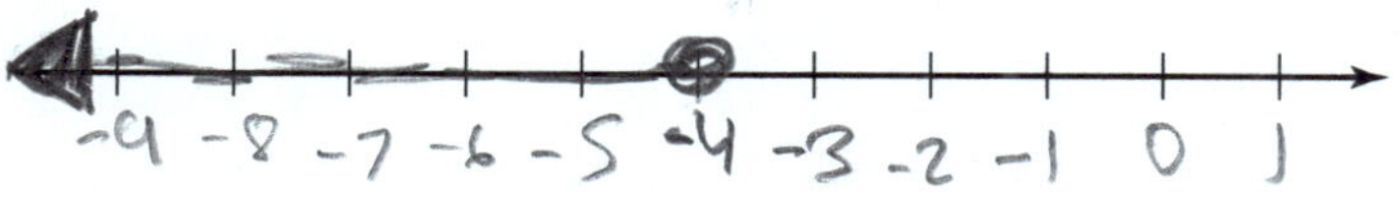

17. $5 < s$

Example 4 Modeling Real Life

7 MTR

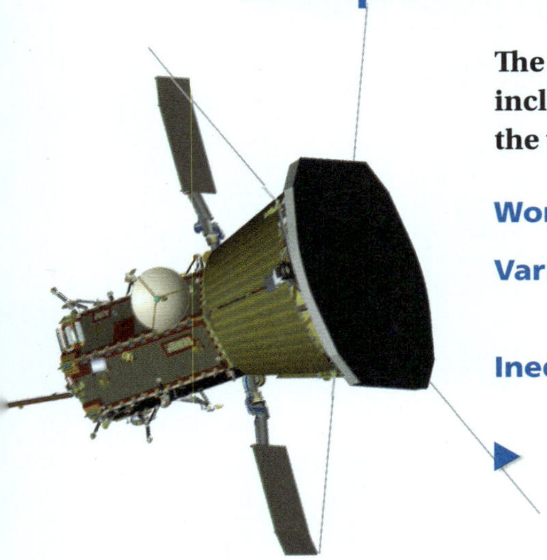

The NASA *Solar Probe Plus* can withstand temperatures up to and including 2600°F. Write and graph an inequality that represents the temperatures the probe can withstand.

Words	temperatures	up to and including	2600°F
Variable	Let t be the temperatures (in degrees Fahrenheit) that the probe can withstand.		
Inequality	t	$\leq$	2600

▶ An inequality is $t \leq 2600$.

Check The graph shows that 2000°F is a solution. Check this in the inequality.

$$t \leq 2600$$

$$2000 \leq 2600 \checkmark$$

In-Class Practice

1 I don't understand yet. | 2 I can do it with help. | 3 I can do it on my own. | 4 I can teach someone else.

18. You need to train for at least 6 hours and 45 minutes to obtain a babysitting license. Describe the amounts of time that you can train and still *not* obtain a license. Use a graph to justify your answer.

19. **Dig Deeper** You control a drone from a floor that is halfway to the top of a building. The farthest away from you that the drone can fly is 200 meters. The building is 380 meters tall. Can the drone reach the top of the building? Explain your reasoning.

20. Each visit to a water park costs \$19.95. An annual pass to the park costs \$89.95. Write an inequality that represents the numbers of times you would need to visit the park for the pass to be a better deal.

GO DIGITAL

7.5 Practice WITH CalcChat® AND CalcView®

Review & Refresh

Solve the equation. Check your solution.

1. $5x = -50$

2. $\frac{y}{9} = 6$

3. $\frac{t}{12} = 4$

4. $8x = 72$

Find the distance between the points.

5. $(2, 8), (6, 8)$

6. $(-5, 9), (7, 9)$

7. $(-3, 6), (-3, -2)$

8. A stack of boards is 24 inches high. The thickness of each board is $\frac{3}{8}$ inch. How many boards are in the stack?

Divide.

9. $12.3 \div 4.1$

10. $80.45 \div 2.5$

11. $21.2 \div 3.2$

Concepts, Skills, & Problem Solving

UNDERSTANDING INEQUALITY STATEMENTS **Choose a number that makes the statement true. What other numbers make the statement true?** (See Exploration 1.)

12. You are less than 3 miles from home.

13. You need at least \$5 for lunch.

14. You buy more than 2 movie tickets.

15. A game lasts no more than 10 minutes.

GO DIGITAL

WRITING INEQUALITIES **Write the word sentence as an inequality.** (See Example 1.)

16. A number k is less than 10.

▶**17.** A number a is more than 6.

18. A number z is fewer than 2.

19. A number b is at least -3.

20. -12 is at least 3 times a number y.

21. 25 is no more than 7 plus a number p.

22. One plus a number y is no more than -13.

23. A number x divided by 3 is at most 5.

CHECKING SOLUTIONS **Tell whether the given value is a solution of the inequality.** (See Example 2.)

24. $x - 1 \leq 7;\ x = 6$

▶**25.** $y + 5 < 13;\ y = 17$

26. $3z > 6;\ z = 3$

27. $6 \leq \frac{b}{2};\ b = 10$

28. $c + 2 < 4;\ c = 2$

29. $a \leq 0;\ a = -5$

30. $4g + 7 > 2g + 15; g = 3$

31. $5s - s < 15; s = 3$

MATCHING **Match the inequality with its graph.**

32. $x \geq -2$

33. $x < -2$

34. $x > -2$

35. $x \leq -2$

A.
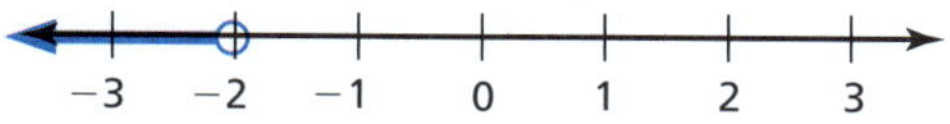

C.
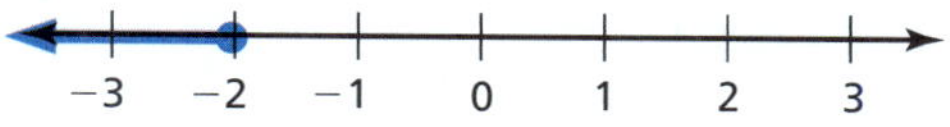

B.
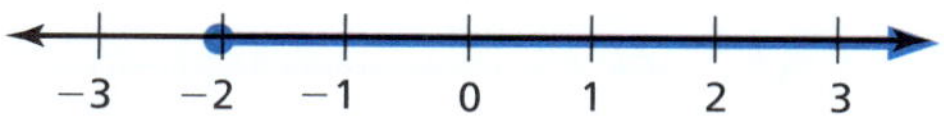

D.
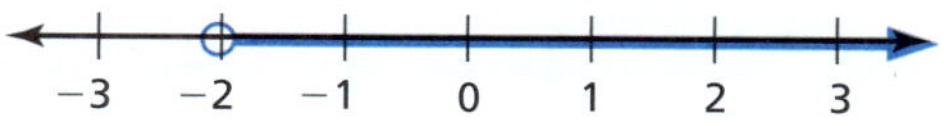

GRAPHING AN INEQUALITY **Graph the inequality on a number line.** (See Example 3.)

36. $a > 4$

▶ **37.** $n \geq 8$

38. $3 \geq x$

39. $y < -7$

40. $x < 9$

41. $-3 \geq c$

42. $m > -5$

43. $0 \leq b$

44. $1 > f$

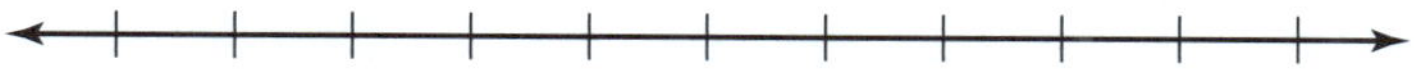

45. $t \geq -2$

46. $p > -15$

47. $12 \geq z$

OPEN-ENDED **Write an inequality and a word sentence that represent the graph.**

48. 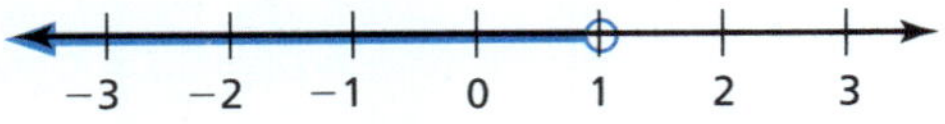

49. 

50. 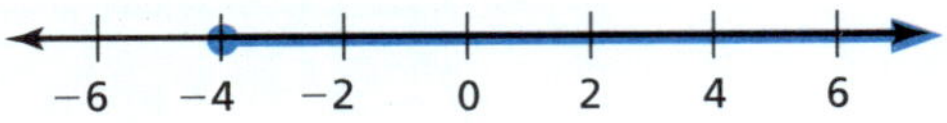

51. 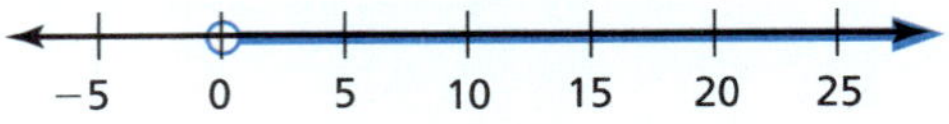

4 MTR **YOU BE THE TEACHER** **Your friend graphs the inequality. Is your friend correct? Explain your reasoning.**

52. 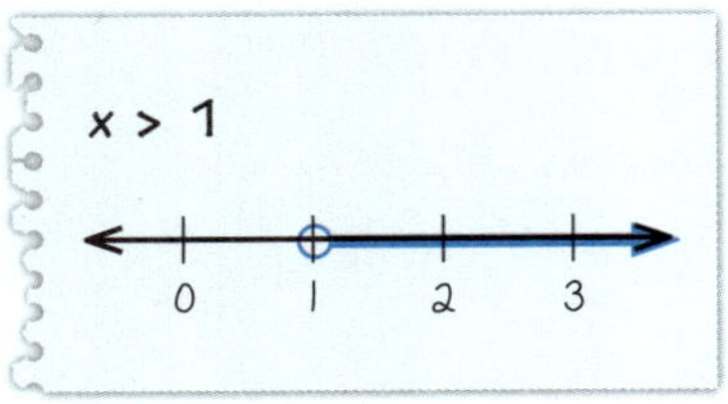

53. 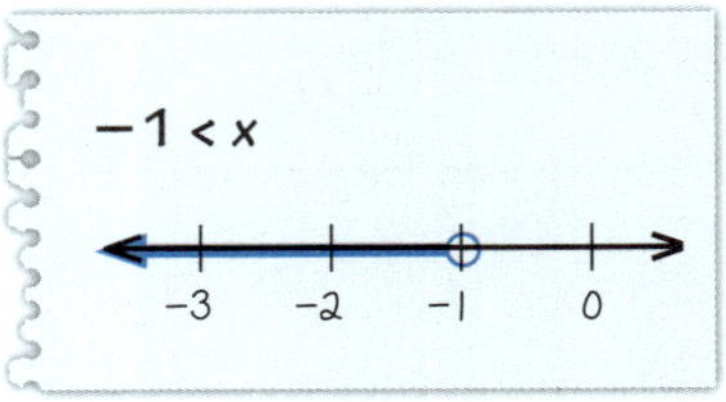

7 MTR **54.** **MODELING REAL LIFE** The world record for the largest collection of flamingo-related items is held by a woman in Ocala, Florida. Her collection contains 793 items. Write and graph an inequality that represents the numbers of items that would set a new world record. (See Example 4.)

7 MTR **55.** **MODELING REAL LIFE** You are allowed to keep at most 3 striped bass while fishing. Each striped bass must be no less than 18 inches long.

a. Write and graph an inequality that represents the numbers of striped bass you are allowed to keep.

b. Write and graph an inequality that represents the lengths of striped bass you are allowed to keep.

56. REASONING You want to buy a necklace and one other item from the list. You have $33.

Item	Price (with tax)
T-shirt	$15
Book	$20
Sunglasses	$13
Necklace	$16

a. Write an inequality that represents the situation.

b. Can the other item be a T-shirt? Explain.

c. Can the other item be a book? Explain.

57. REASONING For a food to be labeled *low sodium,* there must be no more than 140 milligrams of sodium per serving.

Nutrition Facts

4 servings per container

Serving size **½ cup (114g)**

Amount per serving

Calories **90**

	% Daily Value*
Total Fat 3g	**5%**
Saturated Fat 0g	**0%**
Cholesterol 0mg	**0%**
Sodium 300mg	**13%**
Total Carbohydrate 13g	**4%**
Dietary Fiber 3g	**12%**
Sugars 3g	
Protein 3g	

a. Write and graph an inequality that represents the amount of sodium in a low-sodium serving.

b. Write and graph an inequality that represents the amount of sodium in a serving that does *not* qualify as low sodium.

c. Does the food represented by the nutrition facts label qualify as a low-sodium food? Explain.

4 MTR **DISCUSS MATHEMATICAL THINKING** **Determine whether the statement is *always*, *sometimes*, or *never* true. Explain your reasoning.**

58. A number that is a solution of the inequality $x > 5$ is also a solution of the inequality $x \geq 5$.

59. A number that is a solution of the inequality $5 \leq x$ is also a solution of the inequality $x > 5$.

60. **PROBLEM SOLVING** A subway ride costs \$2. A 30-day subway pass costs \$36. Write an inequality that represents the numbers of subway rides you would need to take for the pass to be a better deal.

61. **PROBLEM SOLVING** Fifty people are seated in a commercial airplane. The maximum capacity of the airplane is 425 people. Write an inequality that represents the numbers of additional people who can be seated.

62. **Dig Deeper** The map shows the elevations above sea level for an area of land.

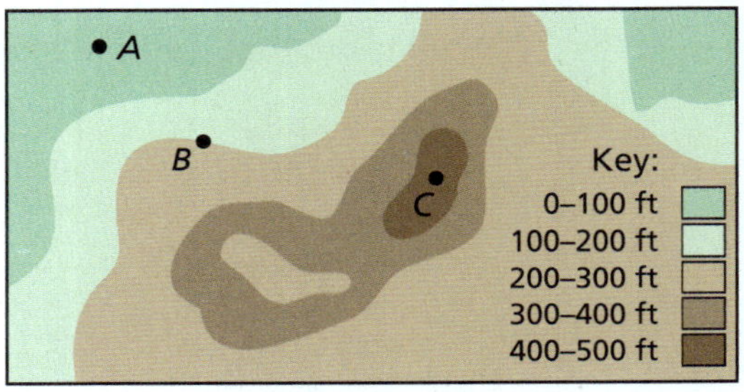

a. Graph the possible elevations of A. Write the set of elevations as two inequalities.

b. Graph the possible elevations of C. How can you write this set of elevations as a single inequality? Explain.

c. What is the elevation of B? Explain.

7.6 Solving Inequalities Using Addition or Subtraction

Learning Target: Write and solve inequalities using addition or subtraction.

Success Criteria:
- I can apply the Addition and Subtraction Properties of Inequality to produce equivalent inequalities.
- I can solve inequalities using addition or subtraction.
- I can apply inequalities involving addition or subtraction to solve real-life problems.

Algebraic Reasoning

MA.7.AR.2.1 Write and solve one-step inequalities in one variable within a mathematical context and represent solutions algebraically or graphically.

Exploration 1 Writing Inequalities

Work with a partner. Use two number cubes on which the odd numbers are negative on one of the number cubes and the even numbers are negative on the other number cube.

- Roll the number cubes. Write an inequality that compares the numbers.
- Roll one of the number cubes. Add the number to each side of the inequality and record your result.
- Repeat the previous two steps five more times.

GO DIGITAL

a. When you add the same number to each side of an inequality, does the inequality remain true? Explain your reasoning.

b. When you subtract the same number from each side of an inequality, does the inequality remain true? Use inequalities generated by number cubes to justify your answer.

c. Use your results in parts (a) and (b) to make a conjecture about how to solve an inequality of the form $x + a < b$ for x.

6 MTR **ASSESS REASONABLENESS**

Use your conjecture to solve $x + 3 < 1$. Does the solution make sense?

7.6 Lesson

Key Ideas

> You can solve inequalities in the same way you solve equations. Use inverse operations to get the variable by itself.

Addition Property of Inequality

Words When you add the same number to each side of an inequality, the inequality remains true.

Numbers

$$\begin{array}{rcr} -4 & < & 3 \\ +2 & & +2 \\ \hline -2 & < & 5 \end{array}$$

Algebra If $a < b$, then $a + c < b + c$.

If $a > b$, then $a + c > b + c$.

Subtraction Property of Inequality

Words When you subtract the same number from each side of an inequality, the inequality remains true.

Numbers

$$\begin{array}{rcr} -2 & < & 2 \\ -3 & & -3 \\ \hline -5 & < & -1 \end{array}$$

Algebra If $a < b$, then $a - c < b - c$.

If $a > b$, then $a - c > b - c$.

These properties are also true for $\leq$ and $\geq$.

Example 1 Solving an Inequality Using Addition

Solve $x - 5 < -3$. Graph the solution.

$x - 5 < -3$	Write the inequality.
$+5 \quad +5$ (Undo the subtraction.)	Addition Property of Inequality
$x < 2$	Simplify.

▶ The solution is $x < 2$.

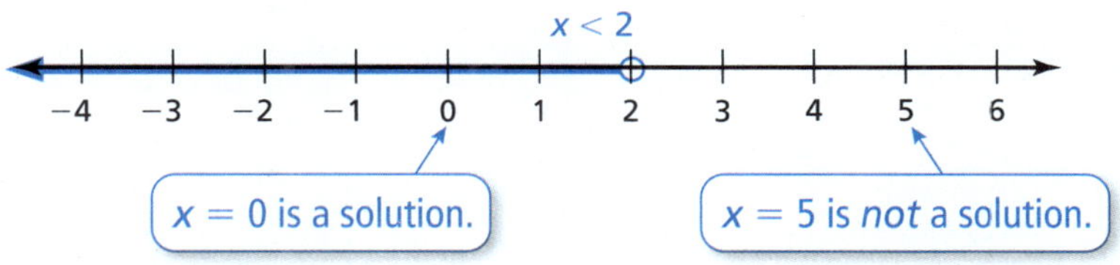

Check

$x = 0$: $0 - 5 \overset{?}{<} -3$

$-5 < -3$ ✓

$x = 5$: $5 - 5 \overset{?}{<} -3$

$0 \not< -3$ ✗

GO DIGITAL

Try It

Solve the inequality. Graph the solution.

1. $y - 6 > -7$

2. $b - 3.8 \leq 1.7$

3. $-\frac{1}{2} > z - \frac{1}{4}$

Example 2 Solving an Inequality Using Subtraction

Solve $13 \leq x + 14$. Graph the solution.

	$13 \leq x + 14$	Write the inequality.
Undo the addition. →	$\underline{-14} \quad \underline{-14}$	Subtraction Property of Inequality
	$-1 \leq x$	Simplify.

Reading

The inequality $-1 \leq x$ is the same as $x \geq -1$.

▶ The solution is $x \geq -1$.

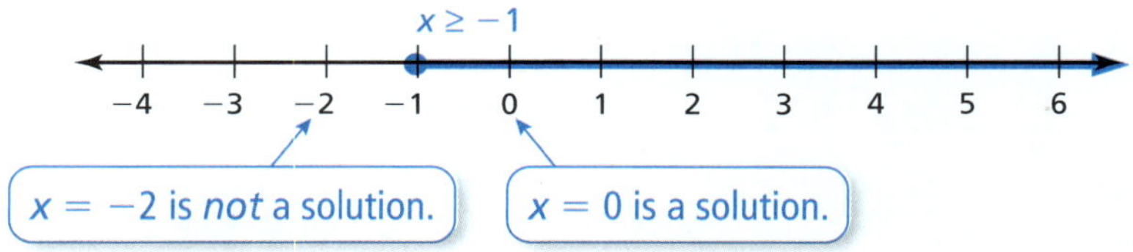

Try It

Solve the inequality. Graph the solution.

4. $w + 3 \leq -1$

5. $8.5 \geq d + 10$

6. $x + \frac{3}{4} > 1\frac{1}{2}$

GO DIGITAL

In-Class Practice

1 I don't understand yet. | 2 I can do it with help. | 3 I can do it on my own. | 4 I can teach someone else.

7. WRITING Are the inequalities $c + 3 > 5$ and $c - 1 > 1$ equivalent? Explain.

4 MTR

8. WHICH ONE DOESN'T BELONG? Which inequality does *not* belong with the other three? Explain your reasoning.

$w + \frac{7}{4} < \frac{3}{4}$	$w - \frac{3}{4} > -\frac{7}{4}$
$w + \frac{7}{4} > \frac{3}{4}$	$-\frac{7}{4} < w - \frac{3}{4}$

SOLVING AN INEQUALITY **Solve the inequality. Graph the solution.**

9. $x - 4 > -6$

10. $z + 4.5 \leq 3.25$

11. $\frac{7}{10} > \frac{4}{5} + g$

12. OPEN-ENDED Write two different inequalities that can be represented using the graph. Justify your answers.

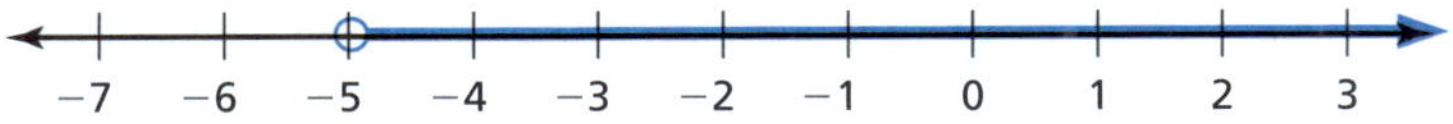

Example 3 Modeling Real Life 7 MTR

To become an astronaut pilot for NASA, a person can be no taller than 6.25 feet. Your friend is 5 feet 9 inches tall. How much can your friend grow and still meet the requirement?

Because the height requirement is given in feet and your friend's height is given in feet and inches, rewrite your friend's height in feet.

Convert 9 inches to feet. Because there are 12 inches in 1 foot and you are converting from a smaller unit to a larger unit, divide 9 by 12.

$$\begin{array}{r} 0.75 \\ 12\overline{)9.00} \\ \underline{84} \\ 60 \\ \underline{60} \\ 0 \end{array}$$

$$5 \text{ ft } 9 \text{ in.} = 5 \text{ ft} + 0.75 \text{ ft} = 5.75 \text{ ft}$$

Use a verbal model to write an inequality that represents the situation.

Verbal Model

Current height (feet)	+	Amount your friend can grow (feet)	≤	Height limit (feet)

Variable Let h be the possible amounts (in feet) your friend can grow.

Inequality $5.75 + h \leq 6.25$

$$\begin{aligned} 5.75 + h &\leq 6.25 && \text{Write the inequality.} \\ -5.75 \quad &\quad -5.75 && \text{Subtraction Property of Inequality} \\ h &\leq 0.5 && \text{Simplify.} \end{aligned}$$

▶ So, your friend can grow no more than 0.5 foot, or $0.5 \times 12 = 6$ inches.

In-Class Practice

1 I don't understand yet. 2 I can do it with help. 3 I can do it on my own. 4 I can teach someone else.

13. **Dig Deeper** A volcanologist rappels 1200 feet into a volcano. He wants to climb out of the volcano in less than 4 hours. He climbs the first 535 feet in 100 minutes. Graph an inequality that represents the average speeds at which he can climb the remaining distance and meet his goal. Justify your answer.

14. You install a mailbox by burying a post as shown. According to postal service guidelines, the bottom of the box must be at least 41 inches, but no more than 45 inches, above the road. Write and interpret two inequalities that describe the possible lengths of the post.

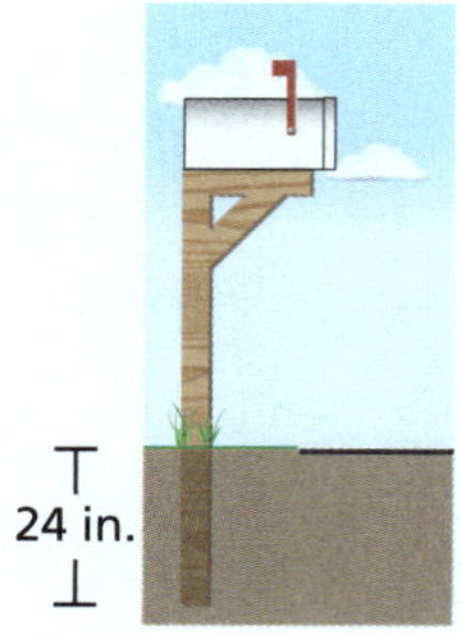

GO DIGITAL

7 MTR **29. MODELING REAL LIFE** A large airboat can hold 30 passengers. Seventeen passengers board the airboat. (See Example 3.)

a. Write and solve an inequality that represents the additional numbers of passengers that can board the airboat.

b. Can 14 more passengers board the airboat? Explain.

GEOMETRY **Find the possible values of x.**

30. The perimeter is less than 28 feet.

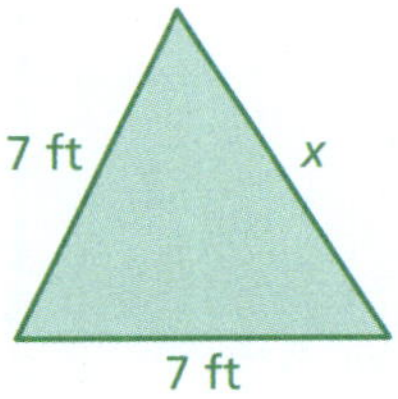

31. The base is greater than the height.

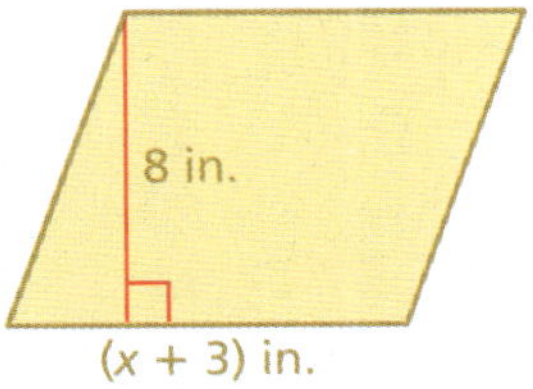

32. The perimeter is less than or equal to 51 meters.

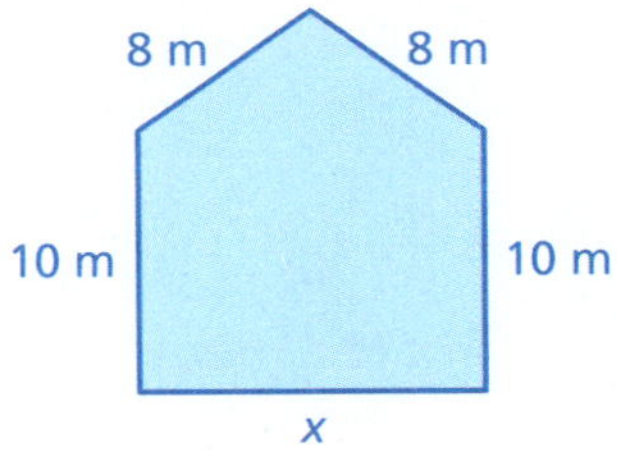

33. REASONING The inequality $d + s > -3$ is equivalent to $d > -7$. What is the value of s?

34. REASONING You can spend up to \$35 at a gift shop at Disney's Animal Kingdom.

a. You want to buy a shirt that costs \$14. Write and solve an inequality that represents the remaining amounts of money you can spend if you buy the shirt.

b. You notice that the shirt is on sale for \$10.50 less than its original price. How does this change your inequality in part (a)?

35. Dig Deeper If items plugged into a circuit use more than 2400 watts of electricity, the circuit overloads. A portable heater that uses 1050 watts of electricity is plugged into the circuit.

Item	Watts
Aquarium	200
Hair dryer	1200
Television	150
Vacuum cleaner	1100

a. Find the additional numbers of watts you can plug in without overloading the circuit.

b. In addition to the portable heater, what two other items in the table can you plug in at the same time without overloading the circuit? Is there more than one possibility? Explain.

36. NUMBER SENSE The possible values of x are given by $x + 8 \leq 6$. What is the greatest possible value of $7x$? Explain your reasoning.

7.7 Solving Inequalities Using Multiplication or Division

Learning Target: Write and solve inequalities using multiplication or division.

Success Criteria:
- I can apply the Multiplication and Division Properties of Inequality to produce equivalent inequalities.
- I can solve inequalities using multiplication or division.
- I can apply inequalities involving multiplication or division to solve real-life problems.

Algebraic Reasoning
MA.7.AR.2.1 Write and solve one-step inequalities in one variable within a mathematical context and represent solutions algebraically or graphically.

Exploration 1 Writing Inequalities

Work with a partner. Use two number cubes on which the odd numbers are negative on one of the number cubes and the even numbers are negative on the other number cube.

- Roll the number cubes. Write an inequality that compares the numbers.
- Roll one of the number cubes. Multiply each side of the inequality by the number and record your result.
- Repeat the previous two steps nine more times.

a. When you multiply each side of an inequality by the same number, does the inequality remain true? Explain your reasoning.

b. When you divide each side of an inequality by the same number, does the inequality remain true? Use inequalities generated by number cubes to justify your answer.

c. Use your results in parts (a) and (b) to make a conjecture about how to solve an inequality of the form $ax < b$ for x when $a > 0$ and when $a < 0$.

4 MTR **CONSTRUCT AN ARGUMENT**

Is $2a \geq a$ true for every possible value of a? Explain your reasoning.

7.7 Lesson

Key Idea

Multiplication and Division Properties of Inequality (Case 1)

Words When you multiply or divide each side of an inequality by the same *positive* number, the inequality remains true.

Numbers

$$-4 < 6 \qquad\qquad 4 > -6$$

$$2 \cdot (-4) < 2 \cdot 6 \qquad\qquad \frac{4}{2} > \frac{-6}{2}$$

$$-8 < 12 \qquad\qquad 2 > -3$$

Algebra If $a < b$ and c is positive, then

$$a \cdot c < b \cdot c \quad \text{and} \quad \frac{a}{c} < \frac{b}{c}.$$

If $a > b$ and c is positive, then

$$a \cdot c > b \cdot c \quad \text{and} \quad \frac{a}{c} > \frac{b}{c}.$$

These properties are also true for $\le$ and $\ge$.

Example 1 Solving an Inequality Using Multiplication

Solve $\frac{x}{5} \le -3$. Graph the solution.

$\frac{x}{5} \le -3$	Write the inequality.
$5 \cdot \frac{x}{5} \le 5 \cdot (-3)$	Multiplication Property of Inequality
$x \le -15$	Simplify.

Undo the division. → $5 \cdot \frac{x}{5} \le 5 \cdot (-3)$

▶ The solution is $x \le -15$.

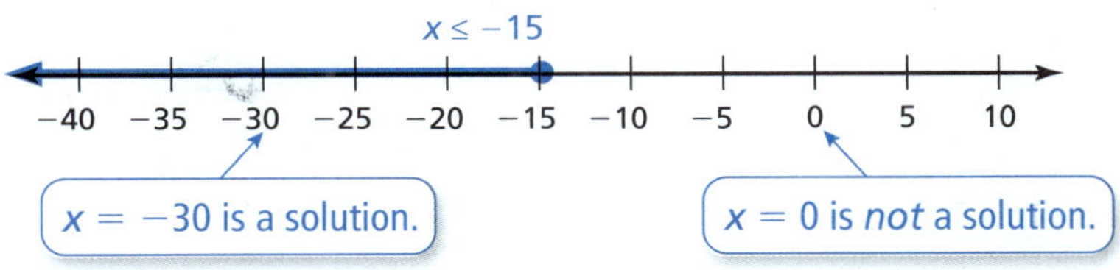

GO DIGITAL

Try It Solve the inequality. Graph the solution.

1. $n \div 3 < 1$

2. $-0.5 \leq \frac{m}{10}$

3. $-3 > \frac{2}{3}p$

Example 2 Solving an Inequality Using Division

Solve $6x > -18$. Graph the solution.

$6x > -18$ — Write the inequality.

Undo the multiplication. → $\frac{6x}{6} > \frac{-18}{6}$ — Division Property of Inequality

$x > -3$ — Simplify.

▶ The solution is $x > -3$.

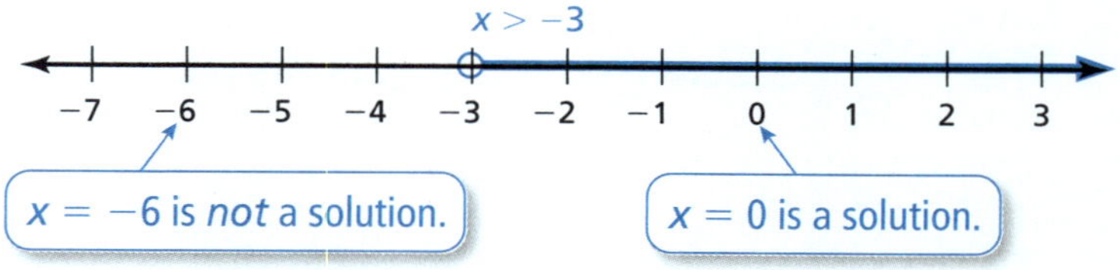

Try It Solve the inequality. Graph the solution.

4. $4b \geq 2$

5. $12k \leq -24$

6. $-15 < 2.5q$

GO DIGITAL

10
0.5

16

Key Idea

Multiplication and Division Properties of Inequality (Case 2)

Words When you multiply or divide each side of an inequality by the same *negative* number, the direction of the inequality symbol must be reversed for the inequality to remain true.

Numbers

$$-4 < 6 \qquad\qquad 4 > -6$$

$$-2 \cdot (-4) > -2 \cdot 6 \qquad\qquad \frac{4}{-2} < \frac{-6}{-2}$$

$$8 > -12 \qquad\qquad -2 < 3$$

Algebra If $a < b$ and c is negative, then

$$a \cdot c > b \cdot c \quad \text{and} \quad \frac{a}{c} > \frac{b}{c}.$$

If $a > b$ and c is negative, then

$$a \cdot c < b \cdot c \quad \text{and} \quad \frac{a}{c} < \frac{b}{c}.$$

These properties are also true for ≤ and ≥.

Common Error

A negative sign in an inequality does not necessarily mean you must reverse the inequality symbol. Reverse the inequality symbol only when you multiply or divide each side by a negative number.

Example 3 Solving an Inequality Using Multiplication

Solve $-\frac{3}{2}n \leq 6$. Graph the solution.

$-\frac{3}{2}n \leq 6$	Write the inequality.
$-\frac{2}{3} \cdot \left(-\frac{3}{2}n\right) \geq -\frac{2}{3} \cdot 6$	Use the Multiplication Property of Inequality. Reverse the inequality symbol.
$n \geq -4$	Simplify.

▶ The solution is $n \geq -4$.

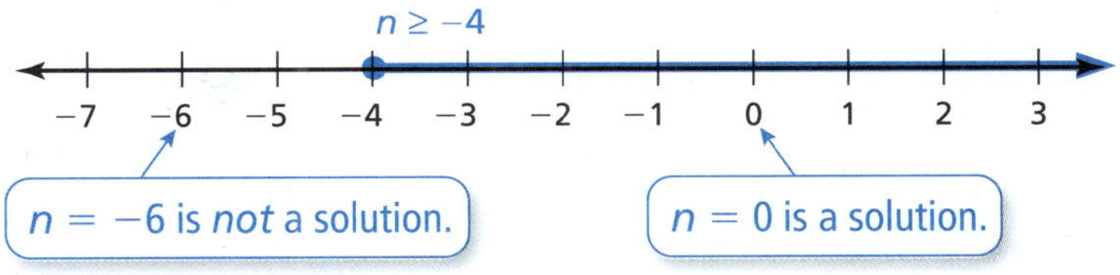

3 MTR REFLECT ON YOUR METHOD

Why do you reverse the inequality symbol when solving in Example 3, but not when solving in Examples 1 and 2?

GO DIGITAL

Try It

Solve the inequality. Graph the solution.

7. $\frac{x}{-3} > -4$

8. $0.5 \le -\frac{y}{2}$

9. $-12 \ge \frac{6}{5}m$

10. $-\frac{2}{5}h \le -8$

In-Class Practice

1 I don't understand yet.	2 I can do it with help.	3 I can do it on my own.	4 I can teach someone else.

11. OPEN-ENDED Write an inequality that you can solve using the Division Property of Inequality where the direction of the inequality symbol must be reversed.

5 MTR **12. STRUCTURE** Explain how solving $4x < -16$ is different from solving $-4x < 16$.

In-Class Practice

1 I don't understand yet. | 2 I can do it with help. | 3 I can do it on my own. | 4 I can teach someone else.

SOLVING AN INEQUALITY **Solve the inequality. Graph the solution.**

13. $6n < -42$

14. $4 \geq -\frac{g}{8}$

15. **WRITING** Are the inequalities $12c > -15$ and $4c < -5$ equivalent? Explain.

$\frac{5}{6} \cdot -\frac{12}{1} =$

Example 4 Modeling Real Life

7 MTR

A submersible descends 37.5 meters per minute from sea level to explore a recent shipwreck. The shipwreck rests on the ocean floor at an elevation deeper than −3810 meters, the elevation of the *Titanic* shipwreck. How long does it take the submersible to reach the shipwreck?

The submersible begins at sea level, where the elevation is 0 feet. Write and solve an inequality to find how long it takes to reach elevations below −3810 meters.

Verbal Model Change in elevation (meters per minute) • Time (minutes) < Elevation of *Titanic* (meters)

Variable Let t be the time (in minutes) that the submersible descends.

Inequality $-37.5 \quad \cdot \quad t \quad < \quad -3810$

The change in elevation is negative because the submersible is descending.

$-37.5t < -3810$ Write the inequality.

$\frac{-37.5t}{-37.5} > \frac{-3810}{-37.5}$ Division Property of Inequality

$t > 101.6$ Simplify.

▶ So, the submersible takes more than 101.6 minutes, or 1 hour, 41 minutes, and 36 seconds, to reach the shipwreck.

GO DIGITAL

In-Class Practice

1 I don't understand yet. 2 I can do it with help. 3 I can do it on my own. 4 I can teach someone else.

16. You want to put up a fence that encloses a triangular region with an area greater than or equal to 60 square feet. Describe the possible values of *c*.

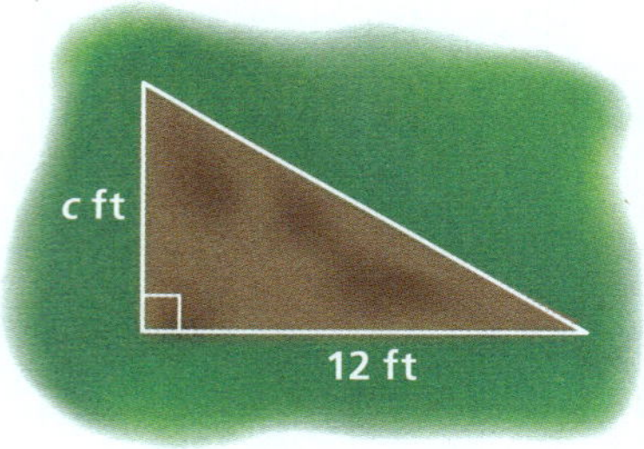

17. A motorcycle rider travels at an average speed greater than 50 miles per hour. Write and solve an inequality to determine how long it will take the motorcycle rider to travel 375 miles. Explain your reasoning.

7.7 Practice WITH CalcChat® AND CalcView®

Review & Refresh

Solve the inequality. Graph the solution.

1. $h + 4 < 6$

2. $c - 5 \geq 4$

3. $\frac{7}{10} \leq n + \frac{4}{5}$

Solve the equation. Check your solution.

4. $-12 = w + 4$

5. $\frac{v}{5} = 3$

6. $3x = 18$

7. $-17 = m - 5$

8. Simplify $\frac{2}{3} + \left(-\frac{5}{7}\right)$.

Concepts, Skills, & Problem Solving

WRITING AN INEQUALITY Write an inequality that compares the given numbers. Does the inequality remain true when you multiply each number in the inequality by 2? by −2? Justify your answers. (See Exploration 1.)

9. $-2; 5$

10. $4; -1$

11. $6; -3$

GO DIGITAL

SOLVING AN INEQUALITY **Solve the inequality. Graph the solution.** (See Examples 1 and 2.)

12. $2n > 20$

13. $\frac{c}{9} \leq -4$

14. $2.2m < 11$

15. $-16 > x \div 2$

16. $\frac{1}{6}w \geq 2.5$

17. $7 < 3.5k$

18. $3x \leq -\frac{5}{4}$

19. $4.2y \leq -12.6$

20. $11.3 > \frac{b}{4.3}$

7 MTR

21. **MODELING REAL LIFE** You earn \$9.20 per hour at your summer job. Write and solve an inequality that represents the numbers of hours you can work to earn enough money to buy a smartphone that costs \$299.

22. **Dig Deeper** You have \$5.60 to buy avocados for a guacamole recipe. Avocados cost \$1.40 each.

a. Write and solve an inequality that represents the numbers of avocados you can buy.

b. Are there infinitely many solutions in this context? Explain.

SOLVING AN INEQUALITY **Solve the inequality. Graph the solution.** (See Example 3.)

23. $-5n \leq 15$

24. $7w > -49$

25. $-\frac{1}{3}h \geq 8$

26. $-9 < -\frac{1}{5}x$

27. $-3y < -14$

28. $-2d \geq 26$

29. $-4.5 > \frac{m}{6}$

30. $\frac{k}{-0.25} \leq 36$

31. $-2.4 > \frac{b}{-2.5}$

YOU BE THE TEACHER **Your friend solves the inequality. Is your friend correct? Explain your reasoning.**

32.

$$\frac{x}{3} < -9$$
$$3 \cdot \frac{x}{3} > 3 \cdot (-9)$$
$$x > -27$$

33.

$$-3m \geq 9$$
$$\underline{+3m} \quad \underline{+3m}$$
$$0 \geq 9 + 3m$$
$$\underline{-9} \quad \underline{-9}$$
$$-9 \geq 3m$$
$$\frac{-9}{3} \geq \frac{3m}{3}$$
$$-3 \geq m$$

WRITING AND SOLVING AN INEQUALITY **Write the word sentence as an inequality. Then solve the inequality.**

34. The quotient of a number g and -4 is at most 5.

35. A number p divided by 7 is less than -3.

36. Six times a number w is at least -24.

37. The product of -2 and a number x is greater than 30.

38. $\frac{3}{4}$ is greater than or equal to a number k divided by -8.

7 MTR

39. **MODELING REAL LIFE** A *cryotherapy* chamber uses extreme cold to reduce muscle soreness. A chamber is currently 0°F. The temperature in the chamber is dropping 2.5°F every second. Write and solve an inequality that represents the numbers of seconds that can pass for the temperature to drop below −20°F. (See Example 4.)

7 MTR

40. **MODELING REAL LIFE** You are moving some of your belongings into a storage facility.

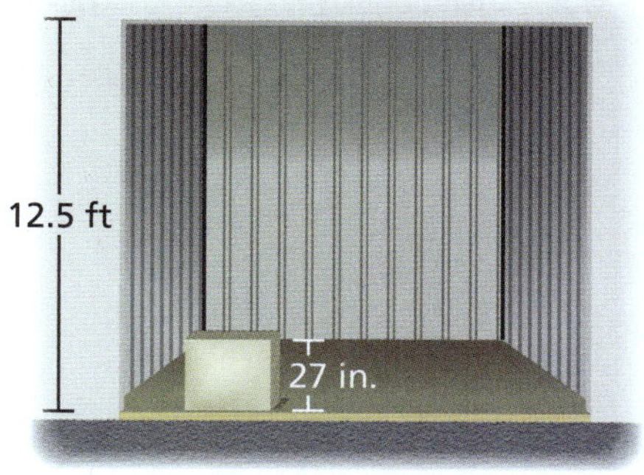

a. Write and solve an inequality that represents the numbers of boxes that you can stack vertically in the storage unit.

b. Can you stack 6 boxes vertically in the storage unit? Explain.

GEOMETRY **Write and solve an inequality that represents *x*.**

41. Area $\geq 120\text{ cm}^2$

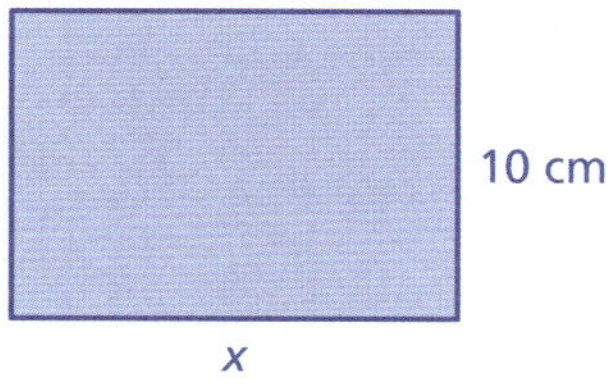

42. Area $< 20\text{ ft}^2$

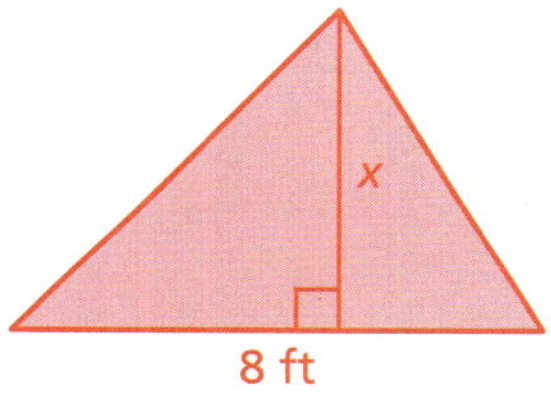

7 MTR

43. **MODELING REAL LIFE** A device extracts no more than 37 liters of water per day from the air. How long does it take to collect at least 185 liters of water? Explain your reasoning.

44. **REASONING** Students in a science class are divided into 6 equal groups with at least 4 students in each group for a project. Describe the possible numbers of students in the class.

7 MTR

45. INVESTIGATE Choose two novels to research.

a. Use the Internet to complete the table below.

b. Use the table to find and compare the average number of copies sold per month for each novel. Which novel do you consider to be the most successful? Explain.

c. Assume each novel continues to sell at the average speed. For what numbers of months will the total number of copies sold exceed twice the current number sold for each novel?

Author	Name of Novel	Release Date	Current Number of Copies Sold
1.			
2.			

46. REASONING When you multiply or divide each side of an inequality by the same negative number, you must reverse the direction of the inequality symbol. Explain why.

3 MTR

ADAPT A PROCEDURE **Describe all numbers that satisfy *both* inequalities. Include a graph with your description.**

47. $4m > -4$ and $3m < 15$

48. $\frac{n}{3} \geq -4$ and $\frac{n}{-5} \geq 1$

49. $2x \geq -6$ and $2x \geq 6$

50. $-\frac{1}{2}s > -7$ and $\frac{1}{3}s < 12$

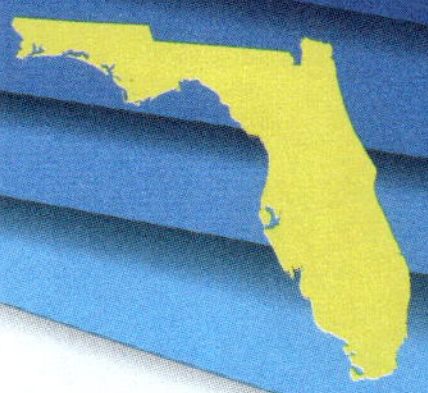

7 Connecting Concepts

5 MTR *Mathematicians who use patterns and structure to help understand and connect mathematical concepts relate previously learned concepts to new concepts.*

1 MTR **1. ANALYZE A PROBLEM** A tornado forms 63,360 feet from a weather station. It travels away from the station at an average speed of 15 miles per hour. How many hours have passed when the tornado is 221,760 feet away from the station?

Using the Problem-Solving Plan

Understand the problem.

You know the total distance between the tornado and the station, and the average speed the tornado is traveling away from the station. You are asked to determine how many hours have passed when the tornado is 221,760 feet away from the station.

Make a plan.

First, find the distance (in feet) the tornado has traveled. Then convert the distance from feet to miles. Finally, write an equation that represents the distance (in miles) the tornado traveled after t hours. Use the equation to find the value of t.

Solve and check.

Use the plan to solve the problem. Then check your solution.

1 case = 12 cans

2. You buy 96 cans of soup to donate to a food bank. The store manager discounts the cost of each case for a total discount of $40. Find the discount for each case of soup when each can of soup costs $2.

3. Your organization has already raised $8000 for cancer research. A business agrees to donate $0.40 for every dollar donated by the community during an additional fundraising event. Write an equation that represents the total amount raised (in dollars). How much money does the community need to donate for the organization to reach its fundraising goal of $15,000?

STEAM Performance Task — Planning the Climb

GO DIGITAL

At the beginning of this chapter, you watched a STEAM Video called "Rock Climbing." You are now ready to complete the STEAM Performance Task, where you will find the average speed of rock climbers. Be sure to use mathematical thinking and reasoning, and the problem-solving plan as you work through the performance task.

7 Chapter Review WITH CalcChat®

Review Vocabulary

Write the definition and give an example of each vocabulary term.

equation, *p. 494*

inequality, *p. 538*

solution set, *p. 539*

solution, *p. 505*

solution of an inequality, *p. 539*

graph of an inequality, *p. 540*

inverse operations, *p. 519*

GO DIGITAL

Graphic Organizers

You can use an **Example and Non-Example Chart** to list examples and non-examples of a concept. Here is an Example and Non-Example Chart for the vocabulary term ***equation***.

Equation

Examples	Non-Examples
$x = 5$	5
$2a = 16$	$2a$
$x + 4 = 19$	$x + 4$
$5 = x + 3$	$x + 3$
$12 - 7 = 5$	$12 - 7$
$\frac{3}{4}y = 6$	$\frac{3}{4}$

Choose and complete a graphic organizer to help you study the concept.

1. inverse operations
2. solving equations using addition
3. solving equations using subtraction
4. solving equations using multiplication
5. solving equations using division
6. solution of an inequality
7. solution set

Chapter Learning Target: Understand equations.

Chapter Success Criteria:
- ◆ I can identify key words and phrases that indicate equality.
- ◆ I can write word sentences as equations or inequalities.
- ■ I can solve equations and inequalities.
- ■ I can use equations or inequalities to model real-life problems.

◆ Surface
■ Deep

Rate your understanding after each section.

1 I don't understand yet. | 2 I can do it with help. | 3 I can do it on my own. | 4 I can teach someone else.

7.1 Writing Equations in One Variable *(pp. 493–502)*

Learning Target: Write equations in one variable and write equations that represent real-life problems.

Write the word sentence as an equation.

1. The product of a number m and -2 is 8.

2. 6 less than a number t is -7.

3. A number m increased by 5 is 7.

4. 8 is the quotient of a number g and 3.

5. The height of the 50-milliliter beaker is one-third the height of the 2000-milliliter beaker. Write an equation you can use to find the height (in centimeters) of the 2000-milliliter beaker.

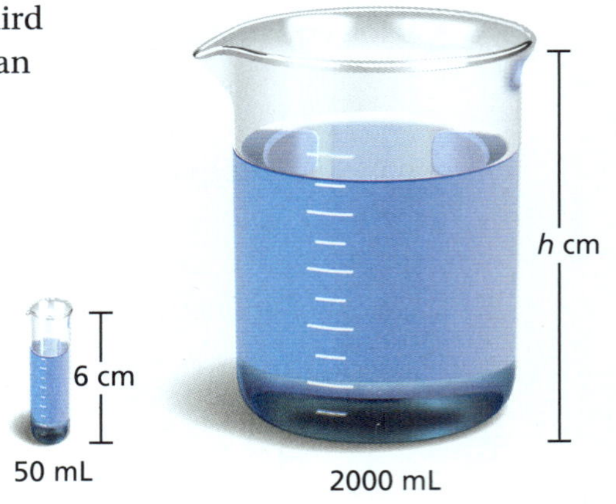

GO DIGITAL

7.2 Finding Unknown Values in Equations *(pp. 503–516)*

Learning Target: Find unknown values in equations.

Tell whether the given value is a solution to the equation.

6. $3p - 10 = 27; p = 12$

7. $32 = -5m + 7; m = -5$

8. $\frac{x}{3} = 7; x = 21$

9. $-4 = 2r - 11; r = 4$

Solve the equation. Explain your method.

10. $y + 5.54 = 11$

11. $-9 = -3 \bullet t$

12. $\frac{4}{9} = \frac{7}{27} \bullet n$

13. $\frac{s}{5} = 3.04$

14. A charity donates $5 in your name for every mile you run during a race. At the end of the race, the charity donates $35 in your name. Use an equation to find how many miles you ran during the race.

7.3 Solving Equations Using Addition or Subtraction *(pp. 517–526)*

Learning Target: Write and solve equations using addition or subtraction.

Solve the equation. Check your solution.

15. $x - 1 = 8$

16. $m + 7 = 11$

17. $-21 = p - 12$

Write the word sentence as an equation. Then solve the equation.

18. 5 more than a number x is -9.

19. 82 is the difference of a number b and 24.

20. Before swallowing a large rodent, a python weighs 152 pounds. After swallowing the rodent, the python weighs 164 pounds. Find the weight of the rodent.

7.4 Solving Equations Using Multiplication or Division (pp. 527–536)

Learning Target: Write and solve equations using multiplication or division.

Solve the equation. Check your solution.

21. $6 \cdot q = 54$

22. $k \div 3 = -21$

23. $\frac{a}{5} = 15$

24. You buy 4 pounds of tomatoes that cost \$3.00 per pound and 2 pounds of sweet potatoes that cost \$2.00 per pound at a farmers' market. You spend $\frac{1}{3}$ of the money in your wallet. How much money is in your wallet before you pay?

7.5 Writing and Graphing Inequalities (pp. 537–548)

Learning Target: Write inequalities and represent solutions of inequalities on number lines.

Write the word sentence as an inequality.

25. A number m is less than 5.

26. -14 is at least twice a number h.

Tell whether $x = 8$ is a solution of the inequality.

27. $\frac{x}{2} \geq 3$

28. $7x - 4x > 24$

29. $5x + 9 > -30 + 11x$

Graph the inequality on a number line.

30. $x < 0$

31. $a \geq 3$

32. $n \leq -1$

33. The speed limit on a road is 35 miles per hour. Write and graph an inequality that represents the legal speeds on the road.

7.6 Solving Inequalities Using Addition or Subtraction *(pp. 549–558)*

Learning Target: Write and solve inequalities using addition or subtraction.

Solve the inequality. Graph the solution.

34. $d + 12 < 19$

35. $t - 4 \leq -14$

36. $-8 \leq z + 6.4$

37. A small cruise ship can hold up to 500 people. There are 115 crew members on board the ship.

a. Write and solve an inequality that represents the additional numbers of people that can board the ship.

b. Can 385 more people board the ship? Explain.

38. Write an inequality that can be solved using the Subtraction Property of Inequality and has a solution of all numbers less than -3.

7.7 Solving Inequalities Using Multiplication or Division *(pp. 559–572)*

Learning Target: Write and solve inequalities using multiplication or division.

Solve the inequality. Graph the solution.

39. $6q < -18$

40. $-\frac{r}{3} \le 6$

41. $-4 > -\frac{4}{3}s$

Write the word sentence as an inequality. Then solve the inequality.

42. The product of -3 and a number p is greater than 21.

43. A number k times 5 is no more than -25.

44. The fraction $\frac{3}{7}$ is less than the quotient of a number m and 4.

45. You are organizing books on a shelf. Each book has a width of $\frac{3}{4}$ inch. Write and solve an inequality for the numbers of books b that can fit on the shelf.

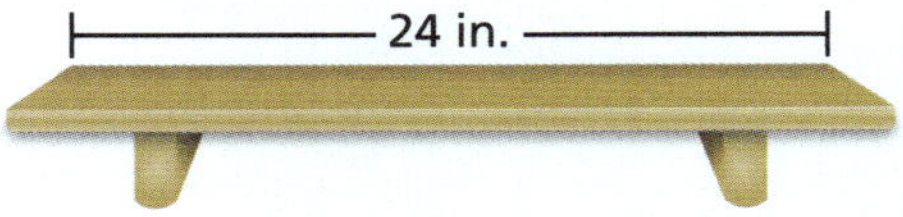

7 Practice Test WITH CalcChat®

1. Write "7 times a number s is 84" as an equation.

Solve the equation. Check your solution.

2. $15 = -7 + b$

3. $v - 6 = 16$

4. $5x = 70$

5. $m \div 6 = -3$

6. The drawbridge shown consists of two identical sections that open to allow boats to pass. Write an equation you can use to find the length s (in feet) of each section of the drawbridge.

7. Each ticket to a school dance is \$4. The total amount collected in ticket sales is \$332. Find the number of students attending the dance.

Write the word sentence as an inequality.

8. A number t is less than 7.

9. 8 is at least 3 times a number p.

Solve the inequality. Graph the solution.

10. $x - 4 > -6$

11. $-\frac{2}{9} + y \leq \frac{5}{9}$

12. $-6z \geq 36$

13. $-5.2 \geq \frac{p}{4}$

GO DIGITAL

14. A soccer team sells T-shirts for a fundraiser. The company that makes the T-shirts charges $10 per shirt. Find the number of shirts sold when the team spends $520 to order the shirts.

15. You hand in 2 homework pages to your teacher. Your teacher now has 32 homework pages to grade. Find the number of homework pages that your teacher originally had to grade.

16. A hurricane has wind speeds that are greater than or equal to 74 miles per hour. Write an inequality that represents the possible wind speeds during a hurricane.

17. You can spend no more than $100 on a party you are hosting. The cost per guest is $8.

a. Write and solve an inequality that represents the numbers of guests you can invite to the party.

b. What is the greatest number of guests that you can invite to the party? Explain your reasoning.

7 B.E.S.T. Test Prep WITH CalcChat®

Cumulative Practice

1. You are making identical fruit baskets using 16 apples, 24 pears, and 32 bananas. What is the greatest number of baskets you can make using all of the fruit?

Ⓐ 2
Ⓑ 4
Ⓒ 8
Ⓓ 16

GRIDDED RESPONSE

2. What is the value of x that makes the equation true?

$$59 + x = 112$$

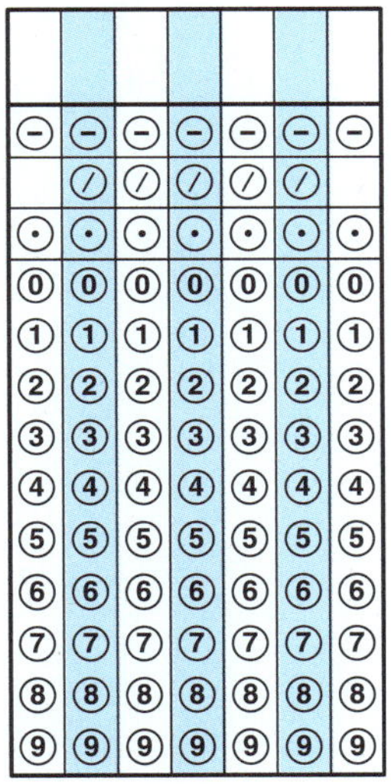

3. Which property is illustrated by the statement?

$$5(a + 6) = 5(a) + 5(6)$$

Ⓐ Associative Property of Multiplication
Ⓑ Commutative Property of Multiplication
Ⓒ Commutative Property of Addition
Ⓓ Distributive Property

GO DIGITAL

4. Which number line shows the graph of $x \geq 5$?

Ⓐ 

Ⓑ 

Ⓒ 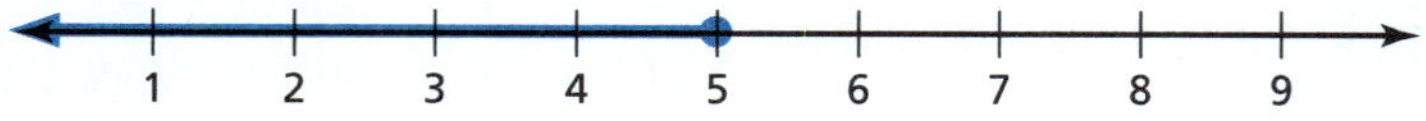

Ⓓ 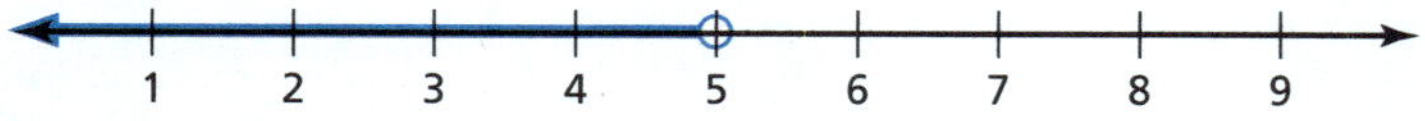

5. What is the value of the expression?

GRIDDED RESPONSE

$$46.8 \div 0.156$$

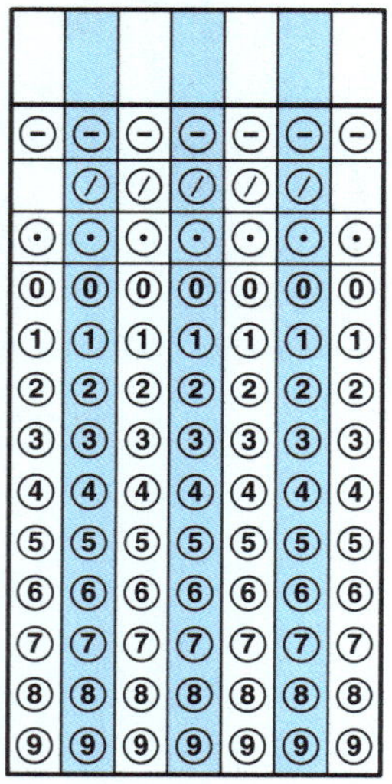

6. Which is equivalent to the expression?

$$9h - 6 + 7h - 5$$

Ⓐ $3h + 2$

Ⓑ $16h + 1$

Ⓒ $2h - 1$

Ⓓ $16h - 11$

7. Which equation represents the word sentence?

The sum of 18 and 5 is equal to 9 less than a number y.

Ⓐ $18 - 5 = 9 - y$

Ⓑ $18 + 5 = 9 - y$

Ⓒ $18 + 5 = y - 9$

Ⓓ $18 - 5 = y - 9$

8. Which graph represents the inequality?

$$-2x \geq -8$$

Ⓐ 

Ⓑ 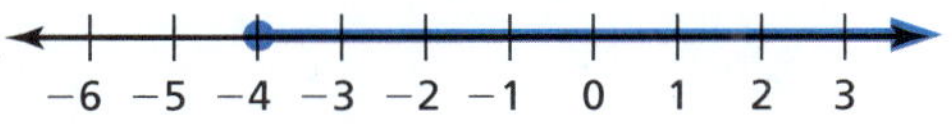

Ⓒ 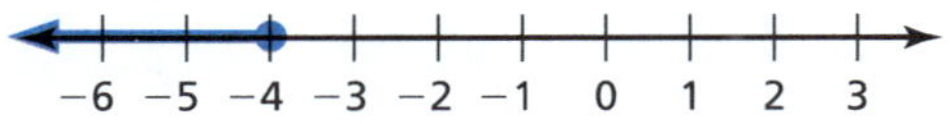

Ⓓ 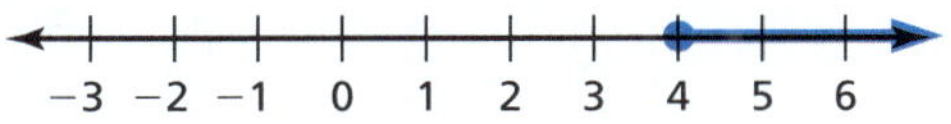

9. The GCF of the terms of the expression $28x - 42$ is [Ⓐ 2 / Ⓑ 7 / Ⓒ 14 / Ⓓ 28]

and the factored form of the expression is [Ⓐ $2(14x - 21)$ / Ⓑ $7(4x - 6)$ / Ⓒ $14(2x - 3)$ / Ⓓ $28(x - 42)$].

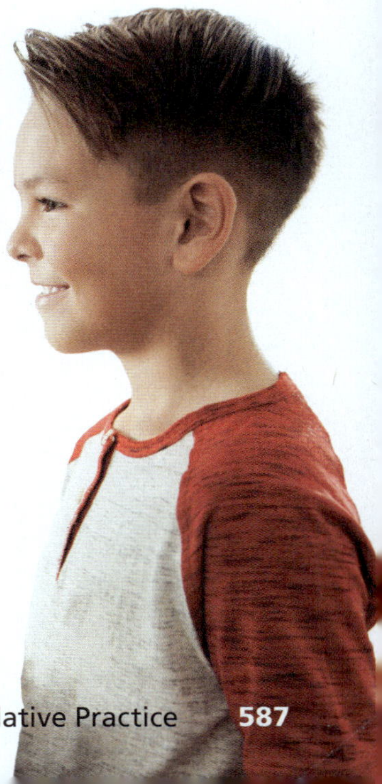

8 Ratios and Rates

Chapter Learning Target:
Understand ratios.

Chapter Success Criteria:

- ◆ I can write and interpret ratios.
- ◆ I can name ratios equivalent to a given ratio.
- ■ I can solve problems using ratios.
- ■ I can apply ratio reasoning to convert units of measure.

◆ Surface ■ Deep

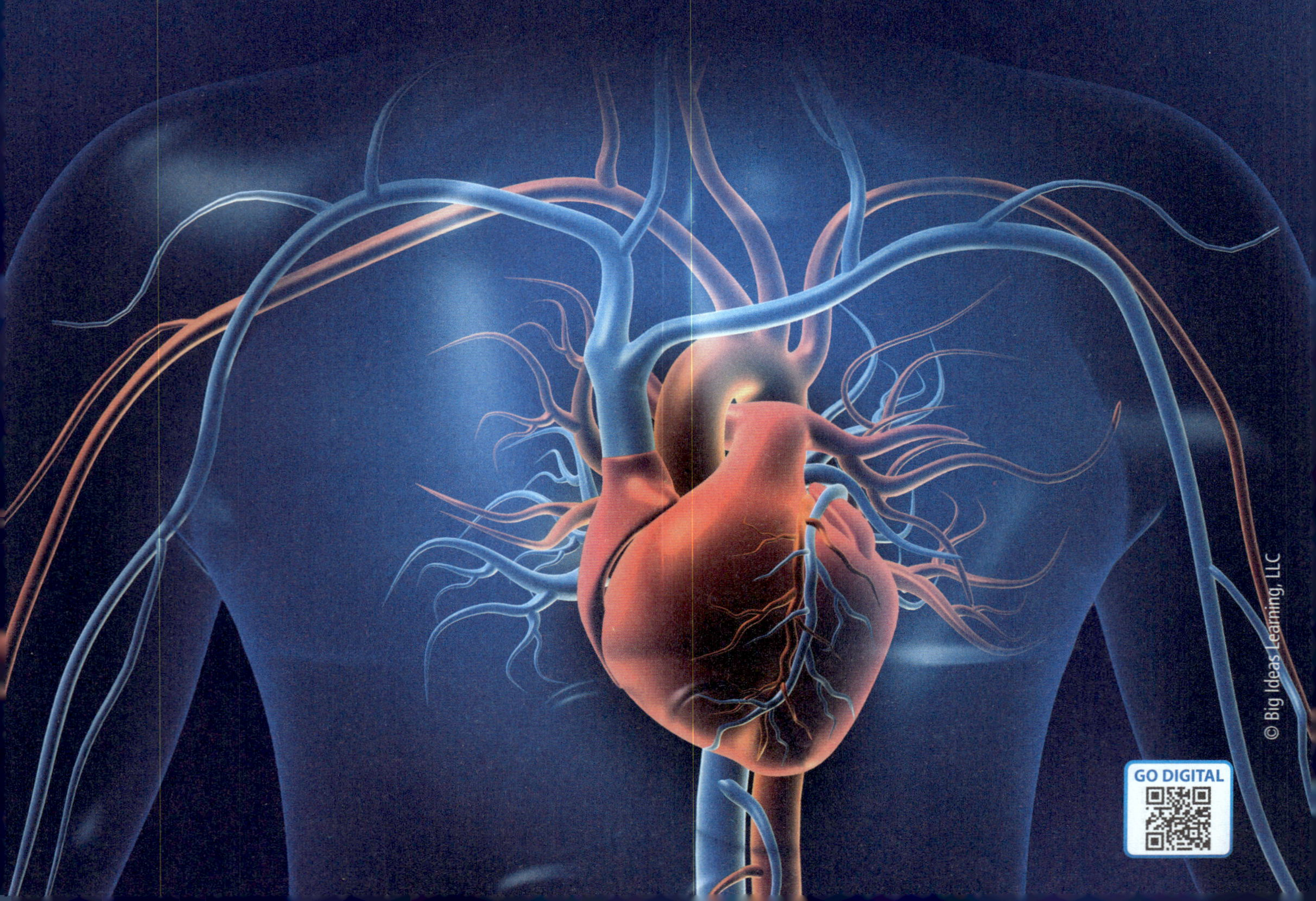

GO DIGITAL

7 MTR *Mathematicians who apply mathematics to real-world contexts connect mathematical concepts to everyday experiences.*

STEAM Video Human Circulatory System

Enid shows Robert a model of the human circulatory system.

3 MTR **1. CHOOSE A METHOD** Enid says the heart pumps about 5 liters of blood each minute. How can you find the amount of blood the heart pumps for any given number of minutes?

2. Explain how you can estimate the blood your heart pumps in one heartbeat.

4 MTR **3. DISCUSS MATHEMATICAL THINKING** The table shows the amounts of blood contained in several different types of blood vessels. How can you make meaningful comparisons of the amounts?

Vessel	Volume
Aorta and large arteries	10.14 fl oz
Small arteries	1.69 cups
Small veins	2.43 qt
Large veins	0.24 gal

STEAM Performance Task Oops! Unit Conversion Mistakes

After completing this chapter, you will be able to use the concepts you learned to answer the questions in the *STEAM Performance Task*. You will analyze and correct unit conversion mistakes in the following real-life situations.

Blood transfusion	**Airplane fuel**
Baby medication	**Zoo enclosure**

GO DIGITAL

Getting Ready for Chapter 8 WITH CalcChat®

Chapter Exploration

1 MTR **Actively Participate in Effortful Learning Collectively**
Work with a partner to prepare for concepts in this chapter.

What portion of the rectangle is red? How did you write your answer?

1. 

2. 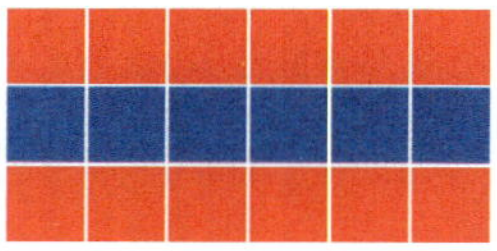

3. 

4. 

5. 

6. 

7. 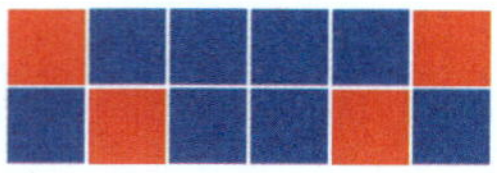

8. 

9. 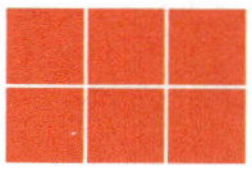

10. In Exercises 1–9, which of the rectangles have the same portion of red tiles? Explain your reasoning.

Use square color tiles to build two different-sized rectangles that represent the description.

11. Five-sixths of the tiles are blue.

12. Three-fourths of the tiles are yellow.

13. Four-fifths of the tiles are green.

14. Five-sevenths of the tiles are red.

7 MTR **15. MODELING REAL LIFE** The soccer committee has 8 girls and 6 boys. The tennis committee has 9 girls and 8 boys. A friend tells you that the tennis committee has a greater portion of girls than the soccer committee. Is your friend correct? Explain. If not, how many boys could you add to the soccer committee so that your friend is correct?

Vocabulary

The following terms are defined in this chapter. Think about what each might mean and record your thoughts.

ratio
rate
equivalent rates
equivalent ratios
unit rate

GO DIGITAL

8.1 Ratios

Learning Target: Understand the concepts of ratios and equivalent ratios.

Success Criteria:
- I can write and interpret ratios using appropriate notation and language.
- I can recognize multiplicative relationships in ratios.
- I can describe how to determine whether ratios are equivalent.
- I can name ratios equivalent to a given ratio.

Algebraic Reasoning

MA.6.AR.3.1 Given a real-world context, write and interpret ratios to show the relative sizes of two quantities using appropriate notation: $\frac{a}{b}$, a to b, or $a:b$ where $b \neq 0$.

MA.6.AR.3.5 Solve mathematical and real-world problems involving ratios, rates and unit rates, including comparisons, mixtures, ratios of lengths and conversions within the same measurement system.

A **ratio** is a comparison of two quantities. Consider two quantities a and b. The ratio $a:b$ indicates that there are a units of the first quantity for every b units of the second quantity.

Exploration 1 Writing Ratios

Work with a partner. A science class has two times as many girls as it has boys.

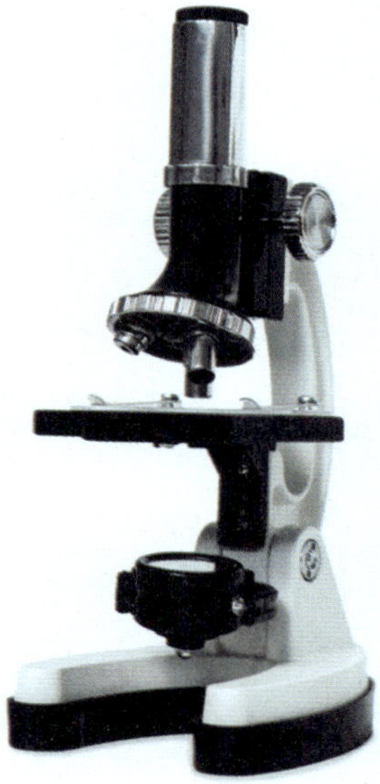

a. The class has 18 students. How many girls are in the class? How many boys are in the class?

b. Compare this class to your class.

c. Write three ratios that you observe in your classroom. Describe what each ratio represents.

2 MTR MODEL A PROBLEM

How can you use a table to represent the relationship between the numbers of girls and boys?

____

____

____

____

____

GO DIGITAL

Exploration 2 Using Ratios in a Recipe

Work with a partner. The ratio of iced tea to lemonade in a recipe is 3 : 1. You begin by combining 3 cups of iced tea with 1 cup of lemonade.

a. You add 1 cup of iced tea and 1 cup of lemonade to the mixture. Does this change the taste of the mixture?

b. Describe how you can make larger amounts without changing the taste.

8.1 Lesson

Key Vocabulary
ratio, *p. 593*
value of a ratio, *p. 594*
equivalent ratios, *p. 594*

Key Idea

Ratio

Words A **ratio** is a comparison of two quantities. Ratios can be part-to-part, part-to-whole, or whole-to-part comparisons. Ratios may or may not include units.

Examples
2 cats *to* 6 dogs
1 cat *for every* 3 dogs
3 dogs *per* 1 cat
3 dogs *for each* cat
3 dogs *out of every* 4 pets
2 cats *out of* 8 pets

Algebra The ratio of a to b can be written as $a : b$.

Reading
Phrases that indicate ratios include *for each, for every,* and *per*.

Example 1 Writing Ratios

You have the coins shown.

a. Write the ratio of pennies to quarters.

6 pennies → 6 to 7 ← 7 quarters

▶ So, the ratio of pennies to quarters is 6 to 7, or 6 : 7.

b. Write the ratio of quarters to dimes.

7 quarters → 7 to 3 ← 3 dimes

▶ So, the ratio of quarters to dimes is 7 to 3, or 7 : 3.

c. Write the ratio of dimes to the total number of coins.

3 dimes → 3 to 16 ← 16 coins

▶ So, the ratio of dimes to the total number of coins is 3 to 16, or 3 : 16.

Try It **Write the indicated ratio using the coins in Example 1.**

1. dimes to pennies

$\frac{\text{3 Dimes}}{\text{6 pennies}} = \frac{\text{1 dime}}{\text{2 pennies}}$

2. quarters to the total number of coins

$\frac{\text{4 quarters}}{\text{16 coins}} = \frac{\text{1 quarter}}{\text{4 coins}}$

GO DIGITAL

The number $\frac{a}{b}$ associated with the ratio $a : b$ is called the **value of the ratio**. It describes the multiplicative relationship between the quantities in a ratio.

Example 2 Writing and Interpreting Ratios

a. The ratio of rubies to diamonds on a ring is 1 : 20. Find and interpret the value of the ratio.

The value of the ratio 1 : 20 is $\frac{1}{20}$. So, the multiplicative relationship is $\frac{1}{20}$.

▶ So, the number of rubies is $\frac{1}{20}$ times the number of diamonds.

b. On another ring, the number of rubies is 4 times the number of diamonds. Write the ratio of rubies to diamonds.

Because the number of rubies is 4 times the number of diamonds, there are 4 rubies per diamond.

▶ So, the ratio of rubies to diamonds is 4 : 1.

Try It

3. An elephant sanctuary contains adult and baby elephants. The ratio of adult elephants to baby elephants is 5 : 1. Find and interpret the value of the ratio.

Key Idea

Equivalent Ratios

Words Two ratios that describe the same relationship are **equivalent ratios**. Two ratios are equivalent when you can multiply each quantity in one ratio by the same positive number to obtain the other ratio. The values of equivalent ratios are equivalent.

Example 1 : 3 and 2 : 6 are equivalent.

$$1:3 \xrightarrow{\times 2,\ \times 2} 2:6 \qquad \frac{1}{3} = \frac{1 \times 2}{3 \times 2} = \frac{2}{6}$$

Algebra Two ratios $a : b$ and $c : d$ are equivalent when there exists a positive number n such that $a \times n = c$ and $b \times n = d$.

GO DIGITAL

Example 3 B.E.S.T. Test Prep: Identifying Equivalent Ratios

Select all the ratios that are equivalent to 8 : 6.

Ⓐ 4 : 3 Ⓑ 24 : 36 Ⓒ 32 : 24 Ⓓ 20 : 12

One way to identify any equivalent ratios is to determine whether you can multiply each quantity in 8 : 6 by the same positive number to obtain any of the given ratios.

Also notice that the values of the equivalent ratios are equal.

$$\frac{8}{6} = \frac{8 \times 0.5}{6 \times 0.5} = \frac{4}{3}$$

$$\frac{8}{6} = \frac{8 \times 4}{6 \times 4} = \frac{32}{24}$$

Choice A: 8 : 6 ($\times$ 0.5, $\times$ 0.5) → 4 : 3
Equivalent

Choice C: 8 : 6 ($\times$ 4, $\times$ 4) → 32 : 24
Equivalent

Choice B: 8 : 6 ($\times$ 3, $\times$ 6) → 24 : 36
Not equivalent

Choice D: 8 : 6 ($\times$ 2.5, $\times$ 2) → 20 : 12
Not equivalent

▶ So, the correct answers are Ⓐ and Ⓒ.

Try It **Determine whether the ratios are equivalent.**

4. 1 : 1 and 6 : 6

yes

5. 1 : 2 and 3 : 4

no

6. 8 : 3 and 6 : 16

$\frac{8}{3}$ no

GO DIGITAL

In-Class Practice

1 I don't understand yet. 2 I can do it with help. 3 I can do it on my own. 4 I can teach someone else.

WRITING AND INTERPRETING RATIOS **Write the ratio. Then find and interpret the value of the ratio.**

7. sharks to dolphins

8. dolphins : animals

IDENTIFYING EQUIVALENT RATIOS **Determine whether the ratios are equivalent. Explain your reasoning.**

9. 2 : 3 and 24 : 36

10. 5 : 7 and 20 : 28

11. 3 : 10 and 9 : 25

4 MTR **12.** **DIFFERENT WORDS, SAME QUESTION** Which is different? Find "both" answers.

What kind of relationship is 2 peaches to 5 pears?	What kind of relationship is 2 pears out of every 5 fruit?
What kind of relationship is 2 pears per 5 peaches?	What kind of relationship is 2 peaches for every 5 pears?

Example 4 Modeling Real Life

7 MTR

You are speedwalking at a pace of 5 meters every 2 seconds. Your friend's pace is 10 meters every 5 seconds. Are you and your friend speedwalking at the same pace? If not, who is faster?

Write a ratio to represent each pace.

Your pace: 5 : 2

Your friend's pace: 10 : 5

To decide whether you and your friend are speedwalking at the same pace, determine whether the ratios are equivalent.

$$\begin{array}{c} 5:2 \\ \times 2 \downarrow \quad \downarrow \times 2.5 \\ 10:5 \end{array}$$

Not equivalent

So, you and your friend are not speedwalking at the same pace. To decide who is faster, use an equivalent ratio to compare your pace to your friend's pace.

Your pace:

$$\begin{array}{c} 5:2 \\ \times 2 \downarrow \quad \downarrow \times 2 \\ 10:4 \end{array}$$

You walk 10 meters every 4 seconds. It takes your friend 5 seconds to walk the same distance.

▶ So, you are speedwalking faster than your friend.

4 MTR

CONSTRUCT AN ARGUMENT

A classmate uses a multiplier of 2.5 for your pace to find 12.5 : 5. Can you use this ratio to decide who is faster? Explain.

GO DIGITAL

In-Class Practice

1 I don't understand yet. 2 I can do it with help. 3 I can do it on my own. 4 I can teach someone else.

13. The ratio of wild hogs to Florida panthers in a swamp forest is 50 : 2. Find and interpret the value of the ratio.

14. You are kayaking at a pace of 63 feet every 12 seconds. Your friend's pace is 21 feet every 3 seconds. Are you and your friend kayaking at the same pace? If not, who is faster?

15. **Dig Deeper** The ratio of Jet Ski rentals to boat rentals at a store is 7 : 2. If the number of boat rentals triples and the number of Jet Ski rentals doubles, then the number of boat rentals is how many times the number of Jet Ski rentals?

8.1 Practice WITH CalcChat® AND CalcView®

Review & Refresh

Solve the inequality. Graph the solution.

1. $-3x \geq 18$

2. $\frac{2}{3}d > 8$

3. $2 \geq \frac{g}{-4}$

Evaluate the expression.

4. $4(9 + 5)$

5. $(6 \times 15) \times 0$

6. $39 + 7 + 1$

7. $5 \times 2 \times 5$

Write an ordered pair corresponding to the point.

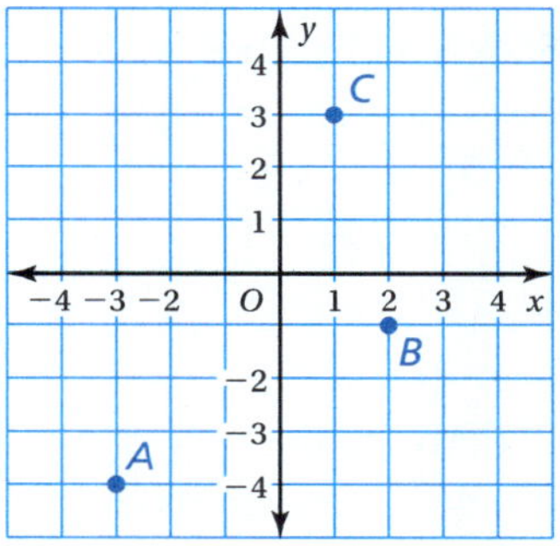

8. Point A

9. Point B

10. Point C

Concepts, Skills, & Problem Solving

USING RATIOS You mix the amounts of iced tea and lemonade shown. Describe how you can make larger amounts without changing the taste. (See Exploration 2.)

11.

12.

GO DIGITAL

WRITING RATIOS Write the ratio. (See Example 1.)

13. frogs to turtles

14. basketballs to soccer balls

15. calculators : pencils

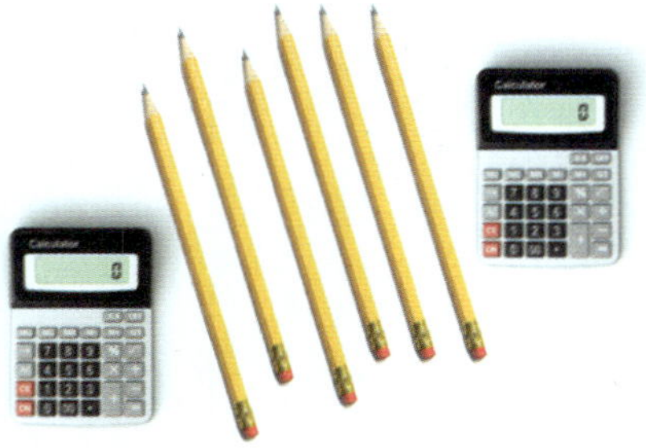

16. shirts : pants

7 MTR **17. MODELING REAL LIFE** Twelve of the 28 students in a class own a dog. What is the ratio of students who own a dog to students who do not?

18. REASONING Name two things that you would like to have in a ratio of 5 : 1 but not in a ratio of 1 : 5. Explain your reasoning.

OPEN-ENDED Describe a real-life relationship that can be represented by the ratio.

19. 1 out of every 7

20. 5 to 26

21. 2 per 5

22. 7 : 1

7 MTR

▶ **23.** **MODELING REAL LIFE** During a given month, the ratio of sunny days to rainy days is 4 : 1. (See Example 2.)

a. Find and interpret the value of the ratio.

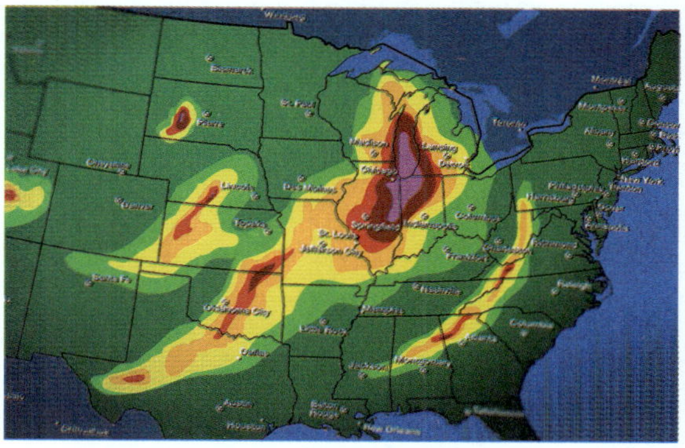

b. In another month, the number of sunny days is 5 times the number of rainy days. Write the ratio of sunny days to rainy days.

IDENTIFYING EQUIVALENT RATIOS **Determine whether the ratios are equivalent.** (See Example 3.)

24. 2 : 3 and 4 : 9

25. 3 : 8 and 9 : 24

26. 1 : 4 and 2 : 6

▶**27.** 5 : 3 and 15 : 12

28. 6 : 10 and 12 : 20

29. 2 : 3 and 4 : 5

30. 28 : 32 and 7 : 8

31. 24 : 100 and 6 : 25

32. 85 : 210 and 340 : 735

5 MTR

STRUCTURE **Fill in the blank so that the ratios are equivalent.**

33. 3 : 9 and 6 : ___

34. 2 : 6 and 8 : ___

35. ___ : 6 and 7 : 2

36. YOU BE THE TEACHER Your friend says that the two ratios are equivalent. Is your friend correct? Explain your reasoning.

> +4 (4 : 8) +4
> 8 : 12
>
> Because you can add 4 to each number in the first ratio to obtain the numbers in the second ratio, the ratios are equivalent.

37. OPEN-ENDED A *non-Newtonian* liquid demonstrates properties of both a solid and a liquid. A recipe for a non-Newtonian liquid calls for 1 cup of water and 2 cups of cornstarch. Find two possible combinations of water and cornstarch that you can use to make a larger batch. Justify your answer.

38. B.E.S.T. Test Prep You are downloading songs to your tablet. The ratio of pop songs to rock songs is 5 : 4. You download 40 pop songs. How many rock songs do you download?

Ⓐ 8 Ⓑ 32 Ⓒ 50 Ⓓ 72

39. PROBLEM SOLVING In the contiguous United States, the ratio of states that border an ocean to states that do not border an ocean is 7 : 9. How many of the states border an ocean?

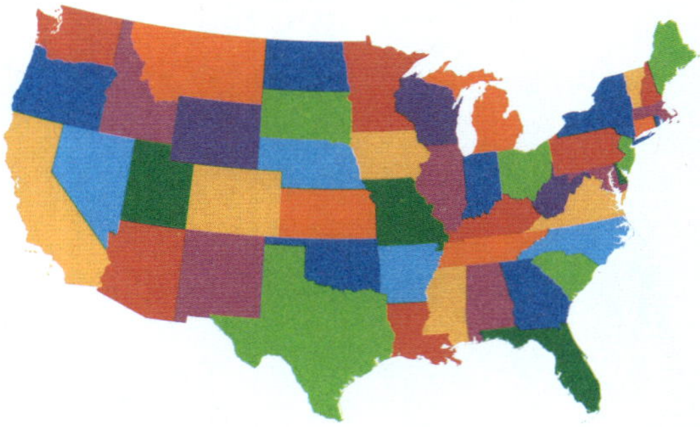

40. REASONING The value of a ratio is $\frac{4}{3}$. The second quantity in the ratio is how many times the first quantity in the ratio? Explain your reasoning.

7 MTR **41. MODELING REAL LIFE** A train moving at a constant speed travels 3 miles every 5 minutes. A car moving at a constant speed travels 12 miles every 20 minutes. Are the vehicles traveling at the same speed? If not, which is faster? (See Example 4.)

1 MTR **42. ANALYZE A PROBLEM** To win a relay race, you must swim 200 yards before your opponent swims 190 yards. You swim at a pace of 50 yards every 40 seconds. Your opponent swims at a pace of 10 yards every 9 seconds. Write two questions that will help you determine who wins the race. Then answer each question and determine who wins the race.

43. Dig Deeper There are 3 boys for every 2 girls in a Latin dance competition. Does it make sense for there to be a total of 9 people in the competition? Explain.

44. **GEOMETRY** Use the blue and green rectangles.

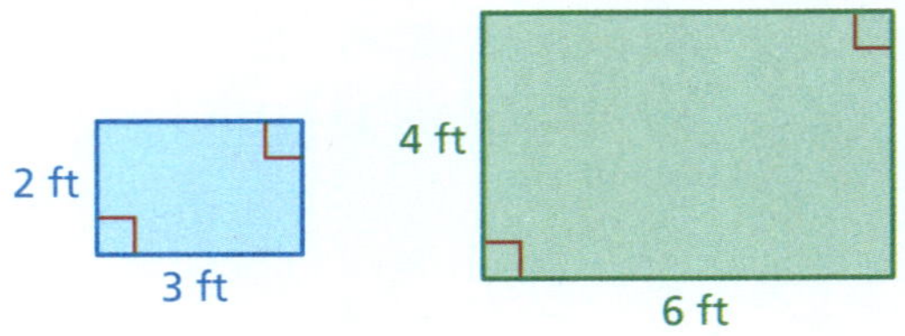

a. Find the ratio of the length of the blue rectangle to the length of the green rectangle. Repeat this for width, perimeter, and area.

b. Compare your ratios in part (a).

5 MTR

45. **STRUCTURE** The ratio of the side lengths of a triangle is 2 : 3 : 4. The shortest side is 15 inches. What is the perimeter of the triangle? Explain.

46. **PROBLEM SOLVING** A restaurant sells tokens that customers use to play games while waiting for their orders.

TOKENS	
1 Token	$1.00
10 Tokens	$5.00
25 Tokens	$10.00
50 Tokens	$25.00
90 Tokens	$40.00

a. Which option is the best deal? Justify your answer.

b. What suggestions, if any, would you give to the restaurant about how it could modify the prices of tokens?

47. **Dig Deeper** There are 12 boys and 10 girls in your gym class. If 6 boys joined the class, how many girls would need to join for the ratio of boys to girls to remain the same? Justify your answer.

8.2 Using Tape Diagrams

Learning Target: Use tape diagrams to model and solve ratio problems.

Success Criteria:
- I can interpret tape diagrams that represent ratio relationships.
- I can draw tape diagrams to model ratio relationships.
- I can find the value of one part of a tape diagram.
- I can use tape diagrams to solve ratio problems.

Algebraic Reasoning

MA.6.AR.3.1 Given a real-world context, write and interpret ratios to show the relative sizes of two quantities using appropriate notation: $\frac{a}{b}$, a to b, or $a : b$ where $b \neq 0$.

MA.6.AR.3.5 Solve mathematical and real-world problems involving ratios, rates and unit rates, including comparisons, mixtures, ratios of lengths and conversions within the same measurement system.

You can use a visual model, called a *tape diagram*, to represent the relationship between two quantities in a ratio.

Exploration 1 Using a Tape Diagram

Work with a partner. The tape diagram models the lengths of two snowboarding trails.

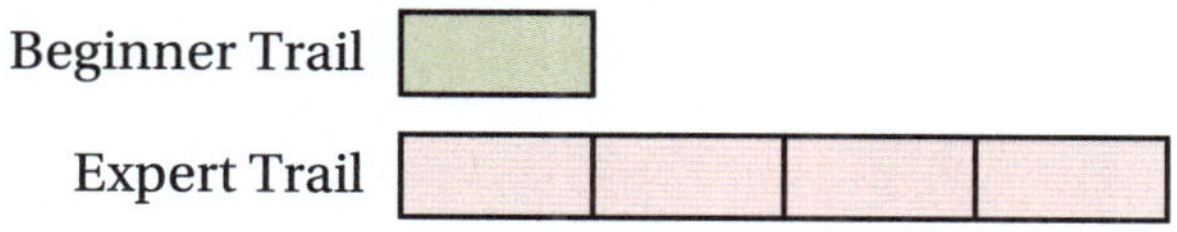

a. Compare the lengths of the two trails.

b. The Beginner Trail is 300 yards long. How long is the Expert Trail. How do you know?

GO DIGITAL

2 MTR **MAKE A CONNECTION**

How does the tape diagram help you make sense of the quantities and the relationship between the quantities?

Exploration 2 Using a Tape Diagram

Work with a partner. The Expert Trail is 5 times longer than the Beginner Trail.

a. Draw a tape diagram to compare the lengths of the two trails.

b. The combined lengths of the two trails is 1500 yards. How long is the Beginner Trail? How long is the Expert Trail?

8.2 Lesson

You can use tape diagrams to represent ratios and solve ratio problems.

Example 1 Interpreting a Tape Diagram

Reading

The tape diagram shows that the ratio of blue monsters to green monsters is 3 : 1.

The tape diagram represents the ratio of blue monsters to green monsters you caught in a game. You caught 10 green monsters. How many blue monsters did you catch?

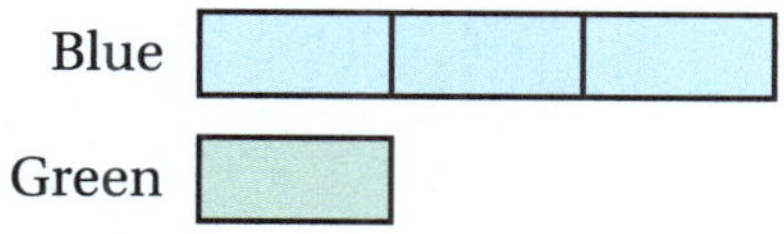

The 1 part for green represents 10 monsters. So, the 3 parts for blue represents $3 \times 10 = 30$ monsters.

▶ You caught 30 blue monsters.

Try It

1. The tape diagram represents the ratio of gifts received to gifts given. You received 4 gifts. How many gifts did you give?

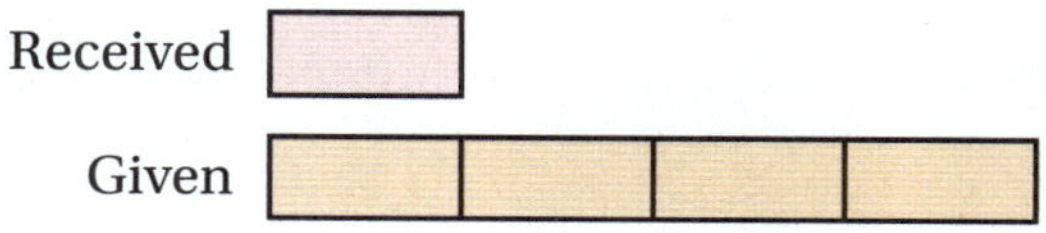

Example 2 Drawing a Tape Diagram

4 MTR COMPARE METHODS

Represent the ratio using counters or other objects. Compare this method with drawing a tape diagram. Which do you prefer?

____

____

____

____

____

____

There are 3 bones in a cat for every 4 bones in a dog. The cat has 240 bones. How many bones does the dog have?

The ratio of bones in the cat to bones in the dog is 3 : 4. Represent the ratio using a tape diagram.

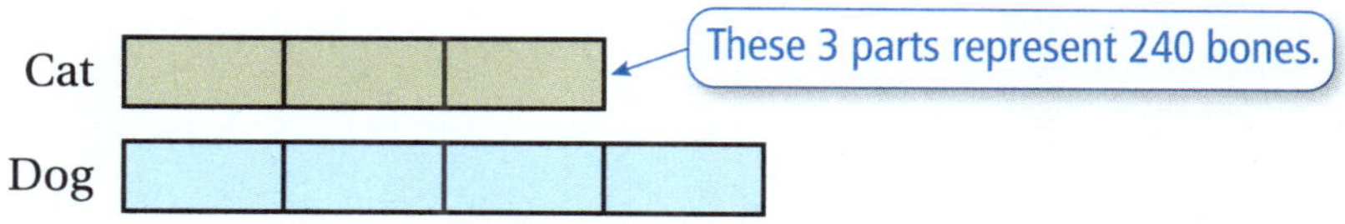

One part represents $240 \div 3 = 80$ bones. So, 4 parts represent $4 \times 80 = 320$ bones.

▶ The dog has 320 bones.

GO DIGITAL

Try It

2. There are 8 bones in a large snake for every 3 bones in a small snake. The small snake has 150 bones. How many bones does the large snake have?

Example 3 Using a Tape Diagram to Solve a Ratio Problem

The ratio of your monthly allowance to your friend's monthly allowance is 5 : 3. The monthly allowances total $40. How much is each allowance?

Represent the ratio 5 : 3 using a tape diagram.

1 part represents $40 ÷ 8 = $5.

5 parts represent $5 \times \$5 = \$25$.

3 parts represent $3 \times \$5 = \$15$.

▶ So, your allowance is $25, and your friend's allowance is $15.

Check Verify that the ratio of allowances is 5 : 3.

5 : 3

×5 ↓ ↓ ×5 ✓

25 : 15

Try It

3. WHAT IF? Repeat Example 3 when the ratio of your monthly allowance to your friend's monthly allowance is 2 to 3.

GO DIGITAL

In-Class Practice

1 I don't understand yet. 2 I can do it with help. 3 I can do it on my own. 4 I can teach someone else.

5 MTR

4. STRUCTURE What ratio is represented by the tape diagram? Can you use the tape diagram to model the ratio 6 : 9? Can you use the tape diagram to model the ratio 8 : 16? Explain your reasoning.

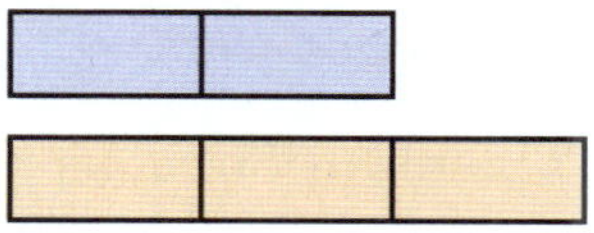

5. REASONING You are given a tape diagram and the total value of the parts. How can you find the value of 1 part?

2 MTR

6. MODEL A PROBLEM Describe two ways that you can represent the ratio 12 : 4 using a tape diagram.

USING A TAPE DIAGRAM **You are given the number of tickets in a bag and the ratio of winning tickets to losing tickets. How many of each kind of ticket are in the bag?**

7. 35 tickets; 1 to 4

8. 80 tickets; 2 : 8

Example 4 Modeling Real Life

In a seven-game basketball series, a team's power forward scores 8 points for every 5 points the center scores. The forward scores 60 more points than the center in the series. How many points does each player score in the series?

The ratio of the forward's points to the center's points is 8 : 5. Represent the ratio using a tape diagram.

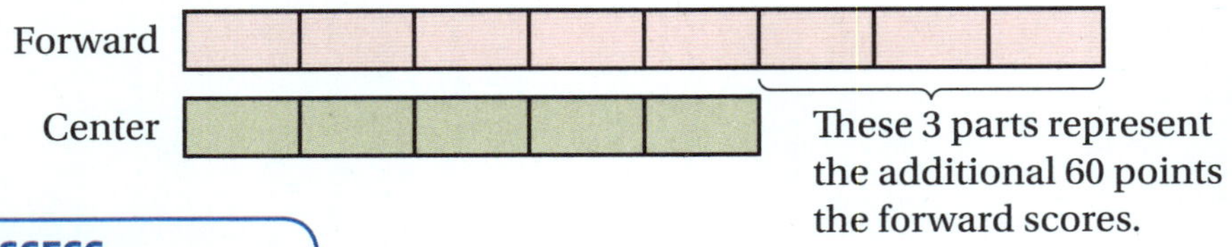

6 MTR ASSESS REASONABLENESS

How can you use the value of a ratio to verify that 8 : 5 and 160 : 100 are equivalent?

____

____

____

1 part represents $60 \div 3 = 20$ points.

8 parts represent $8 \times 20 = 160$ points.

5 parts represent $5 \times 20 = 100$ points.

▶ So, the forward scores 160 points in the series and the center scores 100 points in the series.

In-Class Practice

1 I don't understand yet. 2 I can do it with help. 3 I can do it on my own. 4 I can teach someone else.

9. The tape diagram represents the ratio of the numbers of planets in two different solar systems. There are 8 planets in Solar System B. How many planets are in Solar System A?

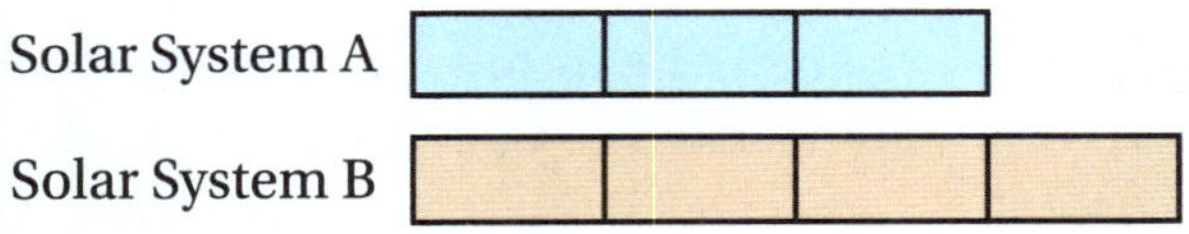

10. You and your friend play an arcade game. You score 5 points for every 9 points that your friend scores. You score 320 points less than your friend. How many points do you each score?

11. **Dig Deeper** Your team wins 18 medals at a track meet that are either gold, silver, or bronze. The ratio of silver medals to total medals is 2 : 9 and the ratio of gold medals to silver medals is 1 : 1. How many of each medal were won by your team?

GO DIGITAL

8.2 Practice WITH CalcChat® AND CalcView®

Review & Refresh

Determine whether the ratios are equivalent.

1. 11 : 4 and 22 : 8

2. 12 : 18 and 2 : 3

3. 56 : 81 and 7 : 9

4. 2 : 12 and 6 : 24

Multiply. Write the answer in simplest form.

5. $\frac{7}{10} \cdot \frac{5}{7}$

6. $2\frac{1}{3} \cdot \frac{3}{4}$

7. $5\frac{3}{8} \cdot 2\frac{1}{2}$

8. Your cousin earns $7.40 per hour working at a grocery store. She works 14.25 hours this week. Find out how much your cousin earns this week.

Concepts, Skills, & Problem Solving

USING A TAPE DIAGRAM **Use the tape diagram in Exploration 1 to answer the question.** (See Exploration 1.)

9. The beginner trail is 200 meters long. How long is the expert trail?

10. The expert trail is 1200 meters long. How long is the beginner trail?

11. The combined length of the trails is 2000 meters. How long is each trail?

12. The expert trail is 750 meters longer than the beginner trail. How long is each trail?

GO DIGITAL

INTERPRETING A TAPE DIAGRAM **The tape diagram represents the ratio of the time you spend tutoring to the time your friend spends tutoring. You tutor for 3 hours. How many hours does your friend spend tutoring?** (See Example 1.)

13.

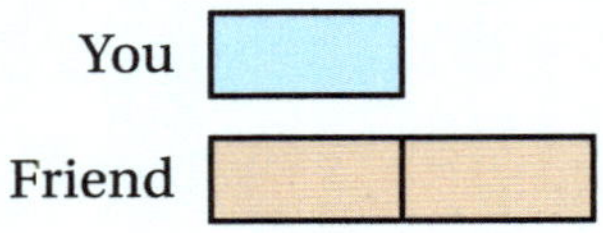

14.

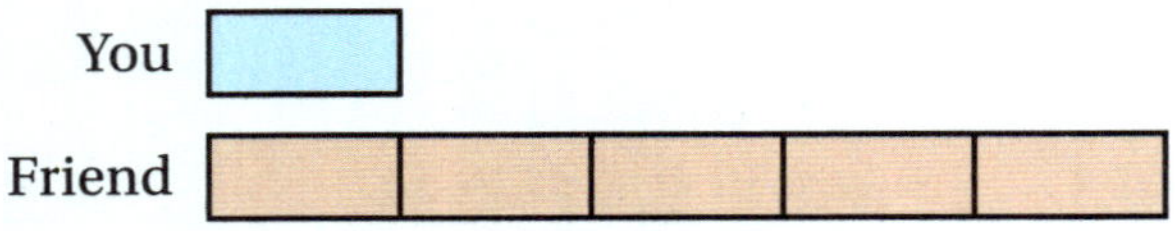

DRAWING A TAPE DIAGRAM **A bag contains red marbles and blue marbles. You are given the number of red marbles in the bag and the ratio of red marbles to blue marbles. Find the number of blue marbles in the bag.** (See Example 2.)

15. 10 red marbles; 5 to 1

16. 3 red marbles; 3 : 7

17. 12 red marbles; 4 : 3

18. 6 red marbles; 2 for every 5

19. 18 red marbles; 6 to 9

20. 12 red marbles; 3 : 4

USING A TAPE DIAGRAM **A bowl contains blueberries and strawberries. You are given the total number of berries in the bowl and the ratio of blueberries to strawberries. How many of each berry are in the bowl?** (See Example 3.)

21. 16 berries; 3 : 1

22. 10 berries; 2 for every 3

23. 12 berries; 1 to 2

24. 20 berries; 4 : 1

25. 48 berries; 9 to 3

26. 46 berries; 11 for every 12

27. PROBLEM SOLVING You separate bulbs of garlic into two groups: one for planting and one for cooking. The tape diagram represents the ratio of bulbs for planting to bulbs for cooking. You use 6 bulbs for cooking. Each bulb has 8 cloves. How many cloves of garlic will you plant?

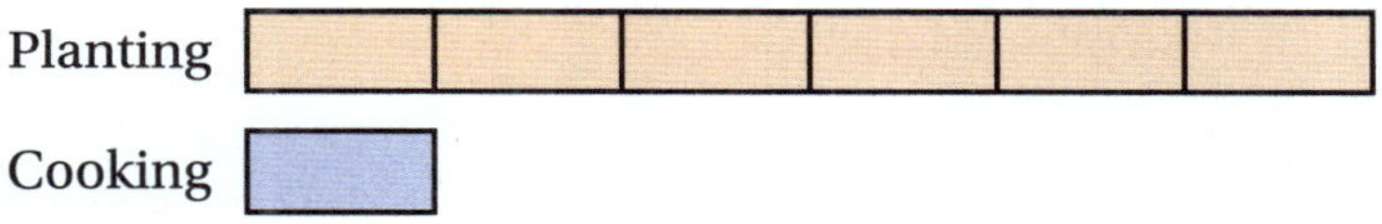

7 MTR **28. MODELING REAL LIFE** Methane gas contains carbon atoms and hydrogen atoms in the ratio of 1 : 4. A sample of methane gas contains 92 hydrogen atoms. How many carbon atoms are in the sample? How many total atoms are in the sample? (See Example 4.)

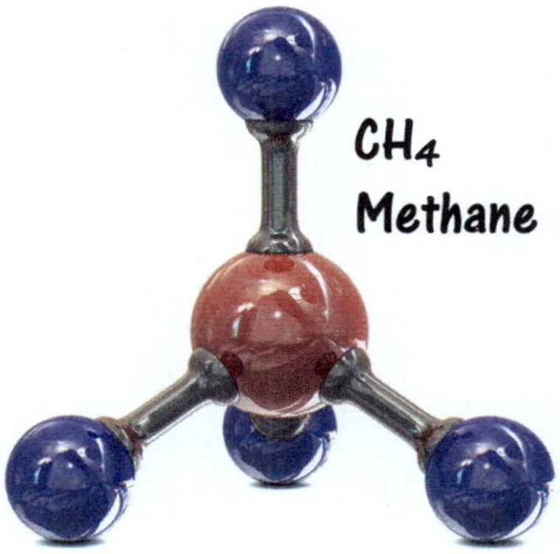

7 MTR

29. **MODELING REAL LIFE** There are 8 more girls than boys in a school play. The ratio of boys to girls is 5 : 7. How many boys and how many girls are in the play?

30. **Dig Deeper** A baseball team sells tickets for two games. The ratio of sold tickets to unsold tickets for the first game was 7 : 3. For the second game, the ratio was 13 : 2. There were 240 unsold tickets for the second game. How many tickets were sold for the first game?

31. **PROBLEM SOLVING** You have $150 in a savings account and you have some cash. The tape diagram represents the ratio of the amounts of money. You want to have twice the amount of money in your savings account as you have in cash. How much of your cash should you deposit into your savings account?

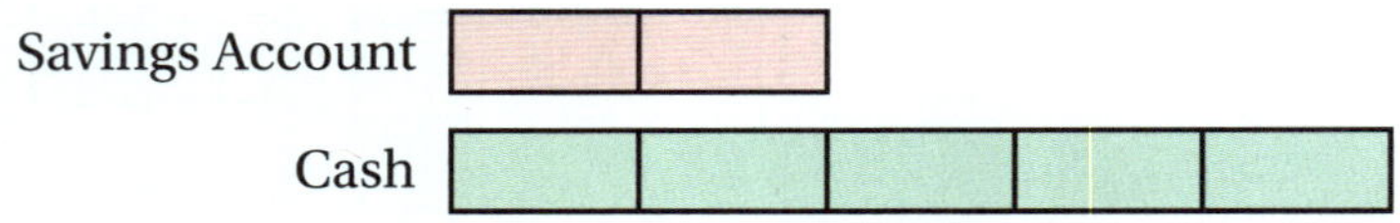

32. **Dig Deeper** A fish tank contains tetras, guppies, and minnows. The ratio of tetras to guppies is 4 : 2. The ratio of minnows to guppies is 1 : 3. There are 60 fish in the tank. How many more tetras are there than minnows? Justify your answer.

8.3 Using Ratio Tables

Learning Target: Use ratio tables to represent equivalent ratios and solve ratio problems.

Success Criteria:
- I can use various operations to create tables of equivalent ratios.
- I can use ratio tables to solve ratio problems.
- I can use ratio tables to compare ratios.

Algebraic Reasoning

MA.6.AR.3.3 Extend previous understanding of fractions and numerical patterns to generate or complete a two- or three-column table to display equivalent part-to-part ratios and part-to-part-to-whole ratios.

MA.6.AR.3.5 Solve mathematical and real-world problems involving ratios, rates and unit rates, including comparisons, mixtures, ratios of lengths and conversions within the same measurement system.

Also MA.6.AR.3.1

Exploration 1 Making a Table of Equivalent Ratios

Work with a partner. You buy milk that contains 180 calories per 2 cups.

a. You measure 2 cups of the milk and pour it into a pitcher. You repeat this four more times. Show the number of calories in a table.

Cups of Milk					
Calories					

b. Describe the relationship between the number of cups and the number of calories.

GO DIGITAL

Exploration 2 Creating a Double Number Line

a. Use the data in Exploration 1 to label the double number line.

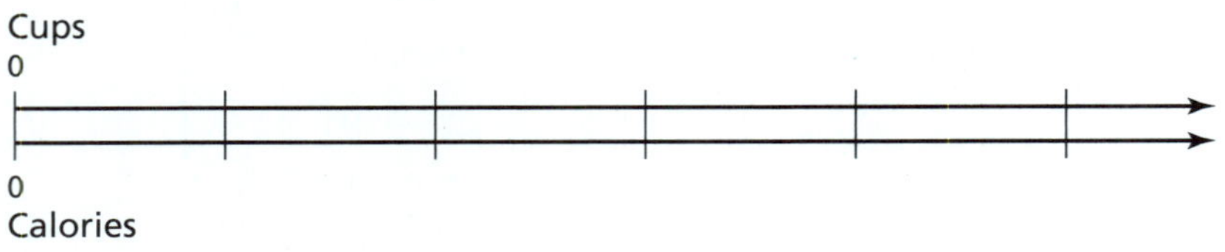

b. Use the double number line to find the number of calories in 3 cups of milk.

c. Use the double number line to find the number of calories in 3.5 cups of milk.

d. Describe another way that you can answer part (b).

4 MTR COMPARE METHODS

Compare your explanations in part (c) with another group's. If they are different, are they both correct?

8.3 Lesson

Key Vocabulary
ratio table, *p. 617*

You can find and organize equivalent ratios in a **ratio table**. You can generate a ratio table by using repeated addition or multiplication.

Example 1 Completing Ratio Tables

Find the missing values in each ratio table. Then write the equivalent ratios.

a.

Triangles	1	2		4
Sides	3		9	

b.

Frogs	4	20		
Toads	6		90	180

a. Because the original ratio is 1 triangle to 3 sides, you can repeatedly add 1 to the first row and repeatedly add 3 to the second row.

+1 +1 +1

Triangles	1	2	3	4
Sides	3	6	9	12

+3 +3 +3

▶ The equivalent ratios are 1 : 3, 2 : 6, 3 : 9, and 4 : 12.

b. You can use multiplication to find the missing values.

×5 ×3 ×2

Frogs	4	20	60	120
Toads	6	30	90	180

×5 ×3 ×2

▶ The equivalent ratios are 4 : 6, 20 : 30, 60 : 90, and 120 : 180.

1 MTR HELP A CLASSMATE
How can you help a classmate understand where to begin when finding missing values in a ratio table?

Try It **Find the missing values in the ratio table. Then write the equivalent ratios.**

1.

Hands	4	8	12	16
People	2	4	6	8

2.

Miles	4	8	24	96
Hours	3	6	18	72

GO DIGITAL

You can also generate a ratio table by using subtraction or division. In summary, you can find equivalent ratios by

- adding or subtracting quantities in equivalent ratios.
- multiplying or dividing each quantity in a ratio by the same number.

Example 2 Completing Ratio Tables

Find the missing values in each ratio table. Then write the equivalent ratios.

a.

Dollars	1			8
Cents	100	300	900	

b.

Meters	3		2	
Minutes	1	2		$\frac{5}{3}$

a. You can use a combination of operations to find the missing values.

$\times 3$ $\times 3$ -1

Dollars	1	3	9	8
Cents	100	300	900	800

$\times 3$ $\times 3$ -100

In Example 2(a), notice that you obtain the fourth column by subtracting the values in the first column from the values in the third column.

$9 - 1 = 8$

$900 - 100 = 800$

▶ The equivalent ratios are 1 : 100, 3 : 300, 9 : 900, and 8 : 800.

b. You can use a combination of operations to find the missing values.

$\times 2$ $\div 3$ $+3$

Meters	3	6	2	5
Minutes	1	2	$\frac{2}{3}$	$\frac{5}{3}$

$\times 2$ $\div 3$ $+1$

▶ The equivalent ratios are $3 : 1$, $6 : 2$, $2 : \frac{2}{3}$, and $5 : \frac{5}{3}$.

Try It **Find the missing values in the ratio table. Then write the equivalent ratios.**

3.

Flowers	1	2	4	6
Petals	5	6	12	15

4.

Students	24	12	48	36
Teachers	2	1	4	3

GO DIGITAL

Example 3 Solving a Ratio Problem

3 MTR REFLECT ON YOUR METHOD

How can you obtain the third column in the table from the first column using one operation?

There are 75 milligrams of sodium in every 12 crackers. You eat 30 crackers. How much sodium do you consume?

Method 1: Use a double number line. Increment the number lines using the original ratio of 75 to 12.

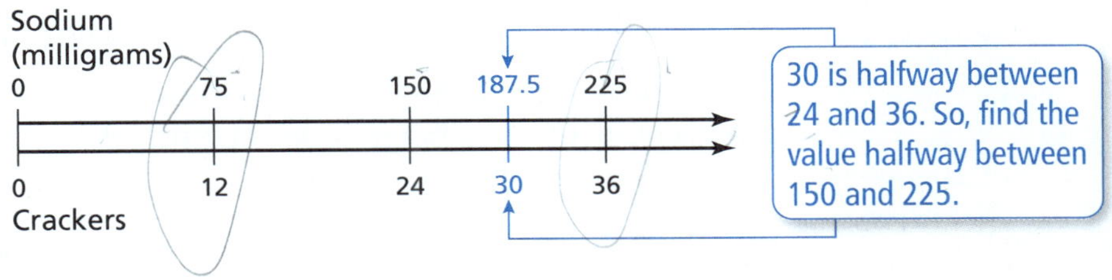

▶ So, you consume 187.5 milligrams of sodium in 30 crackers.

Method 2: Use a ratio table. The ratio of milligrams of sodium to crackers is 75 to 12. Find an equivalent ratio with 30 crackers.

$\div 2$ $\times 5$

Sodium (milligrams)	75	37.5	187.5
Crackers	12	6	30

$\div 2$ $\times 5$

▶ So, you consume 187.5 milligrams of sodium in 30 crackers.

Try It

5. **WHAT IF?** You eat 21 crackers. How much sodium do you consume?

In-Class Practice

1 I don't understand yet. 2 I can do it with help. 3 I can do it on my own. 4 I can teach someone else.

COMPLETING A RATIO TABLE **Find the missing values in the ratio table. Then write the equivalent ratios.**

6.

Fruit	2		6
Vegetables	6	12	

7.

Gnats	2	14		5
Flies	8		28	

8. **WRITING** Explain how creating a ratio table using repeated addition is similar to creating a ratio table using multiplication.

GO DIGITAL

Example 4 Modeling Real Life 7 MTR

You and your friend make colored frosting. You add 3 drops of red food coloring for every 1 drop of blue food coloring. Your friend adds 5 drops of red for every 3 drops of blue. Whose frosting is redder?

Understand the problem.

You are given the numbers of drops of food coloring that you and your friend use to make frosting. You are asked to determine whose frosting is redder.

Make a plan.

Use ratio tables to compare the frostings. Find ratios in which the number of drops of red, the number of drops of blue, or the total number of drops is the same. Then compare the quantities to determine which is redder.

Solve and check.

Create ratio tables for 3 : 1 and 5 : 3 using repeated addition. Include a column for the total number of drops in each frosting.

Your Frosting		
Drops of Red (part)	Drops of Blue (part)	Total Drops (whole)
3	1	4
6	2	8
9	3	12
12	4	16
15	5	20

Your Friend's Frosting		
Drops of Red (part)	Drops of Blue (part)	Total Drops (whole)
5	3	8
10	6	16
15	9	24
20	12	32
25	15	40

When both frostings have 3 drops of blue, your frosting has $9 - 5 = 4$ more drops of red than your friend's frosting.

▶ So, your frosting is redder than your friend's frosting.

Look Back
The tables show that when both frostings have a total of 16 drops, your frosting has 2 more drops of red and 2 fewer drops of blue. So, your frosting is redder. ✓

In-Class Practice

1 I don't understand yet. 2 I can do it with help. 3 I can do it on my own. 4 I can teach someone else.

9. You mix 7 tablespoons of vinegar for every 4 tablespoons of baking soda to produce a chemical reaction. You use 15 tablespoons of baking soda. How much vinegar do you use?

10. You make a carbonated beverage by adding 7 ounces of soda water for every 3 ounces of regular water. Your friend uses 11 ounces of soda water for every 4 ounces of regular water. Whose beverage is more carbonated?

GO DIGITAL

7 MTR **31. MODELING REAL LIFE** You make candles by adding 2 fluid ounces of scented oil for every 22 fluid ounces of wax. Your friend makes candles by adding 3 fluid ounces of the same scented oil for every 37 fluid ounces of wax. Whose candles are more fragrant? Explain your reasoning. (See Example 4.)

7 MTR **32. MODELING REAL LIFE** A mint milkshake contains 1.25 fluid ounces of milk for every 4 ounces of ice cream. A strawberry milkshake contains 1.75 fluid ounces of milk for every 5 ounces of ice cream. Which milkshake is thicker? Explain.

5 MTR **STRUCTURE** **Two whole numbers A and B satisfy the following conditions. Find A and B.**

33. $A + B = 30$

$A : B$ is equivalent to $2 : 3$.

34. $A - B = 18$

$A : B$ is equivalent to $11 : 5$.

Nutrition Facts

8 servings per container

Serving size **1 ounce (28g)**

Amount per serving

Calories **161**

	% Daily Value*
Total Fat 13g	**20%**
Saturated Fat 3g	**13%**
Trans Fat 0g	
Cholesterol 0mg	**0%**
Sodium 4mg	**0%**
Total Carbohydrate 9g	**3%**
Dietary Fiber 1g	**3%**
Total Sugars 1g	
Includes 0g Added Sugars	**0%**
Protein 4g	
Vitamin D 0mcg	0%
Calcium 14.8mg	1%
Iron 2.67mg	15%
Potassium 264mg	8%

* The % Daily Value (DV) tells you how much a nutrient in a serving of food contributes to a daily diet. 2,000 calories a day is used for general nutrition advice.

7 MTR **35. MODELING REAL LIFE** There are 161 calories in 28 grams of dry roasted cashews. You eat 9 cashews totaling 12 grams.

a. Do you think it is possible to find the number of calories you consume? Explain your reasoning.

b. How many cashews are in one serving?

36. REASONING The ratio of three numbers is $4 : 5 : 3$. The sum of the numbers is 54. What are the three numbers?

37. ADAPT A PROCEDURE Seven out of every 8 students surveyed own a bike. The difference between the number of students who own a bike and those who do not is 72. How many students were surveyed?

38. REASONING You and a classmate have an insect collection for science class. You find 5 out of every 9 insects in the collection. You find 4 more insects than your classmate. How many insects are in the collection?

39. PROBLEM SOLVING You earn \$72 for every 8 hours you spend mowing lawns. You earn \$60 for every 5 hours you spend babysitting. For every 3 hours you spend babysitting, you spend 2 hours mowing lawns. You babysit for 15 hours in July. How much money do you earn in July?

40. Dig Deeper You and a friend each have a collection of stamps. Initially, for every 8 stamps you had, your friend had 3. After you give half of your stamps to your friend, your friend now has 18 more stamps than you. Initially, how many more stamps did you have than your friend?

8.4 Rates and Unit Rates

Learning Target: Understand the concept of a unit rate and solve rate problems.

Success Criteria:
- I can find unit rates.
- I can use unit rates to solve rate problems.
- I can use unit rates to compare rates.

Algebraic Reasoning

MA.6.AR.3.2 Given a real-world context, determine a rate for a ratio of quantities with different units. Calculate and interpret the corresponding unit rate.

MA.6.AR.3.5 Solve mathematical and real-world problems involving ratios, rates and unit rates, including comparisons, mixtures, ratios of lengths and conversions within the same measurement system.

Also MA.6.AR.3.1, MA.6.AR.3.3

Exploration 1 Using a Diagram

Work with a partner. The diagram shows a story problem.

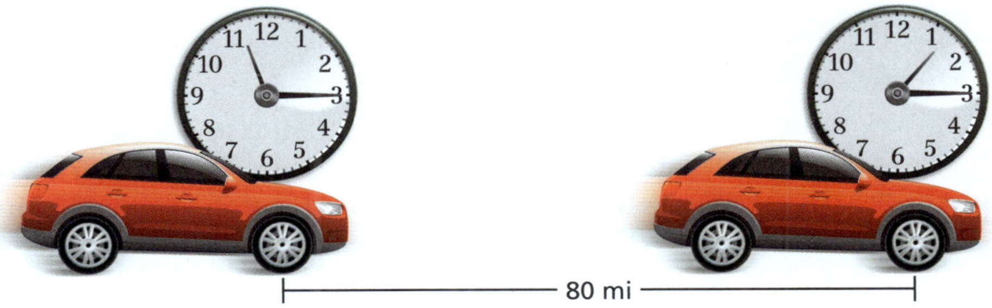

a. What information can you obtain from the diagram?

b. Assuming that the car travels at a constant speed, how far does the car travel in 3.25 hours? Explain your method.

c. Draw a speedometer that shows the speed of the car. How can you use the speedometer to answer part (b)?

GO DIGITAL

1 MTR **ANALYZE A PROBLEM**

What are the units for the speed in part (c)? Why is it important to keep track of units in ratio problems?

Exploration 2 Using Equivalent Ratios

Work with a partner. Count the number of times you can clap your hands in 12 seconds. Have your partner record your results. Then switch roles with your partner and repeat the process.

a. Using your results and your partner's results, write ratios that represent the numbers of claps for every 12 seconds.

b. Use the ratios in part (a) to find the numbers of times you and your partner can clap your hands in 2 minutes, in 2.5 minutes, and in 3 minutes.

8.4 Lesson

Key Vocabulary

rate, *p. 629*
unit rate, *p. 629*
equivalent rates, *p. 629*

A **rate** is a ratio of two quantities using different units. You solved various ratio problems in the previous sections that involved rates. Now you will use *unit rates* to solve rate problems.

Key Idea

Unit Rate

Words A **unit rate** compares a quantity to one unit of another quantity. **Equivalent rates** have the same unit rate.

Numbers You pay \$27 for 3 pizzas.

Rate: \$27 : 3 pizzas

Unit rate: \$9 : 1 pizza

Algebra Rate: a units : b units Unit rate: $\frac{a}{b}$ units : 1 unit

Example 1 Finding Unit Rates

You make fruit juice by adding 4 pints of water for every 2 cups of concentrate.

a. Find the unit rate.

The ratio of pints of water to cups of concentrate is 4 : 2. Divide each quantity by 2 to find the unit rate.

	÷ 2	
Water (pints)	4	2
Concentrate (cups)	2	1
	÷ 2	

▶ The unit rate is 2 pints of water per cup of concentrate.

b. Write the ratio of cups of concentrate to pints of water. Then find the unit rate.

The ratio of cups of concentrate to pints of water is 2 : 4. Divide each quantity by 4 to find the unit rate.

	÷ 4	
Concentrate (cups)	2	$\frac{1}{2}$
Water (pints)	4	1
	÷ 4	

▶ The unit rate is $\frac{1}{2}$ cup of concentrate per pint of water.

For a rate $a : b$, the associated rate $b : a$ has a unit rate of $\frac{b}{a} : 1$. This can be useful when solving rate problems.

GO DIGITAL

Try It

1. WHAT IF? Repeat Example 1 when you add 4 pints of water for every 3 cups of concentrate.

Example 2 Using a Unit Rate to Solve a Rate Problem

A piece of space junk travels 5 miles every 6 seconds.

a. How far does the space junk travel in 30 seconds?

The ratio of miles to seconds is 5 : 6. Divide by 6 to find the unit rate in miles per second. Then multiply each quantity by 30 to find the distance traveled in 30 seconds.

$\div 6$ $\times 30$

Distance (miles)	5	$\frac{5}{6}$	25
Time (seconds)	6	1	30

$\div 6$ $\times 30$

▶ The space junk travels 25 miles in 30 seconds.

b. How many seconds does it take the space junk to travel 2 miles?

In Example 2, notice that you can use one step in each ratio table. Multiply by $\frac{1}{6} \times 30 = 5$ in part (a) and $\frac{1}{5} \times 2 = \frac{2}{5}$ in part (b).

The ratio of seconds to miles is 6 : 5. Divide by 5 to find the unit rate in seconds per mile. Then multiply each quantity by 2 to find the time to travel 2 miles.

$\div 5$ $\times 2$

Time (seconds)	6	$\frac{6}{5}$	$\frac{12}{5}$
Distance (miles)	5	1	2

$\div 5$ $\times 2$

▶ It takes $\frac{12}{5} = 2\frac{2}{5}$ seconds for the space junk to travel 2 miles.

GO DIGITAL

Try It

2. WHAT IF? Repeat Example 2 when the space junk travels 3 miles every 5 seconds.

In-Class Practice

1 I don't understand yet.	2 I can do it with help.	3 I can do it on my own.	4 I can teach someone else.

FINDING UNIT RATES **Write a unit rate for the situation.**

3. 5 revolutions in 50 seconds

4. 1400 words for every 4 pages

4 MTR

5. WHICH ONE DOESN'T BELONG? Which rate does *not* belong with the other three? Explain your reasoning.

8 pounds for every 2 feet

12 pounds per 3 feet

20 pounds per 4 feet

24 pounds for every 6 feet

Example 3 Modeling Real Life

7 MTR

You buy 2 pounds of salmon filets at Store A for \$21.50. Your friend buys 3 pounds of salmon filets at Store B for \$33.75. How much less would you spend for 5 pounds of salmon filets at the store with the better deal?

Use ratio tables to find the cost of 5 pounds of salmon filets at each store.

Find the unit rate at each store.

Find the cost of 5 pounds at each store.

Store A	
Cost (dollars)	Salmon (pounds)
21.50	2
10.75	1
53.75	5

Store B	
Cost (dollars)	Salmon (pounds)
33.75	3
11.25	1
56.25	5

▶ So, you would spend \$56.25 − \$53.75 = \$2.50 less for 5 pounds of salmon filets at Store A.

Check The cost at Store A is \$11.25 − \$10.75 = \$0.50 cheaper per pound than the cost at Store B. Because you are buying 5 pounds, you will pay 5 × \$0.50 = \$2.50 less at Store A. ✓

In-Class Practice

1 I don't understand yet. 2 I can do it with help. 3 I can do it on my own. 4 I can teach someone else.

6. You buy 10 pounds of grass seed at Store A for \$11.50. Your friend buys 15 pounds of grass seed at Store B for \$19.50. How much less would you spend by buying 20 pounds of grass seed at the store with the better deal?

7. A person hikes 4 miles in 2.5 hours. Find the unit rate in miles per hour. Then find the unit rate in hours per mile. How is each unit rate useful in a real-life situation?

8. **Dig Deeper** You can buy 5 bagels for \$7.50. How many bagels can you buy with a \$20 bill? Can you buy at most n times this many bagels with n \$20 bills? Explain your reasoning.

GO DIGITAL

8.5 Solving Proportions

Learning Target: Understand and solve proportions.

Success Criteria:
- I can explain the meaning of a proportion.
- I can explain how to solve a proportion.
- I can use proportions to solve real-life problems.

Algebraic Reasoning
MA.7.AR.3.2 Apply previous understanding of ratios to solve real-world problems involving proportions.

Exploration 1 Using Proportions

Work with a partner. Use one of the methods shown to check your heart rate.

Method 1: Place two fingers on the *radial artery* in your wrist.

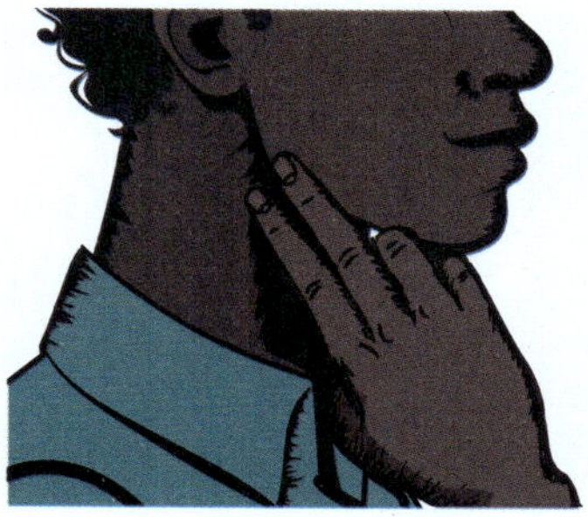

Method 2: Place two fingers on the *carotid artery* in your neck.

1 MTR **HELP A CLASSMATE**
Help your partner understand the definition of *proportion*.

a. Count the number of times your heart beats in 15 seconds. Use this rate to complete the ratio table for different amounts of time.

Heartbeats				
Seconds				

b. Use your ratio table to write several different equations. Explain why you are able to use equations to relate the quantities in the table.

c. Your equations in part (b) are called *proportions*. Use a proportion to find the number of times your heart beats in 1 hour. Explain your method.

GO DIGITAL

8.5 Lesson

Key Vocabulary
proportion, *p. 638*

Key Idea

Proportions

Words A **proportion** is an equation stating that the values of two ratios are equivalent.

Numbers Equivalent ratios: 4 : 5 and 8 : 10

Proportion: $\frac{4}{5} = \frac{8}{10}$

Example 1 Solving a Proportion

Solve $\frac{2}{3} = \frac{x}{12}$.

Method 1: Use mental math.

Think: The product of 3 and what number is 12?

$$\frac{2}{3} = \frac{x}{12}$$

$$3 \times ? = 12$$

Because the product of 3 and 4 is 12, multiply the numerator by 4 to find x.

$$2 \times 4 = 8$$

$$\frac{2}{3} = \frac{x}{12}$$

$$3 \times 4 = 12$$

▶ The solution is $x = 8$.

Method 2: Use multiplication.

$\frac{2}{3} = \frac{x}{12}$ Write the proportion.

$12 \cdot \frac{2}{3} = 12 \cdot \frac{x}{12}$ Multiplication Property of Equality

$8 = x$ Simplify.

▶ The solution is $x = 8$.

3 MTR REFLECT ON YOUR METHOD
How is solving for x in Method 1 similar to completing ratio tables in Section 8.3?

____

____

____

____

____

____

____

GO DIGITAL

Try It

Solve the proportion.

1. $\frac{3}{4} = \frac{p}{24}$

72÷4

18=p

2. $\frac{k}{2} = \frac{27}{6}$

54÷6

K=9

3. $\frac{2}{5} = \frac{10}{b}$

50÷2

b=25

Example 2 B.E.S.T. Test Prep: Writing a Proportion

You mix two parts oil per three parts water to create a suspension. Which proportion can you use to find the number of milliliters of oil you need to create 300 milliliters of the suspension?

Ⓐ $\frac{2}{3} = \frac{300}{x}$ Ⓒ $\frac{2}{3} = \frac{x}{300}$

Ⓑ $\frac{2}{5} = \frac{x}{300}$ Ⓓ $\frac{2}{5} = \frac{300}{x}$

You mix two parts oil per three parts water. So, the ratio of oil to the total amount of liquid is 2 : 5.

Oil (part)	2
Water (part)	3
Total (whole)	5

In 300 milliliters of the suspension, the ratio of the amount x (in milliliters) of oil to the total amount of liquid is $x : 300$. Because the values of the ratios 2 : 5 and $x : 300$ are equal, a proportion is $\frac{2}{5} = \frac{x}{300}$.

▶ The correct answer is Ⓑ.

Try It

4. WHAT IF? Write a proportion that you can use to find the number of milliliters of water in 500 milliliters of the suspension.

GO DIGITAL

In-Class Practice

1 I don't understand yet. | 2 I can do it with help. | 3 I can do it on my own. | 4 I can teach someone else.

5. **VOCABULARY** Explain what a proportion represents.

SOLVING A PROPORTION **Solve the proportion. Explain your method.**

6. $\frac{k}{4} = \frac{35}{28}$

7. $\frac{7}{10} = \frac{g}{30}$

8. **REASONING** Use the ratio table to write at least three different proportions. Explain your reasoning.

White Blood Cells	1	2	3
Red Blood Cells	600	1200	1800

4 MTR

9. **WHICH ONE DOESN'T BELONG?** Which proportion does *not* belong with the other three? Explain your reasoning.

$\frac{2}{3} = \frac{x}{24}$ $\frac{3}{2} = \frac{24}{x}$ $\frac{24}{3} = \frac{x}{2}$ $\frac{24}{2} = \frac{x}{3}$

Example 3 Modeling Real Life

7 MTR

The record for the most touchdown passes thrown in a 16-game season in the NFL is 55. A quarterback throws 39 touchdown passes in the first 12 games of a 16-game season. At this rate, will the quarterback break the record?

Understand the problem.

You know that there are 16 games in the season. You are given the number of passes thrown in the first 12 games. You are asked to determine whether the quarterback will throw more than 55 touchdown passes at this rate.

Make a plan.

The ratio of touchdown passes thrown to games played is 39 : 12. After 16 games, the ratio is x : 16. Assuming that the quarterback continues to throw touchdown passes at the current rate, you can use a proportion to find the value of x for which $\frac{39}{12}$ and $\frac{x}{16}$ are equal.

Solve and check.

$$\frac{39}{12} = \frac{x}{16} \qquad \text{Write the proportion.}$$

$$16 \cdot \frac{39}{12} = 16 \cdot \frac{x}{16} \qquad \text{Multiplication Property of Equality}$$

$$52 = x \qquad \text{Simplify.}$$

▶ At the current rate, the quarterback will throw 52 touchdown passes. So, the quarterback will not break the record.

Another Method

You can use a ratio table to solve the problem.

÷ 3   × 4

Touchdown Passes	39	13	52 ✓
Games	12	4	16

÷ 3   × 4

GO DIGITAL

In-Class Practice

1 I don't understand yet. 2 I can do it with help. 3 I can do it on my own. 4 I can teach someone else.

10. You have $50 to buy new shirts. Your friend buys 3 shirts at a store for $24. Do you have enough money to buy 7 of the shirts?

11. **Dig Deeper** You add 4 cups of cranberry juice to some grape juice to create 12 cups of a juice mixture. You drink 3 cups of the mixture and then add 3 more cups of cranberry juice to the remaining mixture. How much cranberry juice is in 2 cups of the new mixture?

22. **B.E.S.T. Test Prep** You mix five parts powdered peanut butter per two parts water. Which proportion can you use to find the number of tablespoons of water that you use when you use 8 tablespoons of powdered peanut butter? (See Example 2.)

Ⓐ $\frac{2}{5} = \frac{x}{8}$ Ⓒ $\frac{2}{7} = \frac{x}{8}$

Ⓑ $\frac{5}{2} = \frac{x}{8}$ Ⓓ $\frac{2}{5} = \frac{8}{x}$

23. **WRITING A PROPORTION** A manufacturer mixes two parts copper per one part zinc to make brass. Write a proportion that you can use to determine the amount of zinc needed to make 18 ounces of brass.

5 MTR

24. **STRUCTURE** Complete the proportion. Justify your answer.

$$\frac{4}{\square} = \frac{\square}{10}$$

7 MTR

25. **MODELING REAL LIFE** The cost of a trip with a ride-sharing company is based on the length of the trip. You pay \$7 to travel 4 miles to a movie theater. You have \$13 left to travel 7 miles to your friend's apartment. Do you have enough money? (See Example 3.)

7 MTR

26. **MODELING REAL LIFE** A mountain climber melts 16 cups of snow to produce 1.6 cups of water.

a. How much snow does the climber need to melt to produce 8 cups of water?

b. What do you notice about the ratio of snow to water? How can you use this observation to solve part (a) without writing and solving a proportion?

27. **PROBLEM SOLVING** Roll a ball across the floor and record the amount of time that it takes the ball to travel 10 feet. Then determine the speed of the ball in miles per hour. (*Hint:* 1 mile = 5280 feet and 1 hour = 3600 seconds.)

28. **Dig Deeper** The ratio of red paint to yellow paint in a paint mixture is 1 : 3. You have 12 cups of the paint mixture. You want to modify the shade of the mixture by adding red paint until the ratio of red paint to yellow paint is 1 : 2. Find the number of cups of red paint that you should add to the mixture.

8.6 Converting Measures within Systems

Learning Target: Use ratio reasoning to convert units of measure within a system.

Success Criteria:
- I can write conversion facts as unit rates.
- I can convert units of measure using ratio tables.
- I can convert units of measure using conversion factors.
- I can convert rates using conversion factors.

Algebraic Reasoning

MA.6.AR.3.2 Given a real-world context, determine a rate for a ratio of quantities with different units. Calculate and interpret the corresponding unit rate.

MA.6.AR.3.5 Solve mathematical and real-world problems involving ratios, rates and unit rates, including comparisons, mixtures, ratios of lengths and conversions within the same measurement system.

Also MA.6.AR.3.1, MA.6.AR.3.3

Exploration 1 Converting Units of Measure

5 MTR RELATE CONCEPTS

Explain why relationships between capacities form ratios.

____

____

____

____

Work with a partner.

a. The diagram below represents the relationships among gallons, quarts, pints, and cups. Write at least four unit rates represented in the diagram.

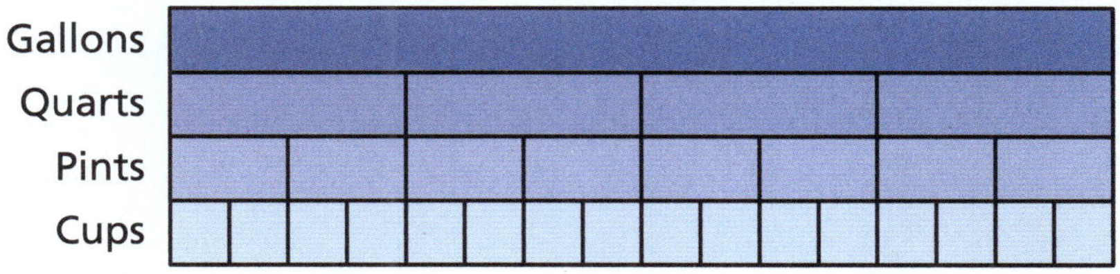

b. Complete the ratio table. How many cups are equivalent to 4 quarts?

Cups						
Quarts	1	2	3	4	5	6

GO DIGITAL

c. A recipe calls for 1.5 quarts of milk, but you only have a 1-cup measuring cup. How can you measure the correct amount?

d. How can you use a ratio table and one operation to find the number of cups in 1.5 quarts?

Cups	4	
Quarts	1	1.5

e. What is the value of any ratio of cups to quarts? How can you use this value to find the number of cups in any number of quarts?

8.6 Lesson

Key Vocabulary

U.S. customary system, *p. 649*
metric system, *p. 649*
conversion factor, *p. 650*
unit analysis, *p. 650*

The **U.S. customary system** is a system of measurement that contains units for length, capacity, and weight. The **metric system** is a decimal system of measurement, based on powers of 10, that contains units for length, capacity, and mass.

You can use unit rates and ratio tables to convert measures within a system.

Example 1 Converting Measures

Convert 36 quarts to gallons.

For a list of conversion facts, see the Mathematics Reference Sheet in the back of this book.

Because 1 gallon = 4 quarts, there are 4 quarts per gallon, and $\frac{1}{4}$ gallon per quart. You can use either of these unit rates to find an equivalent rate with 36 quarts.

Method 1: Create a ratio table using the unit rate 4 quarts per gallon. Multiply each quantity by 9 to find the number of gallons in 36 quarts.

× 9

Quarts	4	36
Gallons	1	9

× 9

▶ So, 36 quarts is 9 gallons.

Method 2: Create a ratio table using the unit rate $\frac{1}{4}$ gallon per quart. Multiply each quantity by 36 to find the number of gallons in 36 quarts.

× 36

Gallons	$\frac{1}{4}$	9
Quarts	1	36

× 36

▶ So, 36 quarts is 9 gallons.

Another Method Use a double number line.

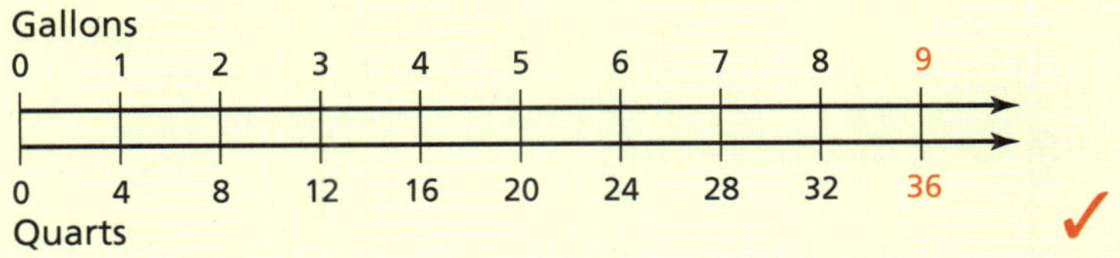

Try It

1. Convert 48 feet to yards.

2. Convert 17 kilograms to grams.

GO DIGITAL

Another way to convert units of measure is to multiply by one or more *conversion factors*.

Key Idea

> Because the quantities in a conversion factor are equal, conversion factors are equal to 1. So, you can multiply a quantity by a conversion factor and not change its value.

Conversion Factor

A **conversion factor** is a rate in which the two quantities are equal. When using conversion factors, write rates using fraction notation.

	Relationship	***Conversion Factors***
Example	1 ft = 12 in.	1 ft per 12 in., or $\frac{1\text{ ft}}{12\text{ in.}}$
		12 in. per 1 ft, or $\frac{12\text{ in.}}{1\text{ ft}}$

You can use **unit analysis** to decide which conversion factor will produce the appropriate units. You can "cross out" a unit that appears in both a numerator and a denominator of a product.

Example 2 Using Conversion Factors

a. Convert 5 pounds to ounces.

Use a conversion factor.

1 lb = 16 oz

$$5\text{ lb} = 5\text{ lb} \times \frac{16\text{ oz}}{1\text{ lb}} = 80\text{ oz}$$

▶ So, 5 pounds is 80 ounces.

b. Convert 10 milliliters to liters.

Use a conversion factor.

1 L = 1000 mL

$$10\text{ mL} = 10\text{ mL} \times \frac{1\text{ L}}{1000\text{ mL}} = 0.01\text{ L}$$

▶ So, 10 milliliters is 0.01 liter.

4 MTR COMMUNICATE CLEARLY

In part (b), explain how you know to use a conversion factor of $\frac{1\text{ L}}{1000\text{ mL}}$ rather than $\frac{1000\text{ mL}}{1\text{ L}}$.

____

____

____

____

GO DIGITAL

Try It

3. Convert 8 kilometers to meters.

4. Convert 40 cups to gallons.

Example 3 Using Conversion Factors to Convert Rates

a. Convert 5 yards per second to yards per minute.

Write 5 yards per second as a fraction. Then use a conversion factor.

1 min = 60 sec

$$5 \text{ yards per second} = \frac{5 \text{ yd}}{1 \cancel{\text{sec}}} \times \frac{60 \cancel{\text{sec}}}{1 \text{ min}} = \frac{300 \text{ yd}}{1 \text{ min}}$$

▶ So, 5 yards per second is 300 yards per minute.

b. Convert 12 kilometers per minute to meters per second.

Write 12 kilometers per minute as a fraction. Then use conversion factors.

1 min = 60 sec   1 km = 1000 m

$$12 \text{ kilometers per minute} = \frac{12 \cancel{\text{km}}}{1 \cancel{\text{min}}} \times \frac{1 \cancel{\text{min}}}{60 \text{ sec}} \times \frac{1000 \text{ m}}{1 \cancel{\text{km}}} = \frac{200 \text{ m}}{1 \text{ sec}}$$

▶ So, 12 kilometers per minute is 200 meters per second.

GO DIGITAL

Try It

5. Convert 20 liters per minute to liters per second.

6. Convert 15 feet per second to yards per minute.

In-Class Practice

1 I don't understand yet. 2 I can do it with help. 3 I can do it on my own. 4 I can teach someone else.

4 MTR

7. **DIFFERENT WORDS, SAME QUESTION** Which is different? Find "both" answers.

Convert 25 grams to milligrams.	How many milligrams are in 25 grams?
Find the number of grams in 25 milligrams.	Twenty-five grams equals how many milligrams?

CONVERTING MEASURES **Complete the statement.**

8. $\frac{20 \text{ m}}{\text{h}} = \frac{\square \text{ cm}}{\text{h}}$

9. $\frac{12 \text{ ft}}{\text{sec}} = \frac{\square \text{ yd}}{\text{min}}$

Example 4 Modeling Real Life

7 MTR

A runner's goal is to complete a mile in 4 minutes or less. The runner's speed is 20 feet per second. Does the runner meet the goal? If not, how much faster (in feet per second) must the runner be to meet the goal?

To meet the goal, the runner must complete 1 mile in 4 minutes or less. The minimum speed required is 1 mile per 4 minutes.

To compare this speed to the runner's speed of 20 feet per second, convert the minimum speed of 1 mile per 4 minutes to feet per second.

1 min = 60 sec

1 mi = 5280 ft

$$\frac{1\text{ mi}}{4\text{ min}} = \frac{1\text{ mi}}{4\text{ min}} \times \frac{1\text{ min}}{60\text{ sec}} \times \frac{5280\text{ ft}}{1\text{ mi}} = \frac{22\text{ ft}}{1\text{ sec}}$$

▶ So, the runner did not meet the goal because a speed of 20 feet per second is below the minimum speed of 22 feet per second. The runner must be 22 − 20 = 2 feet per second faster to meet the goal.

Check Verify that the distances traveled in 4 minutes at the runner's speed and in 4 minutes at the additional speed have a sum of 1 mile.

At runner's speed

$$4\text{ min} \times \frac{60\text{ sec}}{1\text{ min}} \times \frac{20\text{ ft}}{1\text{ sec}} = 4800\text{ ft}$$

At additional speed

$$4\text{ min} \times \frac{60\text{ sec}}{1\text{ min}} \times \frac{2\text{ ft}}{1\text{ sec}} = 480\text{ ft}$$

The sum of the distances is 4800 + 480 = 5280 feet, or 1 mile. ✓

GO DIGITAL

In-Class Practice

1 I don't understand yet. **2** I can do it with help. **3** I can do it on my own. **4** I can teach someone else.

10. The speed of sound in air is about 1225 kilometers per hour. The speed of sound in seawater is about 1520 meters per second. Does sound travel faster in air or in seawater? How many times faster?

11. A race car driver's goal is to complete a 1000-kilometer auto race in 4 hours or less. The driver's average speed is 4200 meters per minute. Does the driver meet the goal? If not, how much faster (in meters per minute) must the driver be to meet the goal?

7 MTR

34. **MODELING REAL LIFE** The Daytona International Speedway is 4 kilometers long. Find the length of the speedway in meters.

CONVERTING RATES **Complete the statement.** (See Example 3.)

▶35. $\frac{3 \text{ pt}}{\text{h}} = \frac{\square \text{ c}}{\text{h}}$

36. $\frac{14 \text{ ft}}{\text{sec}} = \frac{\square \text{ in.}}{\text{sec}}$

37. $\frac{48 \text{ oz}}{\text{in.}} = \frac{\square \text{ lb}}{\text{in.}}$

38. $\frac{256 \text{ L}}{\text{lb}} = \frac{\square \text{ L}}{\text{oz}}$

39. $\frac{3 \text{ km}}{\text{min}} = \frac{\square \text{ km}}{\text{h}}$

40. $\frac{2 \text{ qt}}{\text{mL}} = \frac{\square \text{ c}}{\text{L}}$

41. $\frac{6 \text{ L}}{\text{kg}} = \frac{\square \text{ mL}}{\text{g}}$

42. $\frac{0.96 \text{ kg}}{\text{lb}} = \frac{\square \text{ g}}{\text{oz}}$

43. $\frac{7.2 \text{ km}}{\text{h}} = \frac{\square \text{ m}}{\text{sec}}$

4 MTR

YOU BE THE TEACHER **Your friend converts the rate. Is your friend correct? Explain your reasoning.**

44.

$$\frac{8 \text{ gal}}{1 \text{ min}} = \frac{8 \cancel{\text{gal}}}{1 \text{ min}} \times \frac{1 \text{ qt}}{4 \cancel{\text{gal}}} = \frac{2 \text{ qt}}{\text{min}}$$

45.

$$\frac{24 \text{ ft}}{1 \text{ sec}} = \frac{24 \cancel{\text{ft}}}{1 \cancel{\text{sec}}} \times \frac{1 \text{ yd}}{3 \cancel{\text{ft}}} \times \frac{60 \cancel{\text{sec}}}{1 \text{ min}} = \frac{480 \text{ yd}}{\text{min}}$$

COMPARING MEASURES **Complete the statement using <, >, or =.**

46. 2 lb $\square$ 48 oz

47. 130 g $\square$ 13 kg

48. 3 gal $\square$ 24 pt

49. 12 ft $\square$ 4 yd

50. 70 cm $\square$ 0.07 km

51. 0.04 L $\square$ 400 mL

52. $\frac{3 \text{ qt}}{\text{sec}} \square \frac{6 \text{ pt}}{\text{sec}}$

53. $\frac{6 \text{ g}}{\text{min}} \square \frac{120 \text{ mg}}{\text{h}}$

54. $\frac{0.9 \text{ km}}{\text{h}} \square \frac{14 \text{ m}}{\text{min}}$

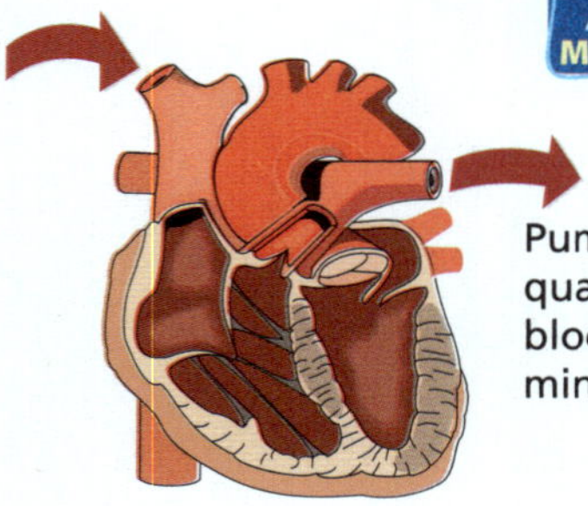

7 MTR **55.** **MODELING REAL LIFE** The diagram shows the number of quarts of blood the human heart pumps per minute. (See Example 4.)

a. How many quarts of blood does the human heart pump per hour?

b. How many cups of blood does the human heart pump per second?

7 MTR **56.** **MODELING REAL LIFE** You are riding on a zip line. Your speed is 15 miles per hour. What is your speed in feet per second?

7 MTR **57.** **MODELING REAL LIFE** You and your friend live the same distance from school. You walk at a speed of 2.5 miles per hour. Your friend walks at a speed of 227 feet per minute. You both leave at the same time. Who arrives at school first? Explain how you found your answer.

CONVERTING MEASURES **Complete the statement.**

58. $\dfrac{36\text{ m}}{\text{min}} = \dfrac{\square\text{ km}}{\text{h}} = \dfrac{\square\text{ mm}}{\text{sec}}$

59. $\dfrac{\square\text{ yd}}{\text{h}} = \dfrac{12\text{ ft}}{\text{min}} = \dfrac{\square\text{ in.}}{\text{sec}}$

60. **PROBLEM SOLVING** Boston, Massachusetts, and Buffalo, New York, are hit by snowstorms that last 3 days. Boston accumulates snow at a rate of 1.5 feet every 36 hours. Buffalo accumulates snow at a rate of 0.01 inch every minute. Which city accumulates more snow in 3 days? How much more snow?

61. **REASONING** The table shows the swimming speeds of several dolphins.

a. Which dolphin is the fastest? Which is the slowest?

Bottlenose Dolphin

Dolphin	Speed
Pacific Whitesided Dolphin	17 mi/h
Orca	50.6 ft/sec
Bottlenose Dolphin	89,760 ft/h
Common Dolphin	2076.8 ft/min

b. The Pacific Spotted Dolphin can reach a swimming speed of up to 24.7 miles per hour. Can the Pacific Spotted Dolphin swim faster than all the dolphins shown in the table? Explain.

62. PERSEVERE You travel 1232 feet every minute on a bicycle.

a. The speed limit is 15 miles per hour. Is your speed less than or equal to the speed limit? Justify your answer.

b. What is your pace in minutes per mile?

c. You are 8 miles from your house at 6:00 P.M. If you continue to travel at this speed, do you reach your house in time for dinner at 6:30 P.M.?

63. Dig Deeper You and your friend are racing remote control cars. Your car travels at a constant speed of 0.4 kilometer per minute. Your friend's car travels at a constant speed of 8 meters per second.

a. The race is 100 meters long. Which car wins the race?

b. You decide that you want the race to have a closer finish. Where could you position the cars at the start to make the race more competitive? Justify your answer.

8 Connecting Concepts

5 MTR *Mathematicians who use patterns and structure to help understand and connect mathematical concepts relate previously learned concepts to new concepts.*

1 MTR **1. ANALYZE A PROBLEM** You mix water, glue, and borax in the ratio of 3 : 1 : 2 to make slime. How many gallons of each ingredient should you use to make 0.75 gallon of slime?

Using the Problem-Solving Plan

Understand the problem. You know the ratio of the ingredients in the slime and that you are making 0.75 gallon of slime. You are asked to find the number of gallons of each ingredient needed to make 0.75 gallon of slime.

Make a plan. Represent the ratio 3 : 1 : 2 using a tape diagram. Because there are 6 parts that represent 0.75 gallon, divide 0.75 by 6 to find the value of one part of the tape diagram. Then use the value of one part to find the number of gallons of each ingredient you should use.

Solve and check. Use the plan to solve the problem. Then check your solution.

2. You buy yogurt cups and frozen fruit bars for a party. Yogurt cups are sold in packages of six. The ratio of the number of yogurt cups in a package to the number of frozen fruit bars in a package is 3 : 2. What are the least numbers of packages you should buy in order to have the same numbers of yogurt cups and frozen fruit bars?

3. The greatest common factor of two whole numbers is 9. The ratio of the greater number to the lesser number is 6 : 5. What are the two numbers? Justify your answer.

STEAM Performance Task

Oops! Unit Conversion Mistakes

GO DIGITAL

At the beginning of this chapter, you watched a STEAM Video called "Human Circulatory System." You are now ready to complete the STEAM Performance Task, where you will identify mistakes in unit conversions. Be sure to use mathematical thinking and reasoning, and the problem-solving plan as you work through the performance task.

8 Chapter Review WITH CalcChat®

Review Vocabulary

Write the definition and give an example of each vocabulary term.

ratio, *p. 593*

rate, *p. 629*

U.S. customary system, *p. 649*

value of a ratio, *p. 594*

unit rate, *p. 629*

metric system, *p. 649*

equivalent ratios, *p. 594*

equivalent rates, *p. 629*

conversion factor, *p. 650*

ratio table, *p. 617*

proportion, *p. 638*

unit analysis, *p. 650*

GO DIGITAL

Graphic Organizers

You can use a **Definition and Example Chart** to organize information about a concept. Here is an example of a Definition and Example Chart for the vocabulary term ***ratio***.

Ratio: a comparison of two quantities. Ratios can be part-to-part, part-to-whole, or whole-to-part comparisons.

- Example: 4 to 5
- Example: 2 : 5
- Example: teachers : students

Choose and complete a graphic organizer to help you study the concept.

1. value of a ratio
2. equivalent ratios
3. tape diagram
4. ratio table
5. rate
6. unit rate
7. conversion factor

Chapter Learning Target: Understand ratios.

Chapter Success Criteria:
- ◆ I can write and interpret ratios.
- ◆ I can name ratios equivalent to a given ratio.
- ■ I can solve problems using ratios.
- ■ I can apply ratio reasoning to convert units of measure.

◆ Surface
■ Deep

Rate your understanding after each section.

1	2	3	4
I don't understand yet.	I can do it with help.	I can do it on my own.	I can teach someone else.

8.1 Ratios *(pp. 591–604)*

Learning Target: Understand the concepts of ratios and equivalent ratios.

Write the ratio.

1. butterflies : caterpillars

2. maracas : sombreros

3. The ratio of hydrogen atoms to nitrogen atoms in a container is 2 : 3.

a. Find and interpret the value of the ratio.

b. In another container, the number of hydrogen atoms is 3 times the number of nitrogen atoms. Write the ratio of hydrogen atoms to nitrogen atoms.

Determine whether the ratios are equivalent.

4. 5 : 2 and 30 : 12

5. 4 : 3 and 8 : 7

6. 6 : 4 and 18 : 6

7. 18 : 12 and 3 : 2

8. Write two equivalent ratios that have values of $\frac{5}{7}$.

GO DIGITAL

9. During a chess match, there are 12 pieces left on the board. The ratio of white pieces to black pieces is 2 : 1. How many white pieces are on the board?

10. You run at a pace of 2 miles every 17 minutes. Your friend runs at a pace of 3 miles every 24 minutes. Are you and your friend running at the same pace? If not, who is running faster?

8.2 Using Tape Diagrams *(pp. 605–614)*

Learning Target: Use tape diagrams to model and solve ratio problems.

The tape diagram represents the ratio of the time you spend reading to the time your friend spends reading. You read for 8 hours. How many hours does your friend spend reading?

11.

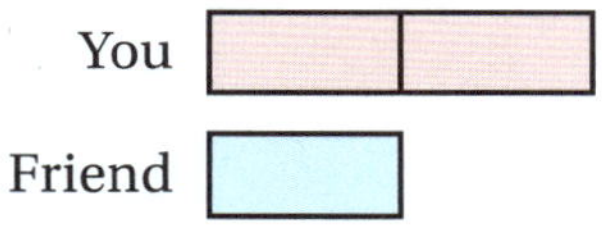

12.

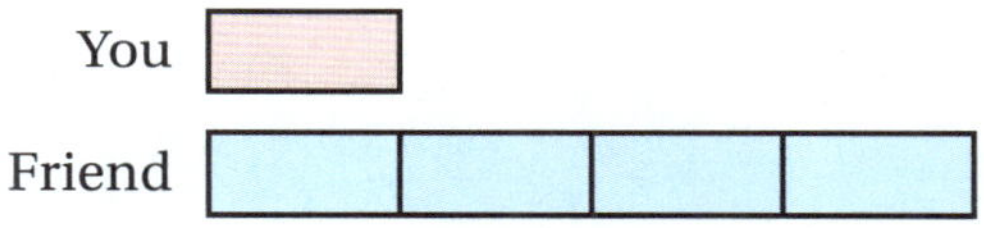

A container has peppermint gum and spearmint gum. You are given the number of pieces of peppermint gum in the container and the ratio of peppermint gum to spearmint gum. Find the number of pieces of spearmint gum in the container.

13. 24 peppermint; 8 to 5

14. 18 peppermint; 2 : 3

15. You perform 7 sit-ups for every 2 pull-ups as part of an exercise routine. You perform 25 more sit-ups than pull-ups. How many sit-ups and how many pull-ups do you perform?

8.3 Using Ratio Tables (pp. 615–626)

Learning Target: Use ratio tables to represent equivalent ratios and solve ratio problems.

Find the missing values in the ratio table. Then write the equivalent ratios.

16.

Levers	6		18
Pulleys	3	6	

17.

Nails	5	10	
Screws	2		6

18. Use all four operations to complete the ratio table. Justify your answer.

Pumpkins Grown	4				
Seeds Planted	12				

19. On New Year's Eve, the Times Square ball is lowered 47 feet every 20 seconds. How long does it take for the ball to be lowered 141 feet?

20. Surf Shop A charges $15 for every 2 hours you rent a surfboard. Surf Shop B charges $25 for every 3 hours. Which shop offers a better deal?

8.4 Rates and Unit Rates (pp. 627–636)

Learning Target: Understand the concept of a unit rate and solve rate problems.

Write a unit rate for the situation.

21. 12 stunts in 4 movies

22. 3600 stitches in 3 minutes

23. A train travels 120 miles in 3 hours. Write two unit rates that describe the relationship between the number of miles and the number of hours the train travels.

24. Mercury orbits the Sun 3 times in 264 days.

a. How many times does Mercury orbit the Sun in 440 days?

b. How many days does it take Mercury to orbit the Sun 8 times?

8.5 Solving Proportions (pp. 637–646)

Learning Target: Understand and solve proportions.

Solve the proportion.

25. $\frac{9}{10} = \frac{d}{40}$

26. $\frac{m}{12} = \frac{9}{4}$

27. $\frac{k}{4} = \frac{7}{28}$

28. $\frac{6}{33} = \frac{2}{y}$

29. After 4 hours, an ultramarathon runner has completed 16 miles of a 100-mile race. At this pace, will the runner complete the race in less than 24 hours? Justify your answer.

8.6 Converting Measures within Systems *(pp. 647–660)*

Learning Target: Use ratio reasoning to convert units of measure within a system.

Complete the statement.

30. 2.5 c = ▭ fl oz

31. 12 ft = ▭ yd

32. 3500 mg = ▭ g

33. $\frac{\$2}{\text{min}} = \frac{\$\square}{\text{h}}$

34. $\frac{13\text{ gal}}{\text{h}} = \frac{\square\text{ qt}}{\text{h}}$

35. $\frac{8\text{ ft}}{\text{h}} = \frac{\square\text{ in.}}{\text{min}}$

36. The distance between two stars increases at a rate of 3 centimeters per month. What is the rate in meters per year?

37. A great white shark's speed is 18 miles per hour. Your swimming speed is 9000 feet per hour. How much faster does the great white shark swim? Write your answer in feet per second.

8 Practice Test WITH CalcChat®

1. Write the ratio of scooters to bikes.

2. Determine whether the ratios 8 : 7 and 15 : 14 are equivalent.

Find the missing values in the ratio table. Then write the equivalent ratios.

3.

Lemons	4		36
Limes	2	6	

4.

Rabbits	2	12	
Hamsters	9		45

5. You travel 224 miles in 4 hours. Find the unit rate.

6. Solve the proportion

$$\frac{h}{32} = \frac{5}{4}.$$

Complete the statement.

7. 4 m = ☐ cm

8. 30 qt = ☐ gal

9. $\frac{10 \text{ gal}}{\text{h}} = \frac{\square \text{ gal}}{\text{wk}}$

10. $\frac{4 \text{ ft}}{\text{sec}} = \frac{\square \text{ in.}}{\text{sec}}$

GO DIGITAL

11. During a baseball season, Team A scores 9 runs for every 7 runs that Team B scores. The total number of runs scored by both teams is 1440. How many runs does each team score?

12. You and your friend mix water and citric acid. You add 3 cups of citric acid for every 16 cups of water. Your friend adds 2 cups of citric acid for every 12 cups of water. Whose mixture is more acidic?

13. Determine which windsurfer is faster. Explain your reasoning.

14. In a rectangle, the ratio of the length to the width is 5 : 2. The length of the rectangle is 13.875 feet greater than the width. What are the perimeter and the area of the rectangle?

8 Review & Refresh WITH CalcChat®
while Building Fluency

Multiply.

1. 3.45×5

2. 4.2×6.37

Divide.

3. $2934 \div 3$

4. $1230 \div 15$

Complete the table.

	Number of Cats	Number of Dogs	Ratio of Cats to Dogs
5.	12	4	:
6.	6		2 : 7

7. Compare the number of baseballs to the number of golf balls using a ratio.

GO DIGITAL

8. Which of the following is *not* equivalent to $\frac{3}{5}$?

Ⓐ $\frac{6}{10}$

Ⓑ $\frac{7}{9}$

Ⓒ $\frac{9}{15}$

Ⓓ $\frac{60}{100}$

9. You and your friend are fishing from a pier in Miami. You catch 4 sailfish in 3 hours and your friend catches 3 sailfish in 2 hours. Are you catching sailfish at the same rate? Explain your answer.

10. Plot 3.467 and 3.47 on the number line. Which number is greater? Explain.

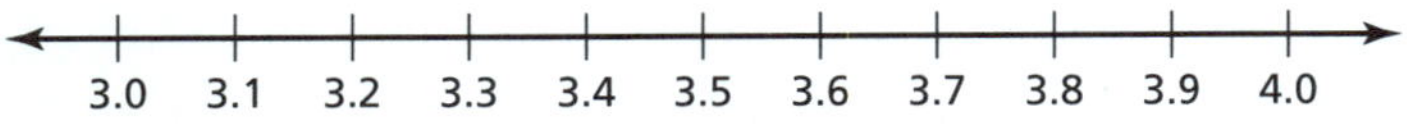

Use the Distributive Property to evaluate the expression. Explain each step.

11. $3(4 + 6)$

12. $(5 - 1) \times 4$

8 B.E.S.T. Test Prep WITH CalcChat®

Cumulative Practice

1. The school store sells 4 pencils for \$0.50. At this rate, what is the cost (in dollars) of 10 pencils?

GRIDDED RESPONSE

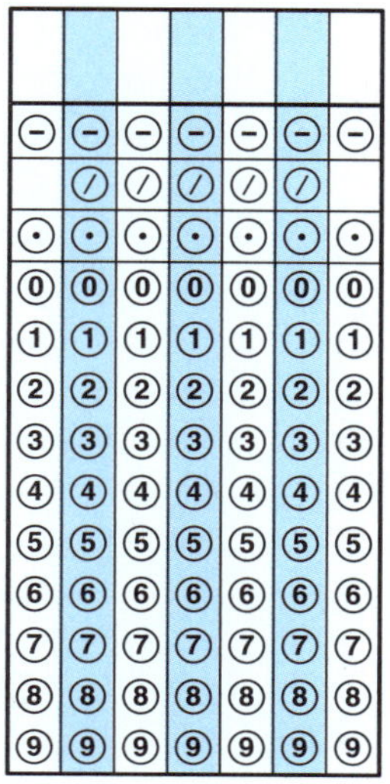

Test-Taking Strategy

Solve Problem before Looking at Choices

2. What is $\frac{1}{4} \times \frac{5}{7}$?

Ⓐ $\frac{5}{11}$ Ⓒ $\frac{5}{28}$

Ⓑ $\frac{6}{11}$ Ⓓ $\frac{3}{14}$

3. What is the least common multiple of 8, 12, and 20?

Ⓐ 24 Ⓒ 60

Ⓑ 40 Ⓓ 120

4. Mole is a sauce typically used in Mexican cuisine. The tape diagram models the ratio of oregano to cocoa powder in a mole recipe. A chef uses 9 teaspoons of oregano. How many teaspoons of cocoa powder does the chef use?

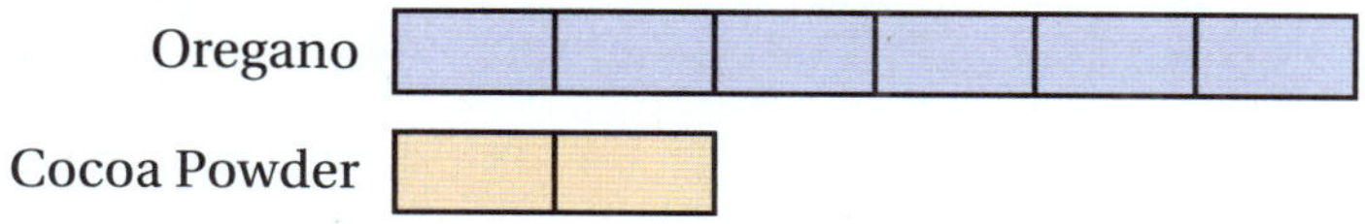

Ⓐ 1.5 teaspoons Ⓒ 9 teaspoons

Ⓑ 3 teaspoons Ⓓ 12 teaspoons

GO DIGITAL

5. What is the value of the expression below?

GRIDDED RESPONSE

$$2 \cdot 4^2 + 3(6 \div 2)$$

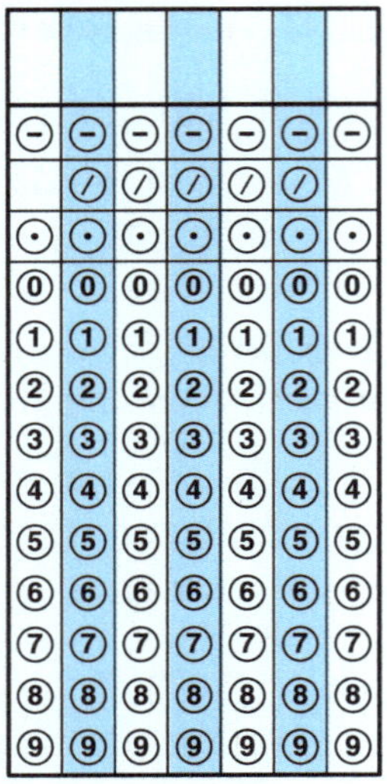

6. Which of the following is a ratio of frogs to snakes?

Ⓐ 4 : 8

Ⓑ 8 : 12

Ⓒ 8 : 4

Ⓓ 4 : 12

7. A factor tree for 14,700 is shown. Which factor of 14,700 is *not* a perfect square?

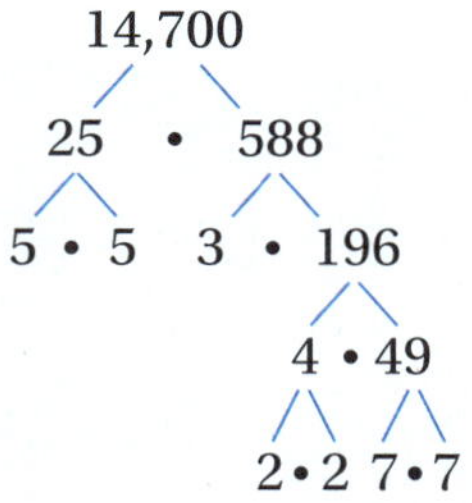

Ⓐ 25

Ⓑ 49

Ⓒ 196

Ⓓ 588

8. When evaluating $5 + 4 \times 2^3$, the first step is to [Ⓐ add 5 and 4 / Ⓑ evaluate 2^3 / Ⓒ multiply 4 and 2 / Ⓓ multiply 2 and 3], and the value of the expression is [Ⓐ 29 / Ⓑ 37 / Ⓒ 72 / Ⓓ 517 / Ⓔ 5832].

9. Which expression is equivalent to 3^5?

Ⓐ $3 \times 3 \times 3 \times 3 \times 3$

Ⓑ 3×5

Ⓒ $5 \times 5 \times 5$

Ⓓ $3 + 3 + 3 + 3 + 3$

9 Percents

Chapter Learning Target:
Understand percents.

Chapter Success Criteria:

- ◆ I can rewrite fractions, decimals, and percents.
- ◆ I can compare and order fractions, decimals, and percents.
- ■ I can use the percent proportion or percent equation to find a percent, a part, or a whole.
- ■ I can apply percents to solve real-life problems.

◆ Surface ■ Deep

Model of DNA Molecule

GO DIGITAL

7 MTR *Mathematicians who apply mathematics to real-world contexts connect mathematical concepts to everyday experiences.*

STEAM Video Chargaff's Rules

Veronica helps Chantell and Molly use forensic evidence to solve a mystery.

DNA contains the genetic information for a living organism. Each DNA molecule is made up of four nucleotide bases: adenine (A), thymine (T), cytosine (C), and guanine (G).

1. Veronica tells Chantell and Molly that the DNA of most mammals contains about 60 percent A and T nucleotides and 40 percent C and G nucleotides. What do you think this means?

2. Use your answer in Question 1 to determine which of the DNA samples in the table is most likely to belong to a mammal. Explain your reasoning.

	Sample 1	Sample 2	Sample 3
A and T Nucleotides	38	61	60
C and G Nucleotides	62	39	100

STEAM Performance Task Genetic Ancestry

After completing this chapter, you will be able to answer the questions in the *STEAM Performance Task*. You will be given the results of your friend's ancestry test.

Native American: $\frac{1}{50}$ **African:** $\frac{30}{40}$ **Mediterranean:** 0.06

Northern European: 0.10 **Southwest Asian:** $\frac{3}{75}$

You will be asked to compare the portions of your friend's ancestry.

GO DIGITAL

Getting Ready for Chapter 9 WITH CalcChat®

Chapter Exploration

1 MTR **Actively Participate in Effortful Learning Collectively**
Work with a partner to prepare for concepts in this chapter.

1. MATCHING Match the "cent" word with its definition.

a. centipede	**b.** centimeter	**c.** centennial	**d.** centiliter
e. centuple	**f.** century	**g.** centenarian	**h.** centigram

J. 100-year anniversary	**K.** one-hundredth of a gram	**L.** multiply by 100
M. bug with many legs	**N.** one-hundredth of a liter	**O.** one-hundredth of a meter
P. 100 years	**Q.** person who is at least 100 years old	

1 MTR **2. HELP A CLASSMATE** Describe a situation where you have seen the word *per* used, and explain its use. Use the description to help a classmate understand what *percent* means.

Represent the shaded portion of the square as (a) a fraction whose denominator is 100, (b) a fraction in simplest form, (c) a decimal, and (d) a percent.

3.
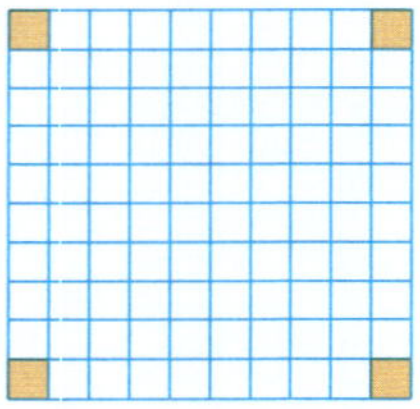

4.
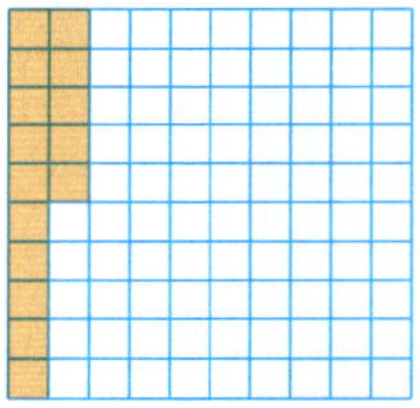

5.
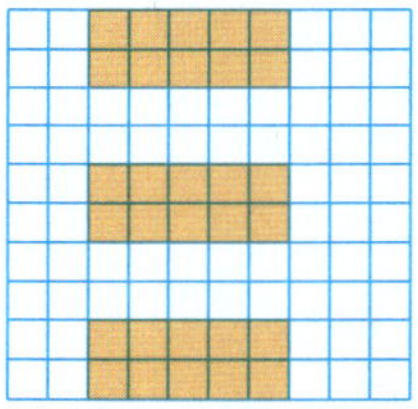

6.

7.
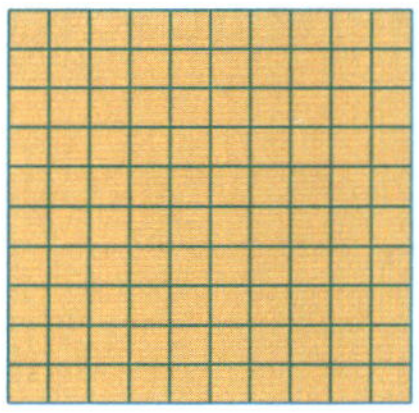

8.
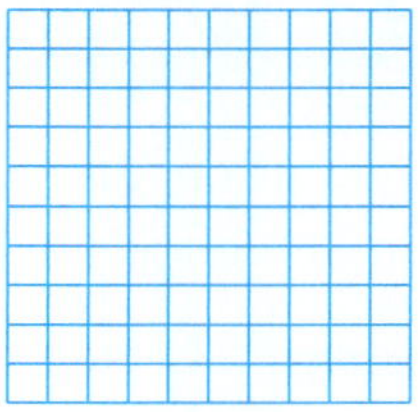

Vocabulary

The following terms are defined in this chapter. Think about what each might mean and record your thoughts.

percent   fraction   decimal

GO DIGITAL

9.1 Percents and Fractions

Learning Target: Write percents as fractions and fractions as percents.

Success Criteria:
- I can draw models to represent fractions and percents.
- I can write percents as fractions.
- I can write equivalent fractions with denominators of 100.
- I can write fractions as percents.

Number Sense and Operations
MA.6.NSO.3.5 Rewrite positive rational numbers in different but equivalent forms including fractions, terminating decimals and percentages.

The Meaning of a Word ▶ Percent

A century is 100 years.

A cent is one hundredth of a dollar.

In Mexico, a centavo is one hundredth of a peso.

Cent means *one hundred*, so **percent** means *per one hundred*. The symbol for percent is %.

Exploration 1 Interpreting Models

2 MTR MAKE A CONNECTION
How does the meaning of *percent* help you determine how to write the percent of each diagram that is shaded?

____

____

____

____

____

____

____

Work with a partner. Write a percent, a fraction, and a ratio shown by each model. How are percents, fractions, and ratios related?

a.

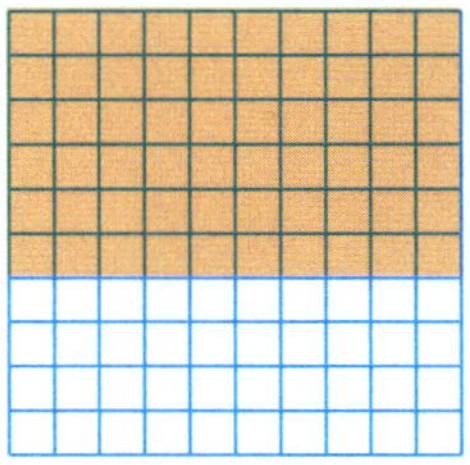

b.

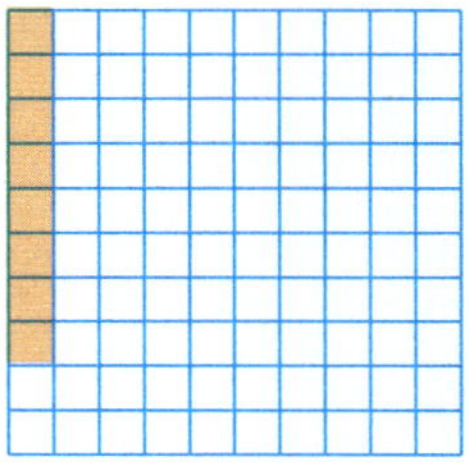

c.

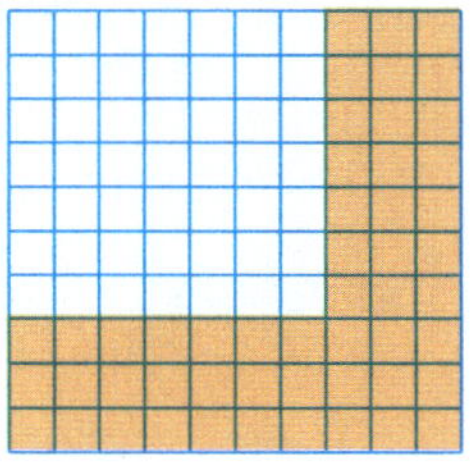

d.

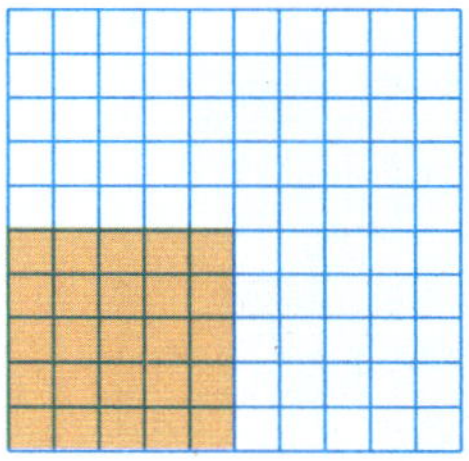

GO DIGITAL

9.1 Lesson

Key Vocabulary
percent, *p. 680*

Key Idea

Writing Percents as Fractions

Words A **percent** is the value of a part-to-whole ratio where the whole is 100. So, you can write a percent as a fraction with a denominator of 100. The symbol % is used to denote a percent.

Numbers $60\% = 60 \text{ out of } 100 = \frac{60}{100}$

part → 60 (in 60%); per one hundred (whole) → %

part → 60; per → fraction bar; one hundred (whole) → 100

Algebra $n\% = \frac{n}{100}$

Example 1 Writing Percents as Fractions

Fractions and percents that are equivalent represent the same number using different notations.

a. Write 35% as a fraction in simplest form.

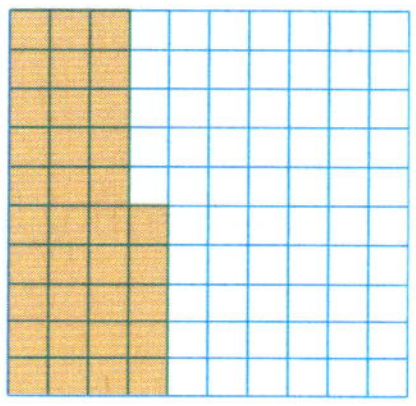

$35\% = \frac{35}{100}$ Write as a fraction with a denominator of 100.

$= \frac{7}{20}$ Simplify.

▶ So, $35\% = \frac{7}{20}$.

b. Write 100% as a fraction in simplest form.

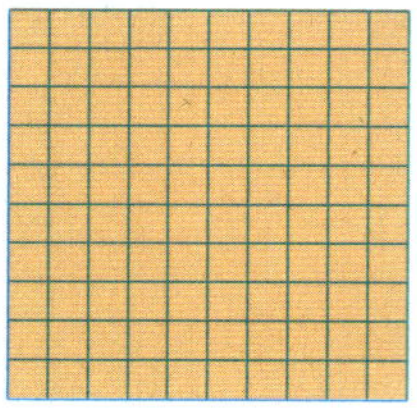

$100\% = \frac{100}{100}$ Write as a fraction with a denominator of 100.

$= 1$ Simplify.

▶ So, $100\% = 1$.

c. Write 174% as a mixed number in simplest form.

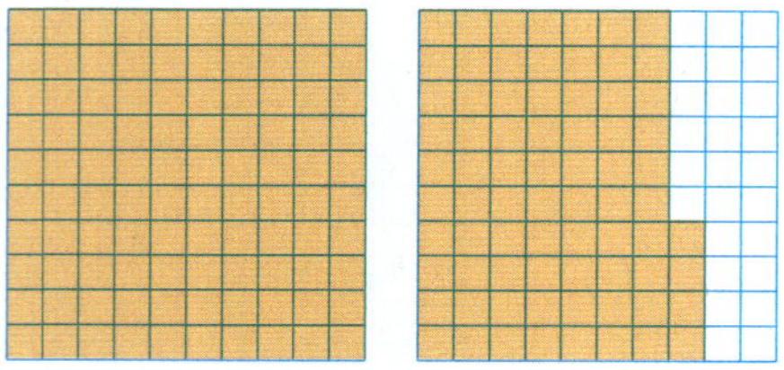

$174\% = \frac{174}{100}$ Write as a fraction with a denominator of 100.

$= \frac{87}{50}$, or $1\frac{37}{50}$ Simplify.

▶ So, $174\% = 1\frac{37}{50}$.

GO DIGITAL

Try It

Write the percent as a fraction or mixed number in simplest form.

1. 5%

$\frac{5}{100} = \frac{1}{20}$

2. 168%

$\frac{168}{100} = \frac{42}{25}$

3. 36%

$\frac{36}{100} = \frac{9}{25}$

4. 83%

$\frac{83}{100}$

Key Idea

Writing Fractions as Percents

Words Write an equivalent fraction with a denominator of 100. Then write the numerator with the percent symbol.

Numbers $\frac{1}{4} = \frac{1 \times 25}{4 \times 25} = \frac{25}{100} = 25\%$

Example 2 Writing Fractions as Percents

$\frac{3}{50} = \frac{6}{100}$

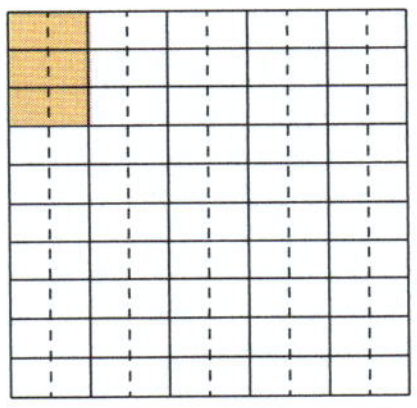

a. Write $\frac{3}{50}$ as a percent.

$$\frac{3}{50} = \frac{3 \times 2}{50 \times 2} = \frac{6}{100} = 6\%$$

Because $50 \times 2 = 100$, multiply the numerator and denominator by 2. Write the numerator with a percent symbol.

b. Write $\frac{9}{1000}$ as a percent.

$$\frac{9}{1000} = \frac{9 \div 10}{1000 \div 10} = \frac{0.9}{100} = 0.9\%$$

Because $1000 \div 10 = 100$, divide the numerator and denominator by 10. Write the numerator with a percent symbol.

19
19
19
19
+ 19

Try It

Write the fraction or mixed number as a percent.

5. $\frac{31}{50}$ ·2

62%

6. $\frac{19}{20}$ $\frac{9}{100}$

95%

7. $\frac{1}{200}$

0.5%

8. $1\frac{1}{2}$

150%

GO DIGITAL

In-Class Practice

1 I don't understand yet. | 2 I can do it with help. | 3 I can do it on my own. | 4 I can teach someone else.

9. **WRITING A FRACTION** Write 40% as a fraction in simplest form. Draw a model that represents this fraction.

10. **WRITING A PERCENT** Write $\frac{9}{20}$ as a percent.

4 MTR

11. **WHICH ONE DOESN'T BELONG?** Which number does *not* belong with the other three? Explain your reasoning.

$\frac{10}{100}$	10%	$\frac{1}{10}$	0.01

12. **OPEN-ENDED** Write three different fractions that are less than 40%.

13. **NUMBER SENSE** Can $1\frac{1}{4}$ be written as a percent? Explain.

Example 3 Modeling Real Life

7 MTR

South Atlantic United States

A hurricane affects 6 out of 8 states in the South Atlantic portion of the United States. What percent of the South Atlantic states are affected by the hurricane?

Write a fraction that represents the portion of the South Atlantic states affected by the hurricane. Then convert the fraction to a percent.

$$\frac{\text{number of states affected}}{\text{total number of South Atlantic states}} = \frac{6}{8}$$

The denominator of 8 is not a factor of 100. But 4 is a factor of both 8 and 100. So, first write an equivalent fraction whose denominator is 4. Then convert that fraction to a percent.

$\frac{6}{8} = \frac{3}{4}$ Divide the numerator and denominator by 2.

$= \frac{75}{100}$ Multiply the numerator and denominator by 25.

$= 75\%$ Write the numerator with a percent symbol.

▶ So, 75% of the South Atlantic states are affected by the hurricane.

Check Reasonableness Notice that $\frac{6}{8} = \frac{6 \times 12}{8 \times 12} = \frac{72}{96}$. Because $\frac{72}{96} > \frac{72}{100}$, you know that $\frac{6}{8} > 72\%$. The answer is reasonable because $75\% > 72\%$. ✓

GO DIGITAL

In-Class Practice

1 I don't understand yet. 2 I can do it with help. 3 I can do it on my own. 4 I can teach someone else.

14. You and a friend fill balloons with water. Two out of every 25 balloons pop while they are being filled. What percent of the balloons do not pop while they are being filled?

15. During the month of April, it rains 2 days for every 3 days that it does not rain. What percent of the days in April does it rain?

16. **Dig Deeper** There are 100 students in a band. Ninety percent of the students are either 12 or 13 years old. The number of 13-year-olds is 125% of the number of 12-year-olds. The rest of the students are 14 years old. Write the portion of the band for each age as a fraction and a percent.

9.1 Practice WITH CalcChat® AND CalcView®

Review & Refresh

Complete the statement.

1. 3 km = ___ m
2. 7 gal = ___ c
3. 54 in. = ___ yd
4. 4 mi = ___ ft
5. 6.5 L = ___ mL
6. 45 lb = ___ oz

Evaluate the expression.

7. $(35 + 7) + 2$
8. 13×1
9. $5 \times 17 \times 2$
10. $3(5 + 2)$
11. $2 + 13 + 28$
12. $(12 \times 3) \times 4$

Use the table to write the ratio.

13. birch to willow
14. trees : oak
15. oak : willow
16. birch to trees

Tree	Oak	Birch	Willow
Number	4	3	1

Concepts, Skills, & Problem Solving

USING TOOLS **Use a 10-by-10 grid to model the percent.** (See Exploration 1.)

17. 10%
18. 55%
19. 45%

GO DIGITAL

WRITING PERCENTS AS FRACTIONS **Write the percent as a fraction or mixed number in simplest form.** (See Example 1.)

20. 45% ▶**21.** 90% **22.** 15% **23.** 7%

24. 34% **25.** 79% **26.** 77.5% **27.** 188%

28. 8% **29.** 224% **30.** 0.25% **31.** 0.4%

4 MTR **32.** **YOU BE THE TEACHER** Your friend writes 225% as a fraction. Is your friend correct? Explain your reasoning.

$$225\% = \frac{225}{100} = \frac{9}{4}$$

WRITING FRACTIONS AS PERCENTS **Write the fraction or mixed number as a percent.** (See Example 2.)

33. $\frac{1}{10}$ **34.** $\frac{1}{5}$ ▶**35.** $\frac{11}{20}$ **36.** $\frac{1}{400}$

37. $\frac{2}{25}$ **38.** $\frac{27}{50}$ **39.** $\frac{3}{250}$ **40.** $\frac{18}{25}$

41. $1\frac{17}{20}$ **42.** $2\frac{41}{50}$ **43.** $3\frac{1}{200}$ **44.** $4\frac{7}{500}$

1 MTR **45.** **HELP A CLASSMATE** Your classmate wants to write $\frac{14}{25}$ as a percent. Explain how your classmate can complete the work.

$$\frac{14}{25} = \frac{14 \times 4}{25 \times 4} = ?$$

7 MTR **46.** **MODELING REAL LIFE** During a 10-year period, 6 out of 30 Major League Baseball teams won the World Series. What percent of Major League Baseball teams won the World Series during the 10-year period? (See Example 3.)

7 MTR **47.** **MODELING REAL LIFE** A doctor conducts an experiment to test new treatments for a medical condition. Of the 16 volunteers in the experiment, 4 do not receive any treatment. What percent of the volunteers do not receive any treatment?

48. **REASONING** Of the students in your class, 12% are left-handed and the rest are right-handed. What fraction of the students are left-handed? Are there more right-handed or left-handed students? Explain.

49. **NUMBER SENSE** You have 125% of the tickets required for a prize. What fraction of the required tickets do you have? Do you need more tickets for the prize? Explain.

FINDING PERCENTS **Find the percent.**

50. 3 is what percent of 8?

51. 13 is what percent of 16?

52. 9 is what percent of 16?

53. 33 is what percent of 40?

7 MTR **54.** **MODELING REAL LIFE** A survey asked students to choose their favorite social media website.

a. What fraction of the students chose Website A?

b. What percent of the students chose Website C?

Social Media Website	Number of Students
Website A	35
Website B	13
Website C	22
Website D	10

55. **Dig Deeper** The percent of the total area of the United States that is in each of four states is shown.

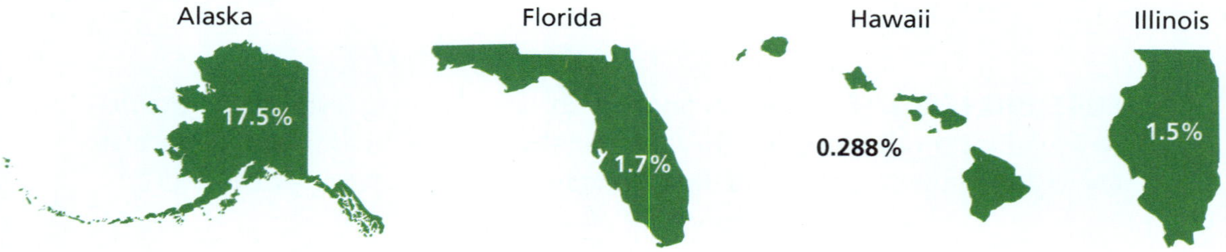

a. Write the percents as fractions in simplest form.

b. Compared to the map of Florida, is the map of Alaska the correct size? Explain your reasoning.

7 MTR **c.** **INVESTIGATE** Which of the 50 U.S. states are larger than Illinois?

56. **PROBLEM SOLVING** A school fundraiser raised 120% of its goal last year and 25% of its new goal this year. Did the fundraiser raise more money this year? Explain your reasoning.

2 MTR 57. **MODEL A PROBLEM** How can you use a 10-by-10 grid to model $\frac{1}{2}$%?

58. **REASONING** Write $\frac{1}{12}$ as a percent. Explain how you found your answer.

9.2 Percents and Decimals

Learning Target: Write percents as decimals and decimals as percents.

Success Criteria:
- I can draw models to represent decimals.
- I can explain why the decimal point moves when multiplying and dividing by 100.
- I can write percents as decimals.
- I can write decimals as percents.

Number Sense and Operations

MA.6.NSO.3.5 Rewrite positive rational numbers in different but equivalent forms including fractions, terminating decimals and percentages.

Exploration 1 Interpreting Models

Work with a partner. Write a percent and a decimal shown by each model. How are percents and decimals related?

a.

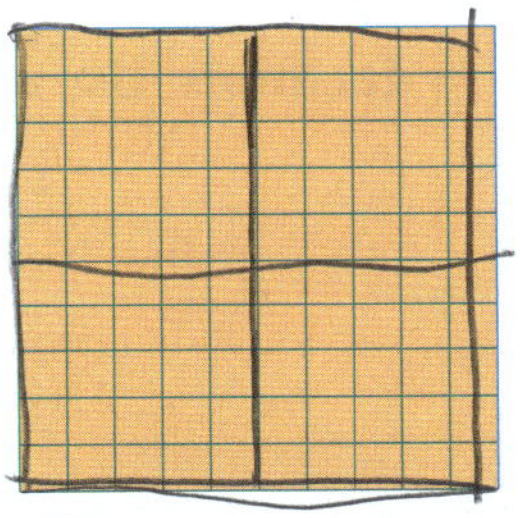

b.

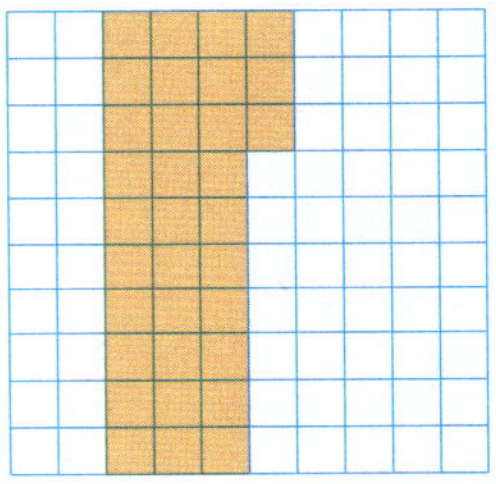

c.

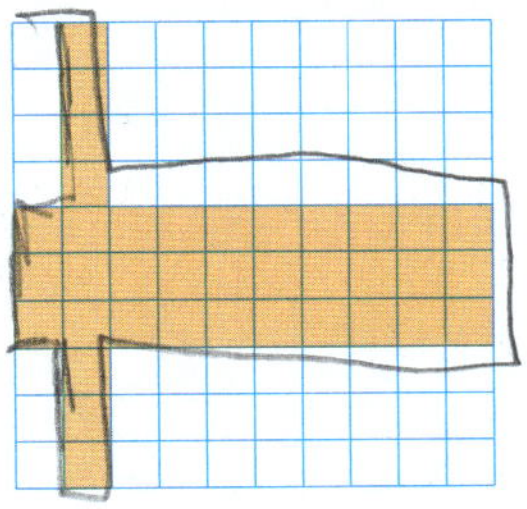

d.

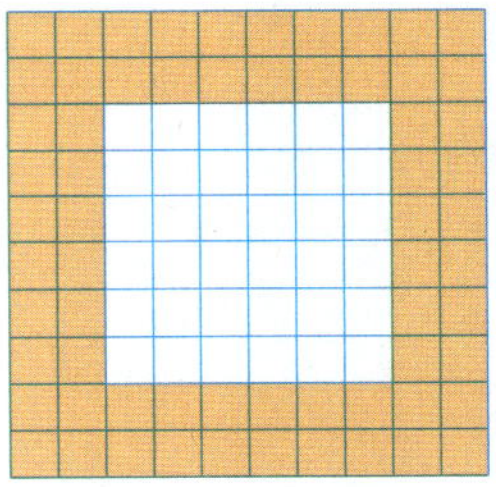

e.

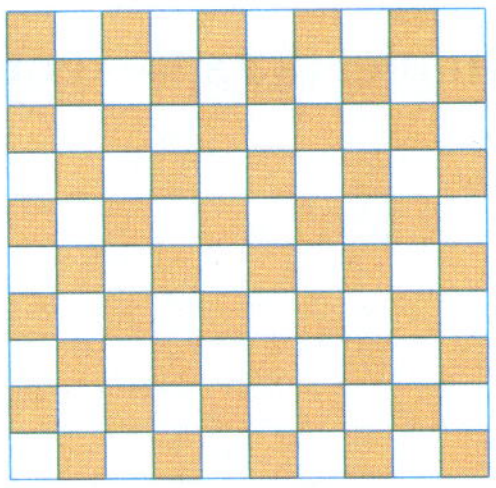

f.

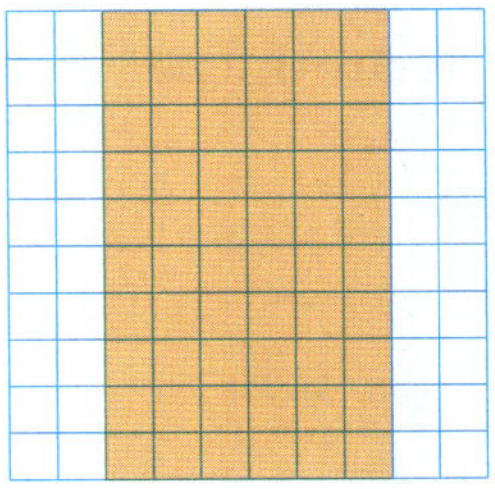

g.

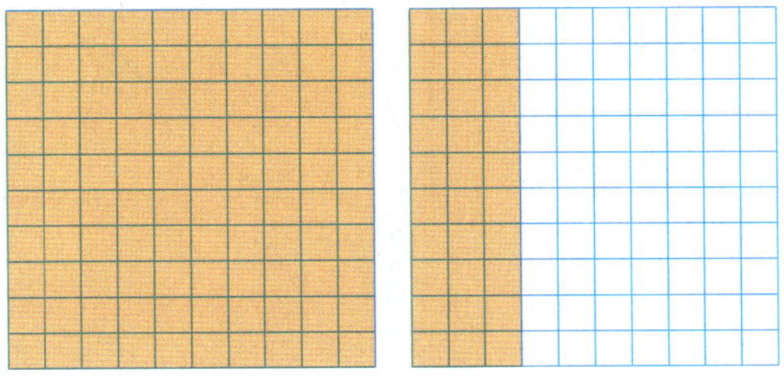

5 MTR USE A PATTERN

How does the decimal point move when you rewrite a percent as a decimal and a decimal as a percent? Explain why this occurs.

____

____

____

____

____

GO DIGITAL

9.2 Lesson

Key Idea

Writing Percents as Decimals

Words Remove the percent symbol. Then divide by 100, which moves the decimal point two places to the left.

Numbers 23% = 23.% = 0.23

Model

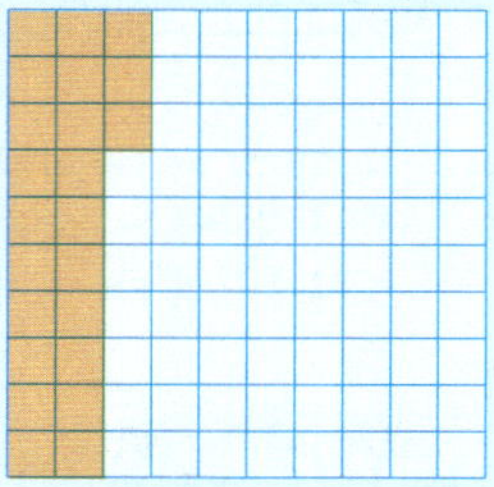

Notice that 23 *percent* is equivalent to 23 *hundredths*.

Example 1 Writing Percents as Decimals

Write (a) 52% and (b) 7% as decimals. Use a model to represent each decimal.

a. 52% = 52.% = 0.52

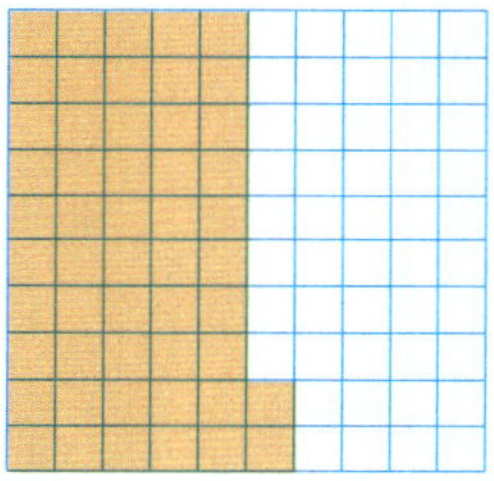

b. 7% = 07.% = 0.07

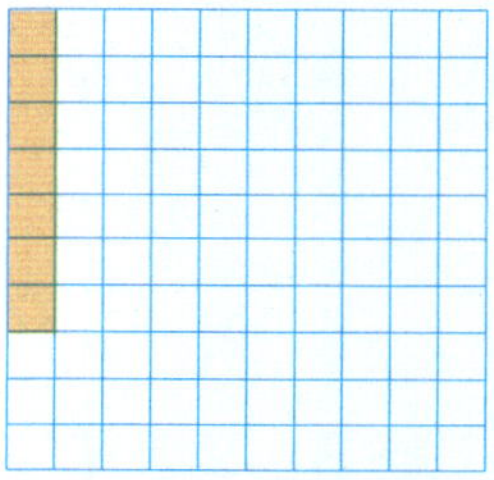

When writing a percent as a decimal, you may need to include one or more zeros so that the digits are in the correct place value positions.

Try It **Write the percent as a decimal. Use a model to represent the decimal.**

1. 24% 0.24

2. 3% .03

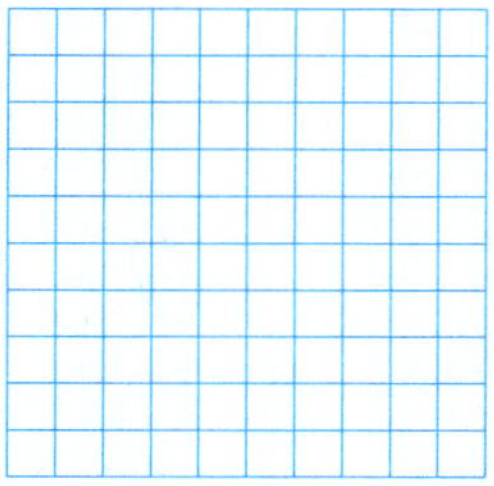

3. 107% 1.07

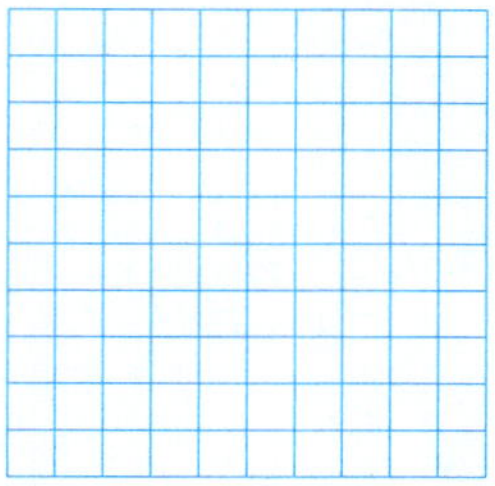

4. 92.5% .925

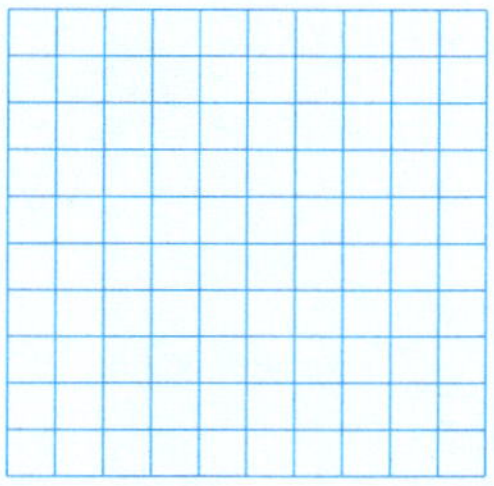

GO DIGITAL

Key Idea

Writing Decimals as Percents

Words Multiply by 100, which moves the decimal point two places to the right. Then add a percent symbol.

Numbers 0.36 = 0.36 = 36%

Model

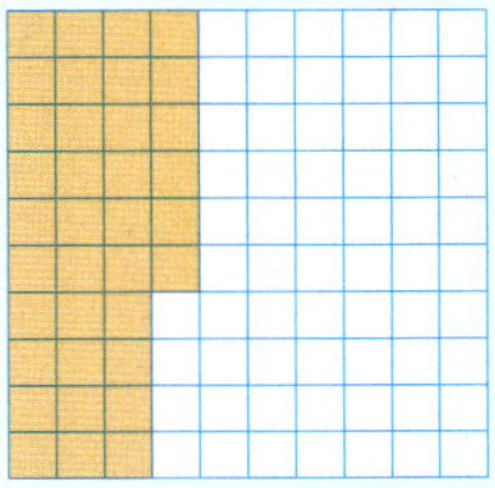

Example 2 Writing Decimals as Percents

a. Write 0.47 as a percent.
0.47 = 0.47 = 47%

b. Write 0.663 as a percent.
0.663 = 0.663 = 66.3%

c. Write 1.8 as a percent.
1.8 = 1.80 = 180%

d. Write 0.009 as a percent.
0.009 = 0.009 = 0.9%

Try It

Write the decimal as a percent.

5. 0.94 — 94%

6. 1.2 — 120%

7. 0.316 — 31.6%

8. 0.005 — 0.5%

Example 3 B.E.S.T. Test Prep: Writing a Fraction as a Percent and a Decimal

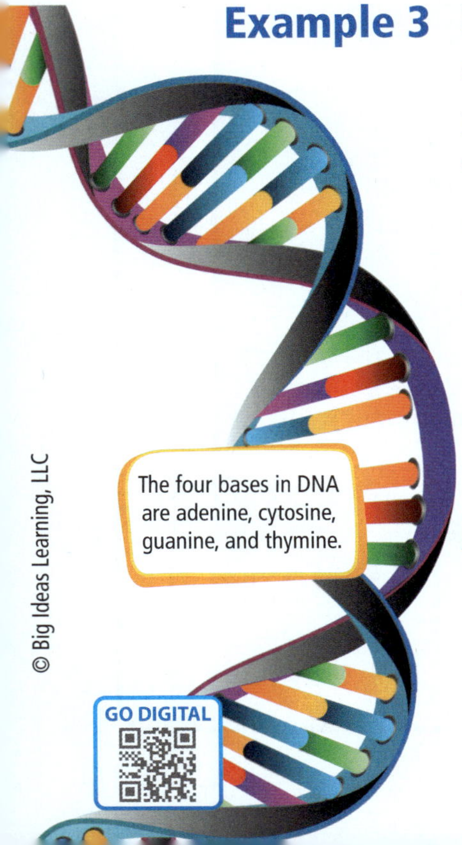

The four bases in DNA are adenine, cytosine, guanine, and thymine.

In a certain sample of DNA, 34 out of 100 bases are guanine. Which of the following is *not* another way of expressing 34 out of 100?

Ⓐ $\frac{17}{50}$ Ⓑ 34% Ⓒ $\frac{7}{20}$ Ⓓ 0.34

Write "34 out of 100" as a fraction, a decimal, and a percent.

$$34 \text{ out of } 100 = \frac{34}{100} \rightarrow = 34\% \quad \text{Eliminate Choice B.}$$
$$\rightarrow = \frac{17}{50} \quad \text{Eliminate Choice A.}$$
$$\rightarrow = 0.34 \quad \text{Eliminate Choice D.}$$

▶ So, the correct answer is Ⓒ.

GO DIGITAL

Try It

9. WHAT IF? In the sample, 28 out of 100 bases are adenine. Write "28 out of 100" as a fraction, a decimal, and a percent.

In-Class Practice

1 I don't understand yet. 2 I can do it with help. 3 I can do it on my own. 4 I can teach someone else.

WRITING PERCENTS AS DECIMALS **Write the percent as a decimal. Use a model to represent the decimal.**

10. 32%

11. 54.5%

12. 108%

WRITING DECIMALS AS PERCENTS **Write the decimal as a percent.**

13. 0.71

14. 0.052

15. 9.66

16. WRITING Explain why the decimal point moves left when dividing a number by 100.

Example 4 Modeling Real Life

The figure shows the portions of ultraviolet (UV) rays reflected by four different surfaces. How many times more UV rays are reflected by water than by sea foam?

The diagram shows that sea foam reflects 25% of UV rays and water reflects $\frac{21}{25}$ of UV rays. First, write 25% and $\frac{21}{25}$ as decimals.

Sea foam: $25\% = 25.\% = 0.25$ **Water:** $\frac{21}{25} = \frac{84}{100} = 0.84$

Next, divide 0.84 by 0.25.

$$0.25\overline{)0.84} \longrightarrow \begin{array}{r} 3.36 \\ 25\overline{)84.00} \\ -75 \\ \hline 9\,0 \\ -7\,5 \\ \hline 1\,50 \\ -1\,50 \\ \hline 0 \end{array}$$

▶ So, water reflects 3.36 times more UV rays than sea foam.

USE ANOTHER METHOD

Show how to solve the problem by first writing 25% as a fraction.

GO DIGITAL

In-Class Practice

1 I don't understand yet. 2 I can do it with help. 3 I can do it on my own. 4 I can teach someone else.

17. Write the amount of occupied space on the computer as a percent.

Volume	Capacity	Free Space
(C:)	150 GB	132 GB

18. *Salinity* is a measure of the salt content of a body of water. One researcher measures the salinity of the Indian Ocean as 3.2%. Another researcher measures the salinity of the Dead Sea as 34%. A bucket of water from the Indian Ocean contains 56 grams of salt. How much salt is contained in the same amount of water from the Dead Sea? Justify your answer.

9.2 Practice WITH CalcChat® AND CalcView®

Review & Refresh

Write the fraction or mixed number as a percent.

1. $\frac{7}{50}$
2. $\frac{2}{5}$
3. $\frac{1}{250}$
4. $5\frac{2}{25}$

You are given the number of marbles in a bag and the ratio of red marbles to green marbles. How many of each kind of marble are in the bag?

5. 40 marbles; 5 to 3
6. 64 marbles; 6 to 2

Multiply. Write the answer in simplest form.

7. $\frac{3}{7} \times \frac{2}{9}$
8. $\frac{5}{12} \times \frac{1}{5}$
9. $4\frac{2}{3} \times \frac{9}{10}$
10. $2\frac{1}{6} \times 3\frac{1}{2}$

Concepts, Skills, & Problem Solving

INTERPRETING MODELS **Write the percent and the decimal shown by the model.** (See Exploration 1.)

11. 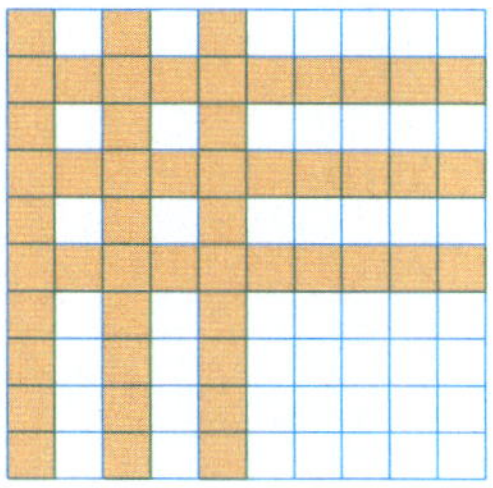

12. 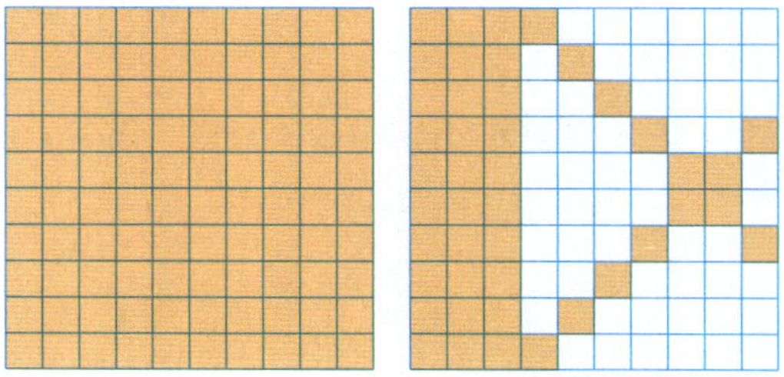

GO DIGITAL

WRITING PERCENTS AS DECIMALS **Write the percent as a decimal.** (See Example 1.)

13. 78%

14. 55%

15. 18.5%

16. 57.4%

17. 33%

18. 9%

19. 47.6%

20. 91.2%

21. 166%

22. 217%

23. 0.6%

24. 0.5%

WRITING DECIMALS AS PERCENTS **Write the decimal as a percent.** (See Example 2.)

25. 0.74

26. 0.52

27. 0.89

28. 0.768

29. 0.99

30. 0.49

31. 0.487

32. 0.128

33. 3.68

34. 5.12

35. 0.007

36. 0.004

4 MTR

37. YOU BE THE TEACHER Your friend writes 0.86 as a percent. Is your friend correct? Explain your reasoning.

$0.86 = 00.86 = 0.0086\%$

7 MTR

38. MODELING REAL LIFE About 0.34 of the length of a cat is its tail. Write this decimal as a percent.

39. B.E.S.T. Test Prep Which of the following is *not* a way to express 70 out of 100? (See Example 3.)

Ⓐ 0.07 Ⓑ $\frac{7}{10}$ Ⓒ $\frac{70}{100}$ Ⓓ 70%

7 MTR

40. **MODELING REAL LIFE** During one month, $\frac{4}{10}$ of Lake Okeechobee is covered in a harmful algal bloom. By the next month, Lake Okeechobee is 90% covered. By how many times does the algal bloom increase? (See Example 4.)

41. **OPEN-ENDED** Write three different decimals that are between 10% and 20%.

WRITING PERCENTS AS FRACTIONS AND DECIMALS **Write the percent as a fraction in simplest form and as a decimal.**

42. 36% **43.** 23.5% **44.** 16.2%

45. **Dig Deeper** The percents of students who travel to school by car, bus, and bicycle are shown for a school of 825 students.

Car: 20%

School bus: 48%

Bicycle: 8%

a. Write the percents as decimals.

b. Write the percents as fractions.

c. What percent of students use another method to travel to school?

7 MTR

d. **INVESTIGATE** Make a bar graph that represents how the students in your class travel to school.

2 MTR

46. MAKE A CONNECTION A running back was the MVP (most valuable player) in 0.14 of the first 50 Super Bowls.

a. What percent of the MVPs were running backs?

b. What fraction of the MVPs were *not* running backs?

3 MTR

47. CHOOSE A METHOD Students in a class were asked to tell their favorite color.

Favorite Color

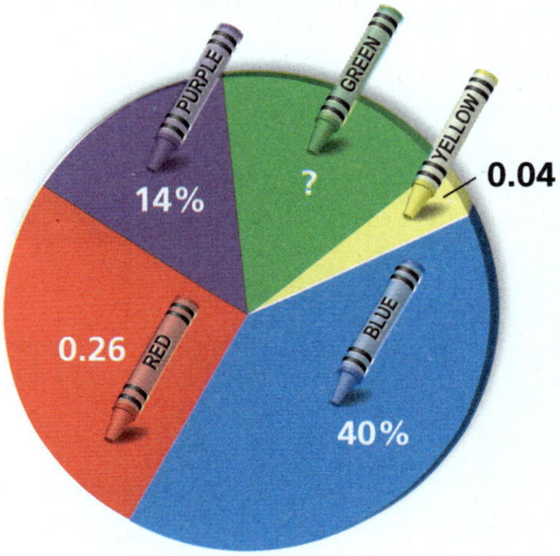

a. What percent said red, blue, or yellow?

b. How many times more students said red than yellow?

c. Use two methods to find the percent of students who said green. Which method do you prefer? Explain.

9.3 Comparing and Ordering Fractions, Decimals, and Percents

Learning Target: Compare and order fractions, decimals, and percents.

Success Criteria:
- I can rewrite a group of fractions, decimals, and percents using the same representation.
- I can explain how to compare fractions, decimals, and percents.
- I can order fractions, decimals, and percents from least to greatest.

Number Sense and Operations

MA.6.NSO.1.1 Extend previous understanding of numbers to define rational numbers. Plot, order and compare rational numbers.

MA.6.NSO.3.5 Rewrite positive rational numbers in different but equivalent forms including fractions, terminating decimals and percentages.

Exploration 1 Using a Number Line to Order Numbers

Work with three partners. Create a number line on the floor. Have your group stand on the number line to represent the four numbers in each list. Use the results to order each list of numbers from least to greatest. How did you know where to stand?

a.

0.25	0.9%	40%	0.5

b.

0%	$\frac{3}{4}$	30%	$\frac{1}{20}$

c.

100%	0.125	75%	$\frac{3}{10}$

GO DIGITAL

d.

12.5%	1.02	$\frac{1}{100}$	25%

e.

0.3	$\frac{1}{8}$	4%	0.75

f.

$\frac{51}{50}$	105%	1.5	$\frac{9}{10}$

3 MTR **SELECT METHODS**

What strategies can you use to determine where to stand?

9.3 Lesson

When comparing and ordering fractions, decimals, and percents, write the numbers as all fractions, all decimals, or all percents.

Example 1 Comparing Fractions, Decimals, and Percents

4 MTR COMPARE METHODS

Compare ordering decimals or percents with ordering fractions. Which do you prefer? Explain.

a. Compare $\frac{3}{20}$ and 16%.

Write $\frac{3}{20}$ as a percent: $\frac{3}{20} = \frac{3 \times 5}{20 \times 5} = \frac{15}{100} = 15\%$.

▶ 15% is less than 16%. So, $\frac{3}{20} < 16\%$.

b. Compare 79% and 0.08.

Write 79% as a decimal: 79% = 79.% = 0.79.

▶ 0.79 is greater than 0.08. So, 79% > 0.08.

Try It **Complete the statement using <, >, or =.**

1. 25% < $\frac{7}{25}$

2. 0.49 < 94%

Example 2 Ordering Fractions, Decimals, and Percents

Order 3%, $\frac{1}{50}$, 0.005, 0.8%, and 0.025 from least to greatest.

One way to order the numbers is to first write each number as a percent.

$\frac{1}{50} = \frac{1 \times 2}{50 \times 2} = \frac{2}{100} = 2\%$ 0.005 = 0.005 = 0.5% 0.025 = 0.025 = 2.5%

Remember

To order numbers from least to greatest, write them as they appear on a number line from left to right.

Graph the percents on a number line.

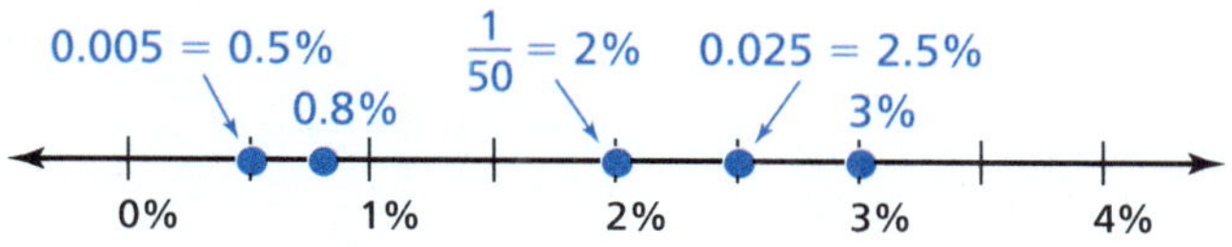

▶ So, the order from least to greatest is 0.005, 0.8%, $\frac{1}{50}$, 0.025, and 3%.

GO DIGITAL

Try It

Order the numbers from least to greatest.

3. $\frac{3}{10}$, 15%, 0.2, $\frac{3}{8}$, 0.09

4. 100%, 0.95, 1.2, $\frac{5}{4}$, 110%

Example 3 Comparing Numbers to Solve a Real-Life Problem

You, your sister, and a friend each take the same number of shots at a soccer goal. You make 72% of your shots, your sister makes $\frac{19}{25}$ of her shots, and your friend makes 0.67 of his shots. Who made the fewest shots?

To determine who made the fewest shots, find who made the least portion of their shots. One way to do this is to first write each number as a decimal.

You: $72\% = 72.\% = 0.72$

Sister: $\frac{19}{25} = \frac{19 \times 4}{25 \times 4} = \frac{76}{100} = 0.76$

Graph the decimals on a number line.

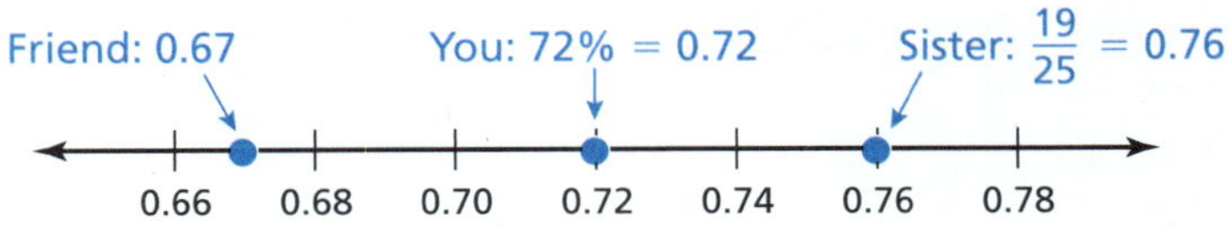

▶ 0.67 is the least number. So, your friend made the fewest shots.

Try It

5. WHAT IF? Your friend makes $\frac{3}{4}$ of his shots. Did your friend make more shots than you? your sister?

GO DIGITAL

In-Class Practice

1 I don't understand yet. | 2 I can do it with help. | 3 I can do it on my own. | 4 I can teach someone else.

6. NUMBER SENSE Complete the table.

Fraction	Decimal	Percent
$\frac{18}{25}$	0.72	
$\frac{17}{20}$		85%
$\frac{13}{50}$		
	0.62	
		45%

3 MTR **7. CHOOSE A METHOD** How would you decide whether $\frac{3}{5}$ or 59% is greater? Explain.

COMPARING NUMBERS **Complete the statement using <, >, or =.**

8. 33% < 0.34

9. 0.8 ___ $\frac{4}{5}$

10. $\frac{9}{50}$ ___ 17%

ORDERING NUMBERS **Order the numbers from least to greatest.**

11. 12%, 0.1, $\frac{4}{25}$

12. 1.35, 125%, $\frac{6}{5}$, 1.5, 130%

$\frac{5}{4}$ $\frac{p}{100}$

Example 4 Modeling Real Life

The map shows the portions of the populations in five Florida counties who identify as Black or African American. List the five counties in order by these portions from least to greatest.

Begin by writing each portion as a fraction, a decimal, and a percent.

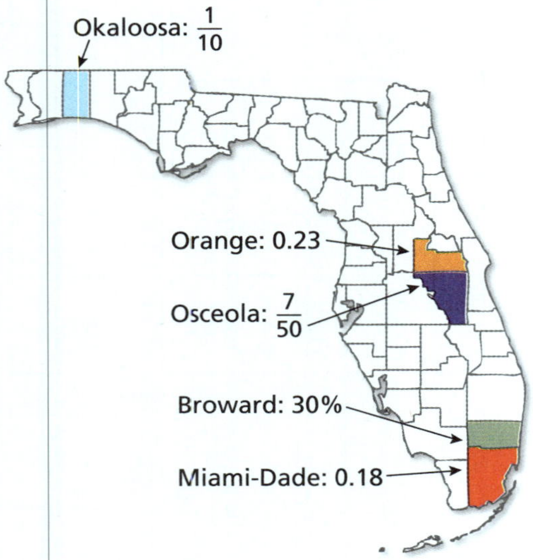

County	Fraction	Decimal	Percent
Broward	$\frac{30}{100}$	0.30	30%
Miami-Dade	$\frac{18}{100}$	0.18	18%
Okaloosa	$\frac{1}{10}$	0.10	10%
Orange	$\frac{23}{100}$	0.23	23%
Osceola	$\frac{7}{50}$	0.14	14%

Graph the percent for each county on a number line.

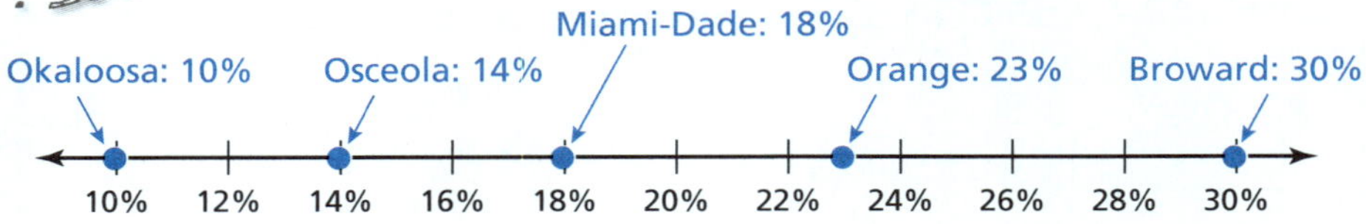

▶ By portion from least to greatest, the counties are Okaloosa, Osceola, Miami-Dade, Orange, and Broward.

In-Class Practice

1 I don't understand yet. 2 I can do it with help. 3 I can do it on my own. 4 I can teach someone else.

Ring	Portion That Is Copper
Red gold	25%
Pink gold	0.2
Rose gold	$\frac{9}{40}$

13. The table shows the portions of copper in three rings. Which ring has the highest portion of copper?

14. **Dig Deeper** The table shows the results of five teams competing in a scavenger hunt. List the five teams in order by the portion of items collected from least to greatest. What is the minimum number of items in the scavenger hunt? Explain.

Team	1	2	3	4	5
Portion Collected	$\frac{3}{4}$	0.8	77.5%	0.825	$\frac{13}{20}$

GO DIGITAL

9.3 Practice WITH CalcChat® AND CalcView®

Review & Refresh

Write the percent as a decimal.

1. 12%

2. 98.3%

3. 4.6%

Find the missing value(s) in the ratio table. Then write the equivalent ratios.

4.

Cashews	5	
Almonds	4	36

5.

Cost (dollars)	2		13.5
Time (minutes)	50	75	

Multiply.

6. $\begin{array}{r} 5.3 \\ \times\ 6 \\ \hline \end{array}$

7. $\begin{array}{r} 18.1 \\ \times\ 9 \\ \hline \end{array}$

8. 12.43×1.51

9. 0.852×6.7

Find the GCF of the numbers.

10. 15, 36

11. 51, 85

12. 176, 704

Concepts, Skills, & Problem Solving

USING A NUMBER LINE **Use a number line to order the numbers from least to greatest.** (See Exploration 1.)

13. 80%, 0.65, $\frac{3}{10}$, 40%

14. 0.27, 20%, $\frac{1}{25}$, 9%

15. $\frac{1}{8}$, 0.25, 15%, $\frac{27}{100}$

GO DIGITAL

COMPARING NUMBERS **Complete the statement using <, >, or =.** (See Example 1.)

16. 0.9 ☐ 95%

▶17. 20% ☐ 0.02

18. $\frac{37}{50}$ ☐ 74%

19. 50% ☐ $\frac{13}{25}$

20. 0.86 ☐ 86%

21. 76% ☐ 0.67

22. 60% ☐ $\frac{5}{8}$

23. 0.12 ☐ 1.2%

24. 17% ☐ $\frac{4}{25}$

25. 140% ☐ 0.14

26. $\frac{3}{8}$ ☐ 30%

27. 70% ☐ $\frac{7}{10}$

ORDERING NUMBERS **Order the numbers from least to greatest.** (See Example 2.)

28. 38%, $\frac{8}{25}$, 0.41

▶29. 68%, 0.63, $\frac{13}{20}$

30. $\frac{43}{50}$, 0.91, $\frac{7}{8}$, 84%

31. 1.6%, $\frac{3}{20}$, 0.015

32. 2.62, $2\frac{2}{5}$, 26.8%, 2.26, 271%

33. $\frac{87}{200}$, 0.44, 43.7%, $\frac{21}{50}$

34. **NUMBER SENSE** You answer 21 out of 25 questions correctly on a test. Do you reach your goal of answering at least 80% of the questions correctly?

7 MTR

35. **MODELING REAL LIFE** The table shows the approximate portions of the world population that live in four countries. Order the countries by population from least to greatest. (See Example 3.)

Country	Brazil	India	Russia	United States
Portion of World Population	2.7%	17.7%	$\frac{19}{1000}$	0.043

3 MTR

MAINTAIN ACCURACY **Order the numbers from least to greatest.**

36. 66.1%, 0.66, $\frac{133}{200}$, 0.667

37. $\frac{111}{500}$, 21%, 0.211, $\frac{11}{50}$

MATCHING **Tell which letter shows the graph of the number.**

38. $\frac{2}{5}$

39. 45.2%

40. 0.435

41. $\frac{89}{200}$

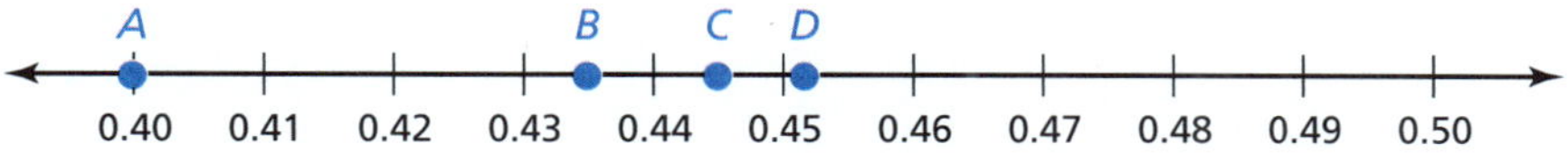

7 MTR

42. MODELING REAL LIFE The Tour de France is a bicycle road race. The whole race is made up of 21 small races called *stages*. The table shows how several stages compare to the whole Tour de France in a recent year. Order the stages by distance from shortest to longest. (See Example 4.)

Stage	1	7	8	17	21
Portion of Total Distance	$\frac{11}{200}$	0.044	$\frac{6}{125}$	0.06	4%

7 MTR

43. MODELING REAL LIFE The table shows the portions of the day that several animals sleep.

a. Order the animals by sleep time from least to greatest.

b. Estimate the portion of the day that you sleep.

c. Where do you fit on the ordered list?

Animal	Portion of Day Sleeping
Dolphin	0.433
Lion	56.3%
Rabbit	$\frac{19}{40}$
Squirrel	$\frac{31}{50}$
Tiger	65.8%

44. NUMBER SENSE Tell what whole number you can substitute for a in each list so the numbers are ordered from least to greatest. If there is none, explain why.

a. $\frac{1}{a}, \frac{a}{20}$, 28%

b. $\frac{3}{a}, \frac{a}{5}$, 75%

9.4 The Percent Proportion

Learning Target: Use the percent proportion to find missing quantities.

Success Criteria:
- I can write proportions to represent percent problems.
- I can solve a proportion to find a percent, a part, or a whole.

Algebraic Reasoning

MA.6.AR.3.4 Apply ratio relationships to solve mathematical and real-world problems involving percentages using the relationship between two quantities.

MA.7.AR.3.1 Apply previous understanding of percentages and ratios to solve multi-step real-world percent problems.

Exploration 1 Using Percent Models

2 MTR MAKE A CONNECTION

What quantities are given in each model? How can you use these quantities to answer the questions in part (b)?

____

____

____

____

____

____

Work with a partner.

a. Complete each model. Explain what each model represents.

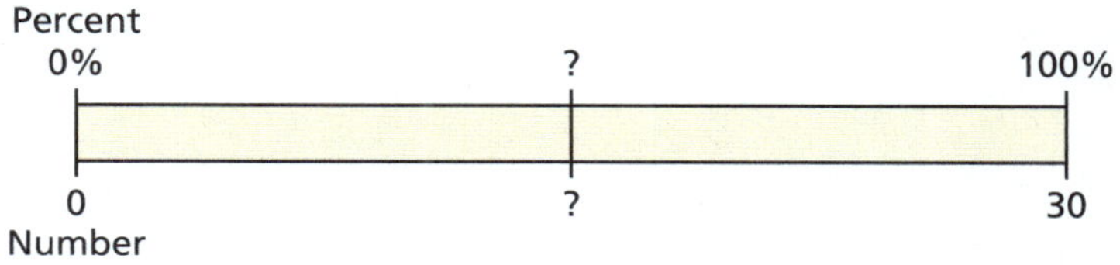

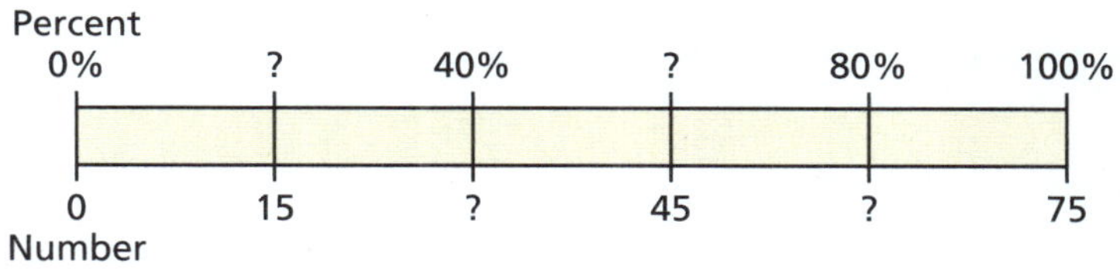

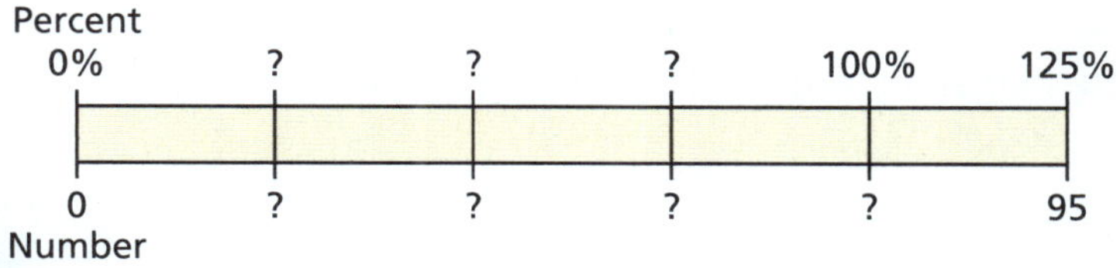

GO DIGITAL

b. Use the models in part (a) to answer each question.

- What number is 50% of 30?

- 15 is what percent of 75?

- 95 is 125% of what number?

c. How can you use ratio tables to check your answers in part (b)? How can you use proportions? Provide examples to support your reasoning.

d. Write a question different from those in part (b) that can be answered using one of the models in part (a). Trade questions with another group and find the solution.

9.4 Lesson

Key Idea

The Percent Proportion

Words You can represent "a is p percent of w" with the proportion

$$\frac{a}{w} = \frac{p}{100}$$

where a is part of the whole w, and $p\%$, or $\frac{p}{100}$, is the percent.

Numbers 3 out of 4 is 75%.

part → $\frac{3}{4} = \frac{75}{100}$ ← percent

whole → 4

In percent problems, the word *of* is usually followed by the whole.

Example 1 Finding a Percent

What percent of 15 is 12?

$\frac{a}{w} = \frac{p}{100}$	Write the percent proportion.
$\frac{12}{15} = \frac{p}{100}$	Substitute 12 for a and 15 for w.
$100 \cdot \frac{12}{15} = 100 \cdot \frac{p}{100}$	Multiplication Property of Equality
$80 = p$	Simplify.

▶ So, 80% of 15 is 12.

Check

Use a model to check your answer.

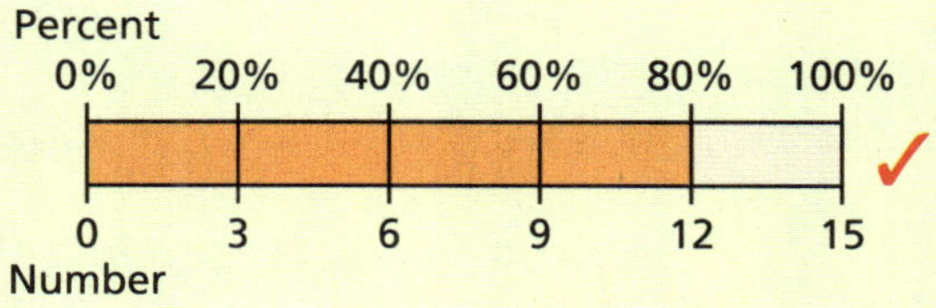

2 MTR USE ANOTHER METHOD

Show how to use a ratio table to find the percent.

____

____

____

____

____

____

____

GO DIGITAL

Try It **Write and solve a proportion to answer the question.**

1. What percent of 5 is 3?

$\frac{3}{5}$ $\frac{p}{100}$ 60%

2. 24 is what percent of 20?

$\frac{24}{20}$ 120%

Example 2 Finding a Part

What number is 0.5% of 200?

$$\frac{a}{w} = \frac{p}{100}$$ Write the percent proportion.

$$\frac{a}{200} = \frac{0.5}{100}$$ Substitute 200 for *w* and 0.5 for *p*.

$$a = 1$$ Multiply each side by 200.

▶ So, 1 is 0.5% of 200.

Try It **Write and solve a proportion to answer the question.**

3. What number is 80% of 60?

48

4. 10% of 40.5 is what number?

$\frac{10}{100} = \frac{x}{40.5}$ 4.05

Example 3 Finding a Whole

150% of what number is 30?

$$\frac{a}{w} = \frac{p}{100}$$ Write the percent proportion.

$$\frac{30}{w} = \frac{150}{100}$$ Substitute 30 for *a* and 150 for *p*.

Use mental math. The product of 30 and 5 is 150. The product of what number and 5 is 100?

$30 \times 5 = 150$

$$\frac{30}{w} = \frac{150}{100}$$

$20 \times 5 = 100$

▶ Because $w = 20$, 150% of 20 is 30.

GO DIGITAL

Try It

Write and solve a proportion to answer the question.

5. 0.1% of what number is 4?

6. $\frac{1}{2}$ is 25% of what number?

In-Class Practice

1 I don't understand yet. **2** I can do it with help. **3** I can do it on my own. **4** I can teach someone else.

7. USING THE PERCENT PROPORTION Write and solve a proportion to determine what percent of 120 is 54.

2 MTR

8. MODEL A PROBLEM Use a model to find 60% of 30.

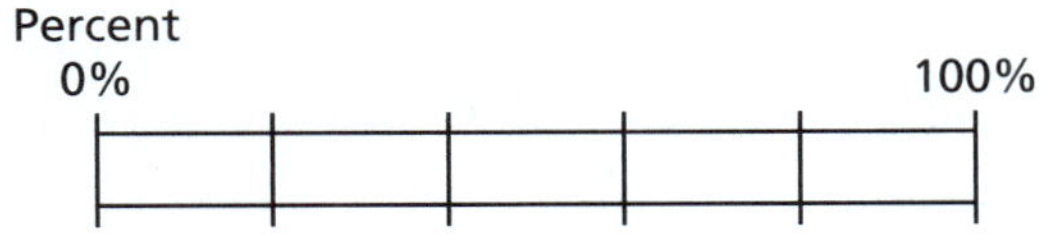

4 MTR

9. WHICH ONE DOESN'T BELONG? Which proportion at the right does *not* belong with the other three? Explain your reasoning.

$\frac{10}{w} = \frac{50}{100}$

$\frac{10}{50} = \frac{p}{100}$

$\frac{10}{20} = \frac{p}{100}$

$\frac{a}{20} = \frac{50}{100}$

Example 4 Modeling Real Life 7 MTR

Tornadoes are classified based on wind speed and resultant damage using the Enhanced Fujita (EF) scale. All the tornadoes in a state in a recent year were EF-0s and EF-1s. The ratio of EF-0s to EF-1s was 7 : 1. What percent of the tornadoes were EF-0s?

Understand the problem.

You are given the ratio of EF-0s to EF-1s and that all the tornadoes were EF-0s and EF-1s. You are asked to find the percent of the tornadoes that were EF-0s.

Make a plan.

The ratio 7 : 1 is a part-to-part ratio. The part that represents EF-0s is 7 and the whole is $7 + 1 = 8$. Use the percent proportion to find the percent of the tornadoes that were EF-0s.

Solve and check.

$$\frac{a}{w} = \frac{p}{100}$$ Write the percent proportion.

$$\frac{7}{8} = \frac{p}{100}$$ Substitute 7 for a and 8 for w.

$$100 \cdot \frac{7}{8} = 100 \cdot \frac{p}{100}$$ Multiplication Property of Equality

$$87.5 = p$$ Simplify.

▶ So, 87.5% of the tornadoes were EF-0s.

Check Reasonableness
The part that represents EF-0 tornadoes, 7, is greater than three-fourths the whole, 8. So, the percent of the tornadoes that were EF-0s should be greater than 75%. Because 87.5% > 75%, the answer is reasonable. ✓

In-Class Practice

1 I don't understand yet. 2 I can do it with help. 3 I can do it on my own. 4 I can teach someone else.

10. An arctic woolly-bear caterpillar lives for 7 years and spends 90% of its life frozen. How many days of its life is the arctic woolly-bear frozen?

11. **Dig Deeper** The table shows the numbers of pictures you upload to a social media website for 5 days in a row. How many total pictures do you upload during the week when 68% of the total pictures are uploaded Monday through Friday?

Day	Pictures Uploaded
Monday	2
Tuesday	2
Wednesday	4
Thursday	1
Friday	8

GO DIGITAL

9.4 Practice WITH CalcChat® AND CalcView®

Review & Refresh

Write the fraction as a decimal and a percent.

1. $\frac{42}{100}$

2. $\frac{7}{1000}$

3. $\frac{9}{20}$

4. $\frac{4}{5}$

Evaluate the expression when $a = -15$ and $b = -5$.

5. $a \div b$

6. $\frac{b + 14}{a}$

7. $\frac{b^2}{a + 5}$

8. What is the solution of $9x = -1.8$?

Concepts, Skills, & Problem Solving

CHOOSE TOOLS **Use a model to answer the question. Use a proportion to check your answer.** (See Exploration 1.)

9. What number is 20% of 80?

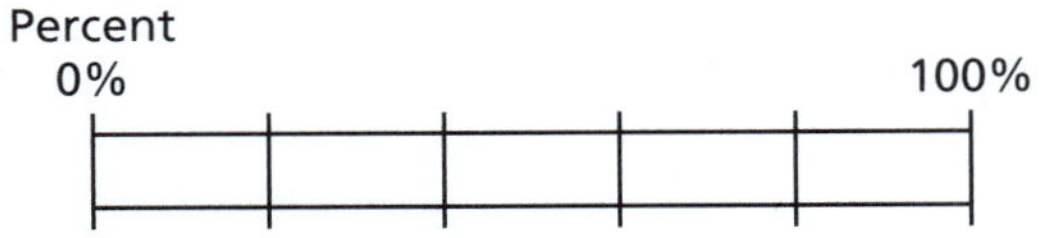

10. 10 is what percent of 40?

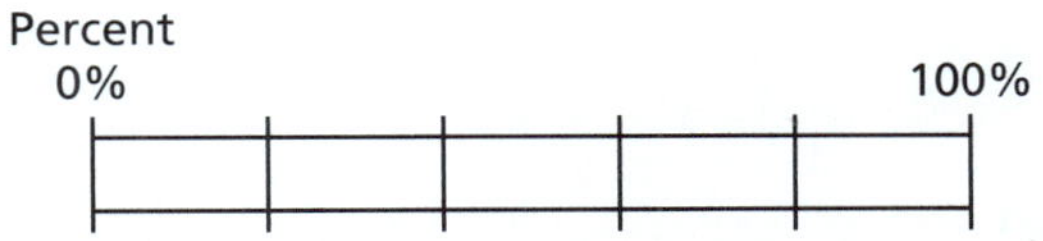

11. 15 is 30% of what number?

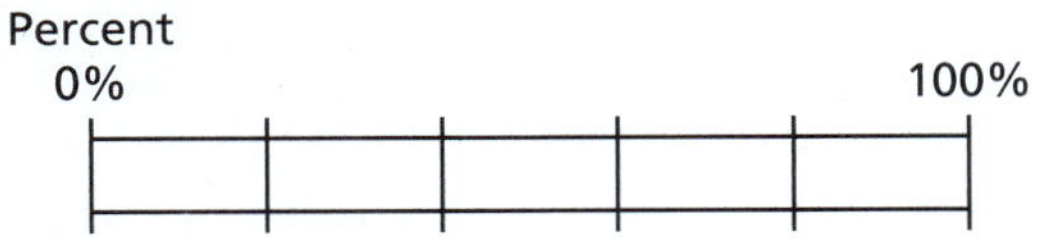

12. What number is 120% of 70?

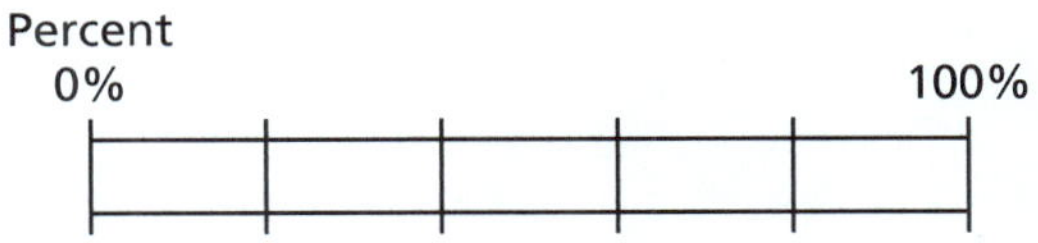

13. 20 is what percent of 50?

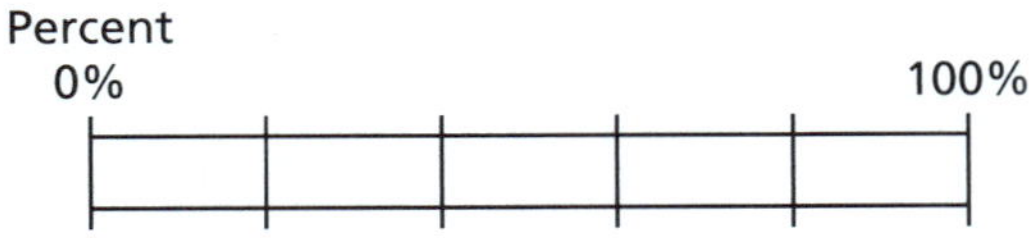

14. 48 is 75% of what number?

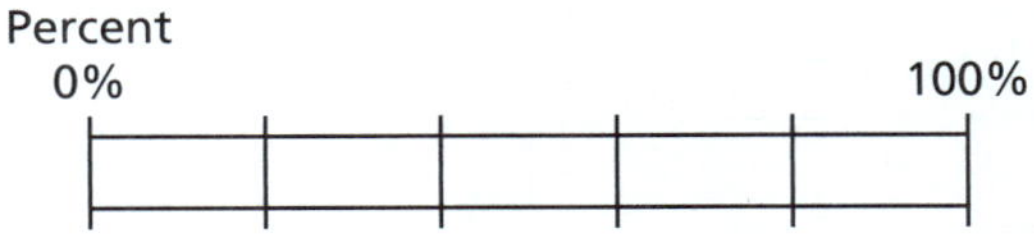

GO DIGITAL

USING THE PERCENT PROPORTION **Write and solve a proportion to answer the question.** (See Examples 1, 2, and 3.)

▶**15.** What percent of 25 is 12?

16. 14 is what percent of 56?

▶**17.** 25% of what number is 5?

18. 36 is 9% of what number?

19. 75% of 124 is what number?

20. 110% of 90 is what number?

▶**21.** What number is 0.4% of 40?

22. 72 is what percent of 45?

4 MTR

23. **YOU BE THE TEACHER** Your friend uses the percent proportion to answer the question below. Is your friend correct? Explain your reasoning.

"40% of what number is 8?"

$$\frac{a}{w} = \frac{p}{100}$$

$$\frac{8}{w} = \frac{40}{100}$$

$$w = 20$$

7 MTR

24. **MODELING REAL LIFE** Of 140 seventh-grade students, 15% earn the Presidential Youth Fitness Award. How many students earn the award?

7 MTR

25. **MODELING REAL LIFE** A salesperson receives a 3% commission on sales. The salesperson receives $180 in commission. What is the amount of sales?

USING THE PERCENT PROPORTION Write and solve a proportion to answer the question.

26. 0.5 is what percent of 20?

27. 14.2 is 28.4% of what number?

28. $\frac{3}{4}$ is 60% of what number?

29. What number is 25% of $\frac{7}{8}$?

7 MTR

30. **MODELING REAL LIFE** Your friend earns $12.50 per hour, which is 125% of her hourly wage last year. How much did your friend earn per hour last year?

7 MTR

▶ **31.** **MODELING REAL LIFE** You board the Iron Gwazi, Florida's tallest and fastest roller coaster. The ratio of empty seats to full seats is 3 : 5. What percent of the seats are full? (See Example 4.)

7 MTR

32. **MODELING REAL LIFE** The bar graph shows the numbers of reserved campsites at a campground for one week. What percent of the reservations were for Friday or Saturday?

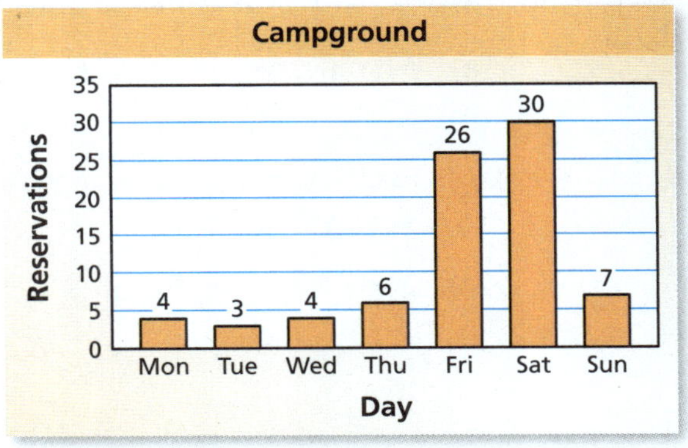

33. PROBLEM SOLVING Your friend displays the results of a survey that asks several people to vote on a new school mascot.

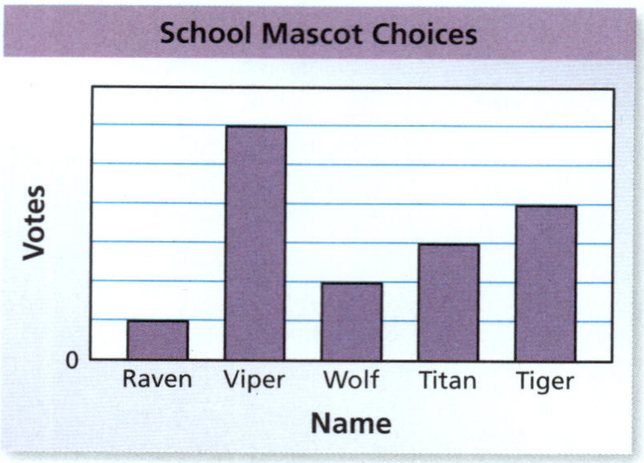

a. What is missing from the bar graph?

b. What percent of the votes does the least popular mascot receive? Explain your reasoning.

c. There are 124 votes total. How many votes does tiger receive?

34. Dig Deeper A quarterback throws 33 passes in the first three quarters of a football game. The ratio of complete passes to incomplete passes during the first three quarters is 6 : 5. He throws 7 passes in the fourth quarter and completes 62.5% of his passes for the entire game. How many passes does the quarterback complete in the fourth quarter? Justify your answer.

35. REASONING 20% of a number is x. What is 100% of the number? Assume $x > 0$.

5 MTR

36. STRUCTURE Answer each question. Assume $x > 0$.

a. What percent of $8x$ is $5x$?

b. What is 65% of $80x$?

9.5 The Percent Equation

Learning Target: Use the percent equation to find missing quantities.

Success Criteria:
- I can write equations to represent percent problems.
- I can use the percent equation to find a percent, a part, or a whole.

Algebraic Reasoning
MA.7.AR.3.1 Apply previous understanding of percentages and ratios to solve multi-step real-world percent problems.

Exploration 1 Using Percent Equations

Work with a partner.

a. The circle graph shows the number of votes received by each candidate during a school election. So far, only half of the students have voted. Find the percent of students who voted for each candidate. Explain your method.

Votes Received by Each Candidate

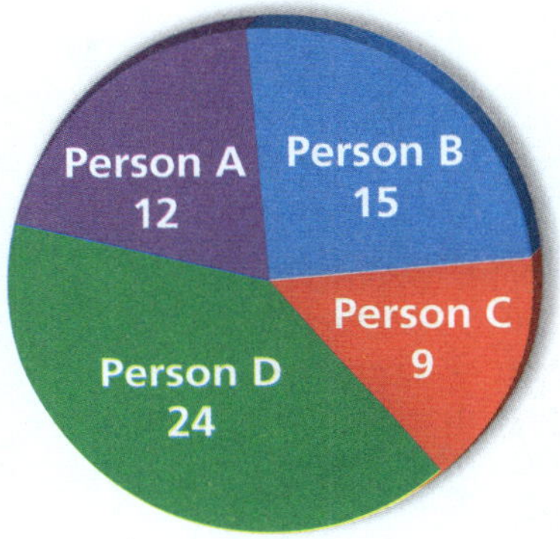

b. You have learned that $\frac{\text{part}}{\text{whole}} = \text{percent}$. Solve the equation for the "part." Explain your reasoning.

Final Results

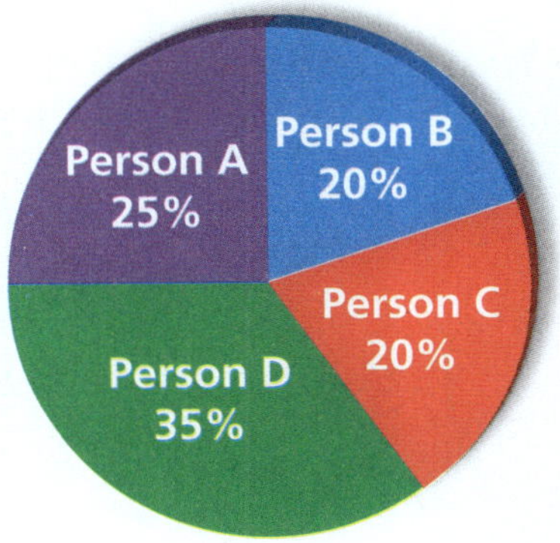

c. The circle graph shows the final results of the election after every student voted. Use the equation you wrote in part (b) to find the number of students who voted for each candidate.

> **5 MTR RELATE CONCEPTS**
> How does the equation you wrote in part (b) compare to the percent proportion? Explain.
>
> ____________________
>
> ____________________
>
> ____________________
>
> ____________________
>
> ____________________
>
> ____________________
>
> ____________________

d. Use a different method to check your answers in part (c). Which method do you prefer? Explain.

GO DIGITAL

9.5 Lesson

Key Idea

The Percent Equation

Words To represent "a is p percent of w," use an equation.

percent

$$a = p\% \cdot w$$

part of the whole — whole

Numbers

$15 = 50\% \cdot 30$

$15 = 0.5 \cdot 30$

$15 = \frac{1}{2} \cdot 30$

Example 1 Finding a Part of a Number

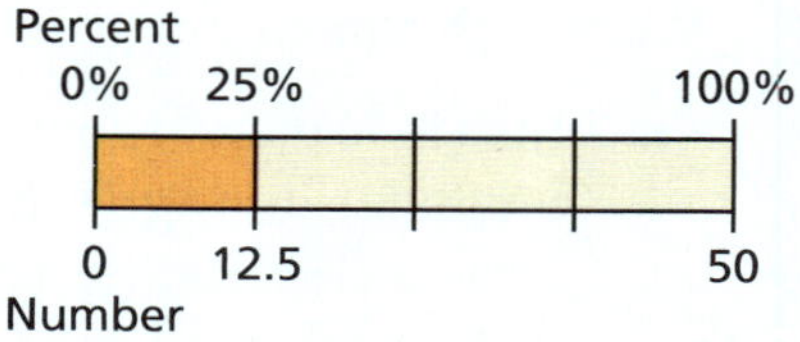

What number is 24% of 50? **Estimate**

$a = p\% \cdot w$ Write the percent equation.

$= \frac{24}{100} \cdot 50$ Substitute $\frac{24}{100}$ for $p\%$ and 50 for w.

$= 12$ Simplify.

▶ So, 12 is 24% of 50. **Reasonable?** $12 \approx 12.5$ ✓

Common Error

Remember to convert a percent to a fraction or a decimal when using the percent equation. For Example 1, write 24% as $\frac{24}{100}$.

Try It **Write and solve an equation to answer the question.**

1. What number is 10% of 20?

2. What number is 150% of 40?

Example 2 Finding a Percent

9.5 is what percent of 25?

$a = p\% \cdot w$ Write the percent equation.

$9.5 = p\% \cdot 25$ Substitute 9.5 for a and 25 for w.

$\frac{9.5}{25} = \frac{p\% \cdot 25}{25}$ Division Property of Equality

$0.38 = p\%$ Simplify.

▶ Because 0.38 equals 38%, 9.5 is 38% of 25.

GO DIGITAL

Try It **Write and solve an equation to answer the question.**

3. 3 is what percent of 600?

4. What percent of 20 is 18?

Example 3 Finding a Whole

2 MTR USE ANOTHER METHOD

Show how to use a ratio table to find the whole.

39 is 52% of what number?

$a = p\% \cdot w$	Write the percent equation.
$39 = 0.52 \cdot w$	Substitute 39 for *a* and 0.52 for *p*%.
$\frac{39}{0.52} = \frac{0.52 \cdot w}{0.52}$	Division Property of Equality
$75 = w$	Simplify.

▶ So, 39 is 52% of 75.

Try It **Write and solve an equation to answer the question.**

5. 8 is 80% of what number?

6. 90 is 180% of what number?

GO DIGITAL

In-Class Practice

1 I don't understand yet. | 2 I can do it with help. | 3 I can do it on my own. | 4 I can teach someone else.

7. **VOCABULARY** Write the percent equation in words.

USING THE PERCENT EQUATION **Write and solve an equation to answer the question.**

8. 14 is what percent of 70?

9. What number is 36% of 85?

10. 9 is 12% of what number?

11. 108 is what percent of 72?

In-Class Practice

1 I don't understand yet. 2 I can do it with help. 3 I can do it on my own. 4 I can teach someone else.

4 MTR **12. DIFFERENT WORDS, SAME QUESTION** Which is different? Find "both" answers.

What number is 20% of 55?	55 is 20% of what number?
20% of 55 is what number?	$0.2 \cdot 55$ is what number?

Example 4 Modeling Real Life

7 MTR

8th Street Cafe

DATE: MAY04 12:45PM
TABLE: 29
SERVER: JANE

Food Total	**27.50**
Tax	**1.65**
Subtotal	**29.15**

TIP: ________

TOTAL: ________

Thank You

You are paying for lunch and receive the bill shown.

a. Find the percent of sales tax on the food total.

Answer the question: \$1.65 is what percent of \$27.50?

$a = p\% \cdot w$ Write the percent equation.

$1.65 = p\% \cdot 27.50$ Substitute 1.65 for a and 27.50 for w.

$0.06 = p\%$ Divide each side by 27.50.

▶ Because 0.06 equals 6%, the percent of sales tax is 6%.

b. You leave a 16% tip on the food total. Find the total amount you pay for lunch.

Answer the question: What tip amount is 16% of \$27.50?

$a = p\% \cdot w$ Write the percent equation.

$= 0.16 \cdot 27.50$ Substitute 0.16 for $p\%$ and 27.50 for w.

$= 4.40$ Multiply.

The amount of the tip is \$4.40.

▶ So, you pay a total of $\$29.15 + \$4.40 = \$33.55$.

GO DIGITAL

In-Class Practice

1 I don't understand yet. 2 I can do it with help. 3 I can do it on my own. 4 I can teach someone else.

13. Water Tank A has a capacity of 550 gallons and is 66% full. Water Tank B is 53% full. The ratio of the capacity of Water Tank A to Water Tank B is 11:15.

a. How much water is in each tank?

b. What percent of the total volume of both tanks is filled with water?

14. **Dig Deeper** A school offers band and chorus classes. The table shows the percents of the 1200 students in the school who are enrolled in band, chorus, or neither class. How many students are enrolled in both classes? Explain.

Class	Enrollment
Band	34%
Chorus	28%
Neither	42%

9.5 Practice WITH CalcChat® AND CalcView®

Review & Refresh

Write and solve a proportion to answer the question.

1. 30% of what number is 6?

2. 42 is what percent of 80?

3. What percent of 36 is 27?

4. What number is 120% of 80?

Find the distance between the two numbers on a number line.

5. -4 and 10

6. $-\frac{2}{3}$ and $\frac{4}{3}$

7. $-5\frac{2}{5}$ and $-1\frac{3}{10}$

8. -4.3 and 7.5

9. There are 160 people in a grade. The ratio of boys to girls is 3 to 5. Write a proportion you can use to find the number of boys.

Concepts, Skills, & Problem Solving

USING PERCENT EQUATIONS The circle graph shows the number of votes received by each candidate during a school election. Find the percent of students who voted for the indicated candidate. (See Exploration 1.)

Votes Received by Each Candidate

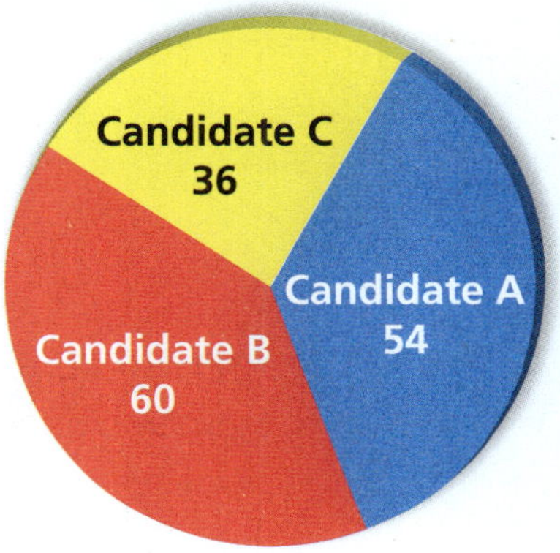

10. Candidate A

11. Candidate B

12. Candidate C

GO DIGITAL

USING THE PERCENT EQUATION **Write and solve an equation to answer the question.** (See Examples 1, 2, and 3.)

▶**13.** 20% of 150 is what number?

14. 45 is what percent of 60?

▶**15.** 35% of what number is 35?

16. 0.8% of 150 is what number?

▶**17.** 29 is what percent of 20?

18. 0.5% of what number is 12?

19. What percent of 300 is 51?

20. 120% of what number is 102?

4 MTR **YOU BE THE TEACHER** **Your friend uses the percent equation to answer the question. Is your friend correct? Explain your reasoning.**

21. What number is 35% of 20?

$$a = p\% \cdot w$$
$$= 0.35 \cdot 20$$
$$= 7$$

22. 30 is 60% of what number?

$$a = p\% \cdot w$$
$$= 0.6 \cdot 30$$
$$= 18$$

7 MTR **23.** **MODELING REAL LIFE** A salesperson receives a 2.5% commission on sales. What commission does the salesperson receive for $8000 in sales? (See Example 4.)

7 MTR

24. **MODELING REAL LIFE** Your school raised 125% of its fundraising goal. The school raised \$6750. What was the goal?

7 MTR

25. **MODELING REAL LIFE** The sales tax on the model rocket shown is \$1.92. What is the percent of sales tax?

26. **B.E.S.T. Test Prep** Florida has about 66 tornadoes each year, which is approximately 5.27% of all tornadoes in the United States annually. Estimate how many tornadoes occur in the United States each year.

Ⓐ 3 Ⓑ 348 Ⓒ 1252 Ⓓ 34,782

5 MTR

STRUCTURE **There were n signers of the Declaration of Independence. The youngest was Edward Rutledge, who was x years old. The oldest was Benjamin Franklin, who was y years old.**

27. x is 25% of 104. What was Rutledge's age?

28. 7 is 10% of y. What was Franklin's age?

29. n is 80% of y. How many signers were there?

30. NUMBER SENSE How can you tell whether a percent of a number will be *greater than, less than,* or *equal to* the number? Give examples to support your answer.

31. PROBLEM SOLVING In a survey, a group of students is asked their favorite sport. Eighteen students choose "other" sports.

a. How many students participate in the survey?

b. How many choose football?

4 MTR **32. DISCUSS MATHEMATICAL THINKING** Tell whether the statement is *true* or *false*. Explain your reasoning.

If W is 25% of Z, then $Z : W$ is $75 : 25$.

33. Dig Deeper At a restaurant, the amount of your bill before taxes and tip is $18.53. A 6% sales tax is applied to your bill, and you want to leave at least a 20% tip, but have only five-dollar bills. You plan to use any change you receive as part of the tip. What is the minimum percent that you can tip? Explain your reasoning.

34. REASONING The table shows your test results in a math class. What score do you need on the last test to earn 90% of the total points on the tests?

Test Score	Point Value
83%	100
91.6%	250
88%	150
?	300

9.6 Percents of Increase and Decrease

Learning Target: Find percents of change in quantities.

Success Criteria:
- I can explain the meaning of percent of change.
- I can find the percent of increase or decrease in a quantity.
- I can find the percent error of a quantity.

Algebraic Reasoning
MA.7.AR.3.1 Apply previous understanding of percentages and ratios to solve multi-step real-world percent problems.

Exploration 1 Exploring Percent of Change

Work with a partner.

Each year in the Columbia River Basin, adult salmon swim upriver to streams to lay eggs.

To go up the river, the adult salmon use fish ladders. But to go down the river, the young salmon must pass through several dams.

At one time, there were electric turbines at each of the eight dams on the main stem of the Columbia and Snake Rivers. About 88% of the young salmon passed through a single dam unharmed.

a. One thousand young salmon passed through a dam. How many passed through unharmed?

GO DIGITAL

b. One thousand young salmon passed through the river basin. How many passed through all 8 dams unharmed?

c. By what percent did the number of young salmon *decrease* when they passed through a single dam?

d. Describe a similar real-life situation in which a quantity *increases* by a constant percent each time an event occurs.

1 MTR ANALYZE A PROBLEM

As the number of dams increases, what should be true about the number of young salmon that passed through unharmed?

9.6 Lesson

Key Vocabulary
percent of change, p. 731
percent of increase, p. 731
percent of decrease, p. 731
percent error, p. 733

A **percent of change** is the percent that a quantity changes from the original amount.

$$\text{percent of change} = \frac{\text{amount of change}}{\text{original amount}}$$

Key Ideas

Percents of Increase and Decrease

When the original amount increases, the percent of change is called a **percent of increase**.

$$\text{percent of increase} = \frac{\text{new amount} - \text{original amount}}{\text{original amount}}$$

When the original amount decreases, the percent of change is called a **percent of decrease**.

$$\text{percent of decrease} = \frac{\text{original amount} - \text{new amount}}{\text{original amount}}$$

Example 1 Finding a Percent of Increase

Day	Hours Online
Saturday	2
Sunday	4.5

The table shows the numbers of hours you spent online last weekend. What is the percent of change in your time spent online from Saturday to Sunday?

The time spent online Sunday is greater than the time spent online Saturday. So, the percent of change is a percent of increase.

$$\text{percent of increase} = \frac{\text{new amount} - \text{original amount}}{\text{original amount}}$$

$$= \frac{4.5 - 2}{2} \quad \text{Substitute.}$$

$$= \frac{2.5}{2} \quad \text{Subtract.}$$

$$= 1.25, \text{ or } 125\% \quad \text{Write as a percent.}$$

▶ So, your time spent online increased 125% from Saturday to Sunday.

GO DIGITAL

Try It **Find the percent of change. Round to the nearest tenth of a percent if necessary.**

1. 10 inches to 25 inches

25 − 10 = 15

15% ↑

2. 57 people to 65 people

65 − 57 = 8 ÷ 57

14.0% ↑

Example 2 Finding a Percent of Decrease

The bar graph shows a softball player's home run totals. What was the percent of change from 2019 to 2020?

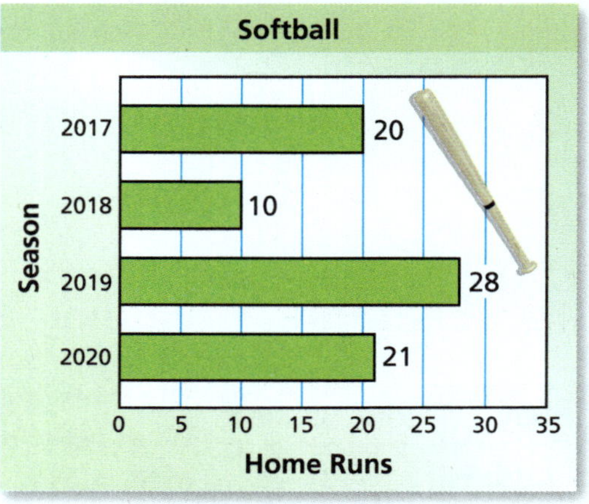

The number of home runs decreased from 2019 to 2020. So, the percent of change is a percent of decrease.

$$\text{percent of decrease} = \frac{\text{original amount} - \text{new amount}}{\text{original amount}}$$

$$= \frac{28 - 21}{28} \qquad \text{Substitute.}$$

$$= \frac{7}{28} \qquad \text{Subtract.}$$

$$= 0.25, \text{ or } 25\% \qquad \text{Write as a percent.}$$

▶ So, the number of home runs decreased 25% from 2019 to 2020.

25% ↓

Try It

3. In Example 2, what was the percent of change from 2017 to 2018?

50% ↓

GO DIGITAL

In-Class Practice

1 I don't understand yet. **2** I can do it with help. **3** I can do it on my own. **4** I can teach someone else.

4. VOCABULARY What does it mean for a quantity to change by $n\%$?

5. NUMBER SENSE Without calculating, determine which situation has a greater percent of change. Explain.

- 5 bonus points added to 50 points
- 5 bonus points added to 100 points

FINDING A PERCENT OF CHANGE **Identify the percent of change as an *increase* or a *decrease*. Then find the percent of change.**

6. 8 feet to 24 feet

7. 300 miles to 210 miles

Key Idea

Percent Error

A **percent error** is the percent that an estimated amount differs from the actual amount.

$$\text{percent error} = \frac{\text{amount of error}}{\text{actual amount}}$$

The amount of error is always positive.

Example 3 Modeling Real Life 7 MTR

Bag A: 15 ounces

Bag B: 16.5 ounces

You fill bags with about 16 ounces of homemade dog treats. The acceptable percent error when filling a bag is 5%. Tell whether each bag is acceptable.

Find the percent error for each bag.

Bag A: The amount of error is $16 - 15 = 1$ ounce.

$$\text{percent error} = \frac{\text{amount of error}}{\text{actual amount}}$$

$$= \frac{1}{16}$$

$$= 0.0625, \text{ or } 6.25\%$$

Bag B: The amount of error is $16.5 - 16 = 0.5$ ounce.

$$\text{percent error} = \frac{\text{amount of error}}{\text{actual amount}}$$

$$= \frac{0.5}{16}$$

$$= 0.03125, \text{ or } 3.125\%$$

▶ Because $6.25\% > 5\%$, Bag A is not acceptable. Because $3.125\% < 5\%$, Bag B is acceptable.

In-Class Practice

1 I don't understand yet. 2 I can do it with help. 3 I can do it on my own. 4 I can teach someone else.

8. In one round of a game, you are asked how many bones are in a human body at birth. If the percent error of your answer is at most 5%, you earn two points. If the percent error is at most 10%, but greater than 5%, you earn one point. You guess 282 bones. The correct answer is 300 bones. How many points do you earn?

9. **Dig Deeper** The manager of a restaurant offers a 20% decrease in price to tennis teams. A cashier applies a 10% decrease and then another 10% decrease. Is this the same as applying a 20% decrease? If so, justify your answer. If not, explain how to achieve a 20% decrease after first applying a 10% decrease.

GO DIGITAL

9.6 Practice WITH CalcChat® AND CalcView®

Review & Refresh

Write and solve an equation to answer the question.

1. What number is 25% of 64?

2. 39.2 is what percent of 112?

3. 5 is 5% of what number?

4. 18 is 32% of what number?

Find the sum. Write fractions in simplest form.

5. $\frac{4}{7} + \left(-\frac{6}{7}\right)$

6. $-4.621 + 3.925$

7. $-\frac{5}{12} + \frac{3}{4}$

Concepts, Skills, & Problem Solving

EXPLORING PERCENT OF CHANGE **You are given the percent of salmon that pass through a single dam unharmed. By what percent does the number of salmon decrease when passing through a single dam?** (See Exploration 1.)

8. 75%

9. 80%

10. 62%

11. 94%

GO DIGITAL

FINDING A PERCENT OF CHANGE **Identify the percent of change as an *increase* or a *decrease*. Then find the percent of change. Round to the nearest tenth of a percent if necessary.**
(See Examples 1 and 2.)

12. 12 inches to 36 inches

▶ **13.** 80 people to 25 people

14. 50 pounds to 35 pounds

▶ **15.** 24 songs to 78 songs

16. 10 gallons to 24 gallons

17. 72 paper clips to 63 paper clips

18. 16 centimeters to 44.2 centimeters

19. 68 miles to 42.5 miles

4 MTR **20.** **YOU BE THE TEACHER** Your friend finds the percent increase from 18 to 26. Is your friend correct? Explain your reasoning.

$$\frac{26 - 18}{26} \approx 0.31 = 31\%$$

7 MTR **21.** **MODELING REAL LIFE** Last week, you finished Level 2 of a video game in 32 minutes. Today, you finish Level 2 in 28 minutes. What is the percent of change?

7 MTR **22.** **MODELING REAL LIFE** You estimate that a baby pig weighs 20 pounds. The actual weight of the baby pig is 16 pounds. Find the percent erro

7 MTR

23. MODELING REAL LIFE A researcher estimates that a fossil is 3200 years old. Using *carbon-14 dating,* a procedure used to determine the age of an object, the researcher discovers that the fossil is 4000 years old.

a. Find the percent error.

b. What other estimate gives the same percent error? Explain your reasoning.

FINDING A PERCENT OF CHANGE **Identify the percent of change as an *increase* or a *decrease*. Then find the percent of change. Round to the nearest tenth of a percent if necessary.**

24. $\frac{1}{4}$ to $\frac{1}{2}$ **25.** $\frac{4}{5}$ to $\frac{3}{5}$ **26.** $\frac{1}{4}$ to $\frac{7}{8}$ **27.** $\frac{5}{4}$ to $\frac{3}{8}$

4 MTR

28. DISCUSS MATHEMATICAL THINKING Explain why a change from 20 to 40 is a 100% increase, but a change from 40 to 20 is a 50% decrease.

6 MTR

29. ASSESS REASONABLENESS The table shows population data for a community.

Year	Population
2014	100,000
2020	138,000

a. What is the percent of change from 2014 to 2020?

b. Your friend predicts that in 2026 the population will be about 190,440 people. Is your friend's prediction reasonable? Explain.

30. **GEOMETRY** Suppose the length and the width of the sandbox are doubled.

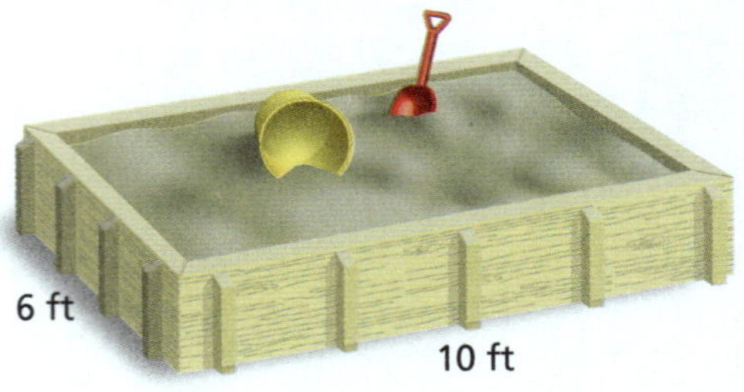

a. Find the percent of change in the perimeter.

b. Find the percent of change in the area.

7 MTR ▶31. **MODELING REAL LIFE** A company fills boxes with about 21 ounces of cereal. The acceptable percent error in filling a box is 6%. Box A contains 20 ounces of cereal, and Box B contains 25 ounces of cereal. Tell whether each box is an acceptable weight. (See Example 3.)

3 MTR 32. **MAINTAIN ACCURACY** Find the percent of change from June to September in the mile-run times shown.

4 MTR 33. **DISCUSS MATHEMATICAL THINKING** A number increases by 10% and then decreases by 10%. Will the result be *greater than*, *less than*, or *equal to* the original number? Explain.

34. **Dig Deeper** Donations to an annual fundraiser are 15% greater this year than last year. Last year, donations were 10% greater than the year before. The amount raised this year is $10,120. How much was raised two years ago?

35. **REASONING** Forty students are in the science club. Of those, 45% are girls. This percent increases to 56% after more girls join the club. How many more girls join?

9.7 Discounts and Markups

Learning Target: Solve percent problems involving discounts and markups.

Success Criteria:
- I can use percent models to solve problems involving discounts and markups.
- I can write and solve equations to solve problems involving discounts and markups.

Algebraic Reasoning
MA.7.AR.3.1 Apply previous understanding of percentages and ratios to solve multi-step real-world percent problems.

Exploration 1 Comparing Discounts

Work with a partner.

a. The same pair of earrings is on sale at three stores. Which store has the best price? Use the percent models to justify your answer.

Store A:
Regular price: \$45
Sale: 40% off

Store B:
Regular price: \$49
Sale: 50% off

Store C:
Regular price: \$39
Sale: 20% off

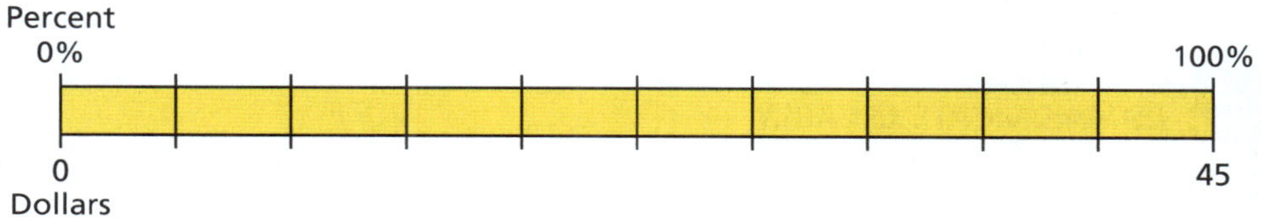

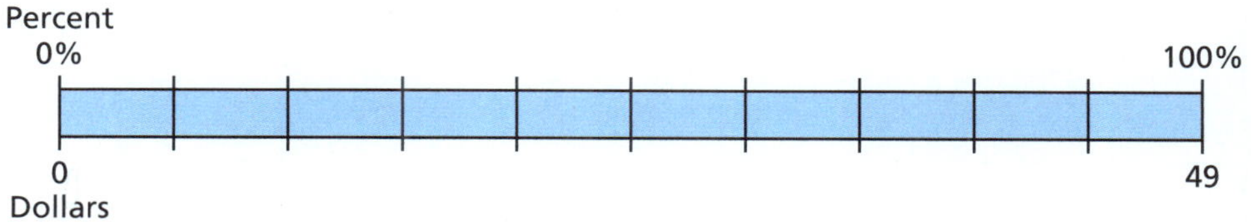

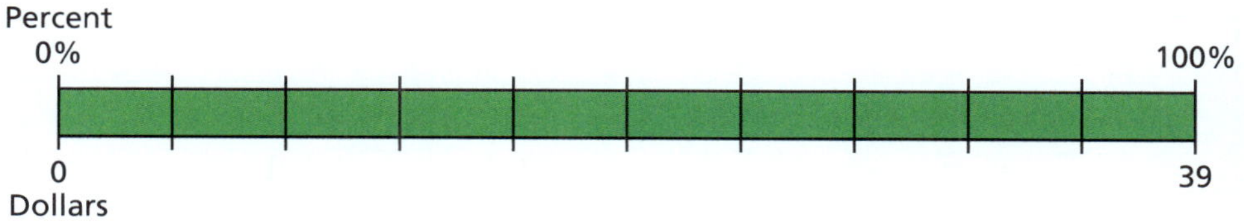

GO DIGITAL

b. You buy the earrings on sale for 30% off at a different store. You pay $22.40. What was the original price of the earrings? Use the percent model to justify your answer.

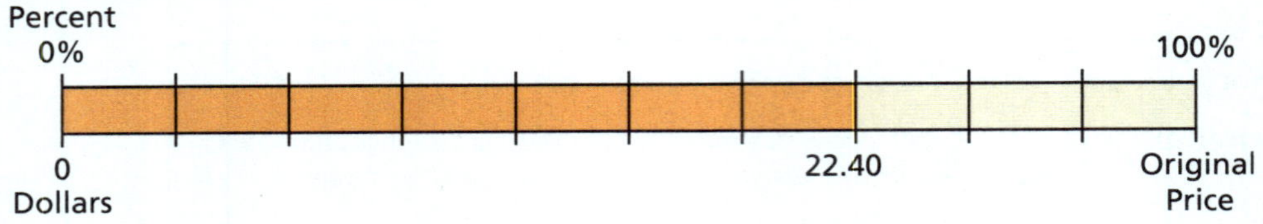

c. You sell the earrings in part (b) to a friend for 60% more than what you paid. What is the selling price? Use a percent model to justify your answer.

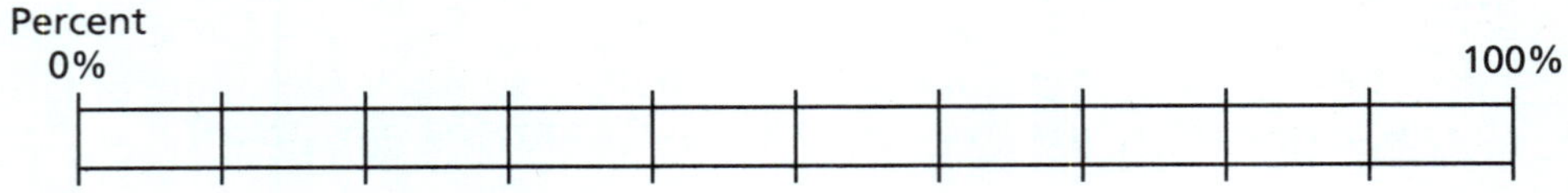

4 MTR COMMUNICATE CLEARLY

Explain to your partner why 30% of the original price in part (b) is not the same as 30% of what you paid.

9.7 Lesson

Key Vocabulary
discount, *p. 741*
markup, *p. 741*

Key Ideas

Discounts

A **discount** is a decrease in the original price of an item.

Markups

To make a profit, stores charge more than what they pay. The increase from what the store pays to the selling price is called a **markup**.

Example 1 Finding a Sale Price

The original price of a video game is $35. What is the sale price?

Method 1: First, find the discount. The discount is 25% of $35.

$a = p\% \cdot w$ Write the percent equation.

$= 0.25 \cdot 35$ Substitute 0.25 for $p\%$ and 35 for w.

$= 8.75$ Multiply.

Next, find the sale price.

Sale price = Original price − Discount

$= 35 - 8.75$

$= 26.25$

▶ So, the sale price is $26.25.

For an item with an original price of p, the price after a 25% discount is $p - 0.25p$, or $0.75p$. So, a 25% discount is the same as paying 75% of the original price.

Method 2: Use the fact that the sale price is $100\% - 25\% = 75\%$ of the original price.

Find the sale price.

Sale price = 75% of $35

$= 0.75 \cdot 35$

$= 26.25$

▶ So, the sale price is $26.25.

Check

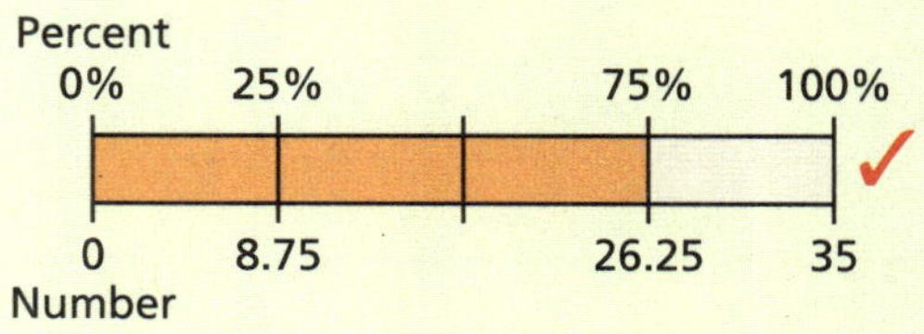

$\frac{25}{35}$ $\frac{p}{100}$

GO DIGITAL

Try It

1. The original price of a skateboard is \$50. The skateboard is on sale for 20% off. What is the sale price?

\$10

Example 2 Finding an Original Price

What is the original price of the cleats?

The sale price is $100\% - 40\% = 60\%$ of the original price.

Answer the question:

33 is 60% of what number?

$a = p\% \cdot w$ Write the percent equation.

$33 = 0.6 \cdot w$ Substitute 33 for a and 0.6 for $p\%$.

$55 = w$ Divide each side by 0.6.

▶ So, the original price of the cleats is \$55.

Check

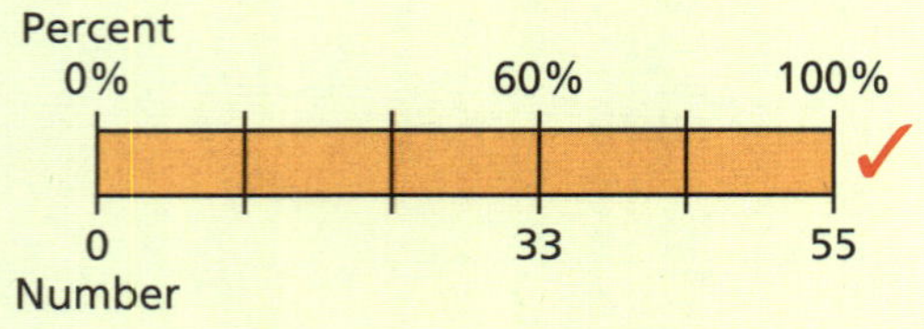

Try It

2. The discount to stream a movie is 50%. It is on sale for \$10. What is the original price to stream the movie?

20 5

GO DIGITAL

In-Class Practice

1 I don't understand yet. | 2 I can do it with help. | 3 I can do it on my own. | 4 I can teach someone else.

1 MTR **3. HELP A CLASSMATE** Explain to a classmate how to find the sale price of an item that has a 15% discount.

FINDING A SALE PRICE **Find the sale price. Use a percent model to check your answer.**

4. A portable table tennis set costs $30 before a 30% discount.

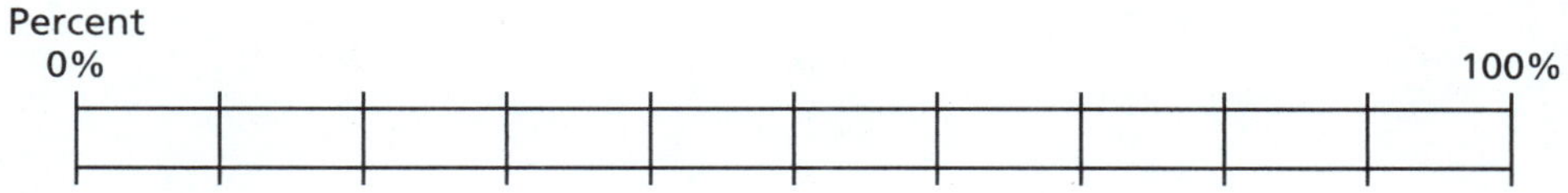

5. The original price of an easel is $70. The easel is on sale for 20% off.

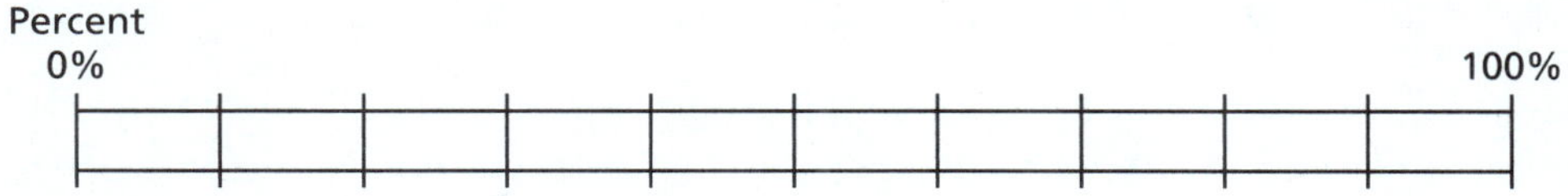

FINDING AN ORIGINAL PRICE **Find the original price. Use a percent model to check your answer.**

6. A bracelet costs $36 after a 25% discount.

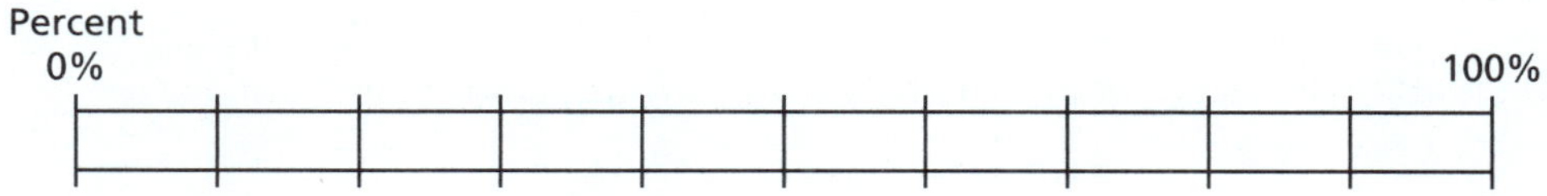

7. The discount on a toy robot is 40%. The toy robot is on sale for $54.

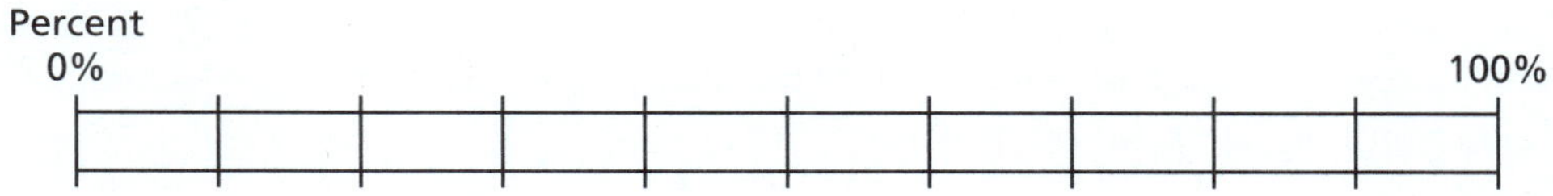

Example 3 Modeling Real Life

7 MTR

A store pays $170 for a bicycle. What is the selling price when the markup is 20%?

Method 1: First, find the markup. The markup is 20% of $170.

$$a = p\% \cdot w$$
$$= 0.20 \cdot 170$$
$$= 34$$

Next, find the selling price.

$$\text{Selling price} = \text{Cost to store} + \text{Markup}$$
$$= 170 + 34$$
$$= 204$$

▶ So, the selling price is $204.

Method 2: Use a ratio table. The selling price is 120% of the cost to the store.

Percent	Dollars
100%	$170
20%	$34
120%	$204

÷ 5 (100% to 20%; $170 to $34), × 6 (20% to 120%; $34 to $204)

▶ So, the selling price is $204.

Check

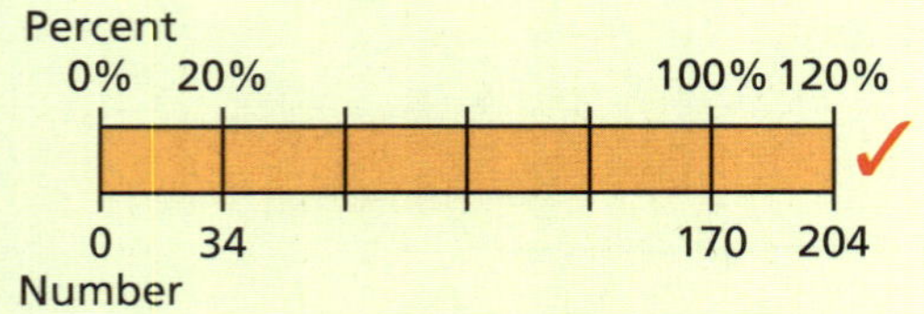

In-Class Practice

1 I don't understand yet. 2 I can do it with help. 3 I can do it on my own. 4 I can teach someone else.

8. You have two coupons for a store. The first coupon applies a $15 discount to a single purchase, and the second coupon applies a 10% discount to a single purchase. You can use only one coupon on a purchase. When should you use each coupon? Explain.

9. A store sells memory cards for $25 each.

a. The markup for each memory card is 25%. How much did the store pay for 50 memory cards?

b. The store offers a discount when a customer buys two or more memory cards. A customer pays $47.50 for two memory cards. What is the percent of discount?

c. How much does a customer pay for three memory cards if the store increases the percent of discount in part (b) by 2%?

GO DIGITAL

4 MTR **27.** **YOU BE THE TEACHER** A store pays \$60 for an item. Your friend finds the selling price when the markup is 20%. Is your friend correct? Explain your reasoning.

> $0.2(\$60) = \$12$
>
> So, the selling price is \$12.

5 MTR **28.** **STRUCTURE** The scooter is being sold at a 10% discount. The original price is shown. Which methods can you use to find the new sale price? Which method do you prefer? Explain.

- Multiply \$42 by 0.9.
- Multiply \$42 by 0.1, then subtract from \$42.
- Multiply \$42 by 0.9, then add to \$42.
- Multiply \$42 by 0.9, then subtract from \$42.

29. **NUMBER SENSE** The original price of an item is p dollars. Is the price of the item with an 18% markup the same as multiplying the original price by 1.18? Use two expressions to justify your answer.

30. **PROBLEM SOLVING** You are shopping for a video game system.

a. At which store should you buy the system?

Store	Cost to Store	Markup
A	\$162	40%
B	\$155	30%
C	\$160	25%

b. Store A has a weekend sale. What discount must Store A offer for you to buy the system there?

31. **Dig Deeper** A pool manager balances the pH level of a pool. The price of a bucket of chlorine tablets is $90, and the price of a pH test kit is $11. The manager uses a coupon that applies a 40% discount to the total cost of the two items. How much money does the pool manager pay for each item?

3 MTR

32. **MAINTAIN ACCURACY** You buy a pair of jeans at a department store.

a. What is the percent of discount to the nearest percent?

b. What is the percent of sales tax to the nearest tenth of a percent?

c. The price of the jeans includes a 60% markup. After the discount, what is the percent of markup to the nearest percent?

33. **PROBLEM SOLVING** You buy a bicycle helmet for $22.26, which includes 6% sales tax. The helmet is discounted 30% off the selling price. What is the original price?

34. **REASONING** A drone that costs $129.50 is discounted 40%. The next month, the sale price is discounted an additional 60%. Is the drone now “free”? If so, explain. If not, find the sale price.

9.8 Simple Interest

Learning Target: Understand and apply the simple interest formula.

Success Criteria:
- I can explain the meaning of simple interest.
- I can use the simple interest formula to solve problems.

Algebraic Reasoning
MA.7.AR.3.1 Apply previous understanding of percentages and ratios to solve multi-step real-world percent problems.

Exploration 1 Understanding Simple Interest

Work with a partner. You deposit $150 in an account that earns 6% *simple interest per year*. You do not make any other deposits or withdrawals. The table shows the balance of the account at the end of each year.

Years	0	1	2	3	4	5	6
Balance	$150	$159	$168	$177	$186	$195	$204

a. Describe any patterns you see in the account balance.

b. How is the amount of interest determined each year?

c. How can you find the amount of simple interest earned when you are given an initial amount, an interest rate, and a period of time?

d. You deposit $150 in a different account that earns simple interest. The table shows the balance of the account each year. What is the interest rate of the account? What is the balance after 10 years?

Years	0	1	2	3
Balance	$150	$165	$180	$195

5 MTR USE A PATTERN
How does the pattern in the balances help you find the simple interest rate?

GO DIGITAL

9.8 Lesson

Key Vocabulary
interest, *p. 750*
principal, *p. 750*
simple interest, *p. 750*

Interest is money paid or earned for using or lending money. The **principal** is the amount of money borrowed or deposited.

Key Idea

Simple Interest

Words **Simple interest** is money paid or earned only on the principal.

Algebra $I = Prt$

- Simple interest → I
- Principal → P
- Annual interest rate (in decimal form) → r
- Time (in years) → t

Reading
An interest rate per year is also called an annual interest rate.

Example 1 Finding a Balance

You deposit \$500 in a savings account. The account earns 3% simple interest per year. What is the balance after 3 years?

To find the balance, calculate the interest and add it to the principal.

$I = Prt$	Write the simple interest formula.
$= 500(0.03)(3)$	Substitute 500 for P, 0.03 for r, and 3 for t.
$= 45$	Multiply.

The interest earned is \$45 after 3 years.

▶ So, the balance is \$500 + \$45 = \$545 after 3 years.

2 MTR MAKE A CONNECTION

Write a formula that you can use to find the total balance B of an account. Explain your reasoning.

____

____

____

____

Try It

1. What is the balance of the account after 9 months?

GO DIGITAL

Example 2 Finding an Annual Interest Rate

You deposit $1000 in an account. The account earns $100 simple interest in 4 years. What is the annual interest rate?

$I = Prt$	Write the simple interest formula.
$100 = 1000(r)(4)$	Substitute 100 for *I*, 1000 for *P*, and 4 for *t*.
$100 = 4000r$	Simplify.
$0.025 = r$	Divide each side by 4000.

▶ So, the annual interest rate of the account is 0.025, or 2.5%.

Try It

2. You deposit $350 in an account. The account earns $17.50 simple interest in 2.5 years. What is the annual interest rate?

Example 3 Finding an Amount of Time

A bank offers three kinds of savings accounts. The simple annual interest rate is determined by the principal. How long does it take an account with a principal of $800 to earn $100 in interest?

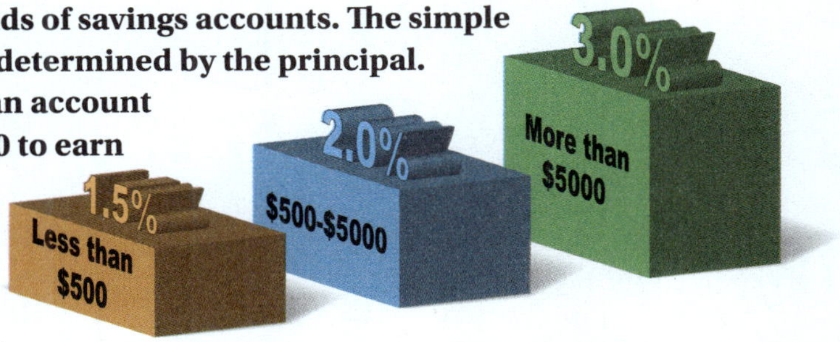

The diagram shows that the interest rate for a principal of $800 is 2%.

$I = Prt$	Write the simple interest formula.
$100 = 800(0.02)(t)$	Substitute 100 for *I*, 800 for *P*, and 0.02 for *r*.
$100 = 16t$	Simplify.
$6.25 = t$	Divide each side by 16.

▶ So, the account earns $100 in interest in 6.25 years.

GO DIGITAL

Try It

3. In Example 3, how long does it take an account with a principal of $10,000 to earn $750 in interest?

In-Class Practice

1 I don't understand yet. 2 I can do it with help. 3 I can do it on my own. 4 I can teach someone else.

4. **VOCABULARY** Explain the meaning of simple interest.

USING THE SIMPLE INTEREST FORMULA Use the simple interest formula.

5. You deposit $20 in a savings account. The account earns 4% simple interest per year. What is the balance after 4 years?

6. You deposit $800 in an account. The account earns $360 simple interest in 3 years. What is the annual interest rate?

7. You deposit $650 in a savings account. How long does it take an account with an annual interest rate of 5% to earn $178.75 in interest?

Example 4 Modeling Real Life

7 MTR

You borrow \$600 to buy a violin. The simple annual interest rate is 15%. You pay off the loan after 2 years of equal monthly payments. How much is each payment?

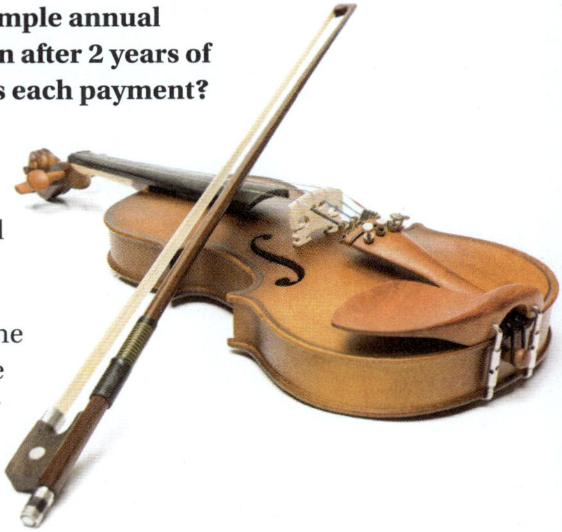

Understand the problem.

You are given the amount and simple annual interest rate of a loan that you pay back in 2 years. You are asked to find the monthly payment.

Make a plan.

Use the simple interest formula to find the interest you pay on the loan. Then divide the total amount you pay by the number of months in 2 years.

Solve and check.

$I = Prt$	Write the simple interest formula.
$= 600(0.15)(2)$	Substitute 600 for P, 0.15 for r, and 2 for t.
$= 180$	Multiply.

You pay \$600 + \$180 = \$780 for the loan.

▶ So, each monthly payment is $\frac{780}{24}$ = \$32.50.

Look Back When you substitute 600 for P and 0.15 for r, you obtain $I = 90t$. This indicates that you pay \$90 in interest each year. So, in 2 years you pay 2(90) = \$180 in interest. ✓

GO DIGITAL

In-Class Practice

1 I don't understand yet. | 2 I can do it with help. | 3 I can do it on my own. | 4 I can teach someone else.

8. You want to deposit \$1000 in a savings account for 3 years. One bank adds a \$100 bonus to your principal and offers a 2% simple annual interest rate. Another bank does not add a bonus, but offers 6% simple interest per year. Which bank should you choose? Explain.

9. Your cousin borrows \$1125 to repair her car. The simple annual interest rate is 10%. She makes equal monthly payments of \$25. How many years will it take to pay off the loan?

10. **Dig Deeper** You borrow \$900 to buy a laptop. You plan to pay off the loan after 5 years of equal monthly payments. After 10 payments, you have \$1200 left to pay. What is the simple annual interest rate of your loan?

23. FINDING AN ACCOUNT BALANCE A savings account earns 5% simple interest per year. The principal is \$1200. What is the balance after 4 years?

24. FINDING AN ANNUAL INTEREST RATE You deposit \$400 in an account. The account earns \$18 simple interest in 9 months. What is the annual interest rate?

25. FINDING AN AMOUNT OF TIME You deposit \$3000 in a CD (certificate of deposit) that earns 5.6% simple annual interest. How long will it take to earn \$336 in interest?

FINDING AN AMOUNT PAID **Find the amount paid for the loan.**

26. \$1500 at 9% for 2 years

27. \$2000 at 12% for 3 years

28. \$2400 at 10.5% for 5 years

29. \$4800 at 9.9% for 4 years

USING THE SIMPLE INTEREST FORMULA **Complete the table.**

	Principal	Annual Interest Rate	Time	Simple Interest
30.	\$12,000	4.25%	5 years	
31.		6.5%	18 months	\$828.75
32.	\$15,500	8.75%		\$5425.00
33.	\$18,000		54 months	\$4252.50

7 MTR

34. MODELING REAL LIFE A family borrows money for a rain forest tour. The simple annual interest rate is 12%. The loan is paid after 3 months. What is the total amount paid for the tour? (See Example 4.)

7 MTR

35. **MODELING REAL LIFE** You deposit \$5000 in an account earning 7.5% simple interest per year. How long will it take for the balance of the account to be \$6500?

7 MTR

36. **MODELING REAL LIFE** You borrow \$1300 to buy a telescope. What is the monthly payment?

37. **REASONING** How many years will it take for \$2000 to double at a simple annual interest rate of 8%? Explain how you found your answer.

38. **Dig Deeper** You take out two loans. After 2 years, the total interest for the loans is \$138. On the first loan, you pay 7.5% simple annual interest on a principal of \$800. On the second loan, you pay 3% simple annual interest. What is the principal for the second loan?

39. **PROBLEM SOLVING** You deposit \$500 in an account that earns 4% simple annual interest. The interest earned each year is added to the principal to create a new principal. Find the total amount in your account after each year for 3 years.

40. **NUMBER SENSE** An account earns r% simple interest per year. Does doubling the initial principal have the same effect on the total interest earned as doubling the amount of time? Justify your answer.

9 Connecting Concepts

5 MTR *Mathematicians who use patterns and structure to help understand and connect mathematical concepts relate previously learned concepts to new concepts.*

1 MTR 1. **ANALYZE A PROBLEM** The table shows the percent of successful shots for each team in a hockey game. A total of 55 shots are taken in the game. The ratio of shots taken by the Blazers to shots taken by the Hawks is 6 : 5. How many goals does each team score?

Team	Percent of Successful Shots
Blazers	10%
Hawks	16%

Using the Problem-Solving Plan

Understand the problem.

You know that 55 shots are taken in a hockey game and that the Blazers take 6 shots for every 5 shots taken by the Hawks. You also know the percent of successful shots for each team.

Make a plan.

Use a ratio table to determine the number of shots taken by each team. Then use the percent equation to determine the number of successful shots for each team.

Solve and check.

Use the plan to solve the problem. Then check your solution.

5 MTR 2. **STRUCTURE** Fill in the blanks with positive numbers so that the sum of the fractions is 37.5% of the first fraction. Justify your answer.

$$\frac{\square}{5} + \left(-\frac{\square}{4}\right)$$

3. The graph shows the distance traveled by a motorcycle on a dirt road. After turning onto a paved road, the motorcycle travels $\frac{1}{5}$ mile every $\frac{1}{4}$ minute. Find the percent of change in the speed of the motorcycle. Round to the nearest tenth of a percent if necessary.

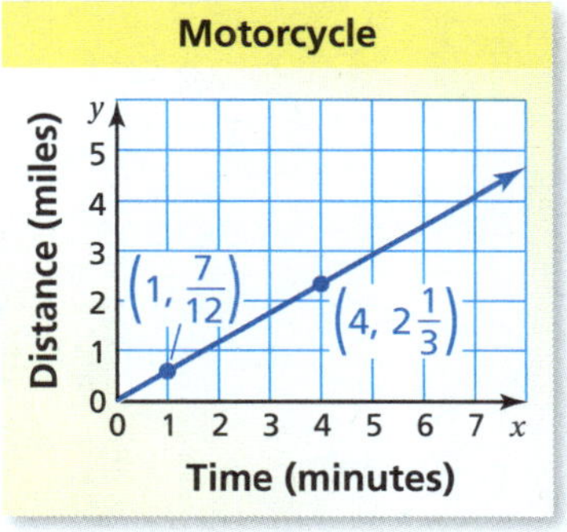

STEAM Performance Task

Genetic Ancestry

GO DIGITAL

At the beginning of this chapter, you watched a STEAM video called "Chargaff's Rules." You are now ready to complete the STEAM Performance Task, where you will analyze the results of a DNA test. Be sure to use mathematical thinking and reasoning, and the problem-solving plan as you work through the performance task.

9 Chapter Review WITH CalcChat®

Review Vocabulary

Write the definition and give an example of each vocabulary term.

percent, *p. 680*

percent error, *p. 733*

interest, *p. 750*

percent of change, *p. 731*

discount, *p. 741*

principal, *p. 750*

percent of increase, *p. 731*

markup, *p. 741*

simple interest, *p. 750*

percent of decrease, *p. 731*

GO DIGITAL

Graphic Organizers

You can use a **Summary Triangle** to explain a concept. Here is an example of a Summary Triangle for ***writing a percent as a decimal***.

Writing a percent as a decimal

Remove the percent symbol. Then divide by 100, which moves the decimal point two places to the left.

$n\% = n \div 100$

Example: $76\% = 0.76$

Choose and complete a graphic organizer to help you study the concept.

1. writing a decimal as a percent

2. comparing and ordering fractions, decimals, and percents

3. the percent proportion

4. the percent equation

5. percent of change

6. discount

7. markup

Chapter Learning Target: Understand percents.

Chapter Success Criteria:
- ◆ I can rewrite fractions, decimals, and percents.
- ◆ I can compare and order fractions, decimals, and percents.
- ■ I can use the percent proportion or percent equation to find a percent, a part, or a whole.
- ■ I can apply percents to solve real-life problems.

◆ Surface ■ Deep

Rate your understanding after each section.

1 I don't understand yet. 2 I can do it with help. 3 I can do it on my own. 4 I can teach someone else.

9.1 Percents and Fractions *(pp. 679–688)*

Learning Target: Write percents as fractions and fractions as percents.

Write the percent as a fraction or mixed number in simplest form.

1. 12% **2.** 88% **3.** 0.8%

4. 127% **5.** 2.5% **6.** 18%

Write the fraction or mixed number as a percent.

7. $\frac{3}{5}$ **8.** $1\frac{18}{25}$ **9.** $1\frac{21}{50}$

10. $\frac{14}{35}$ **11.** $5\frac{7}{25}$ **12.** $\frac{7}{400}$

13. Write a fraction in simplest form that is greater than 43% and less than 47%.

14. Write a percent that is greater than $3\frac{3}{4}$ and less than $3\frac{4}{5}$.

15. Your computer displays the progress of a downloading video. What fraction of the video is downloaded?

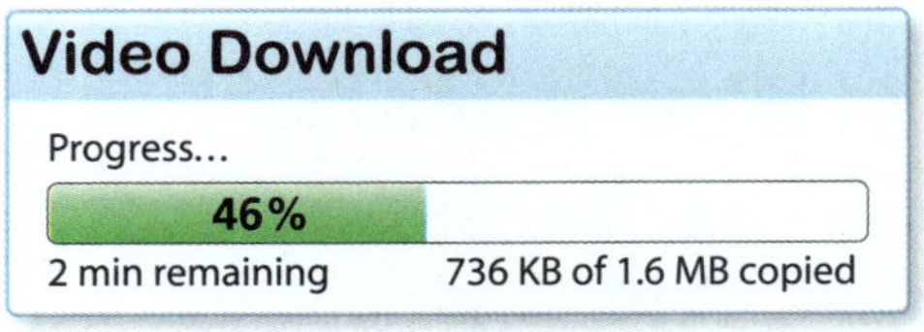

16. You complete 40% of your homework problems before dinner. What fraction of the problems did you complete before dinner?

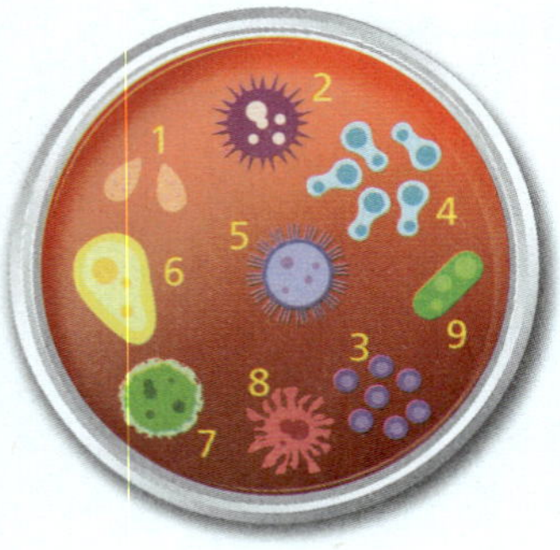

17. There are nine different types of bacteria on the petri dish. What percent of the bacteria on the petri dish is of type 3?

GO DIGITAL

9.2 Percents and Decimals *(pp. 689–698)*

Learning Target: Write percents as decimals and decimals as percents.

Write the percent as a decimal.

18. 76%

19. 6%

20. 17%

21. 0.8%

22. 1.6%

23. 334%

Write the decimal as a percent.

24. 0.15

25. 0.77

26. 0.56

27. 1.06

28. 1.24

29. 0.097

30. Write a decimal that is greater than 6.2% and less than 6.4%.

31. Write a percent that is greater than 0.026 and less than 0.028.

32. On a fishing trip, 38% of the fish that you catch are perch. Write this percent as a decimal.

9.3 Comparing and Ordering Fractions, Decimals, and Percents *(pp. 699–708)*

Learning Target: Compare and order fractions, decimals, and percents.

Complete the statement using <, >, or =.

33. $\frac{1}{2}$ ☐ 52%

34. $\frac{12}{5}$ ☐ 245%

35. 0.46 ☐ 43%

36. 0.023 ☐ 22%

Order the numbers from least to greatest.

37. $\frac{9}{4}$, 220%, 2.15, 218%

38. 0.88, $\frac{7}{8}$, 92%, $\frac{9}{10}$, 0.89

39. Write a percent that is greater than $\frac{13}{25}$ and less than 0.54.

40. The table shows the portions of students in your grade who participate in five activities. List the activities in order by number of students from least to greatest.

Activity	Band	Chorus	Debate	Gymnastics	Theater
Portion of Students	0.14	$\frac{3}{25}$	$\frac{1}{20}$	11%	0.08

9.4 The Percent Proportion *(pp. 709–718)*

Learning Target: Use the percent proportion to find missing quantities.

Write and solve a proportion to answer the question.

41. What percent of 40 is 24?

42. 39 is what percent of 50?

43. What number is 20% of 90?

44. What number is 60% of 30?

45. 70% of what number is 35?

46. 140% of what number is 35?

47. What percent of 60 is 18?

48. What number is 70% of 70?

49. Each cell of a dog contains 78 chromosomes. Exactly 50% of the chromosomes are inherited from the father. How many chromosomes in each cell of the dog are inherited from the father?

50. Of the neighborhoods in a county, the ratio of urban to rural neighborhoods is 5 : 3. Find the percent of rural neighborhoods in the county.

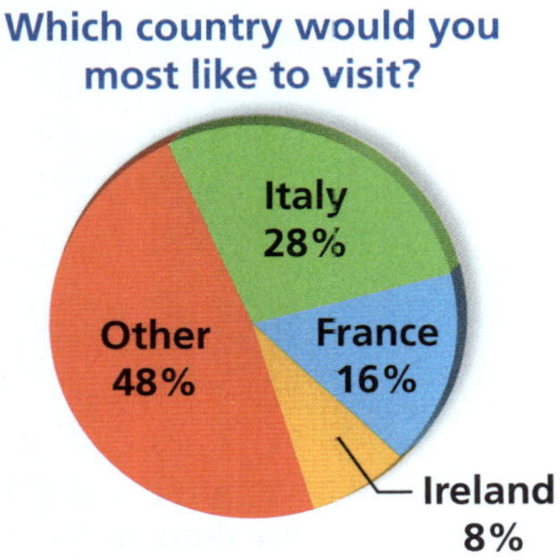

51. The results of a survey are shown at the left. In the survey, 7 students said that they would most like to visit Italy. How many of the students surveyed would most like to visit Ireland?

52. About 29% of Earth's surface is covered by land. The total surface area of Earth is about 510 million square kilometers. What is the area of Earth's surface covered by land?

53. You answer 24 questions on a 100-point test correctly and earn a 96%.

a. All the questions are worth the same number of points. How many questions are on the test? How many points is each question worth?

b. Your friend earns a grade of 76% on the same test. How many questions did your friend answer correctly?

9.5 The Percent Equation *(pp. 719–728)*

Learning Target: Use the percent equation to find missing quantities.

Write and solve an equation to answer the question.

54. What number is 24% of 25?

55. 9 is what percent of 20?

56. 60.8 is what percent of 32?

57. 91 is 130% of what number?

58. 85% of what number is 10.2?

59. 83% of 20 is what number?

60. 15% of the parking spaces at a school are accessible parking spaces. The school has 18 accessible parking spaces. How many parking spaces are there in total?

61. Of the 25 students on a field trip to the Johnson History Museum, 16 bring cameras. What percent of the students bring cameras?

The Johnson History Museum in Palm Beach, Florida, is a nonprofit whose mission is to collect, preserve, and share the history of the county.

9.6 Percents of Increase and Decrease (pp. 729–738)

Learning Target: Find percents of change in quantities.

Identify the percent of change as an *increase* or *decrease*. Then find the percent of change. Round to the nearest tenth of a percent if necessary.

62. 6 yards to 36 yards

63. 100 meals to 52 meals

64. You estimate that a jar contains 68 marbles. The actual number of marbles is 64. Find the percent error.

65. The table shows the numbers of skim boarders at a beach on Saturday and Sunday. What was the percent of change in boarders from Saturday to Sunday?

Day	Number of Skim Boarders
Saturday	12
Sunday	9

9.7 Discounts and Markups (pp. 739–748)

Learning Target: Solve percent problems involving discounts and markups.

Find the sale price or original price.

66. Original price: \$50
Discount: 15%
Sale price: ?

67. Original price: ?
Discount: 20%
Sale price: \$75

68. What is the original price of the tennis racquet?

69. A store pays \$50 for a pair of shoes. The markup is 25%.

a. What is the selling price for the shoes?

b. What is the total cost for a person to buy the shoes, including a 6% sales tax?

9.8 Simple Interest (pp. 749–758)

Learning Target: Understand and apply the simple interest formula.

An account earns simple interest. (a) Find the interest earned. (b) Find the balance of the account.

70. \$300 at 4% for 3 years

71. \$2000 at 3.5% for 4 years

Find the annual interest rate.

72. $I = \$17$, $P = \$500$, $t = 2$ years

73. $I = \$426$, $P = \$1200$, $t = 5$ years

Find the amount of time.

74. $I = \$60$, $P = \$400$, $r = 5\%$

75. $I = \$237.90$, $P = \$1525$, $r = 2.6\%$

76. You deposit \$100 in an account. The account earns \$2 simple interest in 6 months. What is the annual interest rate?

77. Bank A is offering a loan with a simple interest rate of 8% for 2 years. Bank B is offering a loan with a simple interest rate of 6.5% for 3 years.

a. Assuming the monthly payments are equal, what is the monthly payment for the four-wheeler from Bank A? from Bank B?

b. Give reasons for why a person might choose Bank A and why a person might choose Bank B for a loan to buy the four-wheeler. Explain your reasoning.

9 Practice Test WITH CalcChat®

Write the percent as a decimal, or the decimal as a percent. Use a model to represent the number.

1. 0.96%

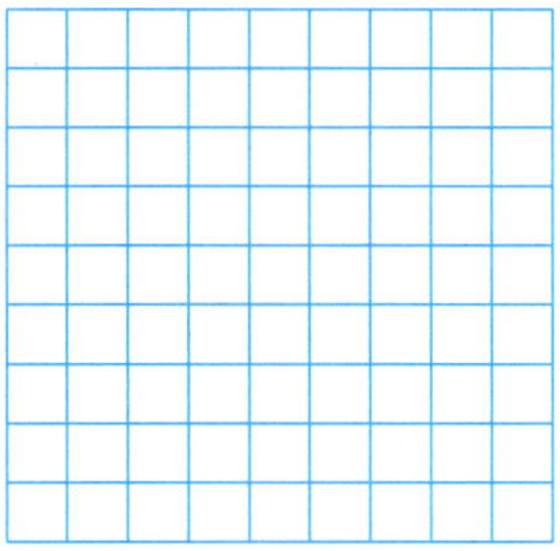

2. 3%

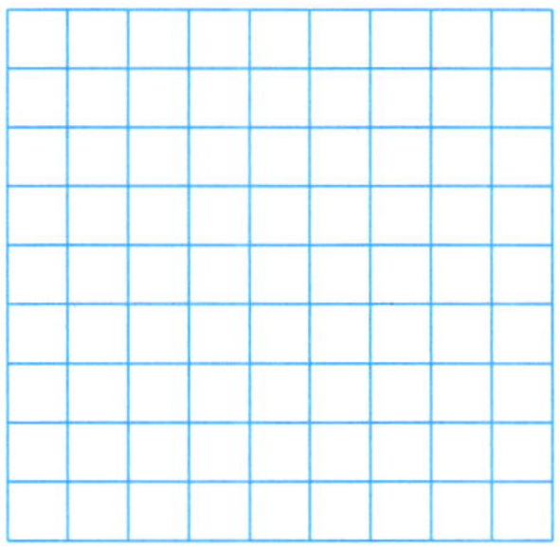

3. 65%

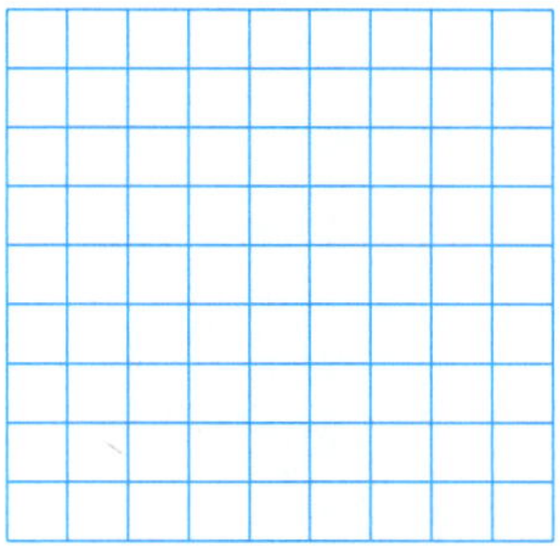

4. 0.42

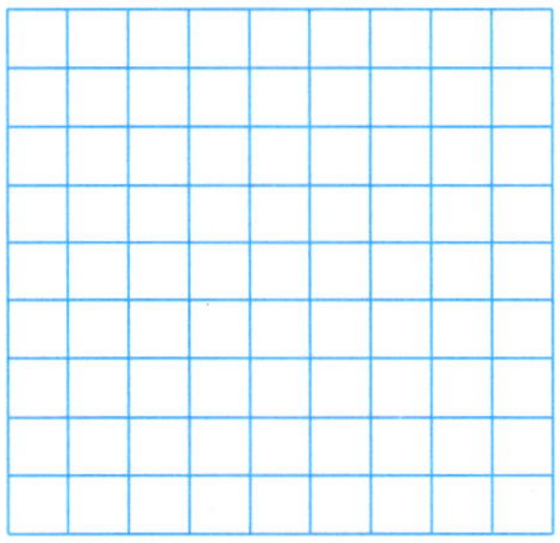

5. 7.88

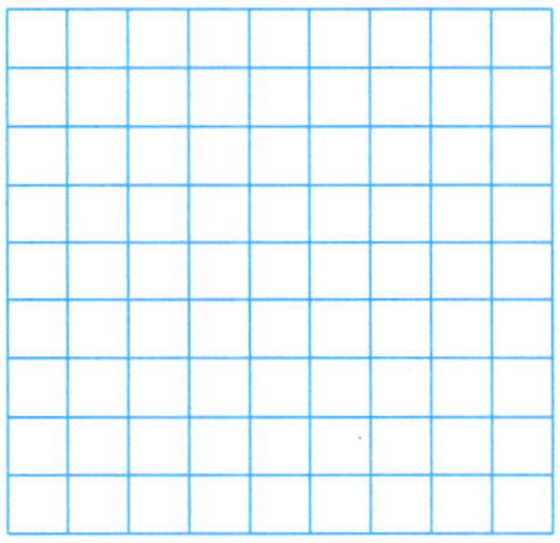

6. 0.58

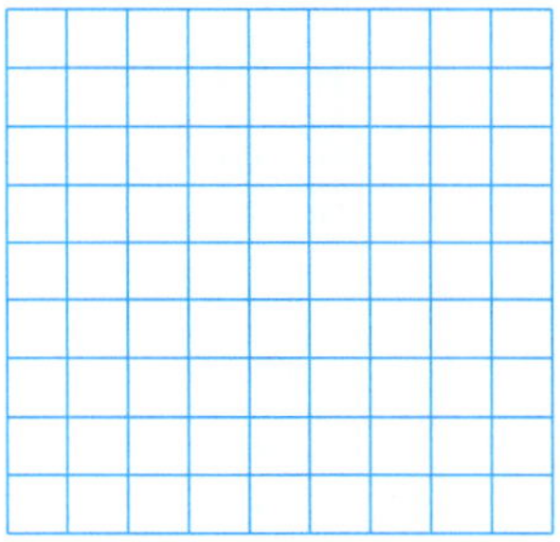

Order the numbers from least to greatest.

7. 85%, $\frac{7}{10}$, 0.74, $\frac{4}{5}$

8. 130%, 1.32, $\frac{6}{5}$, $\frac{5}{4}$, 1.28

Write and solve a proportion or equation to answer the question.

9. What percent of 28 is 21?

10. 64 is what percent of 40?

11. What number is 80% of 45?

12. 0.8% of what number is 6?

Identify the percent of change as an *increase* or *decrease*. Then find the percent of change. Round to the nearest tenth of a percent if necessary.

13. 4 strikeouts to 10 strikeouts

14. $24 to $18

GO DIGITAL

Find the sale price or selling price.

15. Original price: $15
Discount: 5%
Sale price: ?

16. Cost to store: $5.50
Markup: 75%
Selling price: ?

An account earns simple interest. Find the interest earned or the principal.

17. Interest earned: ?
Principal: $450
Interest rate: 6%
Time: 8 years

18. Interest earned: $27
Principal: ?
Interest rate: 1.5%
Time: 2 years

19. The ratio of hours spent at school and not at school on weekdays is 1 : 3. (a) Write the portion of a weekday spent at school as a fraction, a decimal, and a percent. (b) What percent of a week is spent at school if you go to school 4 days that week? Round to the nearest tenth.

20. Research indicates that 90% of the volume of an iceberg is below water. The volume of the iceberg above the water is 160,000 cubic feet. What is the volume of the iceberg below water?

21. You estimate that there are 66 cars in a parking lot. The actual number of cars is 75.

a. Find the percent error.

b. What other estimate gives the same percent error? Explain your reasoning.

9 Review & Refresh WITH CalcChat®
while Building Fluency

Find the absolute value.

1. $|-25|$ 2. $|13|$ 3. $\left|-\frac{2}{5}\right|$

Write the fraction as a decimal and a percent.

4. $\frac{1}{8}$ 5. $\frac{3}{20}$ 6. $\frac{9}{4}$

7. You spend $\frac{1}{8}$ of the hours in a day at soccer practice. How many hours are you *not* at soccer practice that day?

8. During a triathlon, a participant swims 1500 meters, bikes 40 kilometers, and runs 10 kilometers. How many total kilometers does a participant travel during the race?

Use the coordinate plane to answer the following questions.

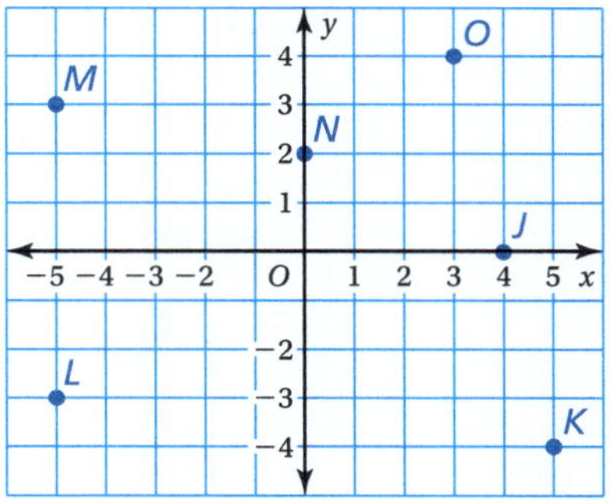

9. Which point corresponds to the ordered pair $(-5, 3)$?

10. What ordered pair corresponds to point L?

11. Which point lies on the x-axis?

12. Which point lies on the y-axis?

GO DIGITAL

Multiply.

13. $3 \cdot 7.8$

14. $-0.6 \cdot 4.2$

15. $15.85 \cdot (-2.2)$

Add or subtract. Write fractions in simplest form.

16. $\frac{1}{5} + \frac{3}{20}$

17. $-\frac{3}{8} + \frac{5}{24}$

18. $\frac{15}{32} - \frac{5}{8}$

19. You are selling T-shirts as a school fundraiser. Any student who raises over $400 during the fundraiser will receive a reward, and each shirt costs $12. Write an inequality that represents the number of T-shirts you must sell to receive the reward.

Find the missing values in the ratio table. Then write the equivalent ratios.

20.

Meters	20	60		220
Minutes	0.6		4.8	

21.

Sugar (cups)	2		8	12
Water (cups)	4.5	13.5		27

9 B.E.S.T. Test Prep WITH CalcChat®
Cumulative Practice

Test-Taking Strategy

Answer Easy Questions First

1. How many pints are in 8 quarts?

Ⓐ 2 pints

Ⓑ 4 pints

Ⓒ 16 pints

Ⓓ 32 pints

2. Which fraction is *not* equivalent to 25%?

Ⓐ $\frac{1}{4}$

Ⓑ $\frac{2}{5}$

Ⓒ $\frac{5}{20}$

Ⓓ $\frac{25}{100}$

3. What is the missing value in the ratio table?

Pairs of Shoes	7		56
Pairs of Boots	2	8	16

4. Which list of numbers is in order from least to greatest?

Ⓐ $0.8, \frac{5}{8}, 70\%, 0.09$

Ⓑ $\frac{5}{8}, 70\%, 0.8, 0.09$

Ⓒ $0.09, \frac{5}{8}, 0.8, 70\%$

Ⓓ $0.09, \frac{5}{8}, 70\%, 0.8$

GO DIGITAL

5. When expressed as a decimal, $\frac{15}{40}$ is [Ⓐ 0.15 Ⓑ 0.375 Ⓒ 0.75 Ⓓ 37.5]. When expressed as a percent, $\frac{15}{40}$ is [Ⓐ 1.5% Ⓑ 15% Ⓒ 3.75% Ⓓ 37.5% Ⓔ 75%].

6. What is 75% of 36?

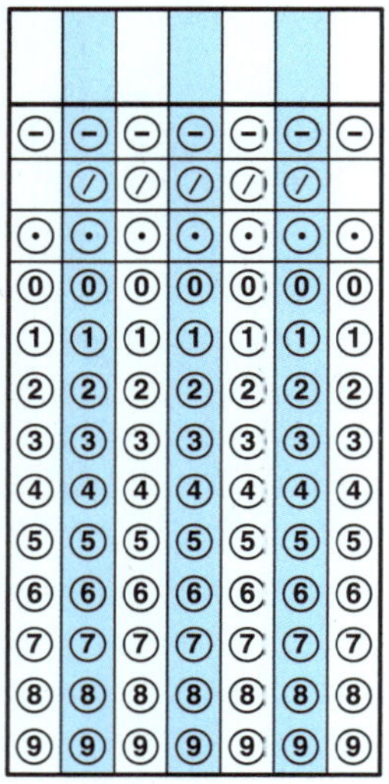

7. Which number is equivalent to $\frac{5}{12} \times \frac{4}{9}$?

Ⓐ $\frac{5}{27}$ Ⓑ $\frac{3}{7}$ Ⓒ $\frac{15}{16}$ Ⓓ $\frac{5}{3}$

8. Find 1.32 ÷ 0.006.

GRIDDED RESPONSE

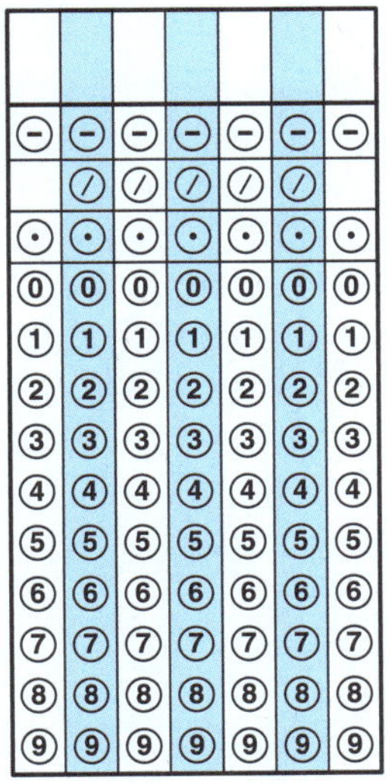

9. Which ratio is equivalent to 4 : 14?

Ⓐ 2 : 12 Ⓑ 10 : 35 Ⓒ 18 : 28 Ⓓ 8 : 18

10. You are comparing the costs of buying bottles of water at the supermarket. Which of the following has the least cost per liter?

Ⓐ 6 one-liter bottles for \$1.80

Ⓑ 1 two-liter bottle for \$0.65

Ⓒ 8 half-liter bottles for \$1.50

Ⓓ 12 half-liter bottles for \$1.98

10

Area, Surface Area, and Volume

Chapter Learning Target:
Understand area and volume of figures.

Chapter Success Criteria:
- ◆ I can find the area of a polygon.
- ◆ I can find the surface area of a solid.
- ■ I can find the volume of a solid.
- ■ I can apply units of measure to solve real-life problems.

◆ Surface ■ Deep

GO DIGITAL

7 MTR *Mathematicians who apply mathematics to real-world contexts connect mathematical concepts to everyday experiences.*

STEAM Video — Packaging Design

In the video, Alex shows Tory how she is designing a special box.

5 MTR

1. **STRUCTURE** Tory says that the length of the design is 47 inches and the width is 16 inches. Show several ways that you can arrange three of the designs on a roll of paper that is 48 inches wide. You can make the length of the paper as long as is needed.

2. Tory says that the cut-out design has an area of 619 square inches. What is the least possible amount of paper that is wasted when you cut out three of the designs?

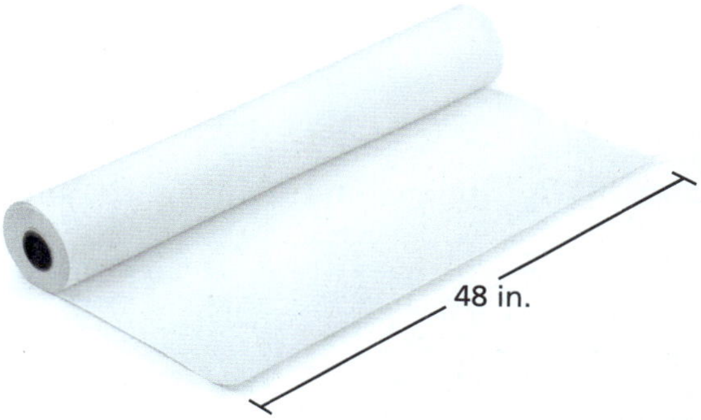

STEAM Performance Task — Maximizing the Volumes of Boxes

After completing this chapter, you will be able to use the concepts you learned to answer the questions in the *STEAM Performance Task*. You will be given the dimensions of a small box and two larger boxes.

Small box: $4\frac{1}{2}$ in. × $4\frac{1}{2}$ in. × 8 in.

Large box 1: $24\frac{1}{4}$ in. × 18 in. × 24 in. **Large box 2:** $20\frac{1}{2}$ in. × $18\frac{1}{2}$ in. × 24 in.

You will be asked to determine how many small boxes can be placed in each larger box.

GO DIGITAL

Getting Ready for Chapter 10 WITH CalcChat®

Chapter Exploration

1 MTR Actively Participate in Effortful Learning Collectively
Work with a partner to prepare for concepts in this chapter.

You can find the area of an unfamiliar figure by considering the areas of figures that compose it. In this chapter, you will use relationships between different figures to reason about their areas.

Rectangle | **Triangles** | **Quadrilaterals**

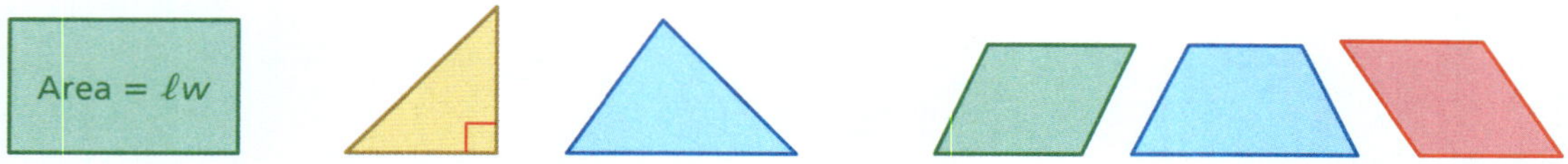

Defined → **Derive formula.** → **Decompose into familiar figures.**

Grades 3, 4, and 5 | *Section 10.1* | *Section 10.2*

Find (a) the dimensions of the figures and (b) the areas of the figures. What do you notice?

1.

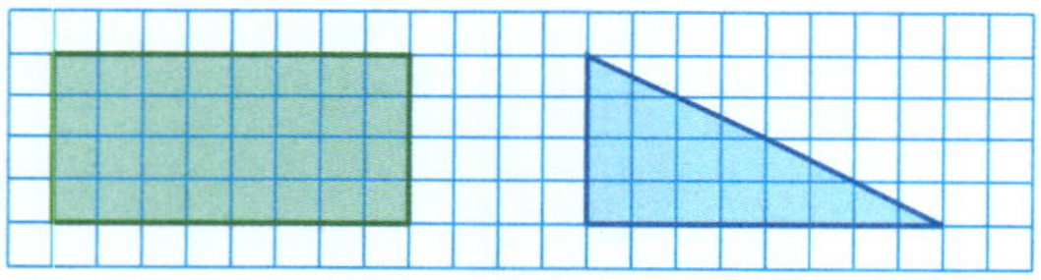

2.

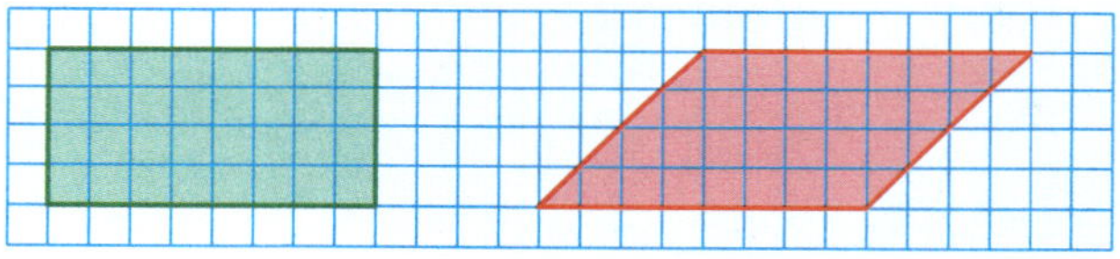

3.

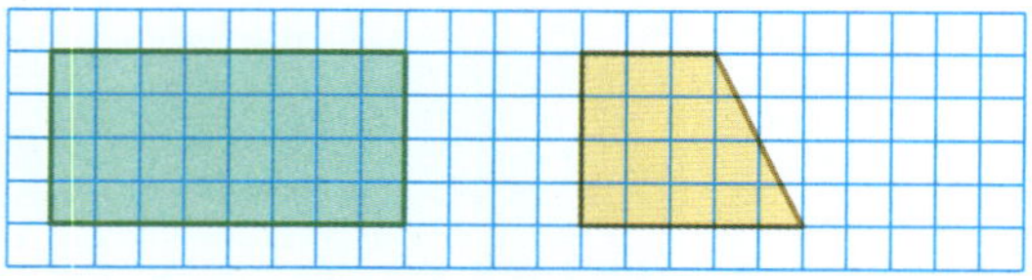

Vocabulary

The following terms are defined in this chapter. Think about what each might mean and record your thoughts.

prism | pyramid | surface area | volume

GO DIGITAL

10.1 Areas of Triangles

Learning Target: Find areas and missing dimensions of triangles.

Success Criteria:
- I can explain how to derive the formula for the area of a triangle.
- I can use a formula to find the area of a triangle.
- I can solve for a missing dimension of a triangle.

Geometric Reasoning
MA.6.GR.2.1 Derive a formula for the area of a right triangle using a rectangle. Apply a formula to find the area of a triangle.

Exploration 1 Deriving the Area Formula of a Triangle

Work with a partner.

a. Find the area of each right triangle created by the diagonal of the rectangle. Explain your method.

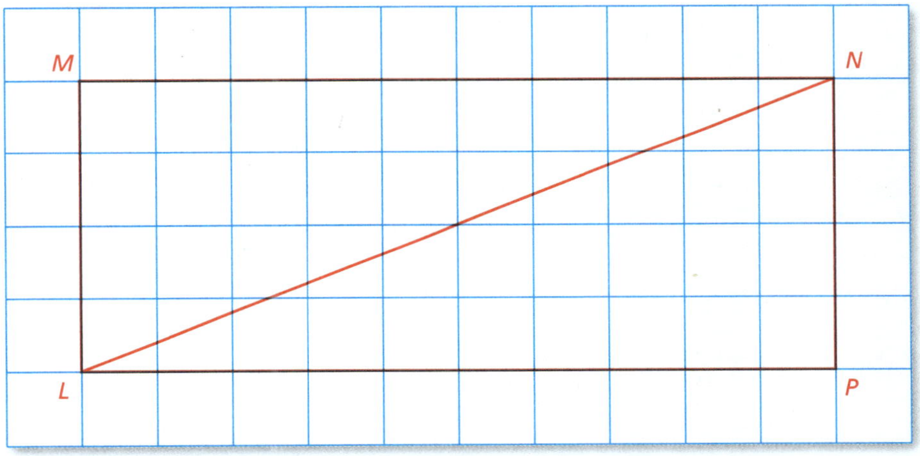

b. Write a formula for the area A of a right triangle with base b and height h. Explain why your formula works for all right triangles.

c. Plot a point Q on the line segment connecting vertices M and N. Find the area of the triangle with vertices L, P, and Q. Explain how you found the area.

GO DIGITAL

d. Draw *any* acute triangle on a piece of centimeter grid paper and find its area. Repeat this process for an obtuse triangle.

e. Do the areas change in parts (c) and (d) when you use different sides as the base? Explain your reasoning.

f. Use your results to write a formula for the area A of any triangle.

3 MTR **SELECT METHODS**

If you use the base and the height to calculate the areas in part (e), how can you estimate the dimensions so that your calculations are accurate?

10.1 Lesson

Key Idea

Area of a Triangle

Words The area A of a triangle is one-half the product of its base b and its height h.

Algebra $A = \frac{1}{2}bh$

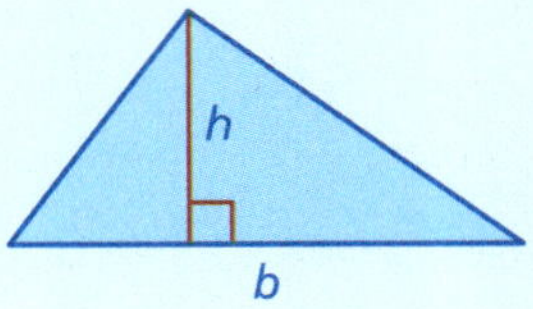

Example 1 Finding Areas of Triangles

Find the area of each triangle.

a.

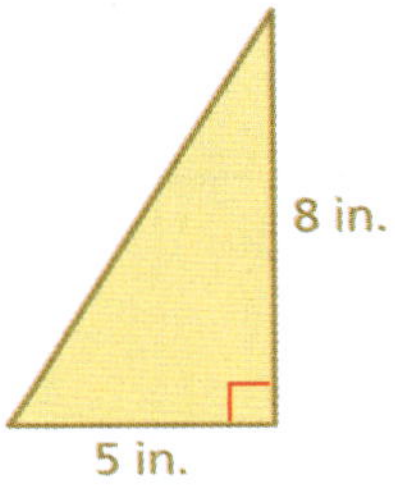

b.

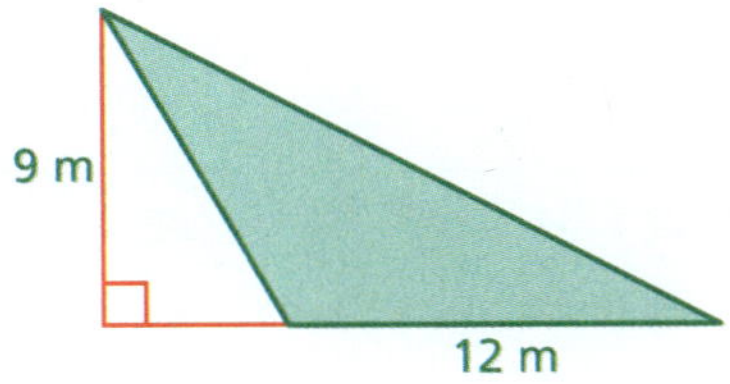

a.		b.
$A = \frac{1}{2}bh$	Write formula.	$A = \frac{1}{2}bh$
$= \frac{1}{2}(5)(8)$	Substitute values.	$= \frac{1}{2}(12)(9)$
$= \frac{1}{2}(40)$	Multiply.	$= 6(9)$
$= 20$	Multiply.	$= 54$

▶ The area of the triangle is 20 square inches.

▶ The area of the triangle is 54 square meters.

The base of a triangle does not have to be horizontal. Any side of a triangle can be the base. In Example 1(a), you could have used 8 inches as the base and 5 inches as the height.

Try It

Find the area of the triangle.

1.

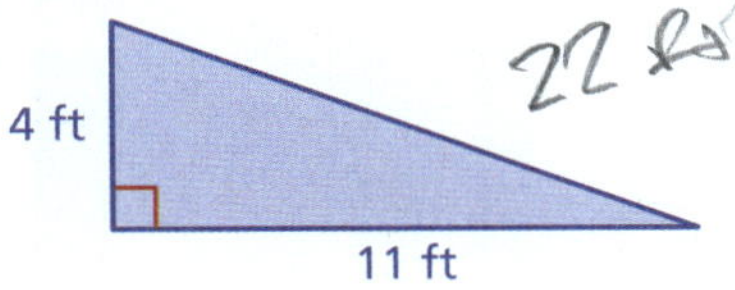

22 ft²

2.

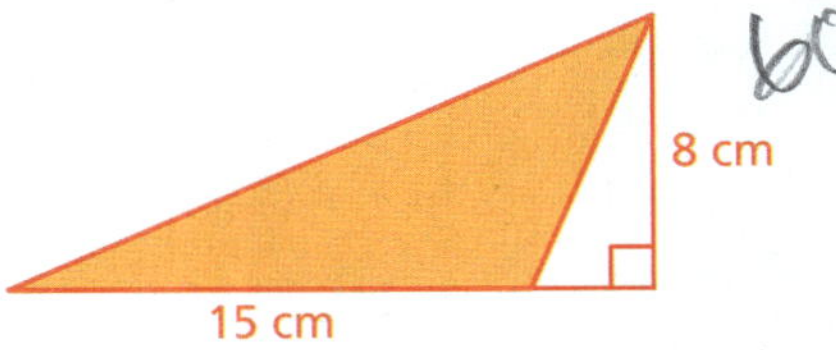

60 cm²

3.

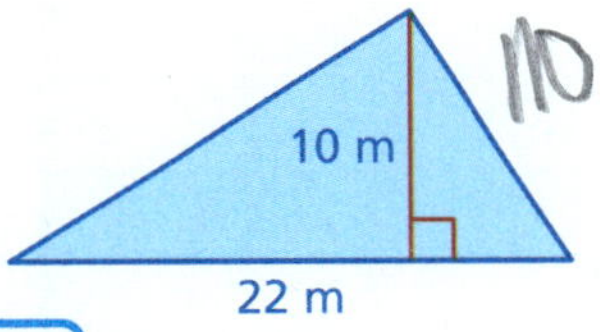

110 m²

4.

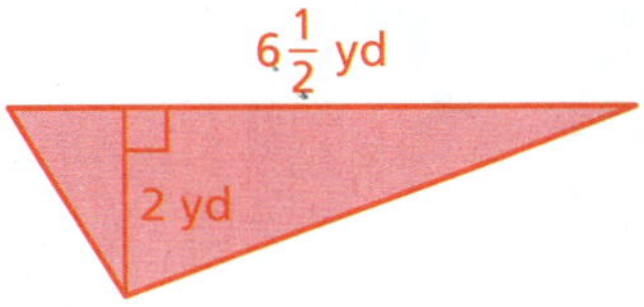

$\frac{13}{2} \cdot \frac{2}{1} = 6.5$ yd²

GO DIGITAL

Example 2 Finding a Missing Dimension

Find the base of the triangle.

Use the formula for the area of a triangle. Substitute for the area and the height, then solve for the base.

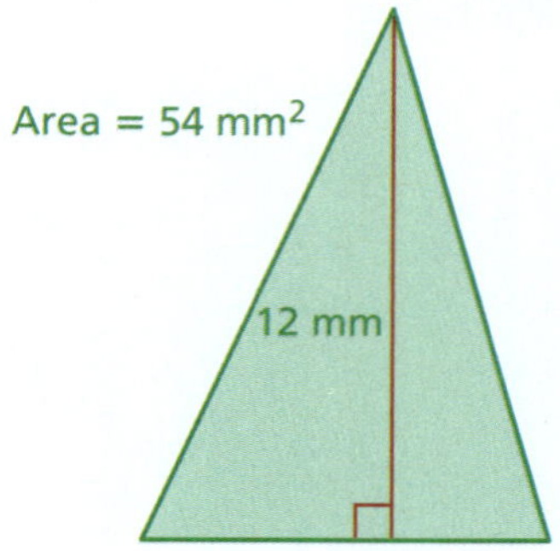

$A = \frac{1}{2}bh$	Write formula for area of a triangle.
$54 = \frac{1}{2}b(12)$	Substitute 54 for *A* and 12 for *h*.
$54 = 6b$	Simplify.
$\frac{54}{6} = \frac{6b}{6}$	Division Property of Equality
$9 = b$	Simplify.

▶ So, the base is 9 millimeters.

4 MTR CONSTRUCT AN ARGUMENT

Write an argument that explains why $\frac{1}{2}b(12)$ and $6b$ are equivalent expressions.

____

____

____

____

____

____

Try It **Find the missing dimension of the triangle.**

5.

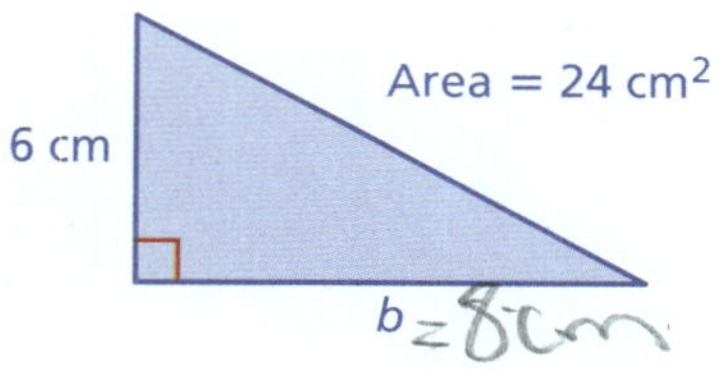

6.

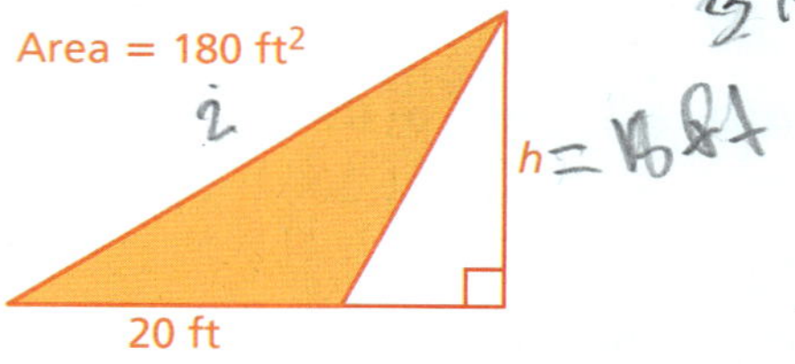

GO DIGITAL

In-Class Practice

1 I don't understand yet. 2 I can do it with help. 3 I can do it on my own. 4 I can teach someone else.

7. FINDING AREA Find the area of the triangle.

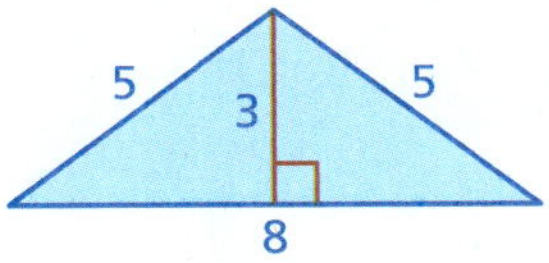

8. WRITING Explain how to use the area of a rectangle to find the area of a triangle.

FINDING A MISSING DIMENSION **Find the missing dimension of the triangle.**

9.

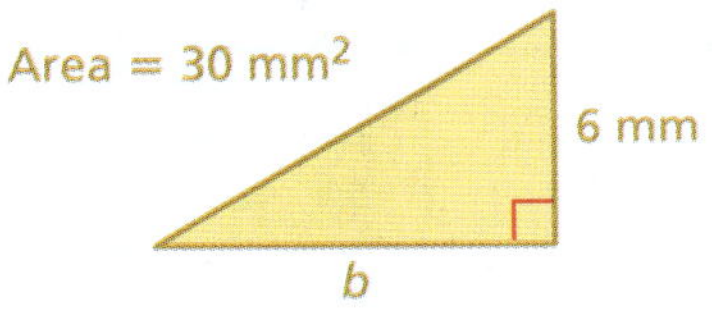

10.

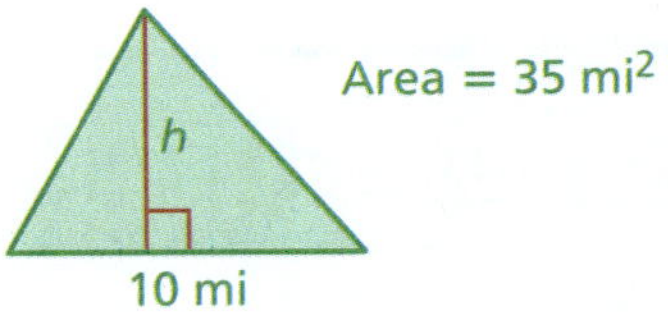

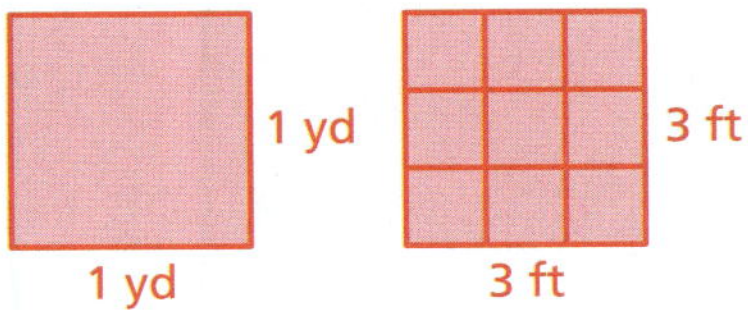

When finding areas, you may need to convert square units. The diagrams at the left show that there are 9 square feet per square yard.

$$1\ \text{yd}^2 = (1\ \text{yd})(1\ \text{yd}) = (3\ \text{ft})(3\ \text{ft}) = 9\ \text{ft}^2$$

You can use a similar procedure to convert other square units.

Example 3 Modeling Real Life

7 MTR

Clearwater Beach is frequently ranked as one of the best beaches in the United States.

You are hang gliding over Clearwater Beach using the hang glider shown. The wingspan of the triangular glider is 36 feet. How many square yards of fabric were used to make the sail?

Find the area of the sail in square feet.

$$A = \frac{1}{2}bh \qquad \text{Write formula for area of a triangle.}$$

$$= \frac{1}{2}(36)(10) \qquad \text{Substitute 36 for } b \text{ and 10 for } h.$$

$$= 18(10) \qquad \text{Multiply } \frac{1}{2} \text{ and 36.}$$

$$= 180 \qquad \text{Multiply.}$$

The area of the sail is 180 square feet. Use a conversion factor to convert to square yards.

$1 \text{ yd}^2 = 9 \text{ ft}^2$

$$180 \text{ ft}^2 = 180 \cancel{\text{ft}^2} \times \frac{1 \text{ yd}^2}{9 \cancel{\text{ft}^2}} = 20 \text{ yd}^2$$

▶ So, 20 square yards of fabric were used to make the sail.

In-Class Practice

1 I don't understand yet. 2 I can do it with help. 3 I can do it on my own. 4 I can teach someone else.

11. A wildlife conservation group buys the 9 square miles of land shown. What is the distance from Point *A* to Point *B*?

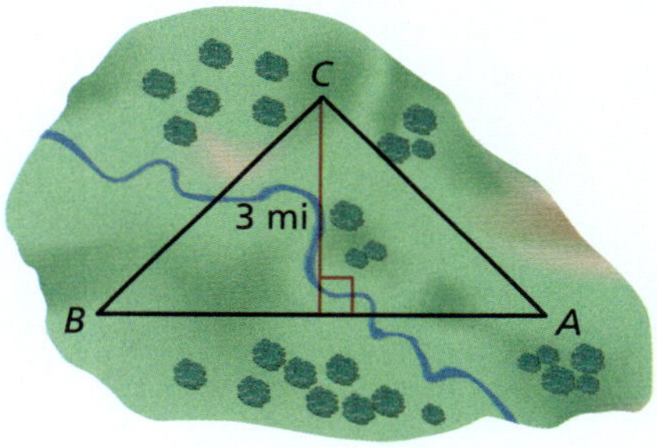

12. **Dig Deeper** You are building a model sailboat. You want the area of the triangular sail to be 150 square inches.

a. Do you need more than 1 square foot of fabric for the sail? Explain.

b. Draw three possible designs for the sail. How can you modify your designs so that the area of the sail is 4 times greater? Justify your answer.

GO DIGITAL

10.1 Practice WITH CalcChat® AND CalcView®

Review & Refresh

An account earns simple interest. Find the interest earned.

1. \$700 at 3% for 4 years

2. \$650 at 2% for 6 years

3. \$480 at 1.5% for 5 years

4. \$1200 at 2.8% for 30 months

Tell which property the statement illustrates.

5. $n \cdot 1 = n$

6. $4 \cdot m = m \cdot 4$

7. $(x + 2) + 5 = x + (2 + 5)$

GO DIGITAL

Concepts, Skills, & Problem Solving

USING TOOLS **Find the area of the triangle by forming a rectangle.** (See Exploration 1.)

8.

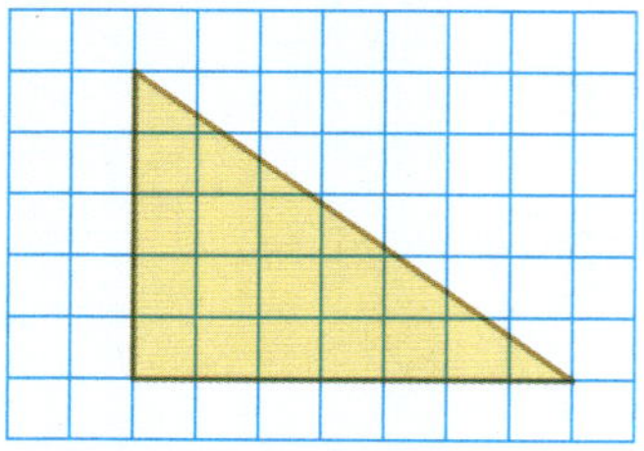

9.

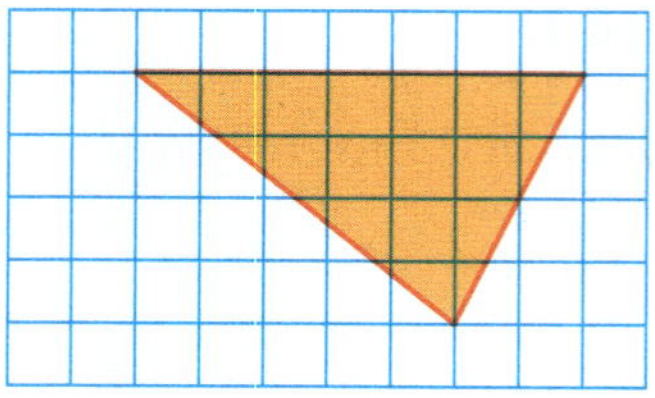

10.

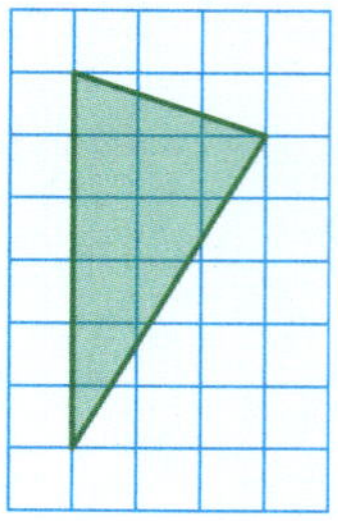

FINDING AREA **Find the area of the triangle.** (See Example 1.)

11.

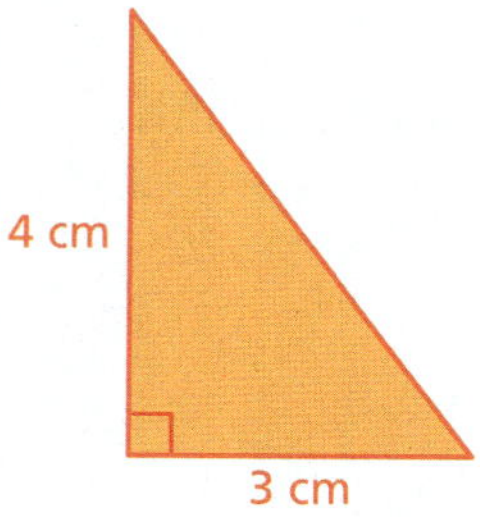

12.

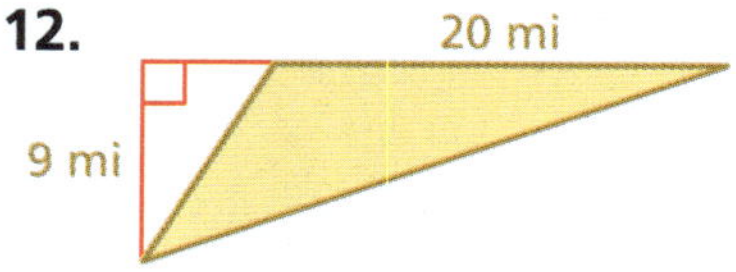

13.

14.

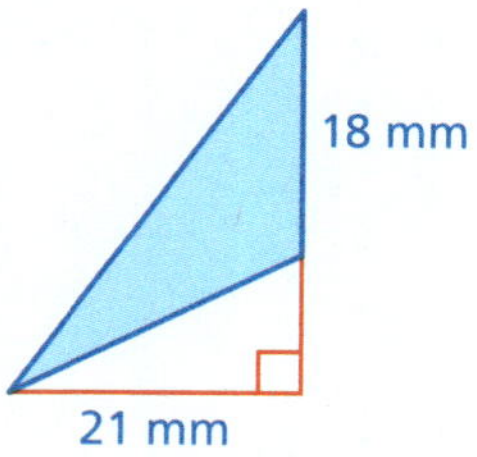

15.

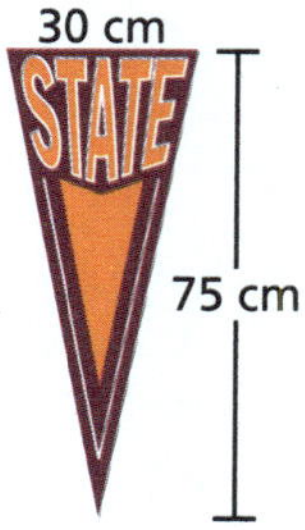

16.

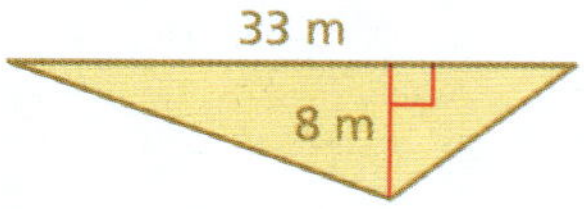

4 MTR

17. YOU BE THE TEACHER Your friend finds the area of the triangle. Is your friend correct? Explain your reasoning.

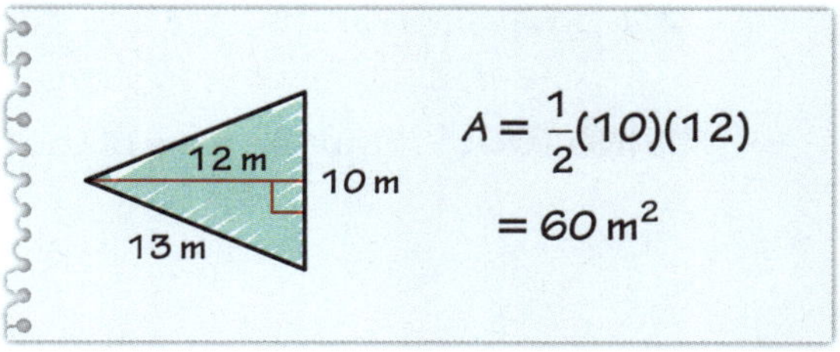

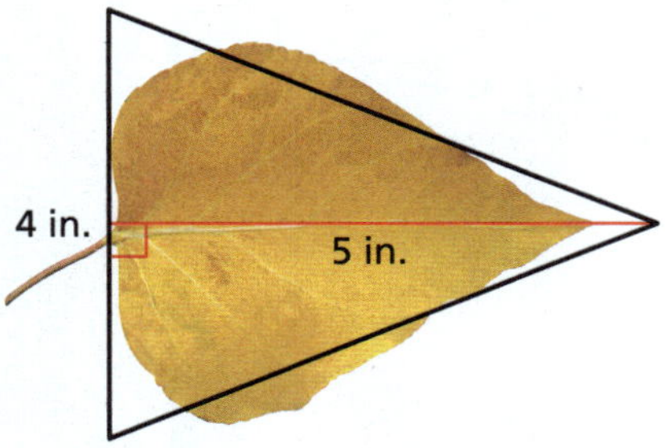

18. B.E.S.T. Test Prep What is the best estimate for the area of the cottonwood leaf?

Ⓐ 5 in. Ⓑ 10 in. Ⓒ 12 in. Ⓓ 20 in.

FINDING A MISSING DIMENSION Find the missing dimension of the triangle. (See Example 2.)

▶**19.** 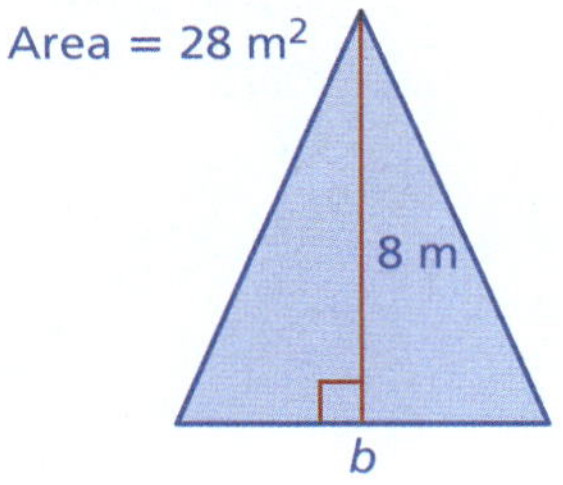

20. 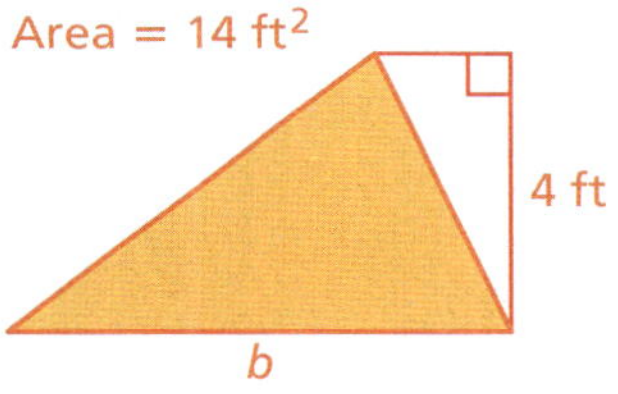

21. 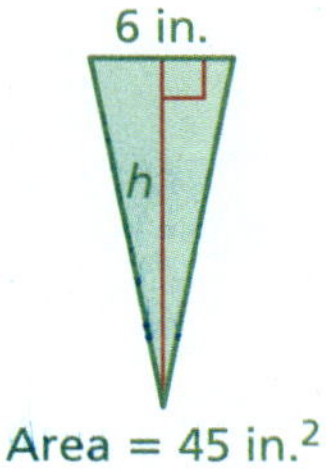

7 MTR

22. MODELING REAL LIFE A shelf has the shape of a triangle. The base of the shelf is 36 centimeters, and the height is 18 centimeters. Find the area of the shelf in square meters. (See Example 3.)

23. **REASONING** The base and the height of Triangle A are one-half the base and one-half the height of Triangle B. How many times greater is the area of Triangle B?

24. **Dig Deeper** There are several basic facial markings on horses. Like fingerprints, no two are precisely the same. A brown horse named Star has a white diamond on its forehead. The diamond is 6 inches tall and 4 inches wide. Is this enough information to determine the area of the diamond? Explain.

25. **OPEN-ENDED** You buy a piece of poster board that is 22 inches wide and 28 inches long to make a sign for a sporting event. You copy the logo shown onto your sign, resizing the logo to be a better fit. Design your poster. Then describe your design process. Estimate the area of the logo on the poster.

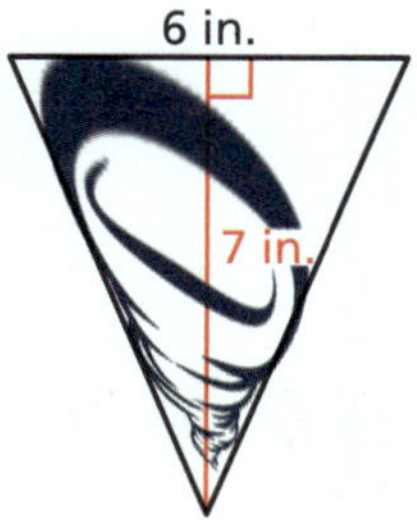

10.2 Areas of Quadrilaterals and Composite Figures

Learning Target: Find areas and missing dimensions of quadrilaterals and composite figures.

Success Criteria:
- I can decompose a figure.
- I can find the area of a quadrilateral.
- I can solve for a missing dimension of a figure.
- I can find the area of a composite figure.

Geometric Reasoning
MA.6.GR.2.2 Solve mathematical and real-world problems involving the area of quadrilaterals and composite figures by decomposing them into triangles or rectangles.

Exploration 1 Finding Areas of Quadrilaterals

Work with a partner.

a. Define each of the following quadrilaterals.

parallelogram trapezoid rhombus

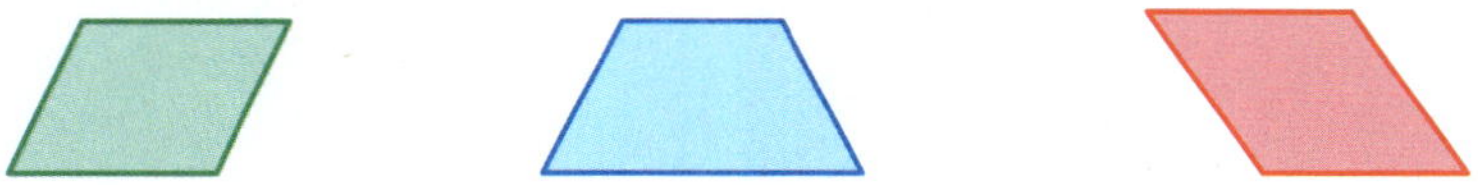

b. Copy the quadrilaterals below on a piece of centimeter grid paper. Find the area of each quadrilateral by separating it into figures with areas you know how to find.

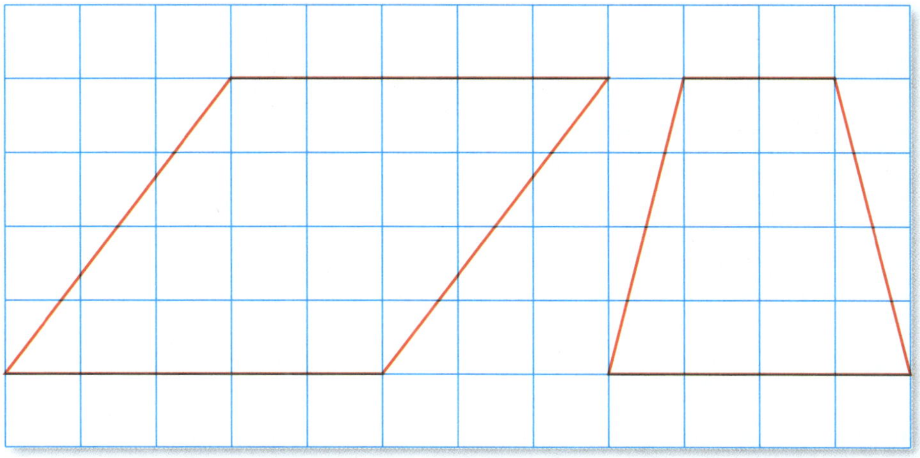

6 MTR ASSESS REASONABLENESS
How can you use the grid to check that your answers are reasonable?

____

____

____

____

GO DIGITAL

c. Do the areas in part (b) change when you *decompose* the quadrilaterals in a different way?

d. Draw *any* type of quadrilateral from part (a) on centimeter grid paper and find its area. Explain your method.

e. How can you find the area of any quadrilateral?

10.2 Lesson

Key Vocabulary
composite figure, *p. 795*

You can find the area of a quadrilateral by separating it into figures with areas you know how to find, such as triangles and rectangles. This is called *decomposition*.

Example 1 Finding Areas of Quadrilaterals

Find the area of each figure.

a.

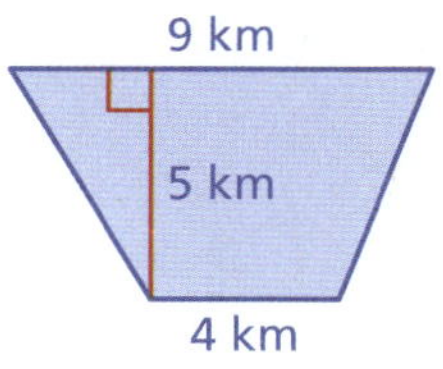

Decompose the trapezoid into a triangle and a rectangle. Find the sum of the areas of the figures.

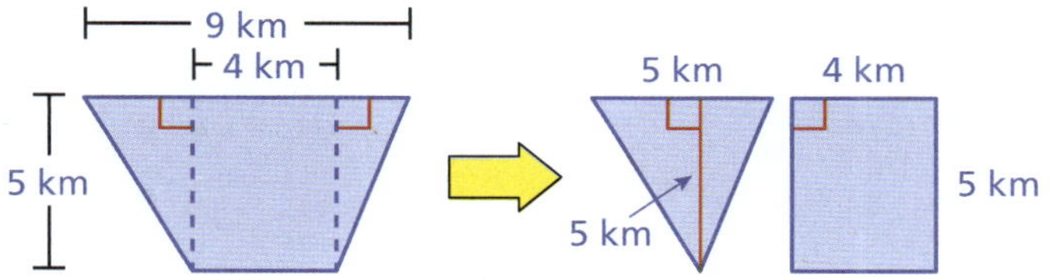

$$A = \frac{1}{2}(5)(5) + 5(4)$$

$$= 12\frac{1}{2} + 20$$

$$= 32\frac{1}{2}$$

▶ The area of the trapezoid is $32\frac{1}{2}$ square kilometers.

b.

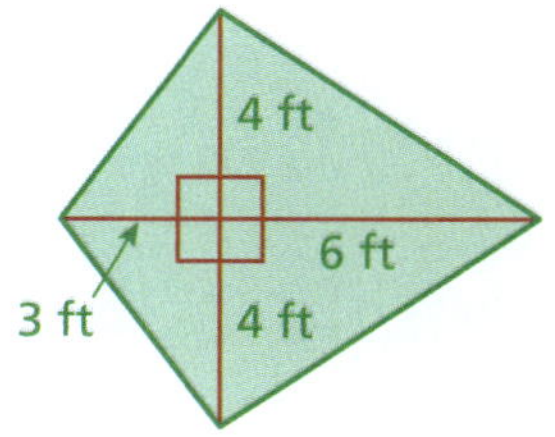

Decompose the figure into two triangles. Find the sum of the areas of the triangles.

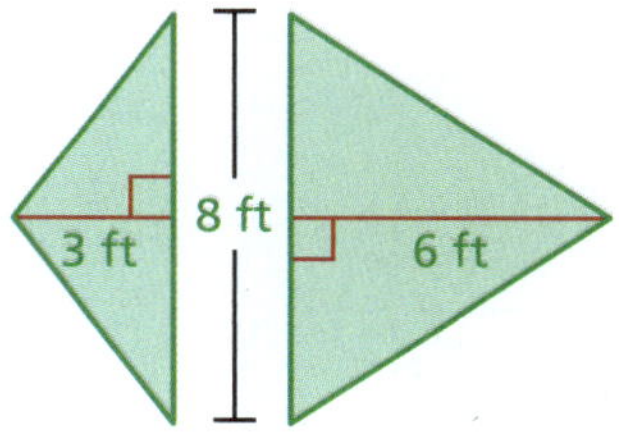

$$A = \frac{1}{2}(8)(3) + \frac{1}{2}(8)(6)$$

$$= 12 + 24$$

$$= 36$$

▶ The area of the figure is 36 square feet.

2 MTR USE ANOTHER METHOD

Can you find the area of the trapezoid in part (a) using only two triangles? Explain.

GO DIGITAL

Try It

Find the area of the figure.

1.

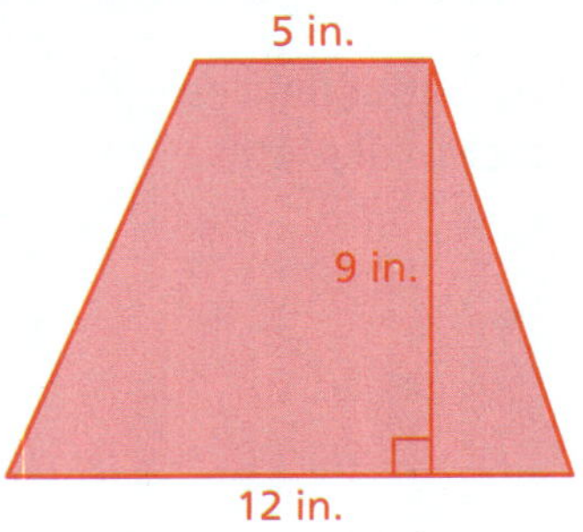

2.

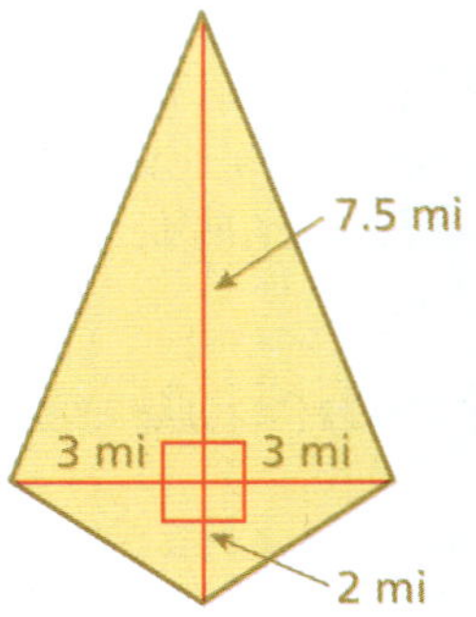

Example 2 B.E.S.T. Test Prep: Finding a Missing Dimension

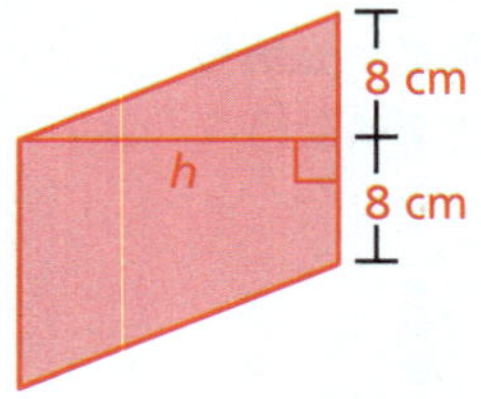

Area = 320 cm^2

What is the height *h* of the parallelogram?

Ⓐ $13\frac{1}{3}$ cm Ⓑ 16 cm Ⓒ 20 cm Ⓓ $26\frac{2}{3}$ cm

Decompose the parallelogram into two identical right triangles and a rectangle. Write expressions that represent the area of each figure.

Area of Rectangle

$A = \ell w$

$= h(8)$

$= 8h$

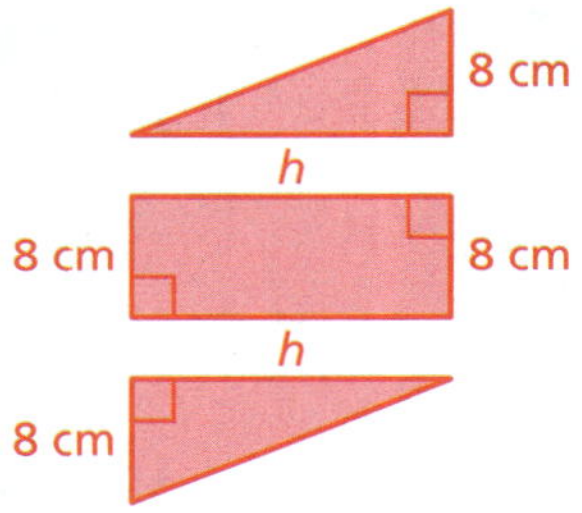

Area of Each Right Triangle

$A = \frac{1}{2}bh$

$= \frac{1}{2}(8)(h)$

$= 4h$

So, the area of the parallelogram can be represented by $A = 8h + 2(4h)$, or $A = 16h$. Substitute the given area, then solve to find h.

$A = 16h$	Write formula for area.
$320 = 16h$	Substitute 320 for A.
$20 = h$	Divide each side by 16.

▶ The height is 20 centimeters. So, the correct answer is Ⓒ.

1 MTR ANALYZE A PROBLEM

How does the height of the parallelogram relate to the rectangle formed by the decomposition?

Try It

3. Find the height h of the trapezoid.

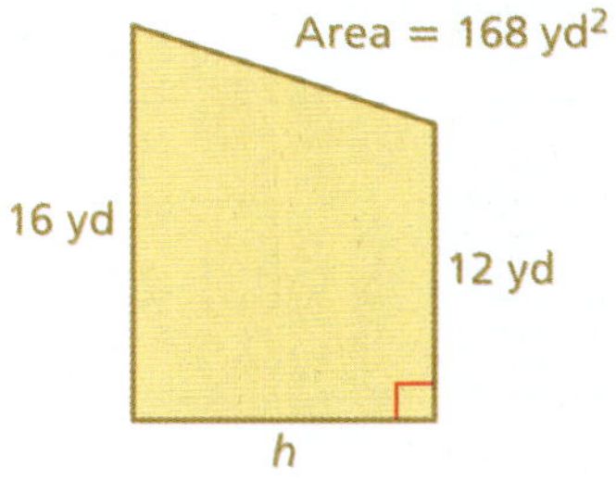

GO DIGITAL

In-Class Practice

1 I don't understand yet. 2 I can do it with help. 3 I can do it on my own. 4 I can teach someone else.

FINDING AREA **Find the area of the figure.**

4.

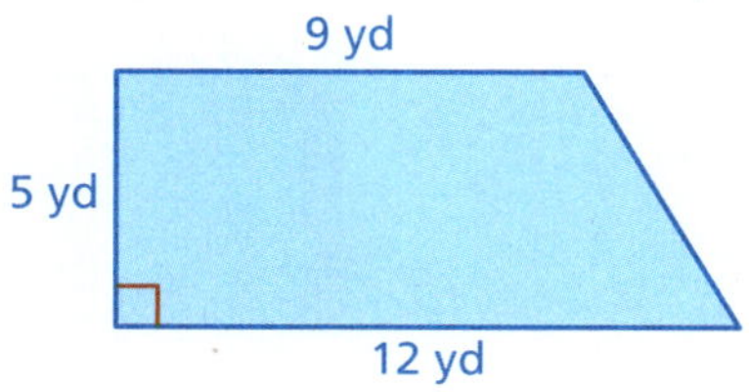

5.

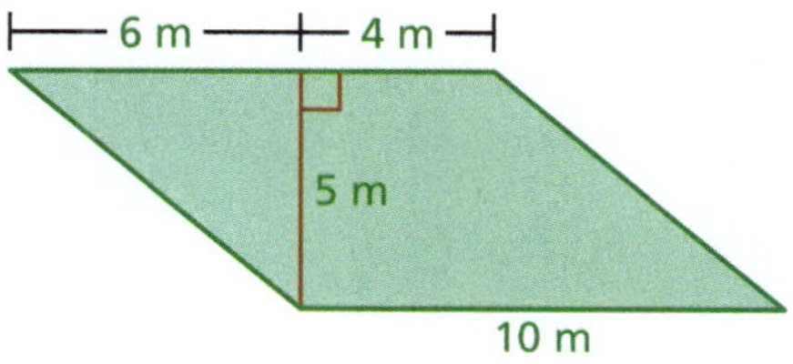

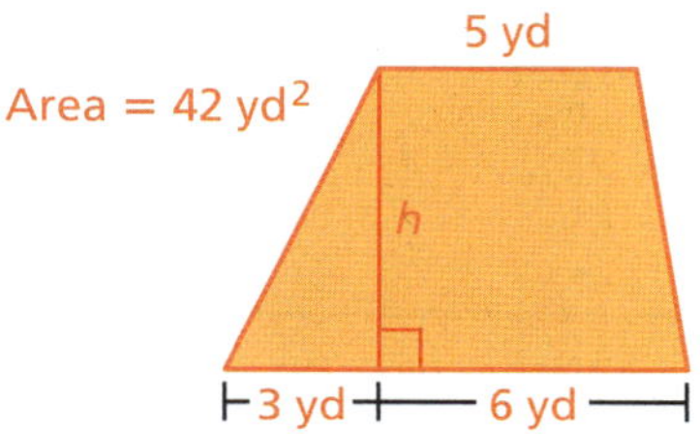

6. **FINDING A MISSING DIMENSION** Find the height h of the trapezoid at the left.

7. **OPEN-ENDED** Draw a trapezoid that has an area of 10 square centimeters. Label the bases and height. Explain your procedure.

A **composite figure** is made up of triangles, squares, rectangles, and other two-dimensional figures. You can use decomposition to find the area of a composite figure.

In-Class Practice

1 I don't understand yet. 2 I can do it with help. 3 I can do it on my own. 4 I can teach someone else.

8. The stained-glass window is made of identical glass panes. The approximate dimensions of one pane are shown. The glass used to make the window costs \$12.50 per square foot. Find the total cost of the glass used to make the window.

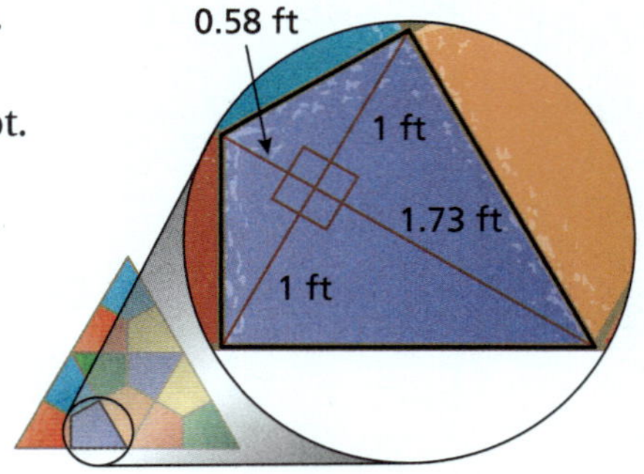

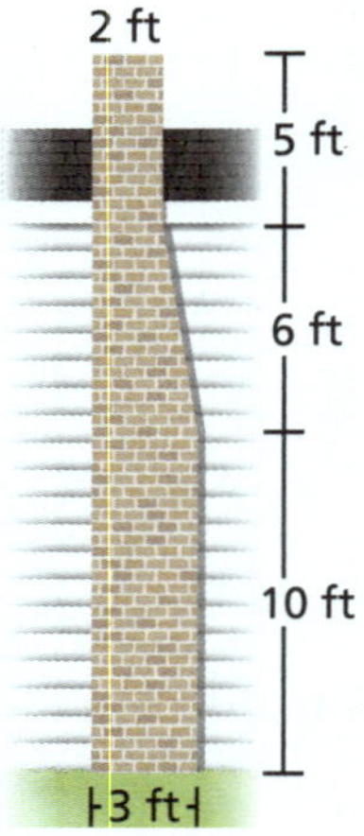

9. Find the area of the side of the chimney. Explain how you found the area.

Example 3 Modeling Real Life 7 MTR

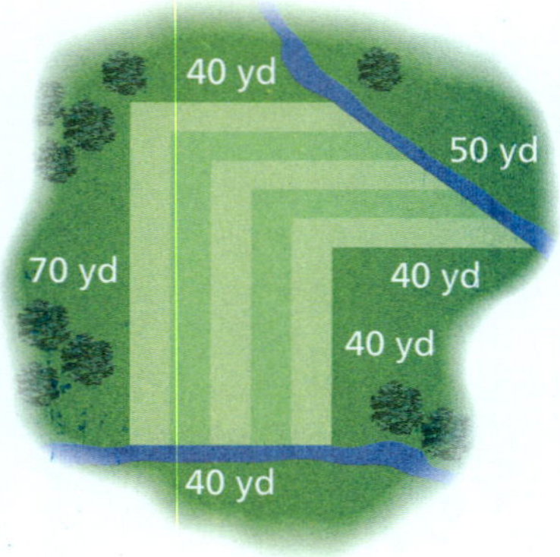

Find the area of the fairway between two streams on a golf course.

There are several ways to use decomposition to find the area of the fairway. One way is to separate the fairway into a rectangle and a right triangle.

Identify each shape and find any missing dimensions. Then find the area of each shape.

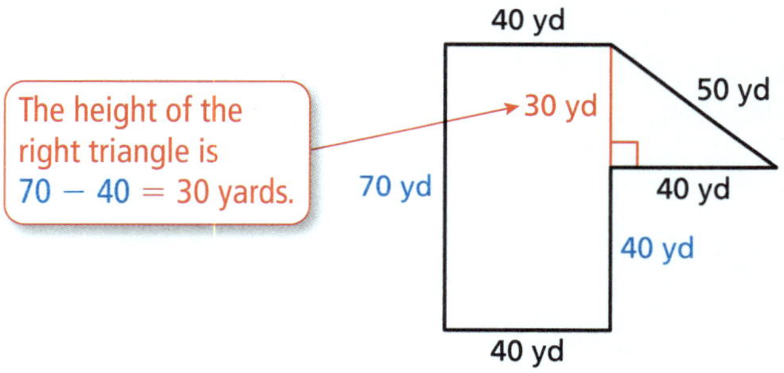

Area of Rectangle	*Area of Right Triangle*
$A = \ell w$	$A = \frac{1}{2}bh$
$= 70(40)$	$= \frac{1}{2}(40)(30)$
$= 2800$	$= 600$

▶ So, the area of the fairway is $2800 + 600 = 3400$ square yards.

GO DIGITAL

10.2 Practice WITH CalcChat® AND CalcView®

Review & Refresh

Find the area of the triangle.

1.

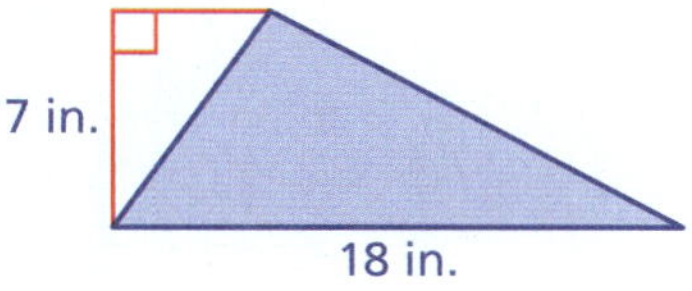

2.

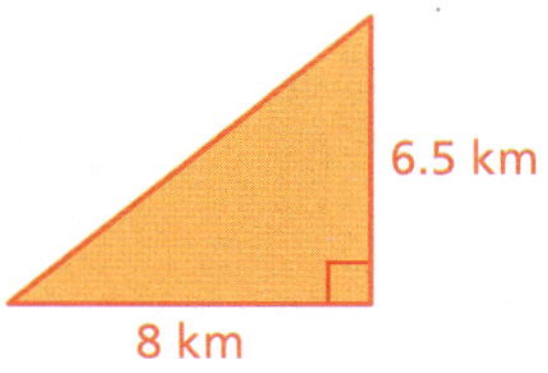

3.

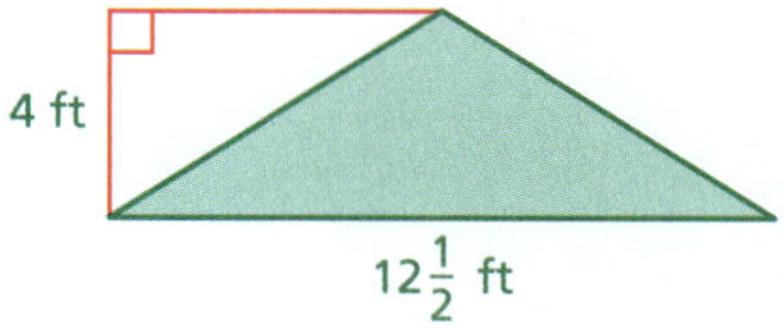

Factor the expression using the GCF.

4. $27 + 81$

5. $42 - 14$

6. $51 + 85$

Order the numbers from least to greatest.

7. $\frac{1}{5}$, 0.23, 22.2%, 0.02, $\frac{11}{50}$

8. 0.77, 7.7%, $\frac{7}{10}$, 0.7%, 0.777

9. 84.4%, $\frac{21}{25}$, 0.86, $\frac{17}{20}$, 0.856

10. 0.999, $\frac{99}{100}$, 9.9%, 0.009, 9%

Concepts, Skills, & Problem Solving

FINDING AREAS OF QUADRILATERALS **Find the area of each figure using two different methods. Explain your methods.** (See Exploration 1.)

11.

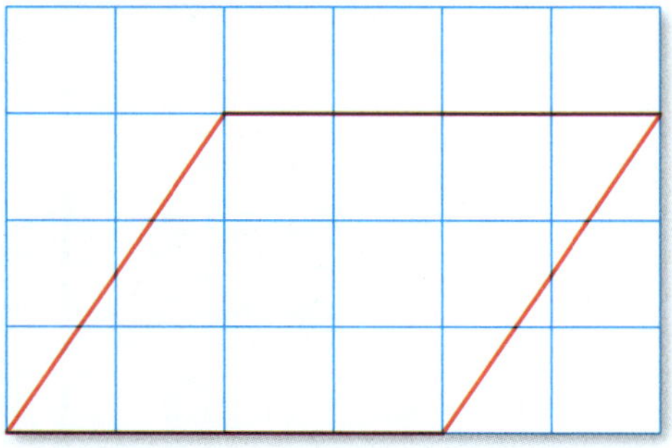

12.

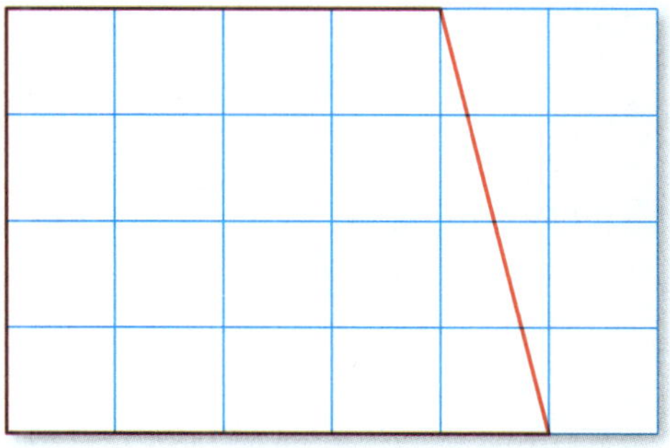

GO DIGITAL

FINDING AREAS OF QUADRILATERALS **Find the area of the figure.** (See Example 1.)

13.

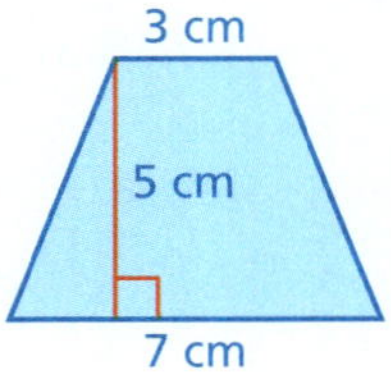

14.

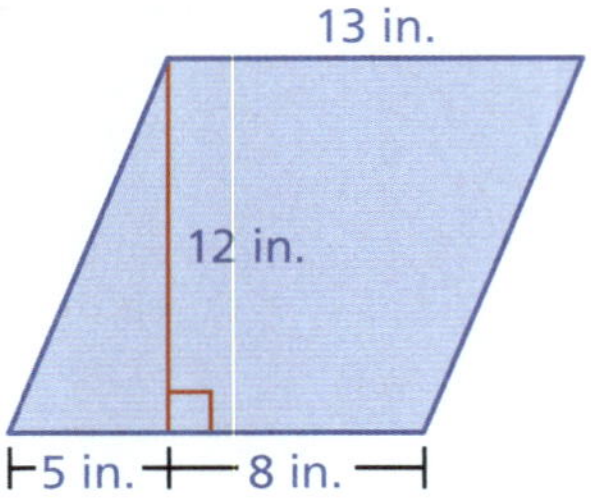

15.

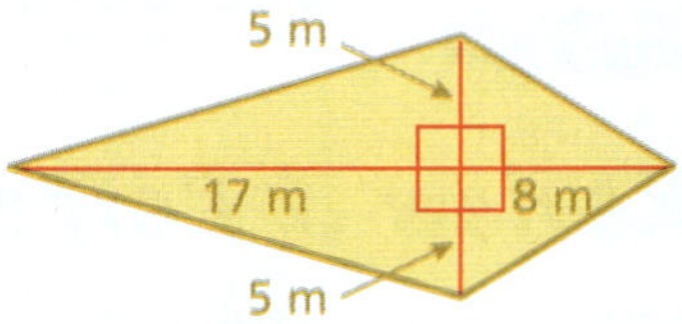

16.

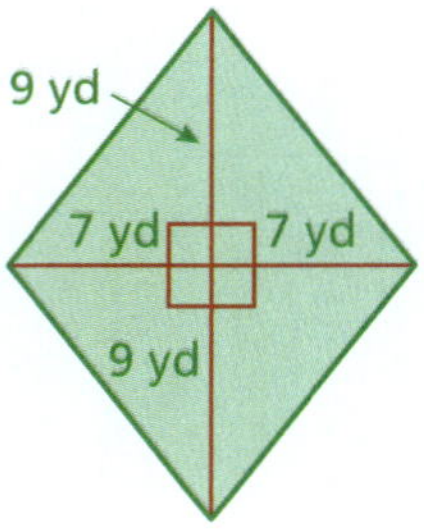

17.

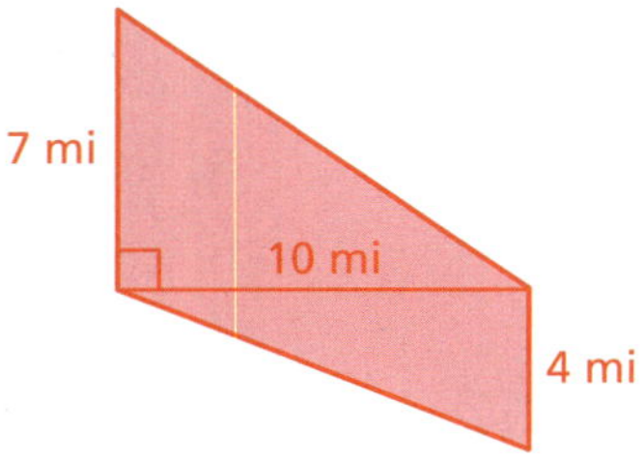

18.

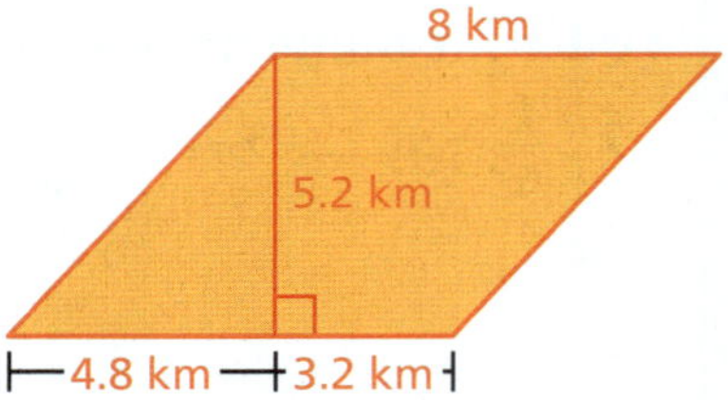

7 MTR

19. **MODELING REAL LIFE** The dance floor is in the shape of a parallelogram. Find the area of the dance floor.

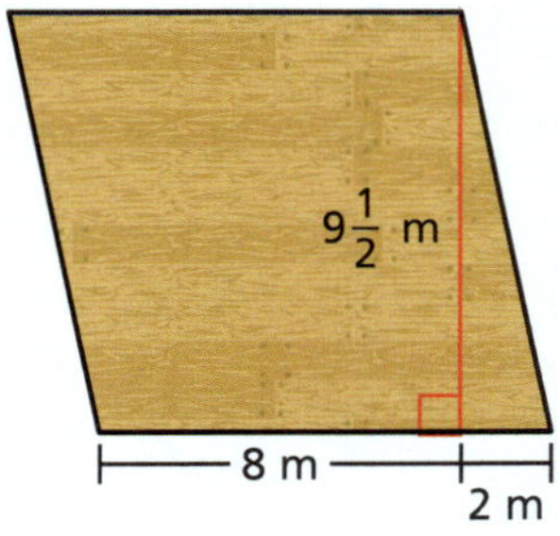

FINDING A MISSING DIMENSION Find the value of *x*. (See Example 2.)

20. Area = $54\ \text{yd}^2$

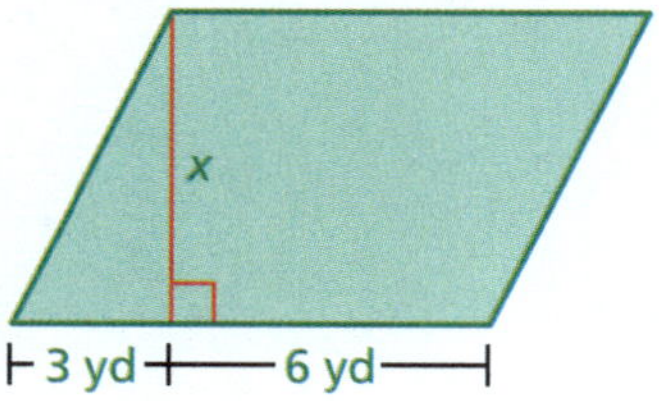

21. Area = $180\ \text{km}^2$

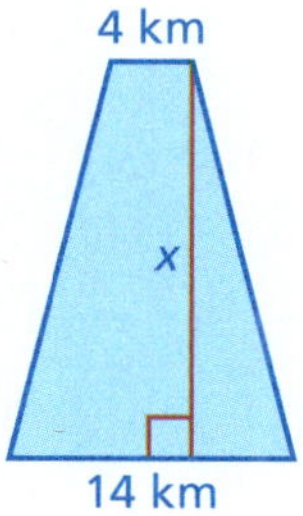

22. Area = $216\ \text{m}^2$

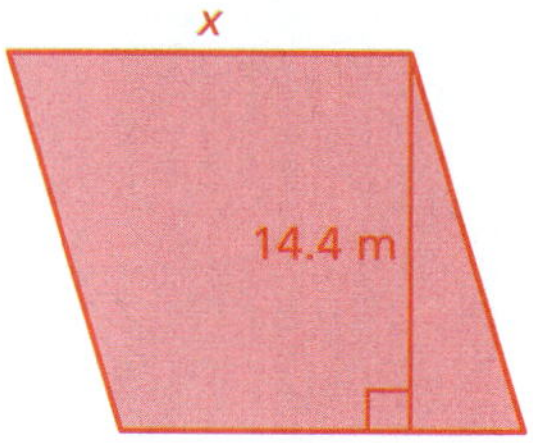

COMPOSITE FIGURES Find the area of the figure. (See Example 3.)

23.

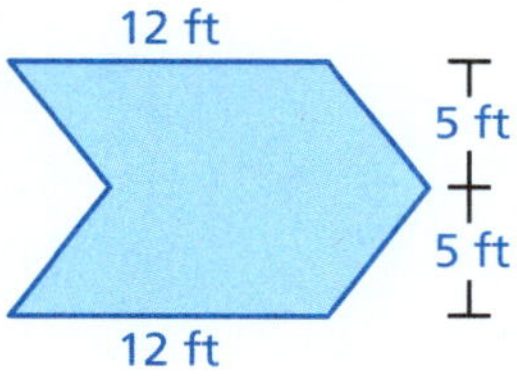

24.

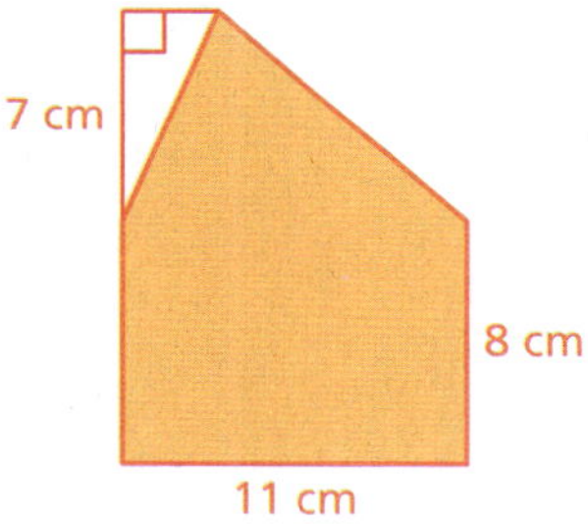

25.

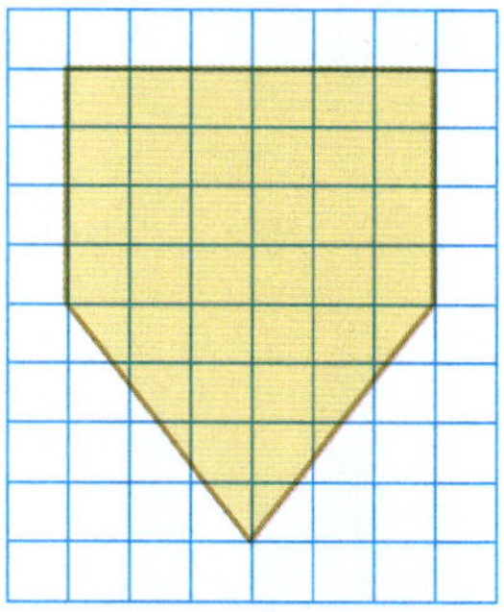

26. B.E.S.T. Test Prep The total area of the trapezoid is 176 square feet. What is the value of x?

Ⓐ 2

Ⓑ 3

Ⓒ 6

Ⓓ 12

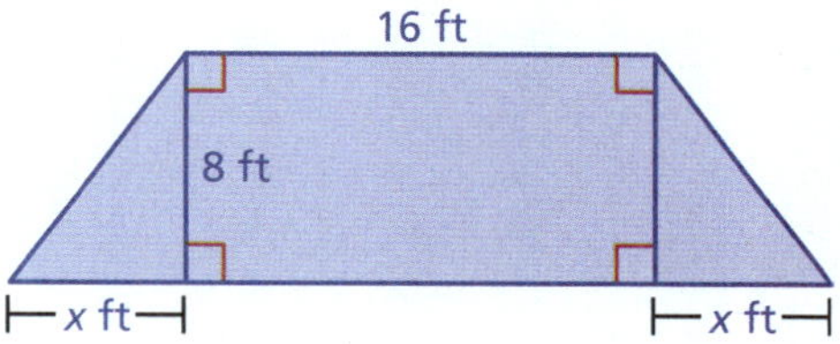

7 MTR

27. MODELING REAL LIFE A restaurant has outdoor patio seating.

a. Find the area of the patio.

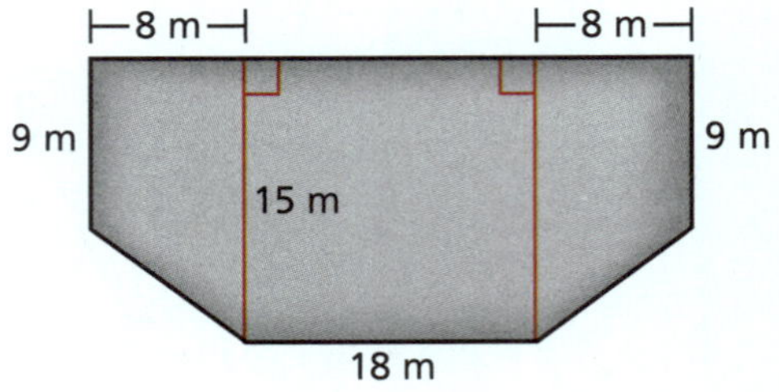

b. The restaurant wants to remodel the patio so that it is rectangular. Describe one way this can be done. Then find the area of the new patio.

5 MTR

28. STRUCTURE Use what you know about finding areas of triangles to write a formula for the area of a rhombus in terms of the lengths of its diagonals.

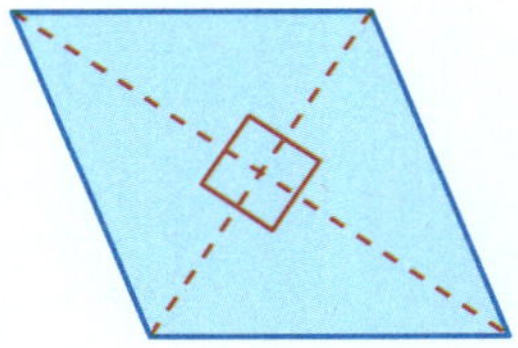

10.3 Areas of Parallelograms

Learning Target: Find areas and missing dimensions of parallelograms.

Success Criteria:
- I can explain how the area of a rectangle is used to find the area of a parallelogram.
- I can use the base and the height of a parallelogram to find its area.
- I can use the area of a parallelogram and one of its dimensions to find the other dimension.

Geometric Reasoning
MA.7.GR.1.1 Apply formulas to find the areas of trapezoids, parallelograms and rhombi.

The formula for the area of a parallelogram can be derived from the definition of the area of a rectangle. Recall that the area of a rectangle is the product of its length ℓ and its width w. The process you use to derive this formula is called *deductive reasoning.*

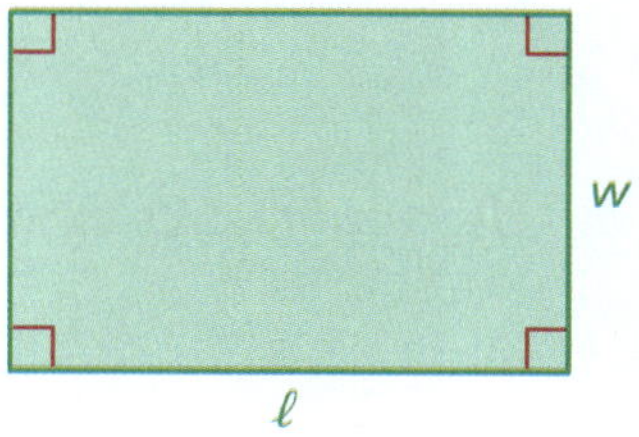

Area = ℓw

Exploration 1 Deriving the Area Formula of a Parallelogram

Work with a partner.

a. Draw *any* rectangle on a piece of centimeter grid paper. Cut the rectangle into two pieces that can be arranged to form a parallelogram. What do you notice about the areas of the rectangle and the parallelogram?

b. Copy the parallelogram below on a piece of centimeter grid paper. Cut the parallelogram and rearrange the pieces to find its area.

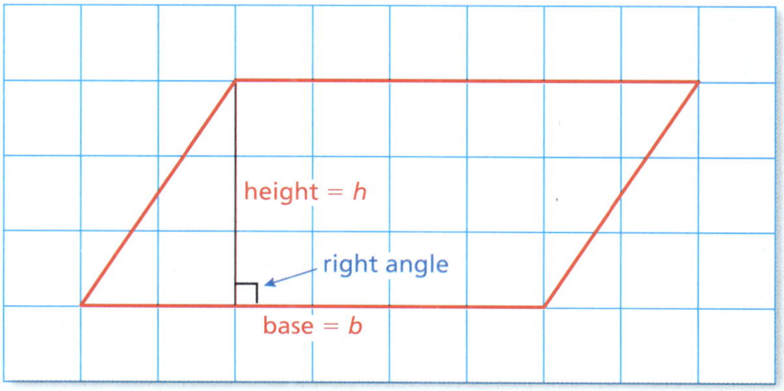

GO DIGITAL

c. Draw *any* parallelogram on a piece of centimeter grid paper and find its area. Does the area change when you use a different side as the base? Explain your reasoning.

d. Use your results to write a formula for the area A of a parallelogram.

5 MTR **DECOMPOSE A PROBLEM**

How does decomposing the parallelogram into other figures help you justify your formula?

10.3 Lesson

The *area* of a polygon is the amount of surface it covers. You can find the area of a parallelogram in much the same way as you can find the area of a rectangle.

The base of a parallelogram does not have to be horizontal. Any side of a parallelogram can be the base.

Key Idea

Area of a Parallelogram

Words The area A of a parallelogram is the product of its base b and its height h.

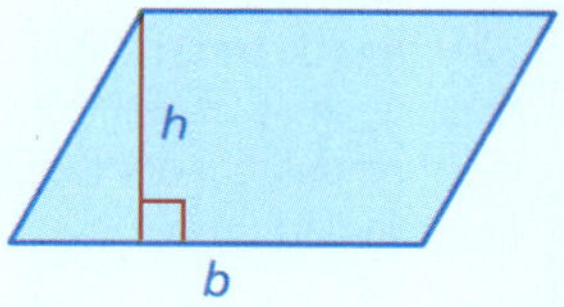

Algebra $A = bh$

Because a rhombus is a parallelogram, you can use the same formula to find the area of a rhombus.

Example 1 Finding Areas of Parallelograms

Find the area of each parallelogram.

a.

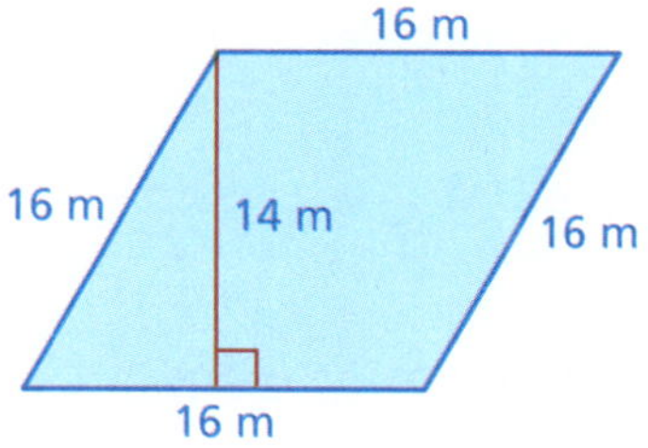

$A = bh$ Write formula.

$= 16(14)$ Substitute values.

$= 224$ Multiply.

▶ The area of the rhombus is 224 square meters.

b.

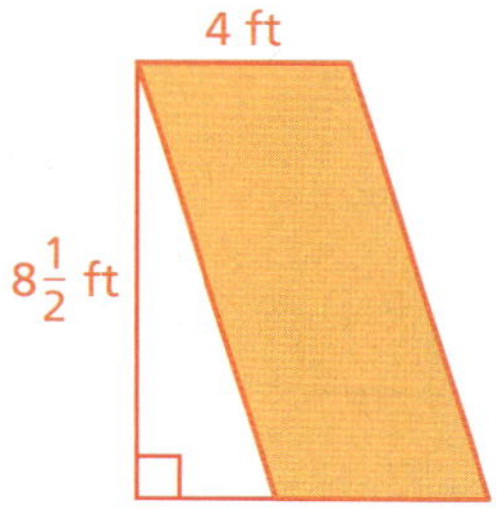

$A = bh$

$= 4\left(8\frac{1}{2}\right)$

$= 34$

▶ The area of the parallelogram is 34 square feet.

Try It **Find the area of the parallelogram.**

1.

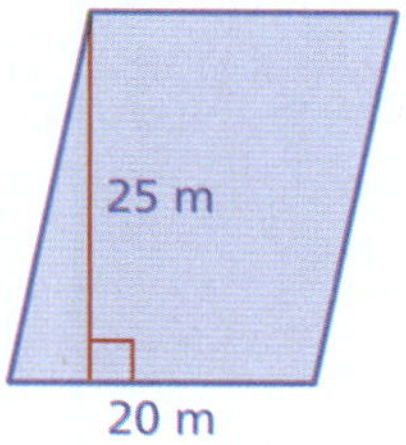

500 m^2

2.

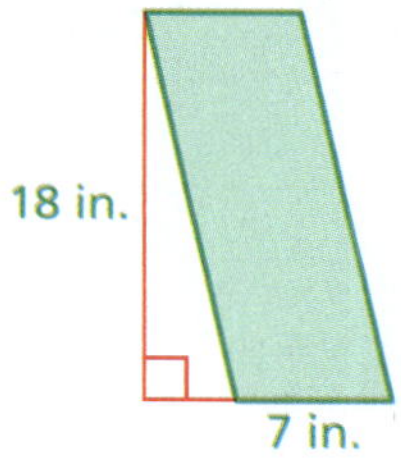

126 in^2

3.

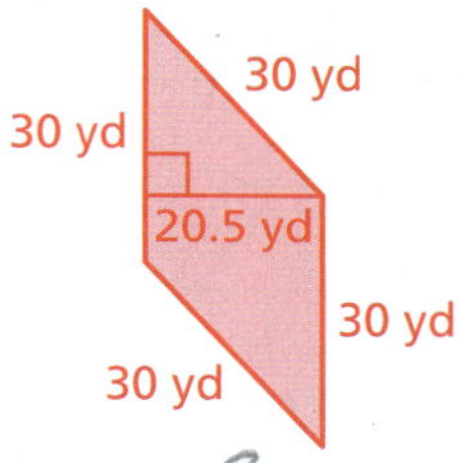

615 yd^2

GO DIGITAL

Example 2 Finding Missing Dimensions of Parallelograms

Find the missing dimension of each parallelogram.

a.

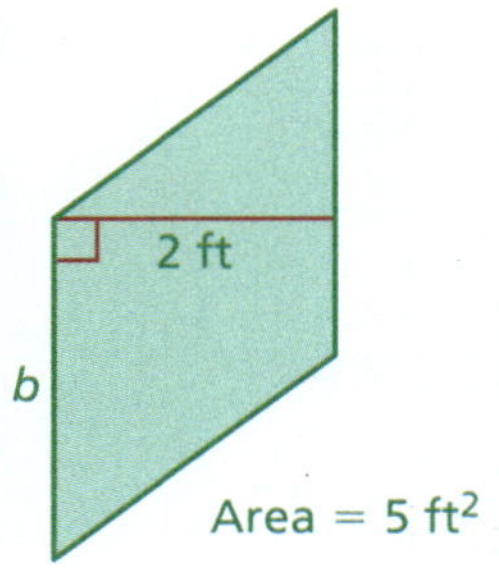

b.

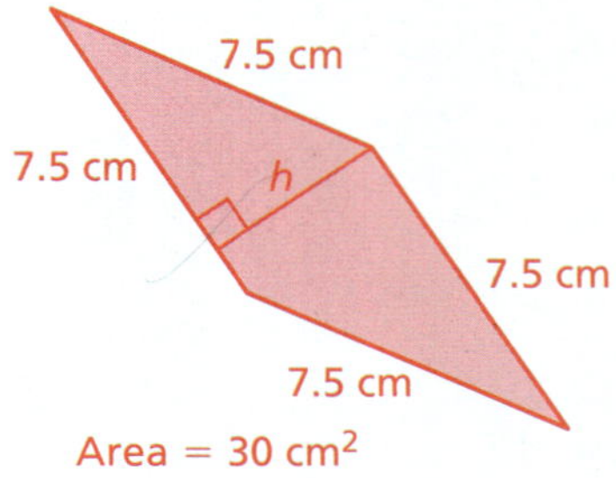

a.		b.
$A = bh$	Write formula.	$A = bh$
$5 = b(2)$	Substitute values.	$30 = 7.5h$
$\frac{5}{2} = \frac{b(2)}{2}$	Division Property of Equality	$\frac{30}{7.5} = \frac{7.5h}{7.5}$
$2\frac{1}{2} = b$	Simplify.	$4 = h$

▶ The base of the parallelogram is $2\frac{1}{2}$ feet.

▶ The height of the rhombus is 4 centimeters.

4 MTR CONSTRUCT AN ARGUMENT

Why can you be certain that the figure in part (b) is a rhombus, but not the figure in part (a)?

____

____

____

____

____

____

____

____

____

____

Try It

4. Find the missing dimension of the parallelogram.

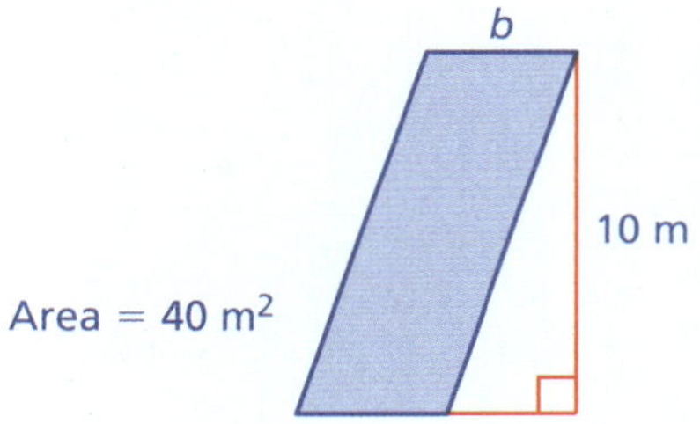

b = 4m

In-Class Practice

1 I don't understand yet. 2 I can do it with help. 3 I can do it on my own. 4 I can teach someone else.

5. WRITING Explain how to use the area of a rectangle to find the area of a parallelogram.

GO DIGITAL

In-Class Practice

1 I don't understand yet. | 2 I can do it with help. | 3 I can do it on my own. | 4 I can teach someone else.

FINDING AREA **Find the area of the parallelogram.**

6.

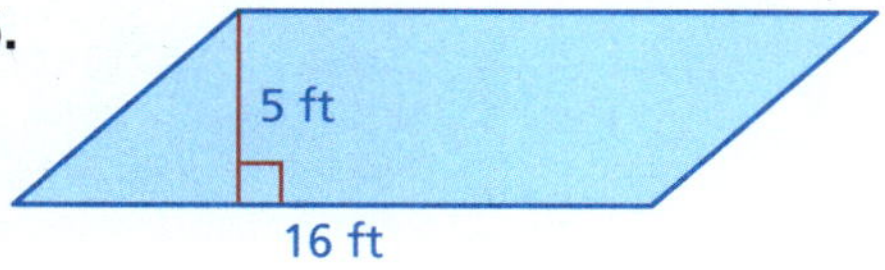

7.

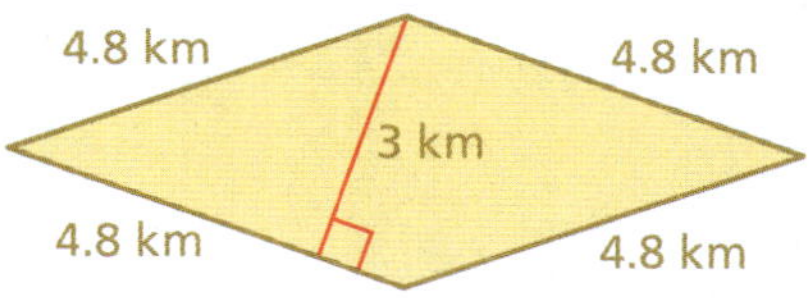

8. REASONING Draw a parallelogram that has a base of 6 inches and an area of 24 square inches. What is the height of the parallelogram?

Example 3 Modeling Real Life

7 MTR

The area of the parallelogram-shaped forest bordered by roads is 99,000 square yards. What is the length of the deer trail?

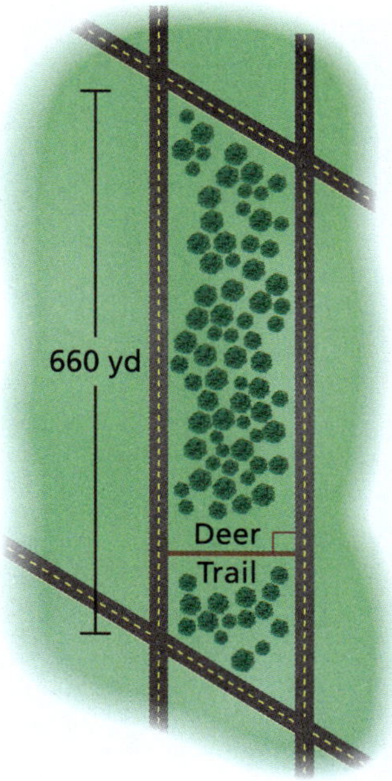

Understand the problem.

You are given the area of a parallelogram-shaped forest, and the map shows the length of one of its bases. You are asked to find the length of the deer trail, which represents the height of the parallelogram.

Make a plan.

Use the formula for the area of a parallelogram. Substitute for the area and the base, and then solve for the height.

Solve and check.

$A = bh$ — Write formula for area of a parallelogram.

$99{,}000 = 660h$ — Substitute 99,000 for *A* and 660 for *b*.

$\frac{99{,}000}{660} = \frac{660h}{660}$ — Division Property of Equality

$150 = h$ — Simplify.

▶ So, the deer trail is 150 yards long.

Check

$A = bh$

$99{,}000 \stackrel{?}{=} 660(150)$

$99{,}000 = 99{,}000$ ✓

GO DIGITAL

In-Class Practice

1 I don't understand yet. 2 I can do it with help. 3 I can do it on my own. 4 I can teach someone else.

9. The side of an office building in Hamburg, Germany, is in the shape of a parallelogram. The area of the side of the building is about 2150 square meters. What is the length x of the portion of the building that extends over the river?

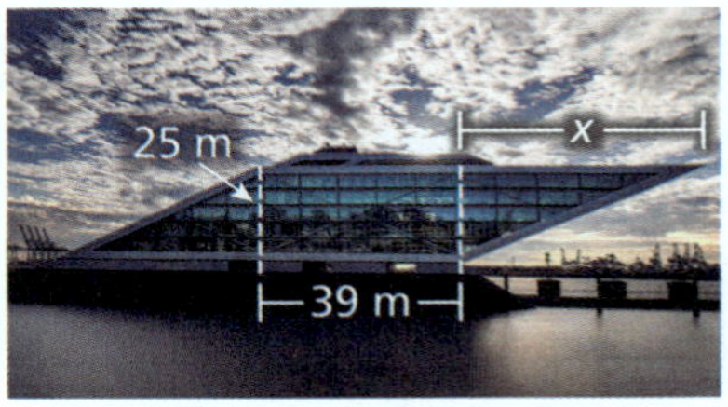

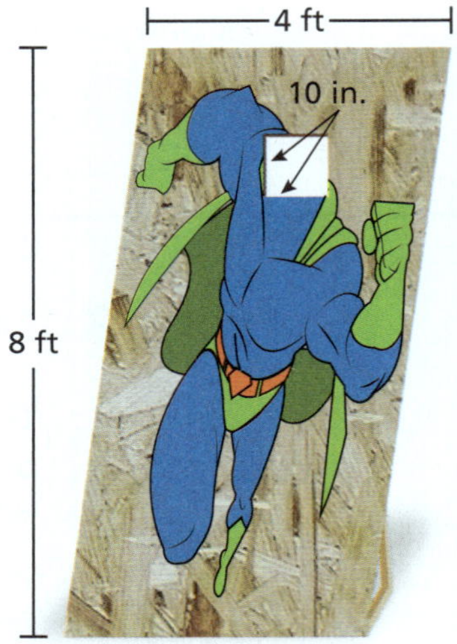

10. You make a photo prop for a school fair. You cut a 10-inch square out of a parallelogram-shaped piece of wood. What is the area of the photo prop?

11. **Dig Deeper** A galaxy contains a rhombus-shaped dust field. The dust field has a side length of 150 miles. What is the greatest possible area of the dust field? Explain your reasoning.

10.3 Practice WITH CalcChat® AND CalcView®

Review & Refresh

Find the area of the figure.

1. 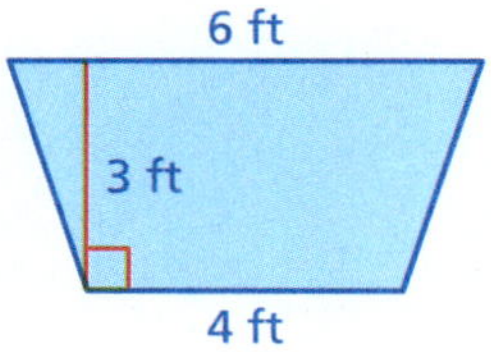

2. 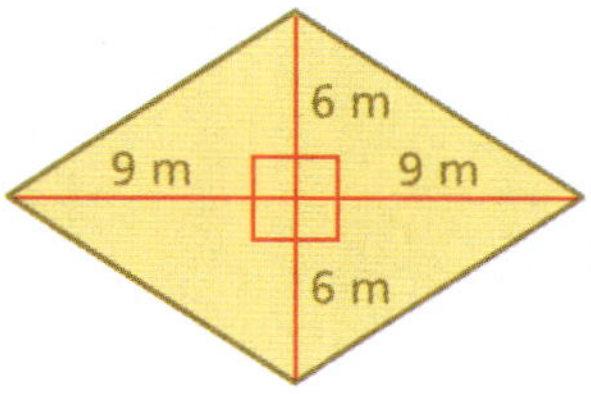

Find the missing values in the ratio table. Then write the equivalent ratios.

3.

Length (inches)	4		12
Width (inches)	3	6	

4.

Red Blood Cells	600	1200	
White Blood Cells	1		3

Concepts, Skills, & Problem Solving

USING TOOLS Rearrange the parallelogram as a rectangle. Then find the area. (See Exploration 1.)

5. 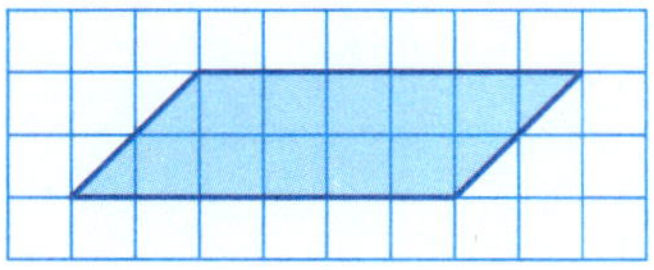

6. 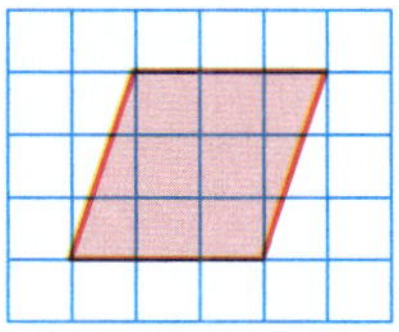

7. 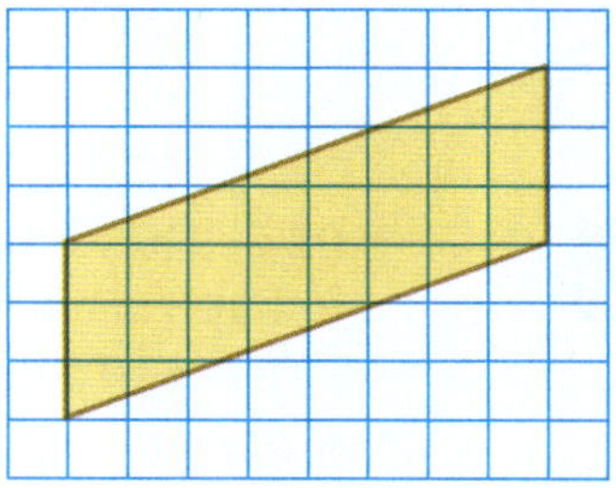

GO DIGITAL

FINDING AREA **Find the area of the parallelogram.** (See Example 1.)

8.

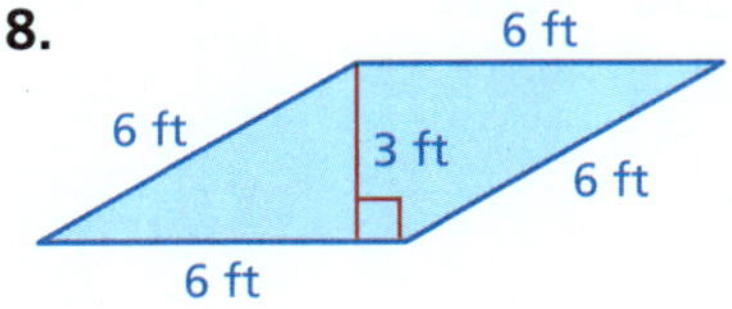

9.

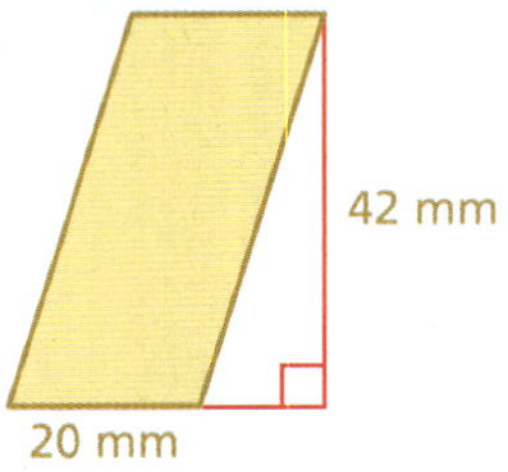

10.

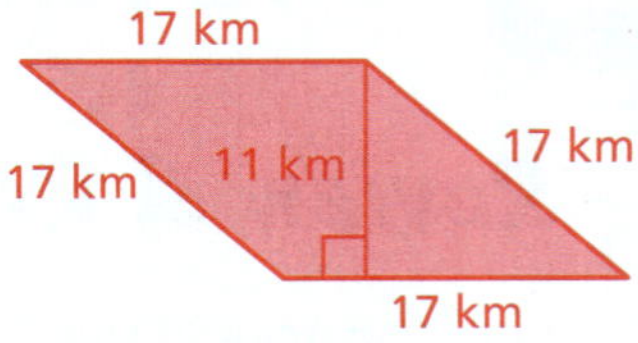

11.

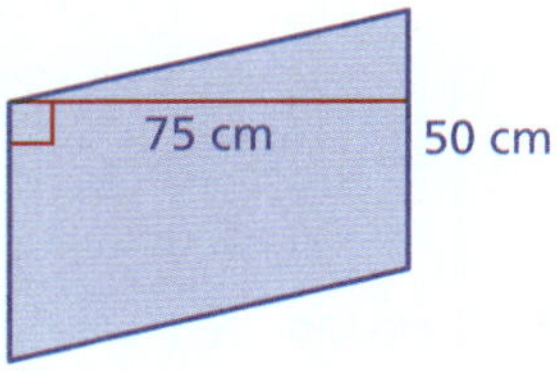

12.

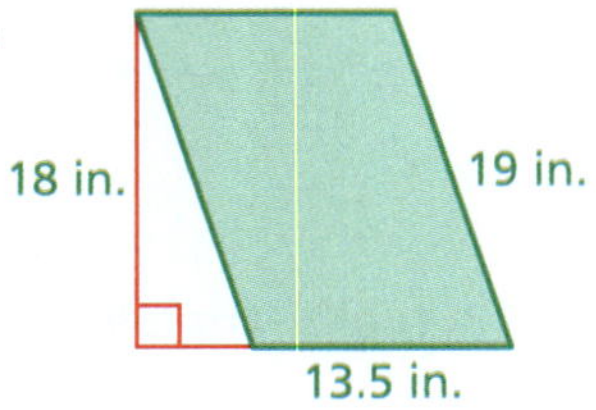

13.

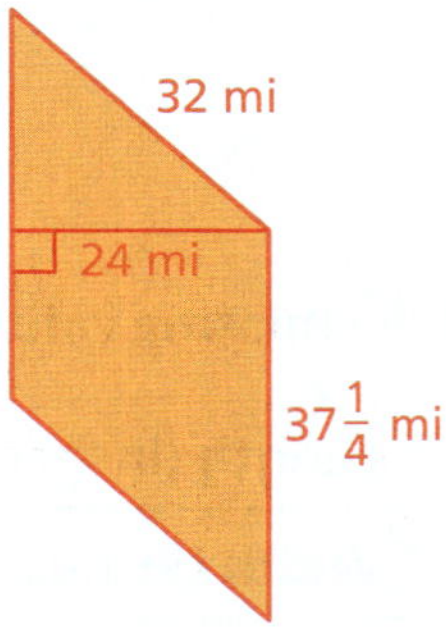

4 MTR **14.** **YOU BE THE TEACHER** Your friend finds the area of the parallelogram. Is your friend correct? Explain your reasoning.

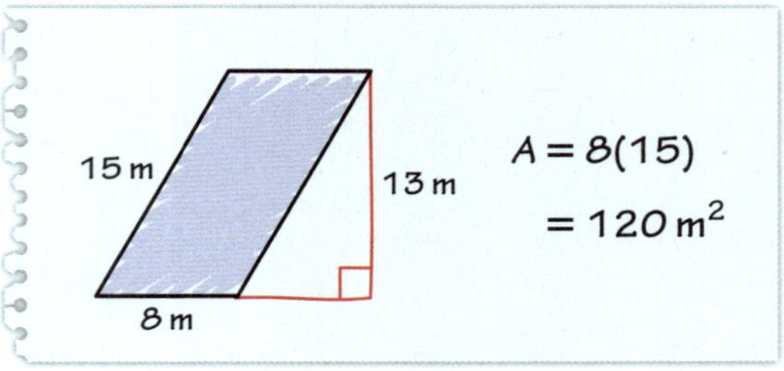

7 MTR **15.** **MODELING REAL LIFE** A ceramic tile in the shape of a parallelogram has a base of 4 inches and a height of 1.5 inches. What is the area of the tile?

FINDING MISSING DIMENSIONS OF PARALLELOGRAMS **Find the missing dimension of the parallelogram.** (See Example 2.)

16.

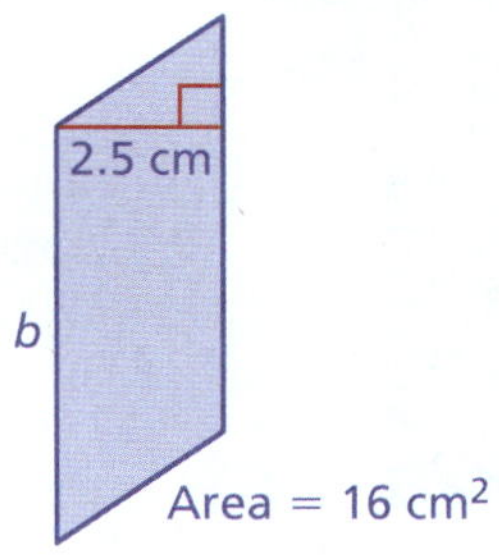

17.

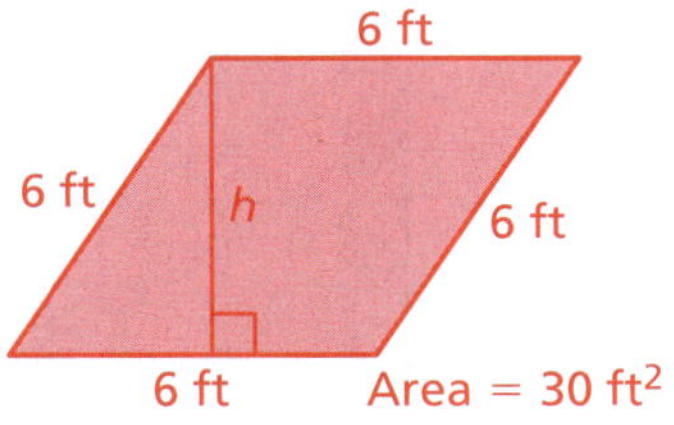

18.

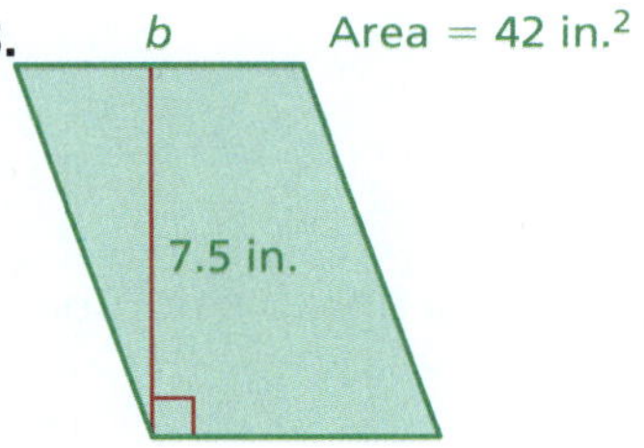

7 MTR

19. **MODELING REAL LIFE** You are sewing a parallelogram-shaped piece of fabric on a quilt. The piece of fabric has an area of 320 square inches. What is the height of the piece of fabric? (See Example 3.)

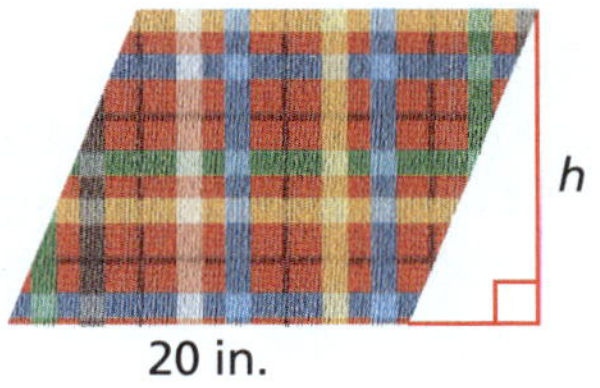

20. **OPEN-ENDED** Your deck has an area of 128 square feet. After adding a section, the area will be $(s^2 + 128)$ square feet. Draw a diagram of how this can happen.

7 MTR

21. MODELING REAL LIFE You use the parallelogram-shaped sponge to create the T-shirt design. The area of the design is 66 square inches. How many times do you use the sponge to create the design? Draw a diagram to support your answer.

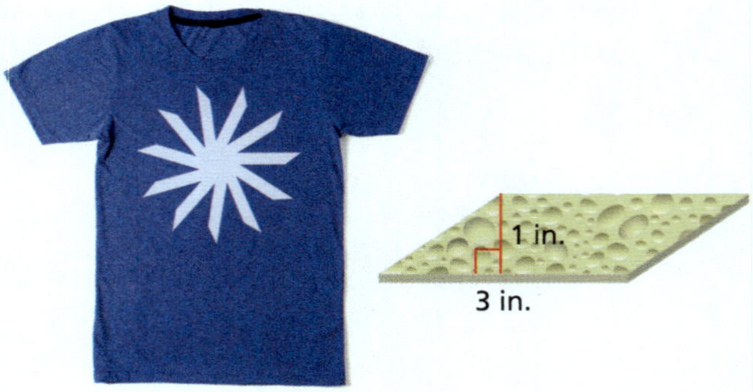

22. Dig Deeper The staircase has three identical parallelogram-shaped panels. The horizontal distance between each panel is 4.25 inches. The area of each panel is 287 square inches. What is the value of x?

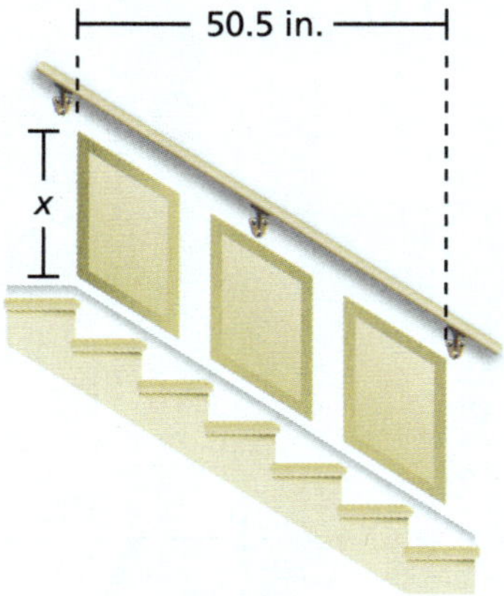

23. REASONING Each dimension of a parallelogram is multiplied by a positive number n. Write an expression for the area of the new parallelogram.

5 MTR

24. STRUCTURE Rearrange the rhombus shown to write a formula for the area of a rhombus in terms of its diagonals.

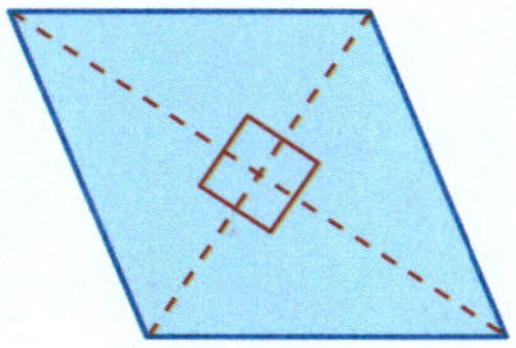

10.4 Areas of Trapezoids and Other Polygons

Learning Target: Find areas of trapezoids and other polygons.

Success Criteria:
- I can decompose trapezoids and other polygons into smaller shapes.
- I can use decomposition to find the area of a figure.
- I can use the bases and the height of a trapezoid to find its area.
- I can use the formula for the area of a trapezoid to find a missing dimension.

Geometric Reasoning

MA.7.GR.1.1 Apply formulas to find the areas of trapezoids, parallelograms and rhombi.

MA.7.GR.1.2 Solve mathematical or real-world problems involving the area of polygons or composite figures by decomposing them into triangles or quadrilaterals.

Also MA.6.GR.2.2

Exploration 1 Deriving the Area Formula of a Trapezoid

Work with a partner.

a. Draw *any* parallelogram on a piece of centimeter grid paper. Cut the parallelogram into two identical trapezoids that have exactly one pair of parallel sides. How can you use the area of the parallelogram to find the area of each trapezoid?

b. Copy the trapezoid below on a piece of centimeter grid paper. Use two different methods to find the area of the trapezoid. Explain your methods.

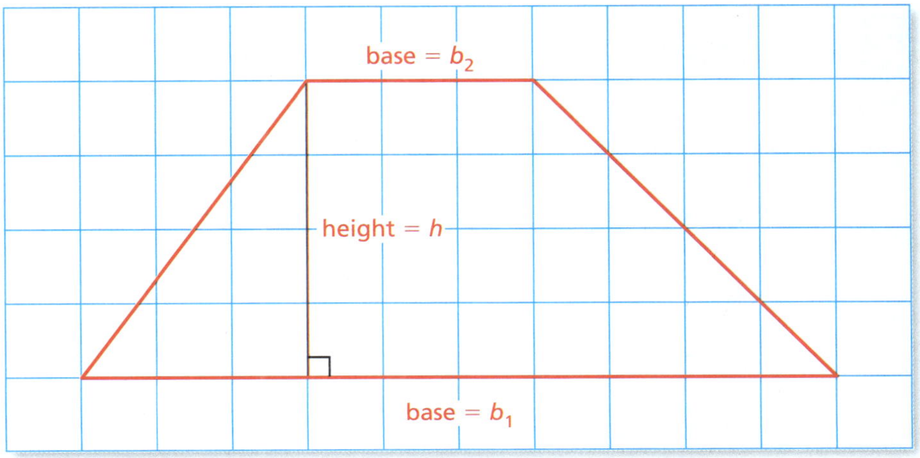

GO DIGITAL

c. Draw *any* trapezoid that has exactly one pair of parallel sides on a piece of centimeter grid paper, and find its area. Explain your choice of method.

d. Use your results to write a formula for the area A of a trapezoid that has exactly one pair of parallel sides. Does this formula work for trapezoids that have two pairs of parallel sides? Explain.

5 MTR **MAKE A PLAN**

How can you use the diagram below to justify the formula you wrote in part (d)?

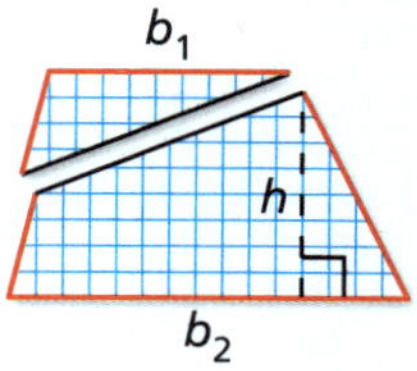

10.4 Lesson

You can use decomposition to find areas of trapezoids and other polygons.

Example 1 Finding Areas of Trapezoids and Other Polygons

Find the area of each figure.

a.

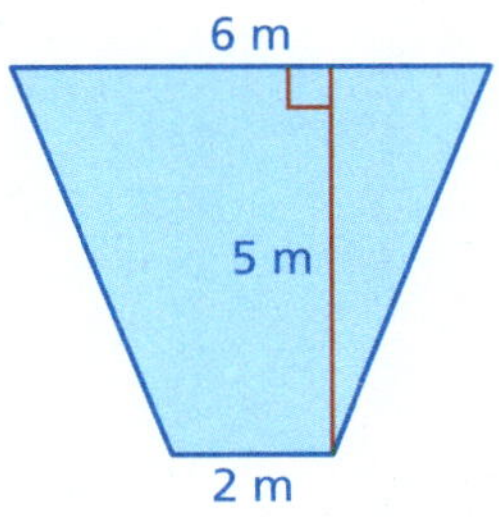

Decompose the trapezoid into a triangle and a rectangle. Find the sum of the areas of the figures.

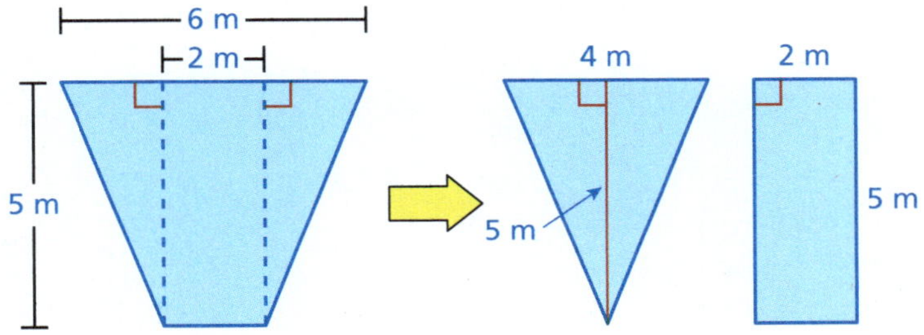

$$A = \frac{1}{2}(4)(5) + 5(2)$$
$$= 10 + 10$$
$$= 20$$

▶ The area of the trapezoid is 20 square meters.

b.

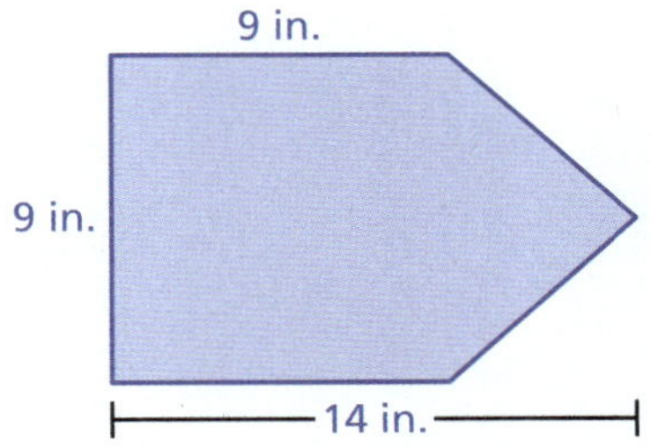

Decompose the pentagon into a square and a triangle. Find the sum of the areas of the figures.

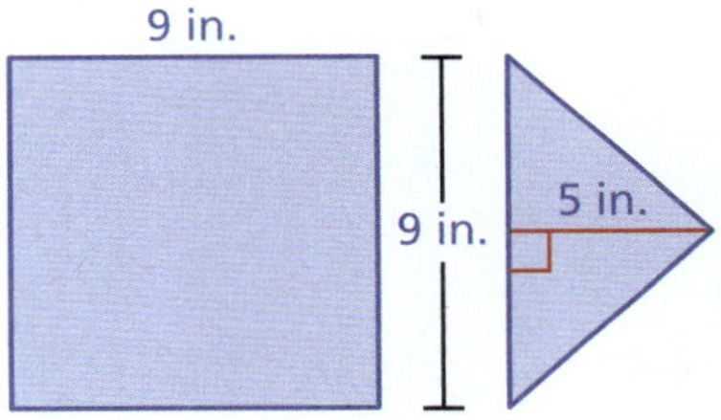

$$A = 9(9) + \frac{1}{2}(9)(5)$$
$$= 81 + 22\frac{1}{2}$$
$$= 103\frac{1}{2}$$

▶ The area of the pentagon is $103\frac{1}{2}$ square inches.

1 MTR HELP A CLASSMATE

Help a classmate find the area in Example 1(a) using a decomposition into two triangles.

____

____

____

____

____

GO DIGITAL

Try It **Find the area of the figure.**

1.

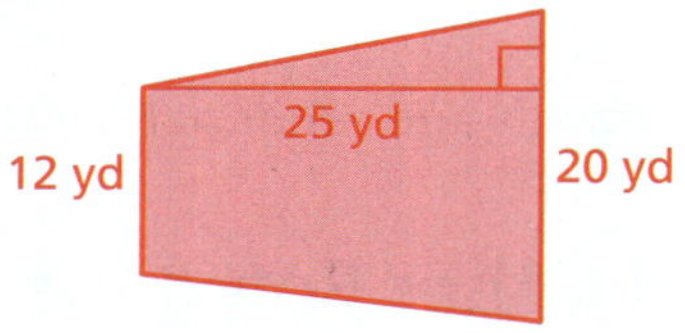

2.

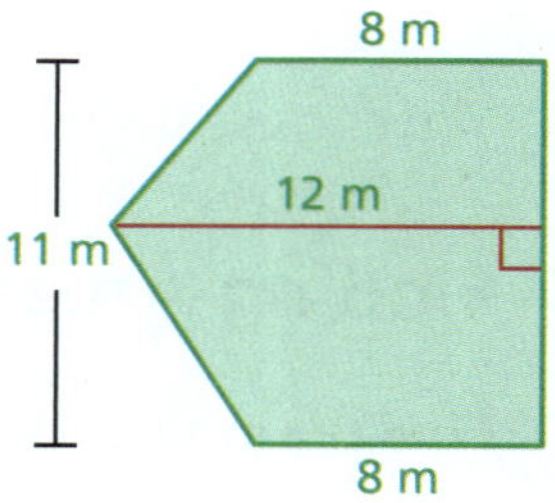

In Example 1(a), you could have used a copy of the trapezoid to form a parallelogram. As you may have discovered in the exploration, this leads to the following formula for the area of a trapezoid.

Key Idea

Area of a Trapezoid

Words The area A of a trapezoid is one-half the product of its height h and the sum of its bases b_1 and b_2.

Algebra $A = \frac{1}{2}h(b_1 + b_2)$

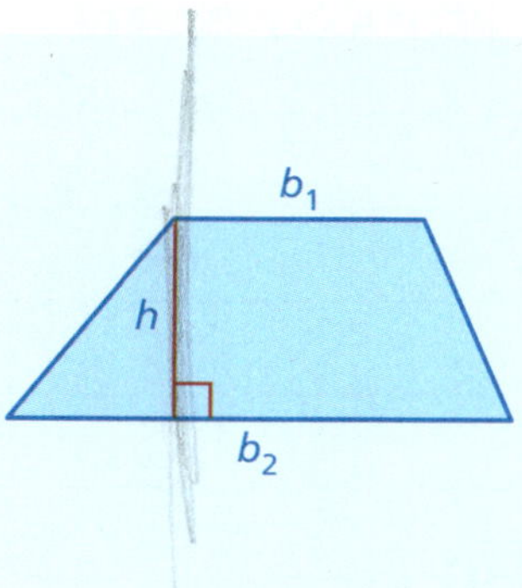

Example 2 Finding Areas of Trapezoids

Find the area of each trapezoid.

a.

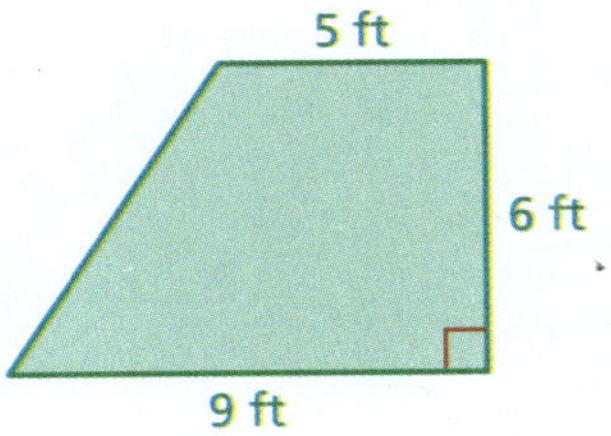

b.

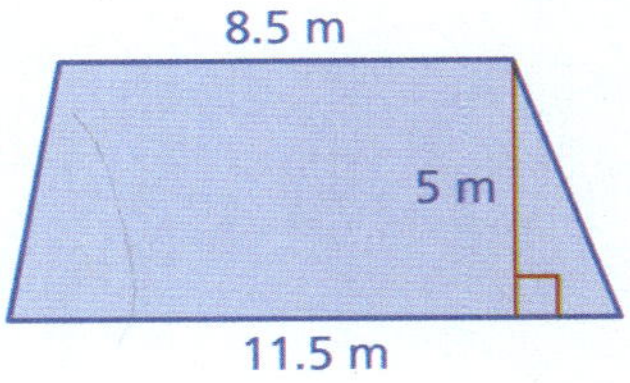

2 MTR USE ANOTHER METHOD

Use decomposition to find the area in Example 2(a).

____

____

____

____

a.

$A = \frac{1}{2}h(b_1 + b_2)$	Write formula.	$A = \frac{1}{2}h(b_1 + b_2)$
$= \frac{1}{2}(6)(5 + 9)$	Substitute.	$= \frac{1}{2}(5)(8.5 + 11.5)$
$= \frac{1}{2}(6)(14)$	Add.	$= \frac{1}{2}(5)(20)$
$= 42$	Multiply.	$= 50$

▶ The area of the trapezoid is 42 square feet.

▶ The area of the trapezoid is 50 square meters.

Try It

Find the area of the trapezoid.

3.

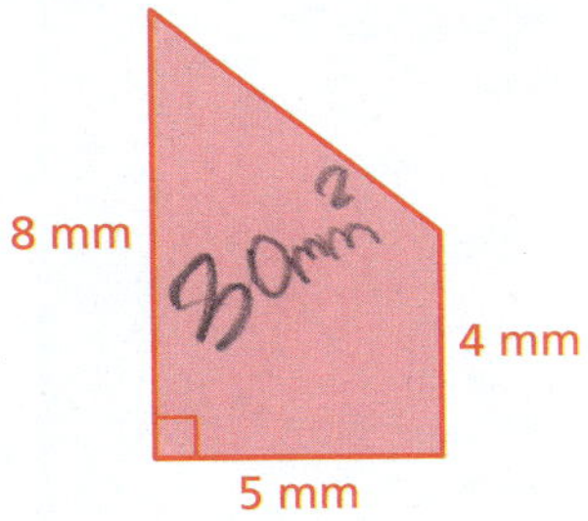

4.

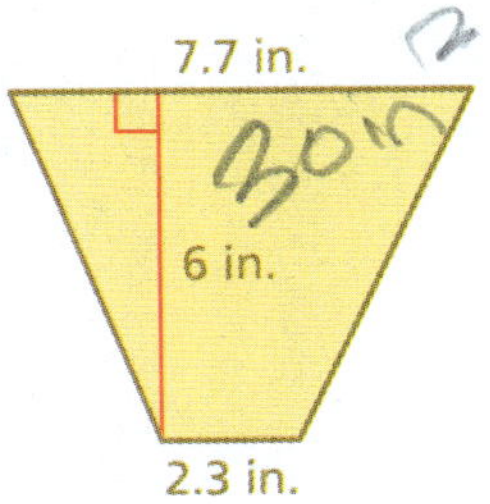

GO DIGITAL

Example 3 Finding the Height of a Trapezoid

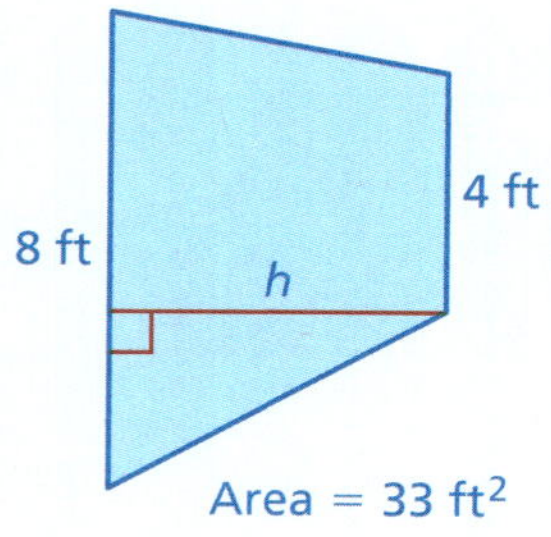

Find the height of the trapezoid.

$A = \frac{1}{2}h(b_1 + b_2)$	Write formula for area of a trapezoid.
$33 = \frac{1}{2}h(4 + 8)$	Substitute 33 for A, 4 for b_1, and 8 for b_2.
$33 = \frac{1}{2}h(12)$	Add.
$33 = 6h$	Simplify.
$\frac{33}{6} = \frac{6h}{6}$	Division Property of Equality
$5\frac{1}{2} = h$	Simplify.

▶ The height of the trapezoid is $5\frac{1}{2}$ feet.

Try It **Find the height of the trapezoid.**

5.

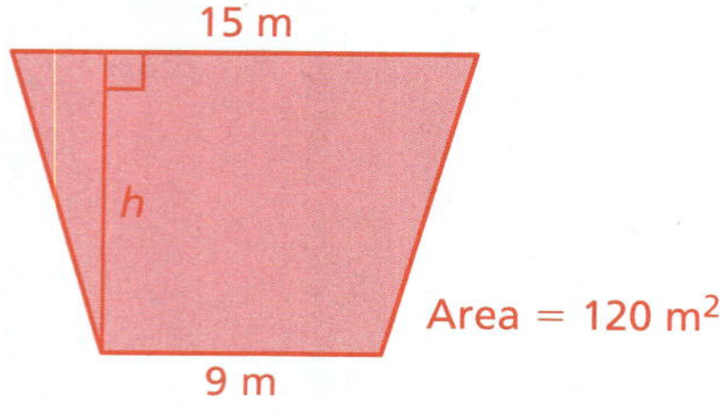

6.

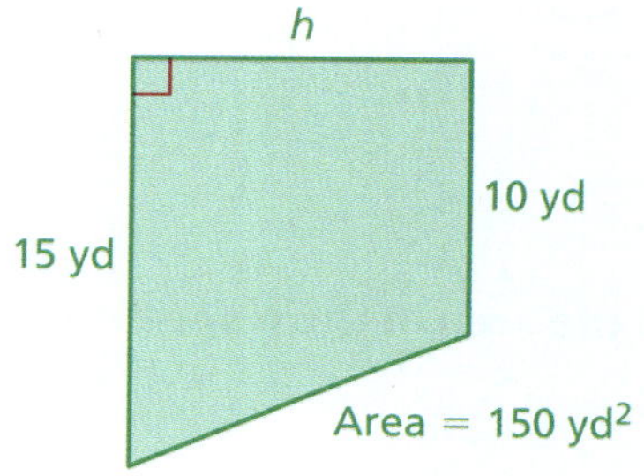

In-Class Practice

1 I don't understand yet. 2 I can do it with help. 3 I can do it on my own. 4 I can teach someone else.

7. **REASONING** What measures do you need to find the area of a trapezoid that has exactly one pair of parallel sides?

GO DIGITAL

In-Class Practice

1 I don't understand yet. 2 I can do it with help. 3 I can do it on my own. 4 I can teach someone else.

FINDING AREA **Find the area of the figure.**

8.

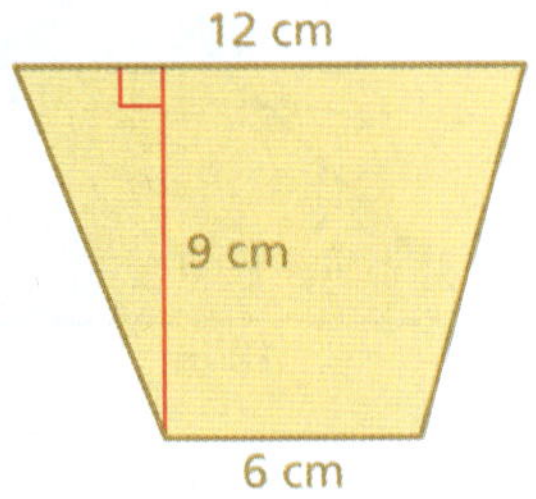

9.

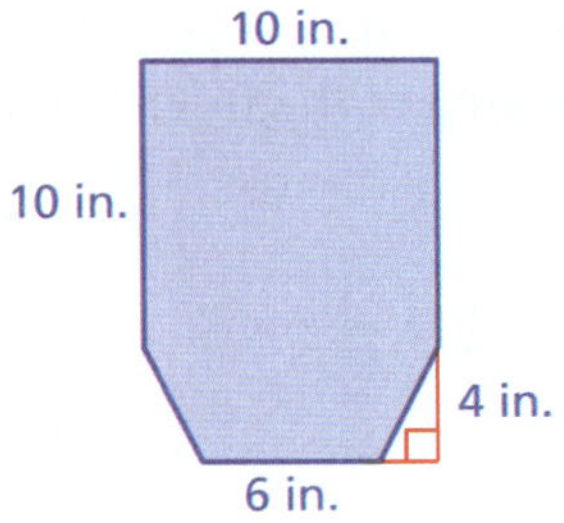

10. **FINDING A HEIGHT** The sum of the bases of a trapezoid is 20 inches. The area of the trapezoid is 50 square inches. What is the height of the trapezoid?

Example 4 Modeling Real Life

7 MTR

You can use a trapezoid to approximate the shape of Polk County, Florida. The trapezoid contains about 200 square miles of water area. The population is about 725,000. About how many people are there per square mile of land?

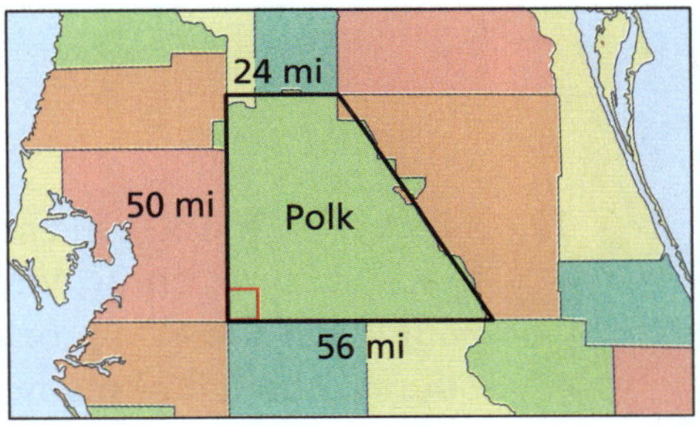

Understand the problem.

You are given the dimensions, the water area, and the population of a county shaped like a trapezoid. You are asked to find the number of people per square mile of land.

Make a plan.

Use the formula for the area of a trapezoid to find the total area of Polk County. Find the difference of the total area and the water area to find the land area. Then divide the population by the land area to find the number of people per square mile of land.

Solve and check.

$A = \frac{1}{2}h(b_1 + b_2)$	Write formula for area of a trapezoid.
$= \frac{1}{2}(50)(24 + 56)$	Substitute 50 for h, 24 for b_1, and 56 for b_2.
$= \frac{1}{2}(50)(80)$	Add.
$= 2000$	Multiply.

The total area of Polk County is about 2000 square miles. The land area is about $2000 - 200 = 1800$ square miles.

▶ So, there are about $\frac{725{,}000 \text{ people}}{1800 \text{ mi}^2} \approx 403$ people per square mile of land.

Check Reasonableness

Round the population to 720,000 to obtain an estimate that is simpler to calculate.

$720{,}000 \div 1800 = 400$

The answer is reasonable because $400 \approx 403$. ✓

GO DIGITAL

In-Class Practice

1 I don't understand yet. 2 I can do it with help. 3 I can do it on my own. 4 I can teach someone else.

11. Dig Deeper An archaeologist estimates that the manuscript shown was originally a rectangle with a length of 20 inches. Estimate the area of the fragment that is missing.

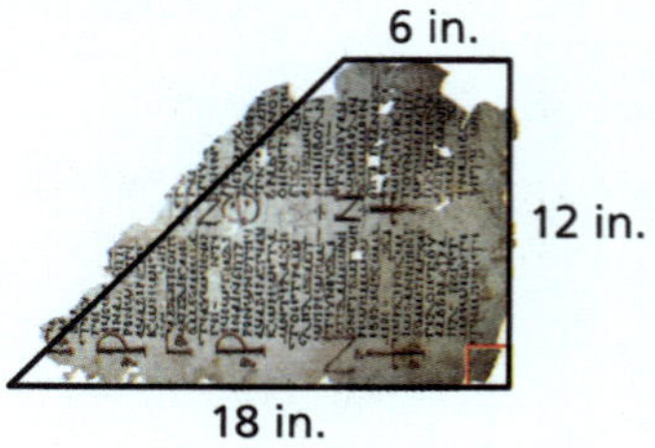

7 MTR **12. INVESTIGATE** You can use a trapezoid to approximate the shape of Scott County, Virginia. Use the Internet to find the population and to approximate the dimensions of Scott County. About how many people are there per square mile?

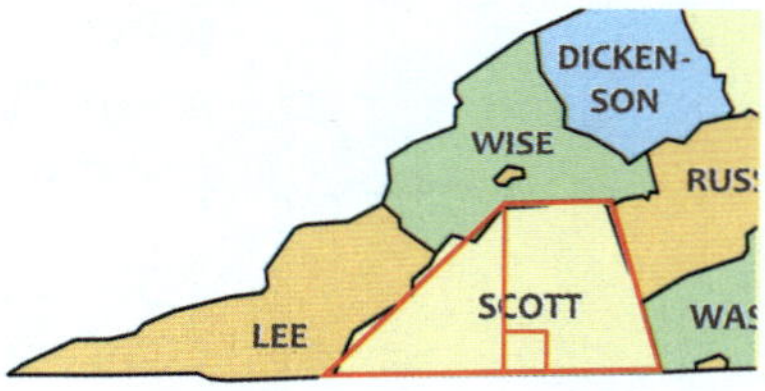

10.4 Practice WITH CalcChat® AND CalcView®

Review & Refresh

Find the area of the parallelogram.

1. 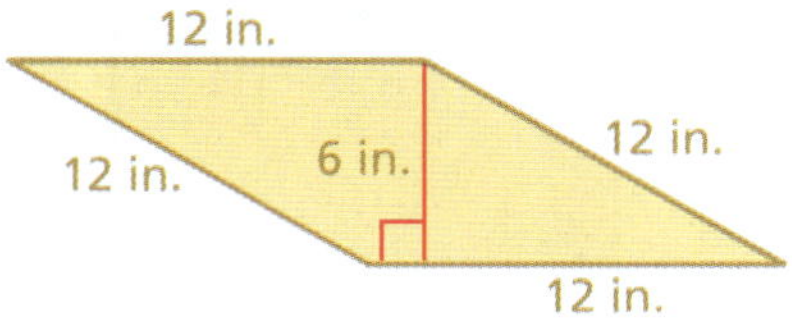

2. 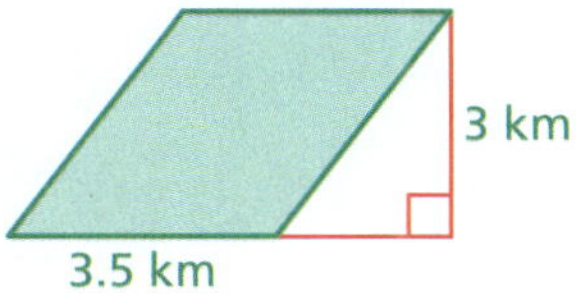

3. 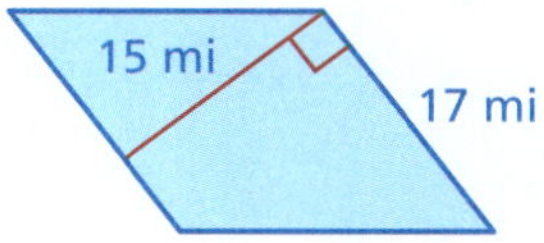

Find the sum or difference. Write fractions in simplest form.

4. $5.6 - 2.9$

5. $-\frac{3}{8} + \frac{1}{8}$

6. $-2\frac{1}{4} - \left(-1\frac{1}{2}\right)$

7. On a normal day, 12 airplanes arrive at an airport every 15 minutes. Write a rate that represents this situation.

GO DIGITAL

Concepts, Skills, & Problem Solving

USING TOOLS Find the area of the trapezoid by forming a parallelogram. (See Exploration 1.)

8.

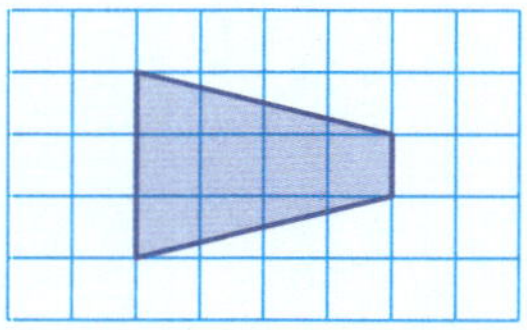

9.

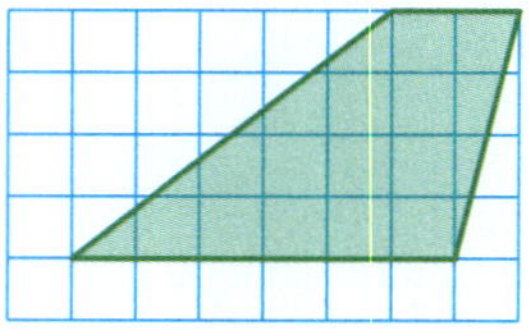

10.

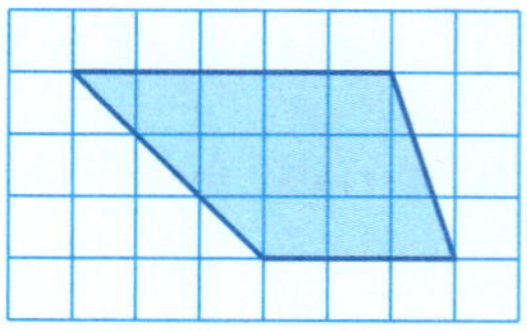

FINDING AREA Find the area of the figure. (See Example 1.)

11.

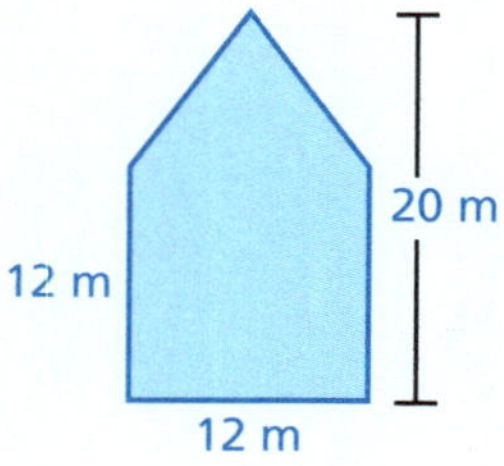

12.

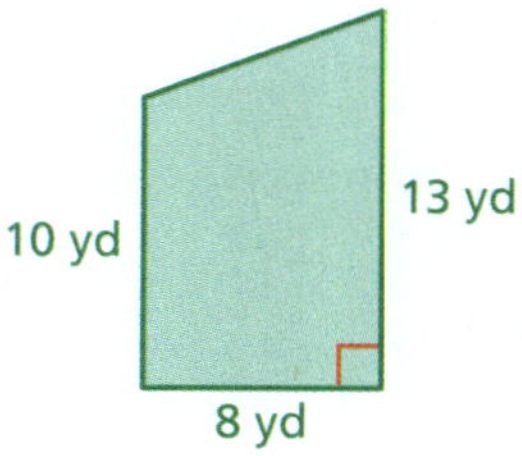

▶ **13.**

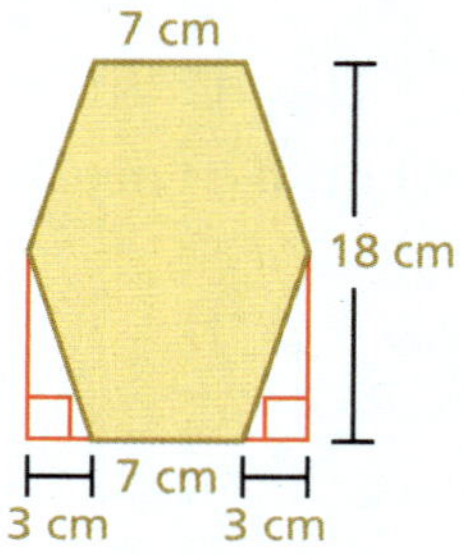

14.

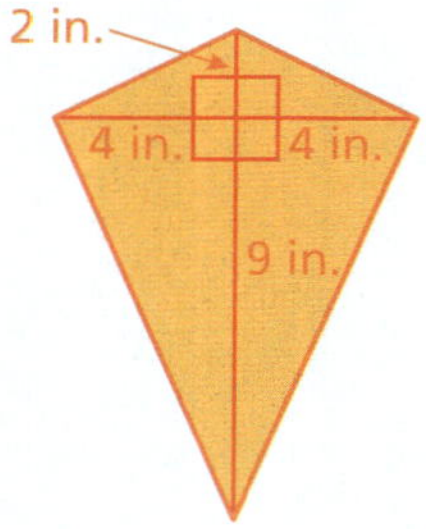

15.

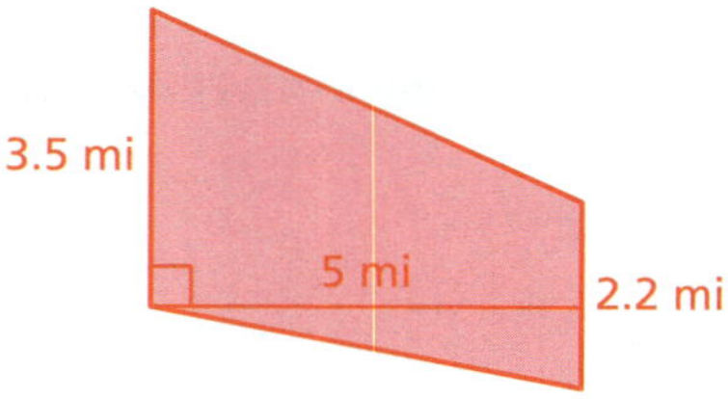

16.

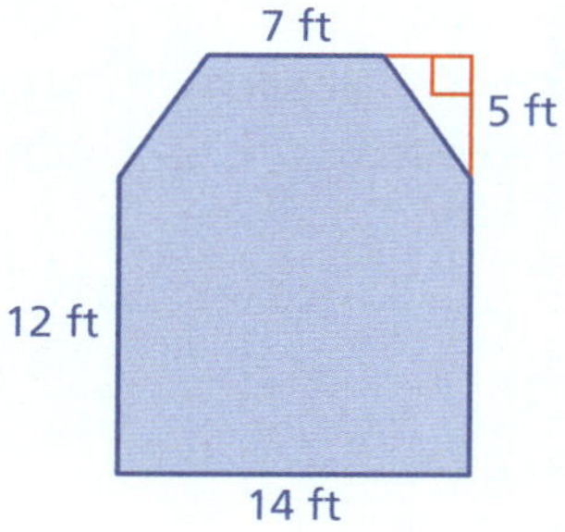

FINDING AREA **Find the area of the trapezoid.** (See Example 2.)

▶ **17.**

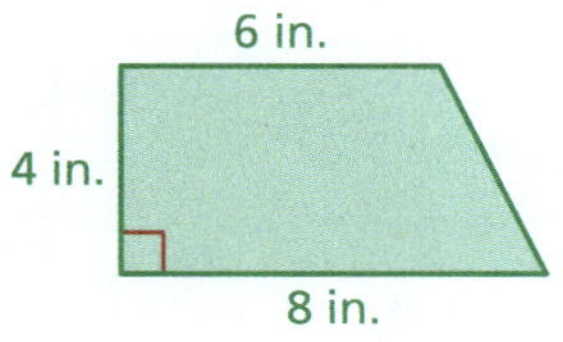

18.

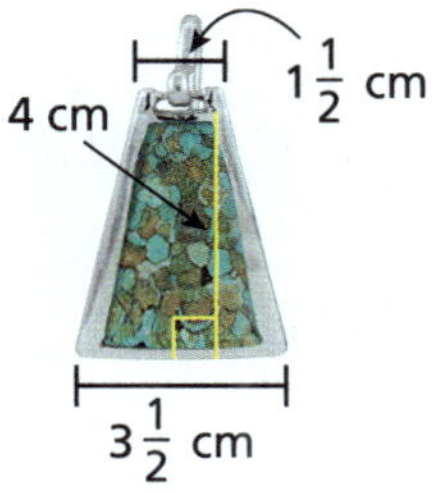

19.

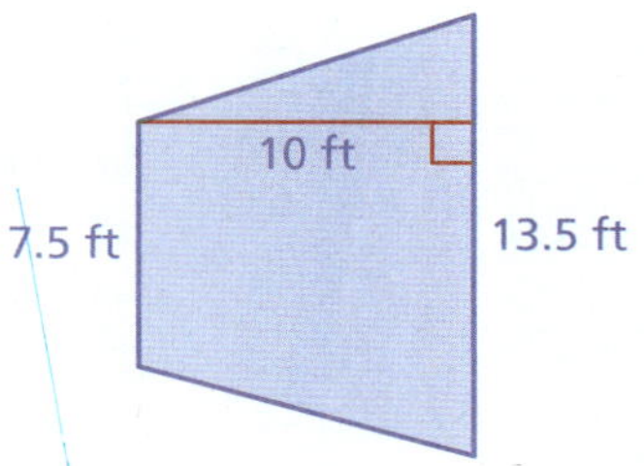

4 MTR **20.** **YOU BE THE TEACHER** Your friend finds the area of the trapezoid. Is your friend correct? Explain your reasoning.

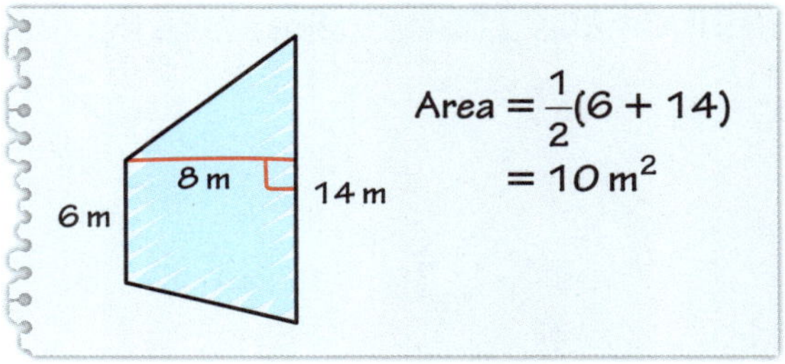

7 MTR **21.** **MODELING REAL LIFE** Light shines through a window. What is the area of the trapezoid-shaped region created by the light?

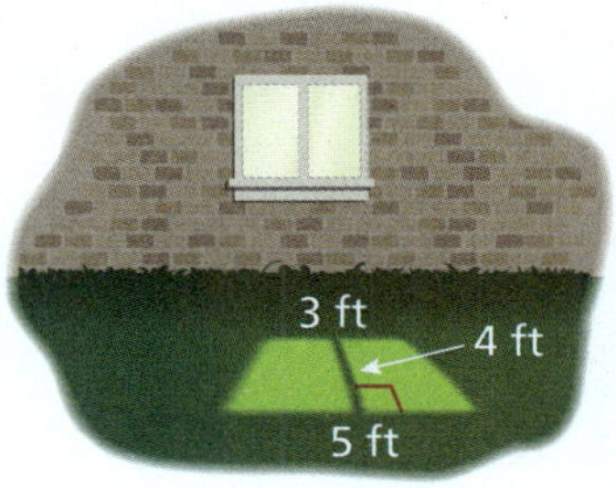

22. **B.E.S.T. Test Prep** What is the area of the figure?

Ⓐ 16 square units

Ⓑ 32 square units

Ⓒ 48 square units

Ⓓ 64 square units

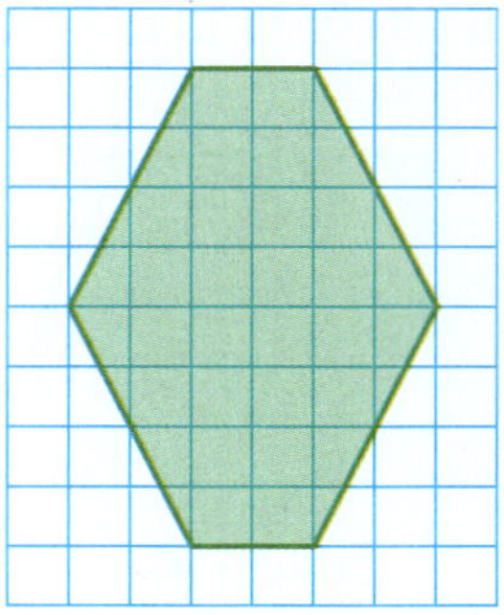

FINDING A HEIGHT **Find the height of the trapezoid.** (See Example 3.)

23. Area = 391 km^2

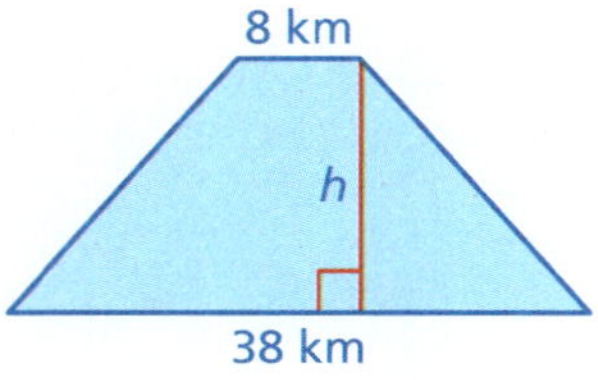

24. Area = 600 yd^2

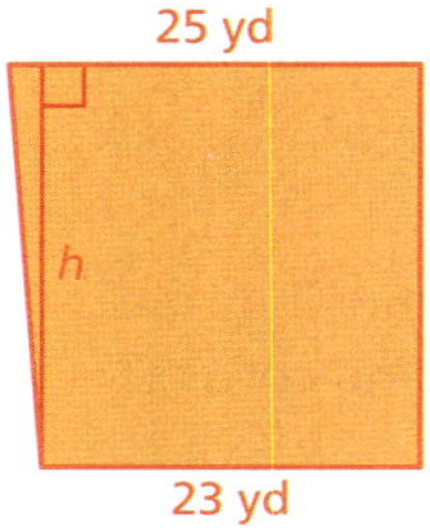

25. Area = $21\frac{1}{5}$ mm^2

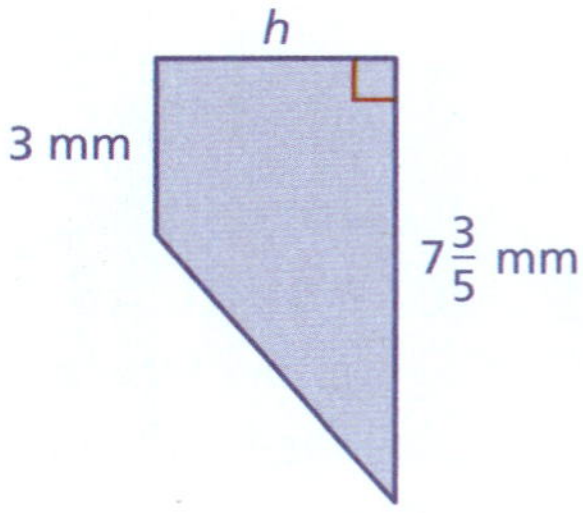

26. **PROBLEM SOLVING** *Discus throw* is a sport that was part of the ancient Greek pentathlon, an event at the ancient Olympic Games. Modern discus throw competitions use large safety nets to protect spectators. Approximate the area the safety net encloses by finding the area of the polygon formed by the net.

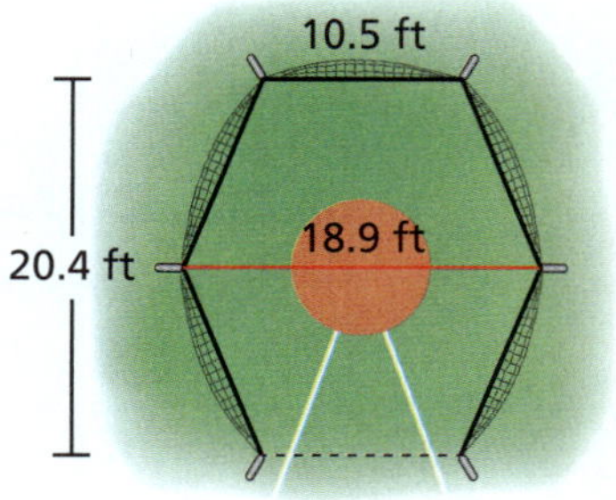

7 MTR

27. MODELING REAL LIFE A trapezoid can be used to approximate the shape of Taylor County, Florida. There are about 189 square miles of water area in Taylor County. The population is about 21,600. About how many people are there per square mile of land? Compare your results to the number of people per square mile of land in Polk County, Florida. (See Example 4.)

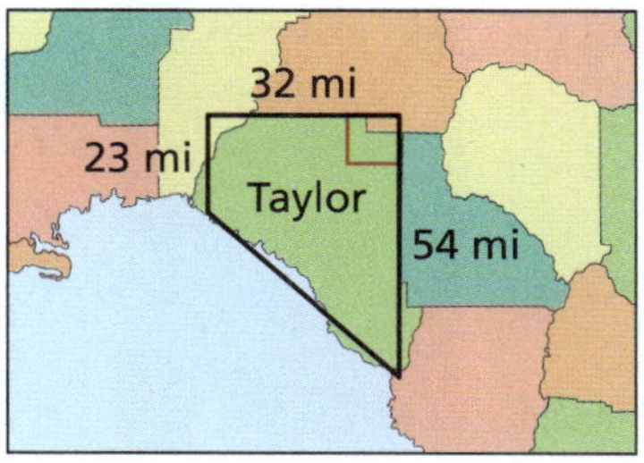

28. OPEN-ENDED The area of the trapezoidal student election sign is 5 square feet. Find two possible values for each base length.

29. REASONING How many times greater is the area of the floor covered by the larger speaker than by the area of the smaller speaker?

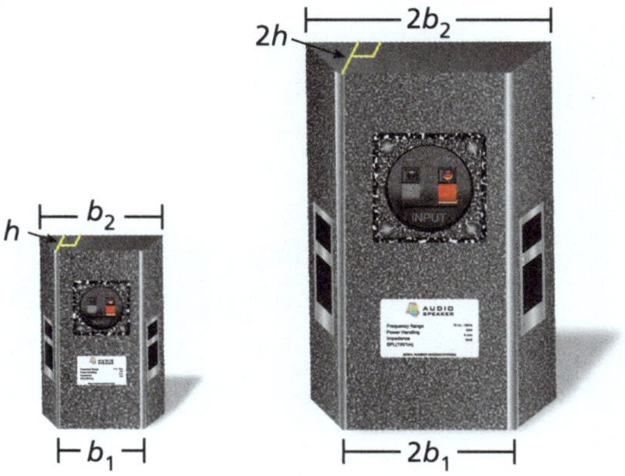

30. REASONING The rectangle and the trapezoid have the same area. What is the length ℓ of the rectangle?

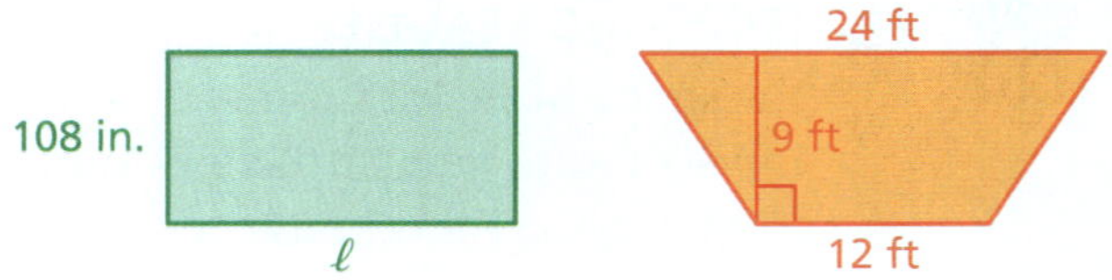

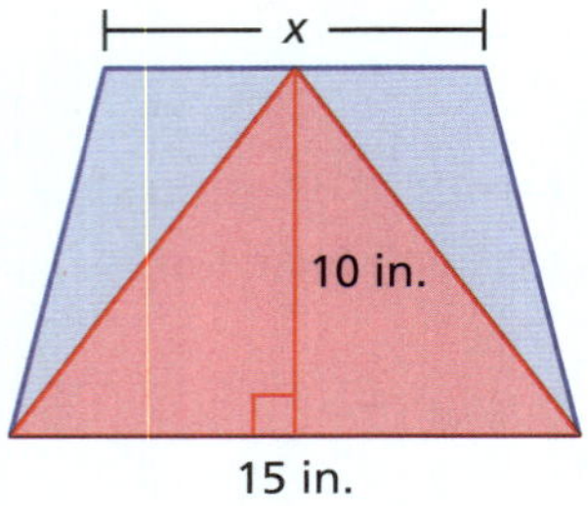

31. PROBLEM SOLVING In the figure shown, the area of the trapezoid is less than twice the area of the triangle. Find the possible values of x. Can the trapezoid have the same area as the triangle? Explain your reasoning.

5 MTR

32. STRUCTURE The quadrilateral shown is called a *kite*. This kite has two pairs of adjacent sides with the same length and opposite sides with different lengths. In Section 10.3 Exercise 24, you wrote a formula for the area of a rhombus in terms of its diagonals.

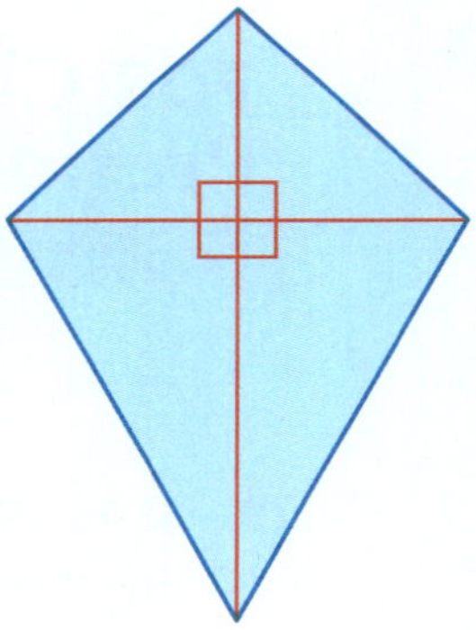

a. Use what you know about finding areas of figures to write a formula for the area of a kite in terms of its diagonals.

b. Are there any similarities between your formula in part (a) and the formula you found in Section 10.3? Explain.

10.5 Three-Dimensional Figures

Learning Target: Describe and draw three-dimensional figures.

Success Criteria:
- I can identify faces, edges, and vertices.
- I can sketch prisms and pyramids.
- I can draw different views of a solid.

Geometric Reasoning

preparing for MA.6.GR.2.3 Solve mathematical and real-world problems involving the volume of right rectangular prisms with positive rational number edge lengths using a visual model and a formula.

preparing for MA.6.GR.2.4 Given a mathematical or real-world context, find the surface area of right rectangular prisms and right rectangular pyramids using the figure's net.

Exploration 1 Exploring Faces, Edges, and Vertices

Work with a partner. Use the rectangular prism shown.

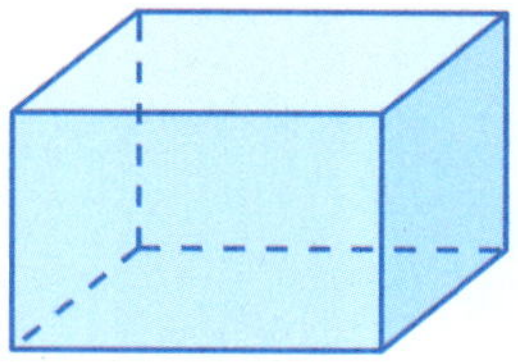

a. Prisms have *faces*, *edges*, and *vertices*. What does each of these terms mean?

b. What does it mean for lines or planes to be parallel or perpendicular in three dimensions? Use drawings to identify one pair of each of the following.

- parallel faces
- perpendicular faces
- parallel edges
- perpendicular edges
- edge parallel to a face
- edge perpendicular to a face

GO DIGITAL

Exploration 2 Drawing Views of a Solid

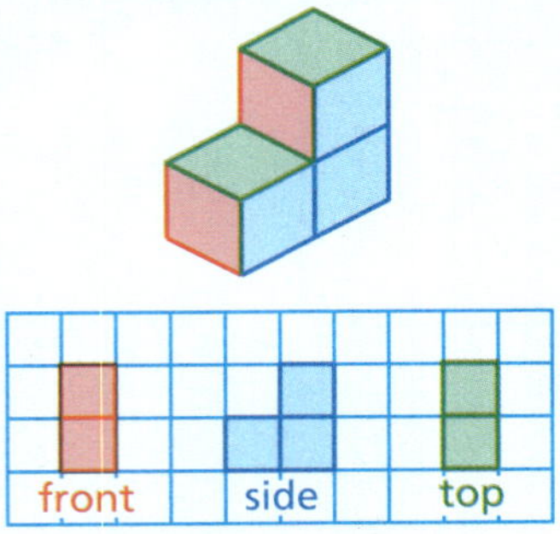

Number of cubes: 3

Work with a partner. Draw the front, side, and top views of each stack of cubes. Then find the number of cubes in the stack.

a.

b.

c.

d.

2 MTR MODEL A PROBLEM

What objects can you use to model the stacks of cubes? How can this help you perform the exploration?

10.5 Lesson

Key Vocabulary
solid, *p. 827*
polyhedron, *p. 827*
face, *p. 827*
edge, *p. 827*
vertex, *p. 827*
prism, *p. 827*
pyramid, *p. 827*

A **solid** is a three-dimensional figure that encloses a space. A **polyhedron** is a solid whose *faces* are all polygons.

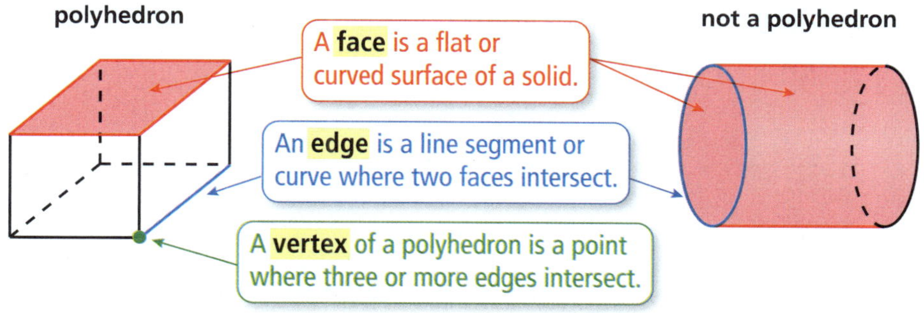

Example 1 Finding the Numbers of Faces, Edges, and Vertices

Find the numbers of faces, edges, and vertices of the solid.

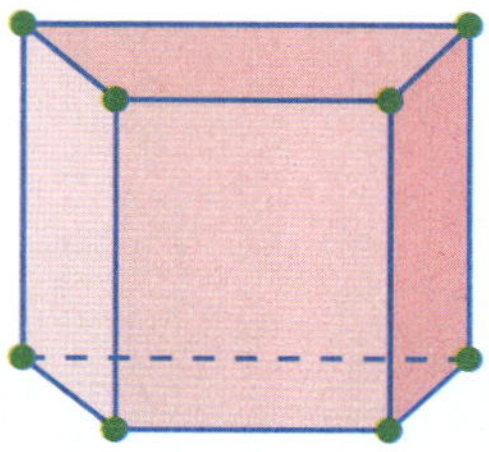

The solid has 1 face on the bottom, 1 face on the top, and 4 faces on the sides.

The faces intersect at 12 different line segments.

The edges intersect at 8 different points.

▶ So, the solid has 6 faces, 12 edges, and 8 vertices.

Try It

1. Find the numbers of faces, edges, and vertices of the solid.

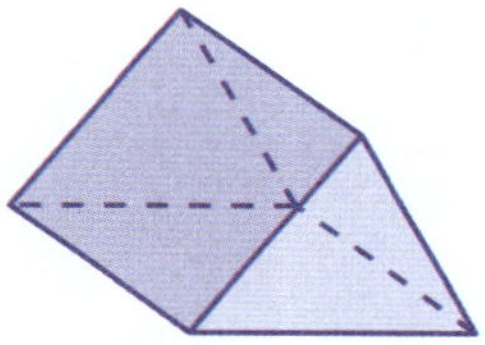

Key Ideas

Prisms

A **prism** is a polyhedron that has two parallel, identical *bases*. The *lateral faces* are parallelograms.

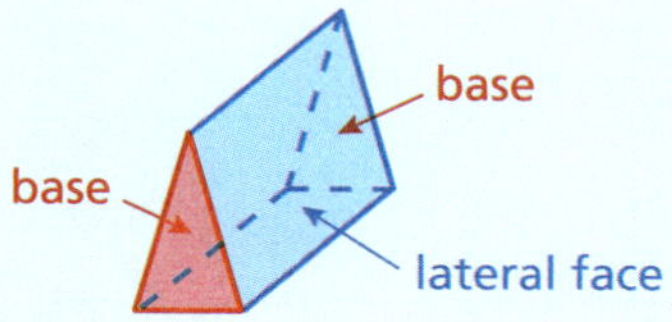

Triangular Prism

Pyramids

A **pyramid** is a polyhedron that has one base. The lateral faces are triangles.

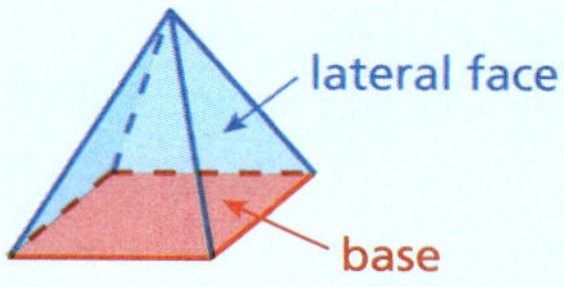

Rectangular Pyramid

The shape of the base tells the name of the prism or the pyramid.

GO DIGITAL

Example 2 Drawing Solids

a. Draw a rectangular prism.

Step 1: Draw identical rectangular bases.

Step 2: Connect corresponding vertices.

Step 3: Change any *hidden* lines to dashed lines.

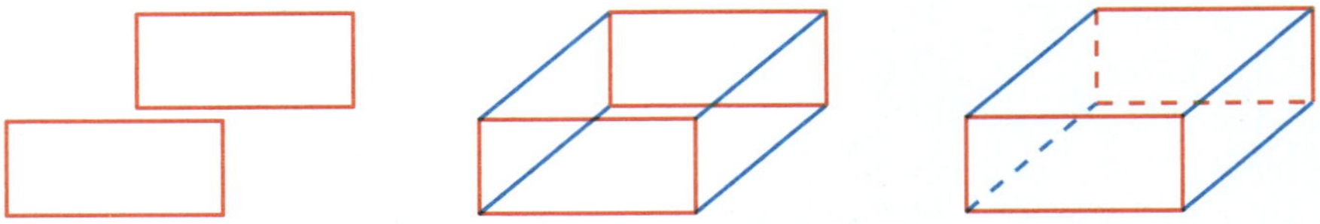

b. Draw a triangular pyramid.

Step 1: Draw a triangular base and a point.

Step 2: Connect the vertices of the triangle to the point.

Step 3: Change any *hidden* lines to dashed lines.

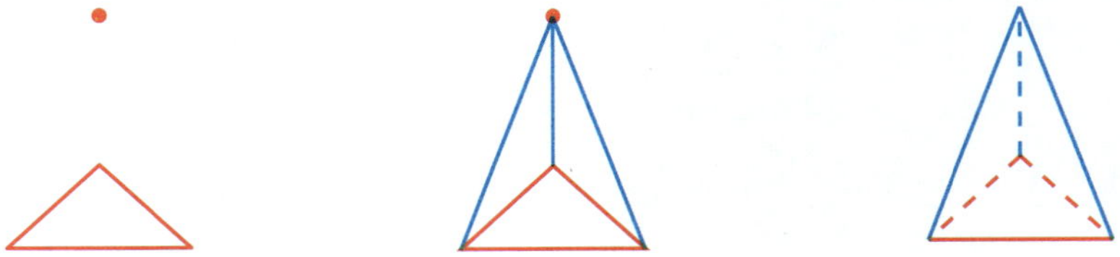

Try It **Draw the solid.**

2. square prism

3. pentagonal pyramid

In-Class Practice

1 I don't understand yet. 2 I can do it with help. 3 I can do it on my own. 4 I can teach someone else.

4. FACES, EDGES, AND VERTICES Find the numbers of faces, edges, and vertices of the solid at the left.

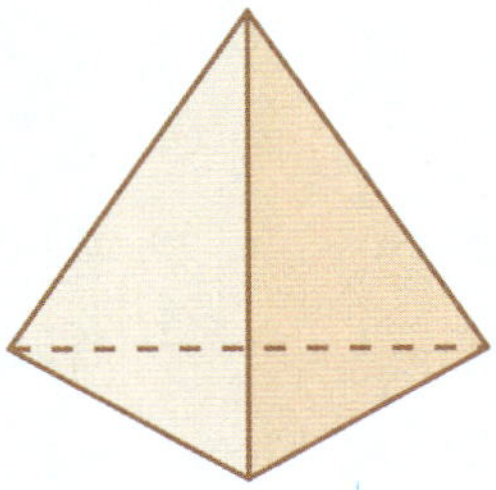

GO DIGITAL

In-Class Practice

1 I don't understand yet. 2 I can do it with help. 3 I can do it on my own. 4 I can teach someone else.

5. DRAWING A SOLID Draw an octagonal prism.

4 MTR **6. WHICH ONE DOESN'T BELONG?** Which figure does *not* belong with the other three? Explain your reasoning.

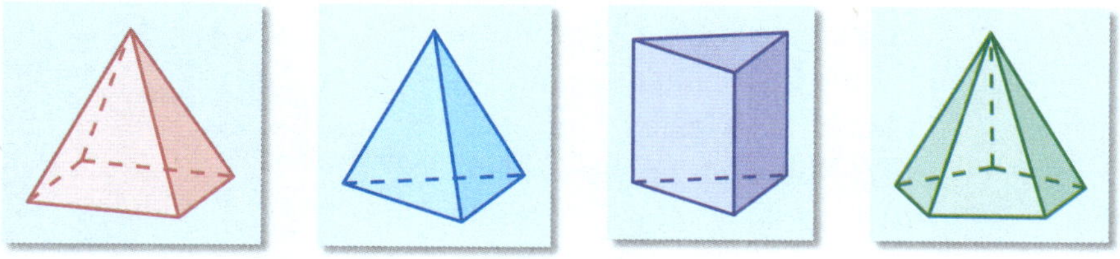

Example 3 Modeling Real Life

7 MTR

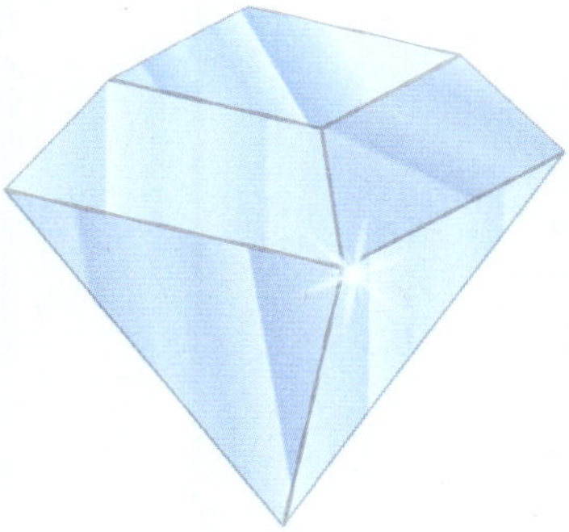

a. Find the numbers of faces, edges, and vertices of the table-cut diamond.

The diamond has 1 face on the top and 8 faces on the sides.

The faces intersect at 16 different line segments.

The edges intersect at 9 different points.

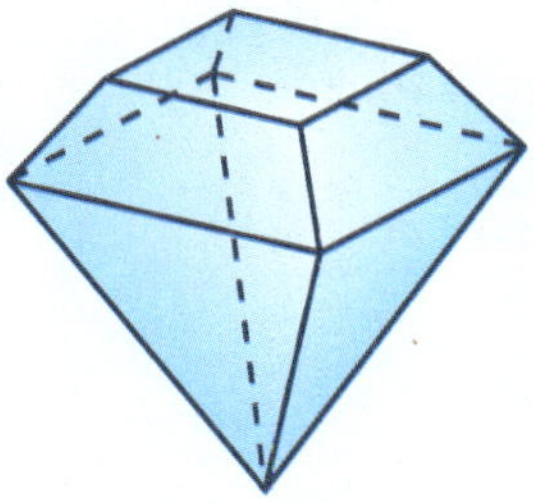

▶ So, the diamond has 9 faces, 16 edges, and 9 vertices.

b. Draw the front, side, and top views of the diamond.

Front view | Side view | Top view

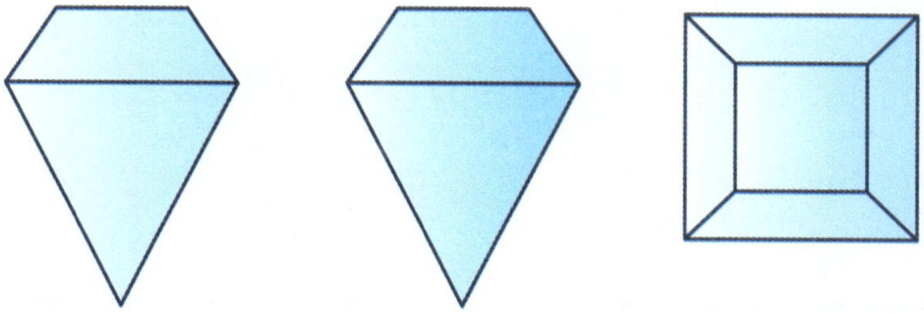

GO DIGITAL

In-Class Practice

1 I don't understand yet. | 2 I can do it with help. | 3 I can do it on my own. | 4 I can teach someone else.

7. The Flatiron Building in New York City is in the shape of a triangular prism. Draw a sketch of the building.

8. The Pyramid of the Niches is in El Tajín, an archaeological site in Veracruz, Mexico. Draw the front, side, and top views of the pyramid. Explain.

9. Use the point-cut diamond shown.

   a. Find the numbers of faces, edges, and vertices of the diamond.

   b. Draw the front, side, and top views of the diamond.

   c. How can a jeweler transform the point-cut diamond into a table-cut diamond as in Example 3?

10.5 Practice WITH CalcChat® AND CalcView®

Review & Refresh

Find the area of the trapezoid.

1. 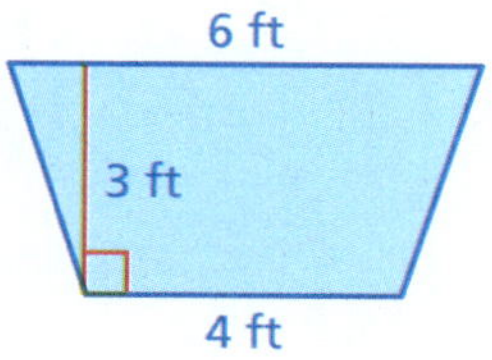

2. 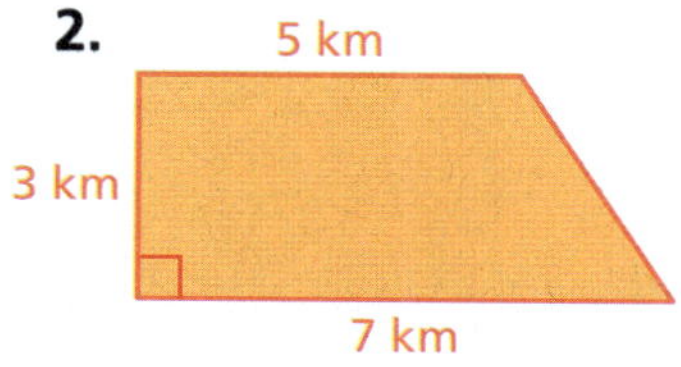

3. 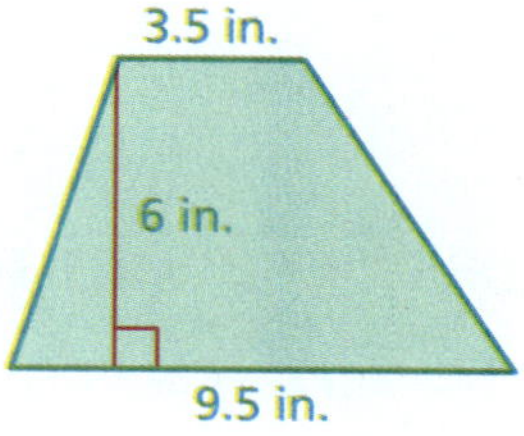

Find the LCM of the numbers.

4. 8, 12

5. 15, 25

6. 14, 24

7. 3, 7, 10

A bucket contains stones and seashells. You are given the number of seashells in the bucket and the ratio of stones to seashells. Find the number of stones in the bucket.

8. 18 seashells; 2 to 1

9. 30 seashells; 4 : 3

10. 40 seashells; 7 : 4

Concepts, Skills, & Problem Solving

DRAWING VIEWS OF A SOLID **Draw the front, side, and top views of the stack of cubes. Then find the number of cubes in the stack.** (See Exploration 2.)

11. 

12. 

13. 

GO DIGITAL

FACES, EDGES, AND VERTICES **Find the numbers of faces, edges, and vertices of the solid.** (See Example 1.)

14. 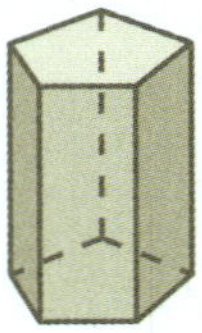

▶**15.** 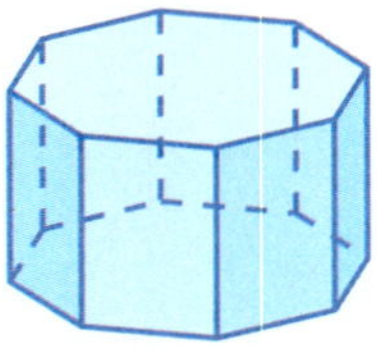

16. 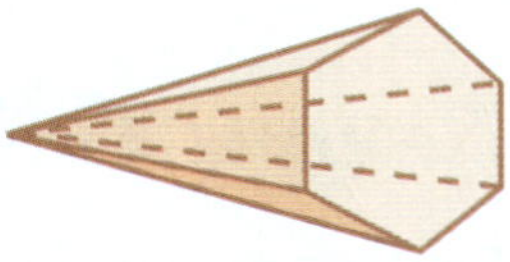

DRAWING SOLIDS **Draw the solid.** (See Example 2.)

17. triangular prism

18. pentagonal prism

▶**19.** rectangular pyramid

20. hexagonal pyramid

7 MTR

21. **MODELING REAL LIFE** The Pyramid of Cestius in Rome, Italy, is in the shape of a square pyramid. Draw a sketch of the pyramid.

7 MTR

22. **INVESTIGATE** Use the Internet to find a picture of the Washington Monument. Describe its shape.

DRAWING VIEWS OF A SOLID **Draw the front, side, and top views of the solid.** (See Example 3.)

23. 

24. 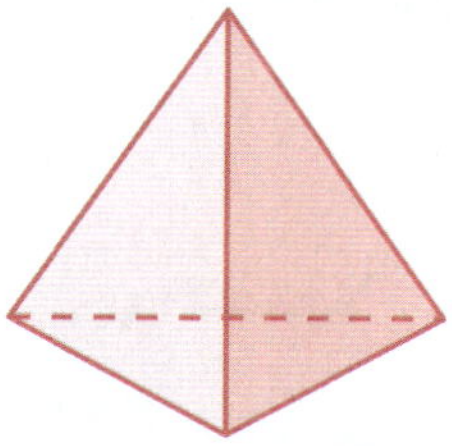

25. 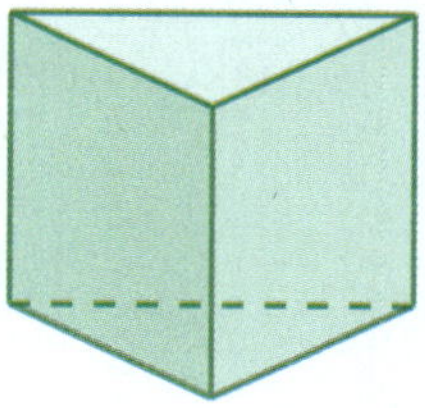

26. 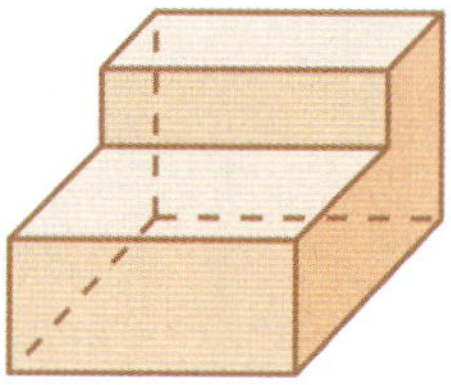

27. 

28. 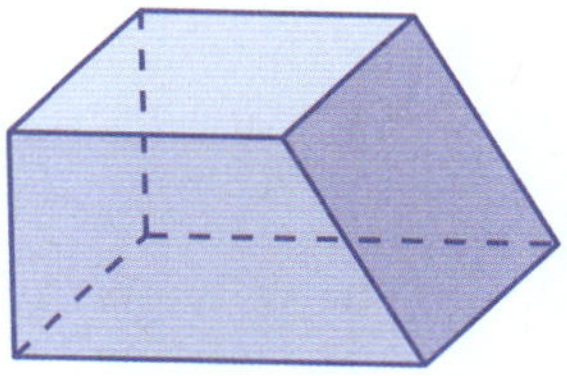

DRAWING SOLIDS **Draw a solid with the following front, side, and top views.**

29. 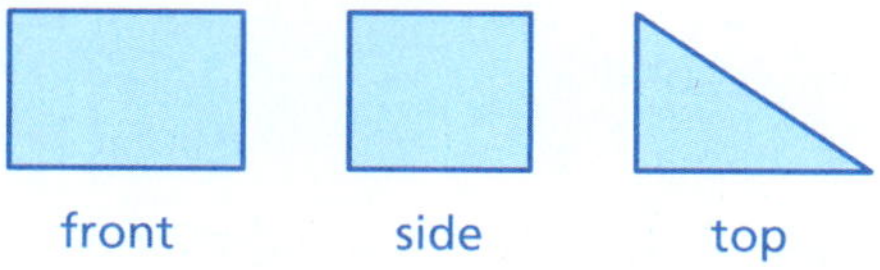

30. 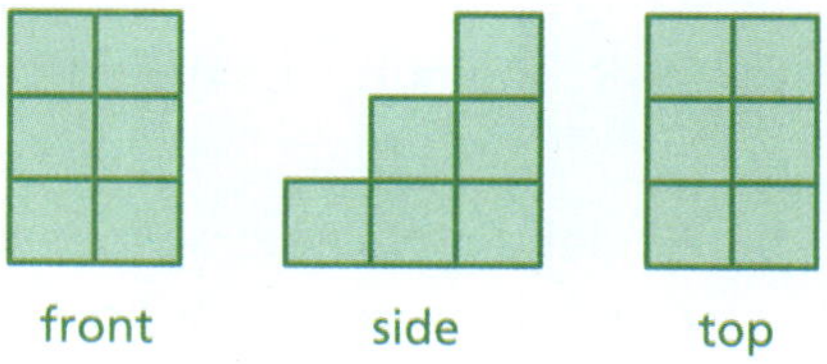

7 MTR

31. MODELING REAL LIFE Design and draw a house. Name the different solids that you can use to make a model of the house.

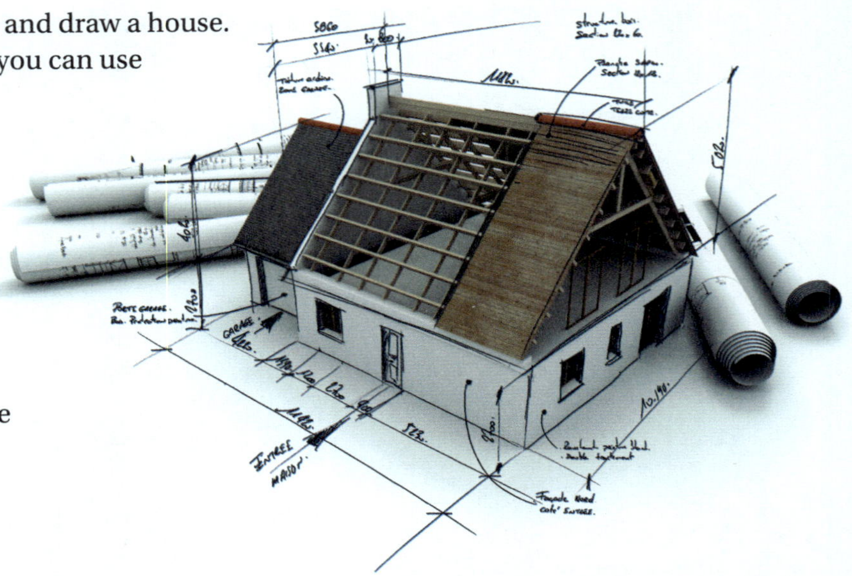

32. Dig Deeper Two of the three views of a solid are shown.

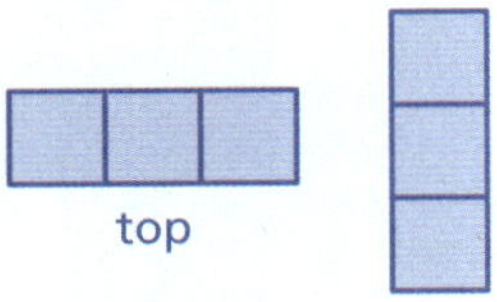

a. What is the greatest number of cubes in the solid?

b. What is the least number of cubes in the solid?

c. Draw the front views of both solids in parts (a) and (b).

33. OPEN-ENDED Draw two different solids with five faces.

a. Write the numbers of vertices and edges for each solid.

b. Explain how knowing the numbers of edges and vertices helps you draw a three-dimensional figure.

34. PROBLEM SOLVING The base of a pyramid has n sides. Find the numbers of faces, edges, and vertices of the pyramid. Explain your reasoning.

10.6 Surface Areas of Rectangular Prisms

Learning Target:	Find surface areas of rectangular prisms.
Success Criteria:	• I can identify a net for a prism. • I can use a net to find the surface area of a rectangular prism. • I can solve real-life problems involving surface areas of rectangular prisms.

Geometric Reasoning
MA.6.GR.2.4 Given a mathematical or real-world context, find the surface area of right rectangular prisms and right rectangular pyramids using the figure's net.

Exploration 1 Using Grid Paper to Construct a Solid

Work with a partner. Copy the figure shown below onto grid paper.

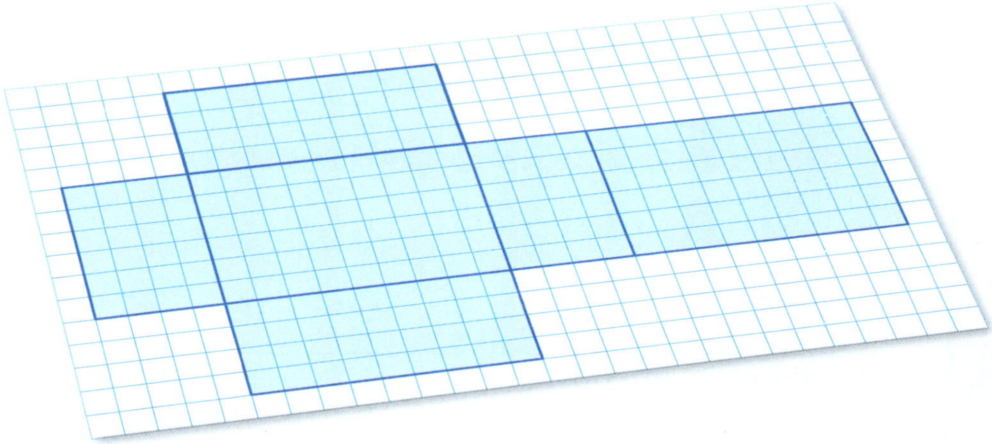

a. Cut out and fold the figure to form a solid. What type of solid does the figure form?

b. What is the area of the entire surface of the solid?

GO DIGITAL

Exploration 2 Finding the Area of the Entire Surface

Work with a partner. Find the area of the entire surface of each solid. Explain your reasoning.

a.

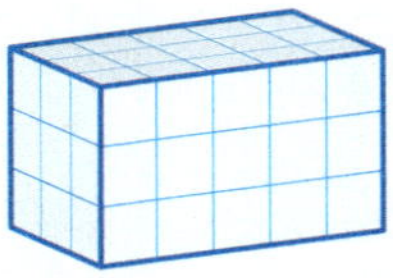

b.

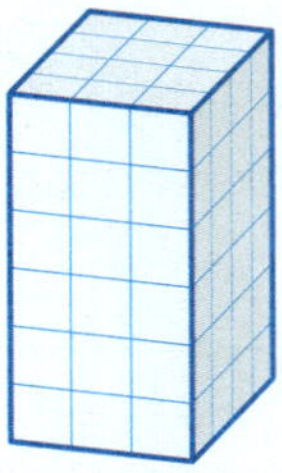

3 MTR **REFLECT ON YOUR METHOD**

When finding the area of the entire surface, what calculations do you repeat?

10.6 Lesson

Key Vocabulary
surface area, *p. 837*
net, *p. 837*

The **surface area** of a solid is the sum of the areas of all of its faces. You can use a two-dimensional representation of a solid, called a **net**, to find the surface area of the solid. Surface area is measured in *square units.*

Key Idea

Net of a Rectangular Prism

A *rectangular prism* is a prism with rectangular bases.

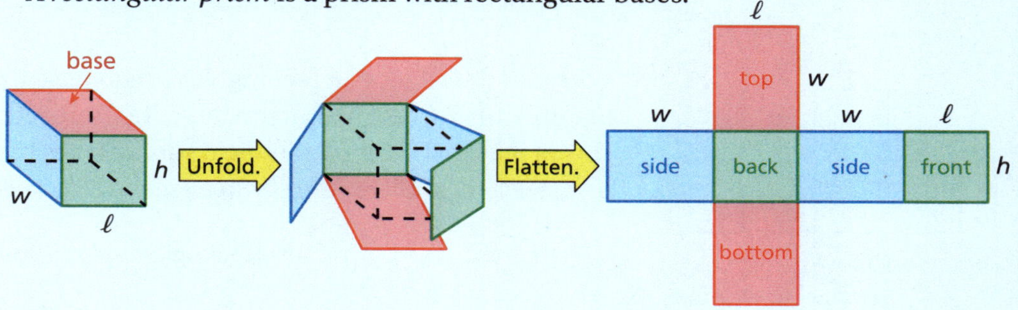

Example 1 Finding the Surface Area of a Rectangular Prism

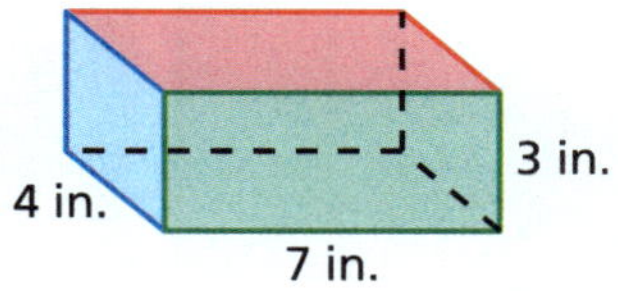

Find the surface area of the rectangular prism.

Use a net to find the area of each face.

Top: $7 \cdot 4 = 28$

Bottom: $7 \cdot 4 = 28$

Front: $7 \cdot 3 = 21$

Back: $7 \cdot 3 = 21$

Side: $4 \cdot 3 = 12$

Side: $4 \cdot 3 = 12$

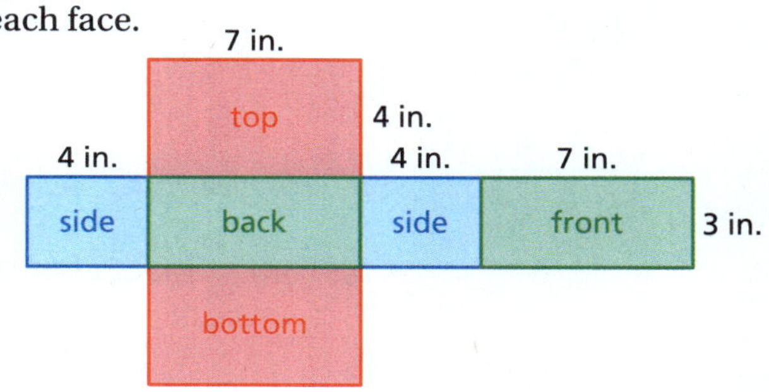

Find the sum of the areas of the faces.

Surface Area = Area of top + Area of bottom + Area of front + Area of back + Area of a side + Area of a side

$S = 28 + 28 + 21 + 21 + 12 + 12 = 122$

▶ So, the surface area is 122 square inches.

GO DIGITAL

Try It Find the surface area of the rectangular prism.

1.

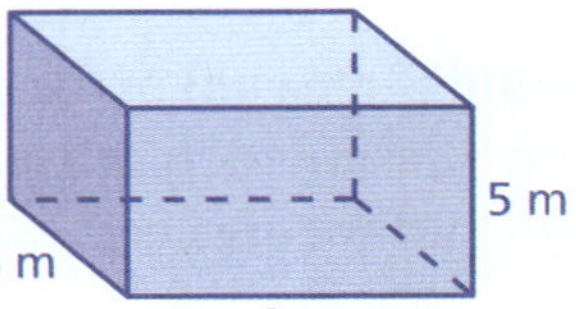

238 m²

2.

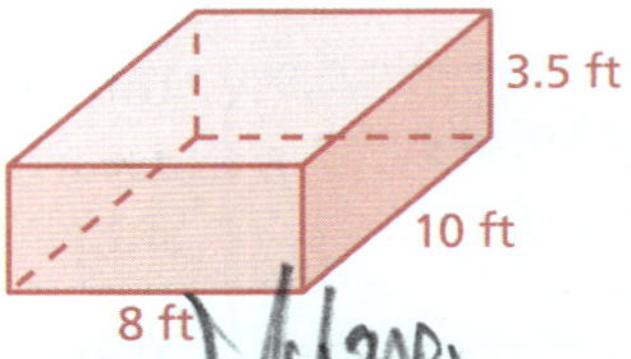

286 ft²

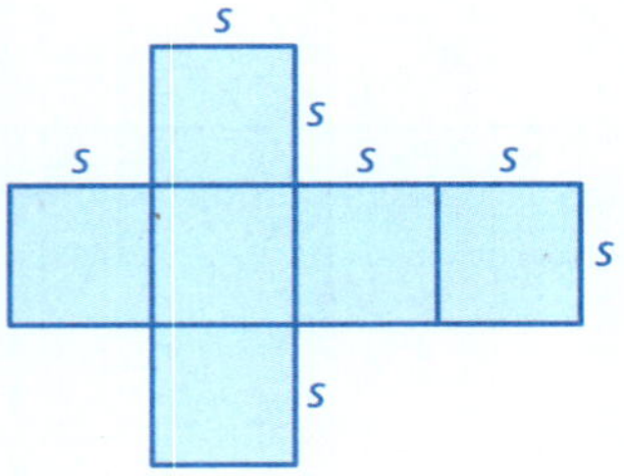

When all the edges of a rectangular prism have the same length s, the rectangular prism is a cube. The net of a cube shows that each of the 6 identical square faces has an area of s^2. So, a formula for the surface area of a cube is

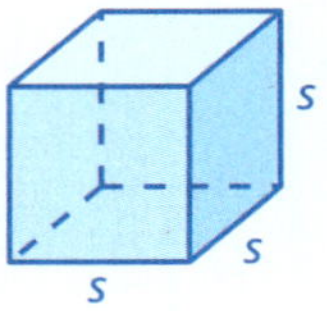

$S = 6s^2$. Formula for surface area of a cube

Example 2 Finding the Surface Area of a Cube

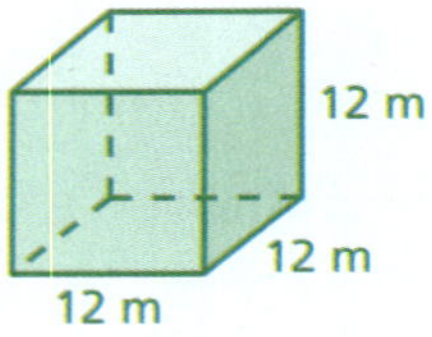

Find the surface area of the cube.

$S = 6s^2$ Write formula for surface area of a cube.

$= 6(12)^2$ Substitute 12 for s.

$= 6(144)$ Evaluate power.

$= 864$ Multiply.

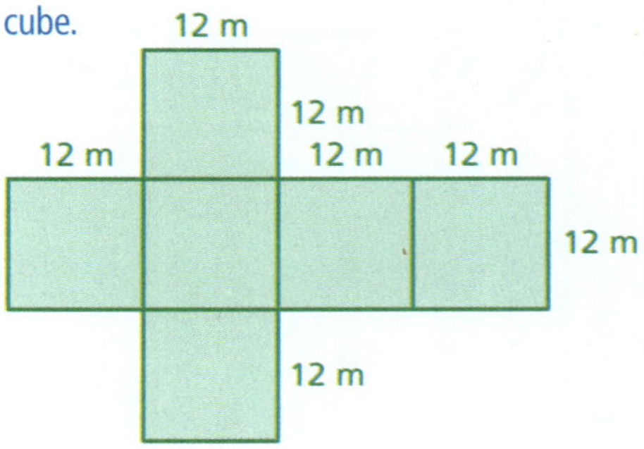

▶ The surface area of the cube is 864 square meters.

Try It Find the surface area of the cube.

3.

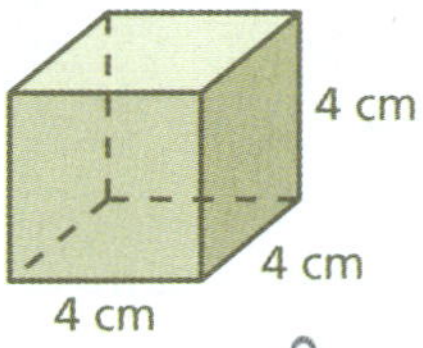

96 cm²

4.

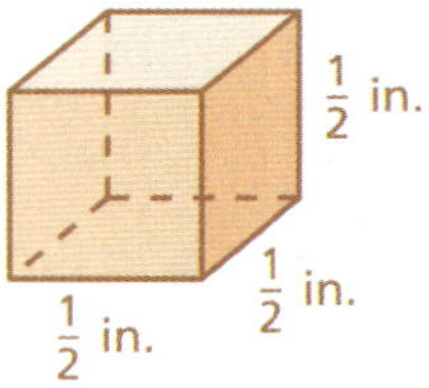

1.5 in²

GO DIGITAL

In-Class Practice

1 I don't understand yet. | 2 I can do it with help. | 3 I can do it on my own. | 4 I can teach someone else.

5. **FINDING SURFACE AREA** Find the surface area of a cube with edge lengths of 9 centimeters.

4 MTR

6. **DIFFERENT WORDS, SAME QUESTION** Which is different? Find "both" answers.

What is the sum of the areas of all of the faces of the prism?

What is the area of the entire surface of the prism?

What is the area of the largest face of the prism?

What is the surface area of the prism?

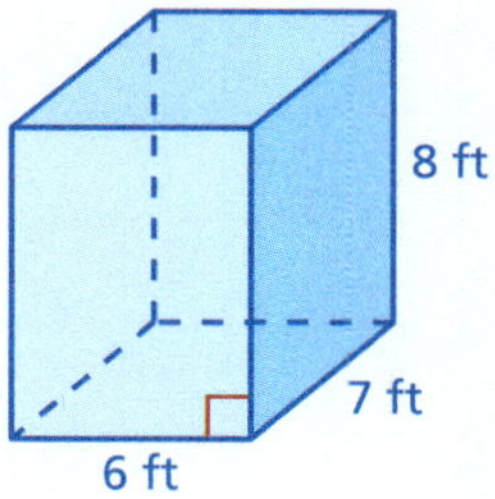

Example 3 Modeling Real Life

7 MTR

Space instruments are often wrapped in gold-colored multi-layer insulation (MLI) to reflect radiation from the Sun. What is the least amount of MLI needed to wrap an instrument in the shape of a rectangular prism with a length of 5 feet, a width of 5 feet, and a height of 3 feet?

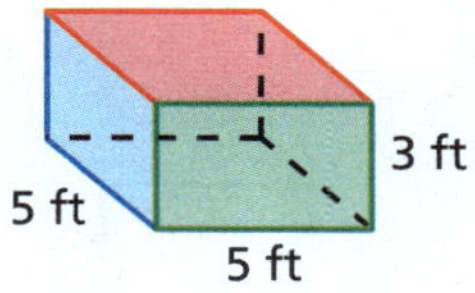

Draw the prism. The least amount of MLI needed is represented by the surface area of the prism. Use a net to find the surface area.

Top: $5 \cdot 5 = 25$

Bottom: $5 \cdot 5 = 25$

Front: $5 \cdot 3 = 15$

Back: $5 \cdot 3 = 15$

Side: $5 \cdot 3 = 15$

Side: $5 \cdot 3 = 15$

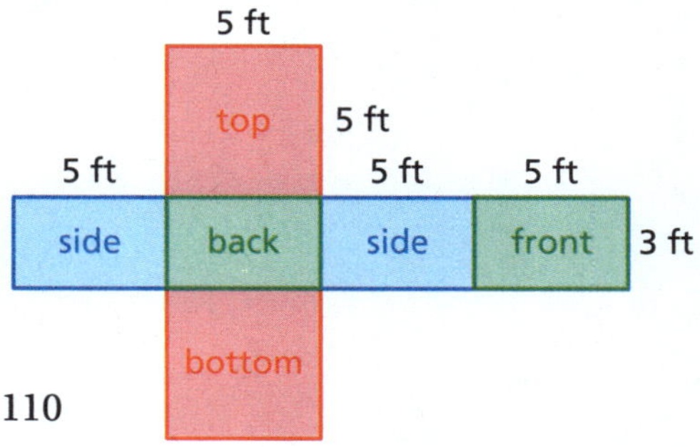

$S = 25 + 25 + 15 + 15 + 15 + 15 = 110$

▶ So, the least amount of MLI needed is 110 square feet.

Check Reasonableness The surface area of a 5 ft × 5 ft × 3 ft prism should be less than the surface area of a 5 ft × 5 ft × 5 ft cube. The cube has a surface area of $6(5)^2 = 150$ square feet. Because $110 \text{ ft}^2 < 150 \text{ ft}^2$, the answer is reasonable. ✓

GO DIGITAL

3 MTR ADAPT A PROCEDURE

What is the surface area of a rectangular prism with length ℓ, width w, and height h? Explain your reasoning.

In-Class Practice

1 I don't understand yet. 2 I can do it with help. 3 I can do it on my own. 4 I can teach someone else.

7. One pint of chalkboard paint covers 60 square feet. What is the least number of pints of paint needed to paint the walls of a room in the shape of a rectangular prism with a length of 15 feet, a width of 13 feet, and a height of 10 feet? Explain.

8. **Dig Deeper** A flexible *metamaterial* is developed for use in robotics and prosthetics. A block of metamaterial is in the shape of a cube with a surface area of 600 square centimeters. What is the edge length of the block of metamaterial?

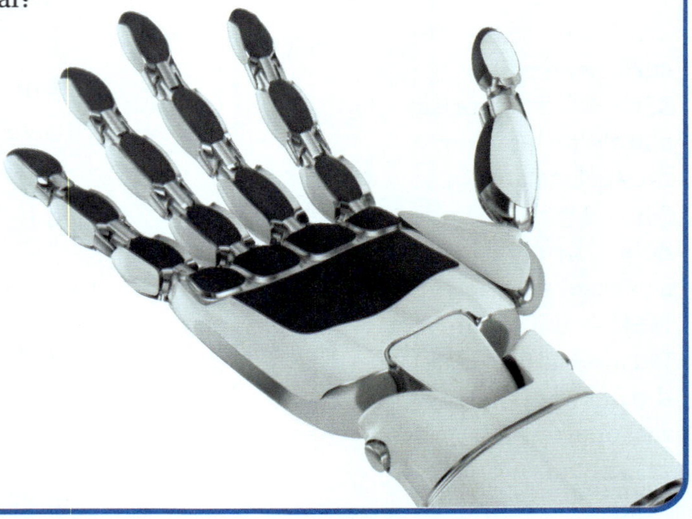

10.6 Practice WITH CalcChat® AND CalcView®

Review & Refresh

Draw the front, side, and top views of the solid.

1. 

2. 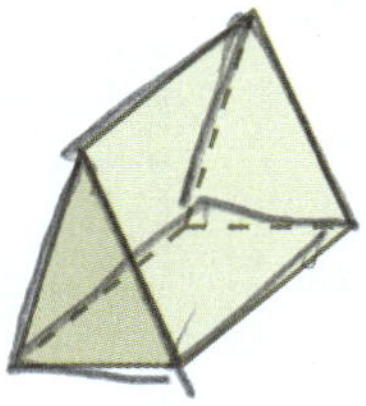

3. 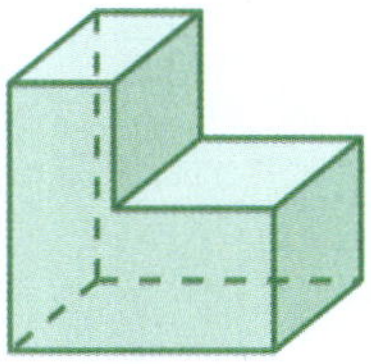

Find the GCF of the numbers.

4. 18, 72

5. 44, 110

6. 78, 315

7. 60, 96, 156

Solve the equation.

8. $s - 5 = 12$

9. $x + 9 = 20$

10. $48 = 6r$

11. $\frac{m}{5} = 13$

GO DIGITAL

Concepts, Skills, & Problem Solving

USING TOOLS **Use a net to find the area of the entire surface of the solid. Explain your reasoning.** (See Exploration 2.)

12. 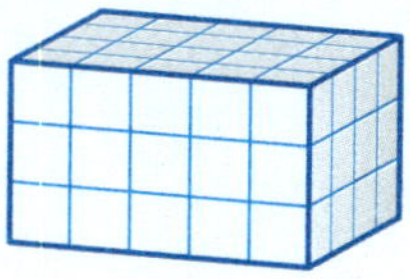

13. 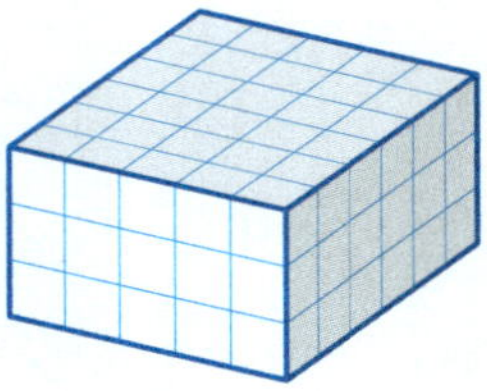

14. 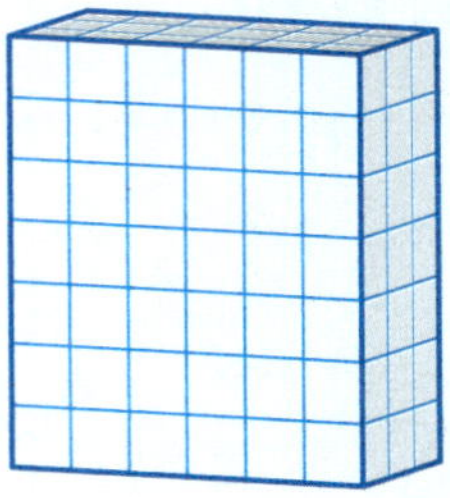

FINDING SURFACE AREA **Find the surface area of the prism represented by the net.**

15. 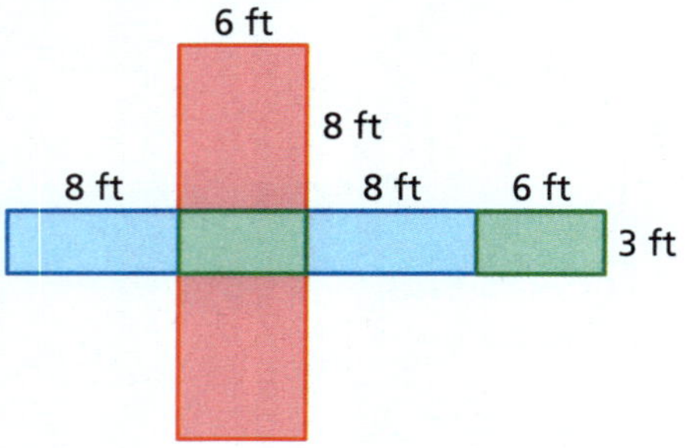

16. 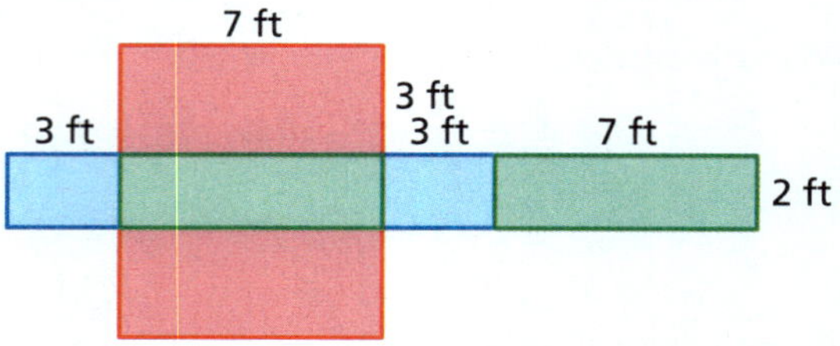

FINDING SURFACE AREA **Find the surface area of the rectangular prism.** (See Example 1.)

▶**17.** 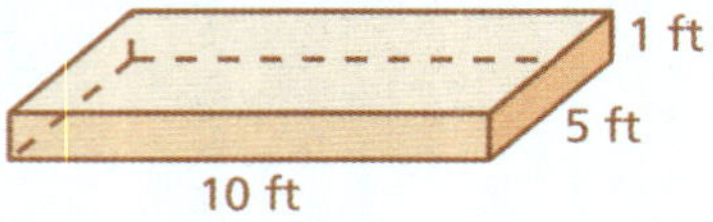

18. 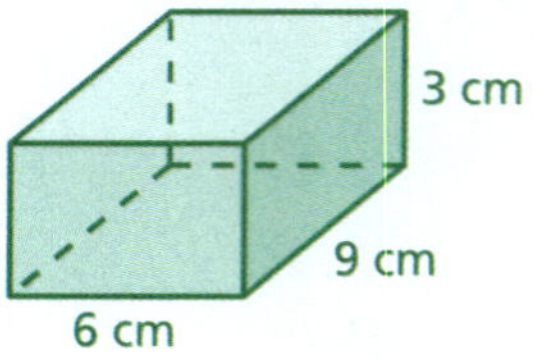

19. 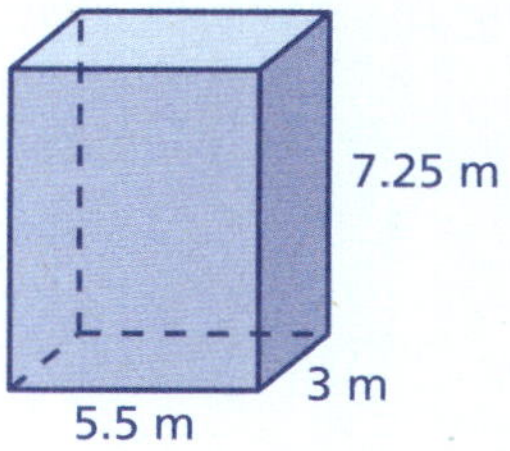

FINDING SURFACE AREA **Find the surface area of the cube.** (See Example 2.)

20.

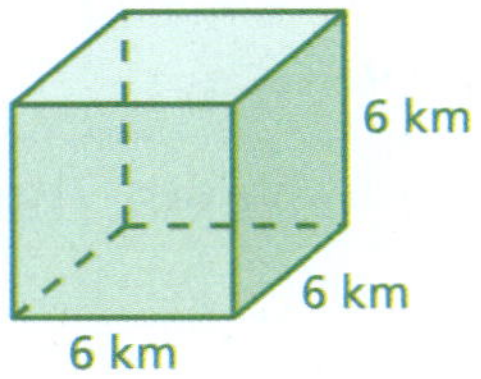

▶21.

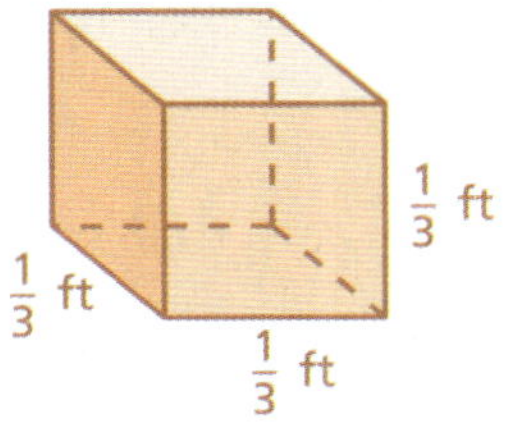

22.

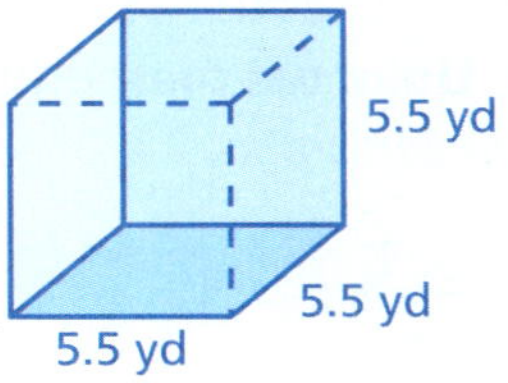

23. **MODELING REAL LIFE** A gift box in the shape of a rectangular prism measures 8 inches by 8 inches by 10 inches. What is the least amount of wrapping paper needed to wrap the gift box? Explain. (See Example 3.)

24. **MODELING REAL LIFE** A piece of dry ice is in the shape of a cube with edge lengths of 7 centimeters. Find the surface area of the dry ice.

25. **PROBLEM SOLVING** A cereal box has the dimensions shown.

a. Find the surface area of the cereal box.

b. The manufacturer decides to decrease the size of the box by reducing each of the dimensions by 1 inch. Find the decrease in surface area.

26. REASONING The material used to make a storage box costs $1.25 per square foot. The boxes have the same volume. Which box might a company prefer to make? Explain your reasoning.

	Length	Width	Height
Box 1	20 in.	6 in.	4 in.
Box 2	15 in.	4 in.	8 in.

27. B.E.S.T. Test Prep Which of the following are nets of a cube? Select all that apply.

Ⓐ 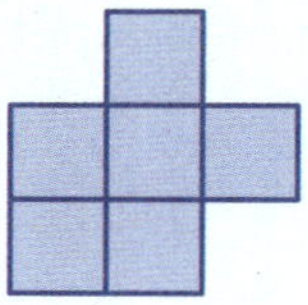

Ⓑ 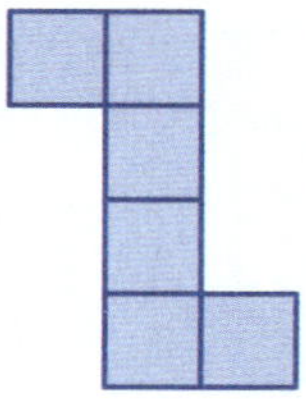

Ⓒ 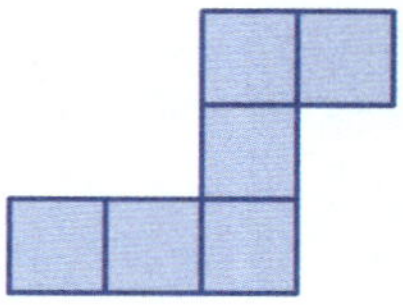

Ⓓ 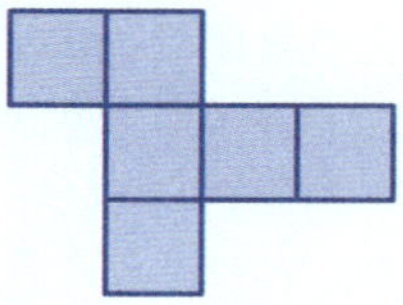

6 MTR

28. ASSESS REASONABLENESS The main portion of the Panorama Tower in Miami can be modeled by a rectangular prism. You know that the tower is 867 feet tall. One of your friends estimates that one side of the prism's base is 100 feet long, and the surface area of the tower is 289,440 square feet. Another friend estimates that one side of the prism's base is 150 feet long, and the surface area of the tower is 1,887,300 square feet. Which estimate do you find more reasonable? Explain your reasoning.

29. Dig Deeper A cube is removed from a rectangular prism. Find the surface area of the figure after removing the cube.

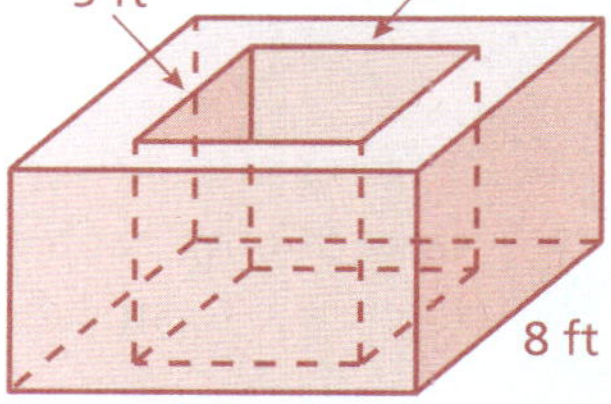

10.7 Surface Areas of Rectangular Pyramids

Learning Target: Find surface areas of rectangular pyramids.

Success Criteria:
- I can identify a net for a pyramid.
- I can use a net to find the surface area of a rectangular pyramid.
- I can solve real-life problems involving surface areas of rectangular pyramids.

Geometric Reasoning

MA.6.GR.2.4 Given a mathematical or real-world context, find the surface area of right rectangular prisms and right rectangular pyramids using the figure's net.

Exploration 1 Using a Net to Construct a Solid

Work with a partner. Copy the net shown below onto grid paper.

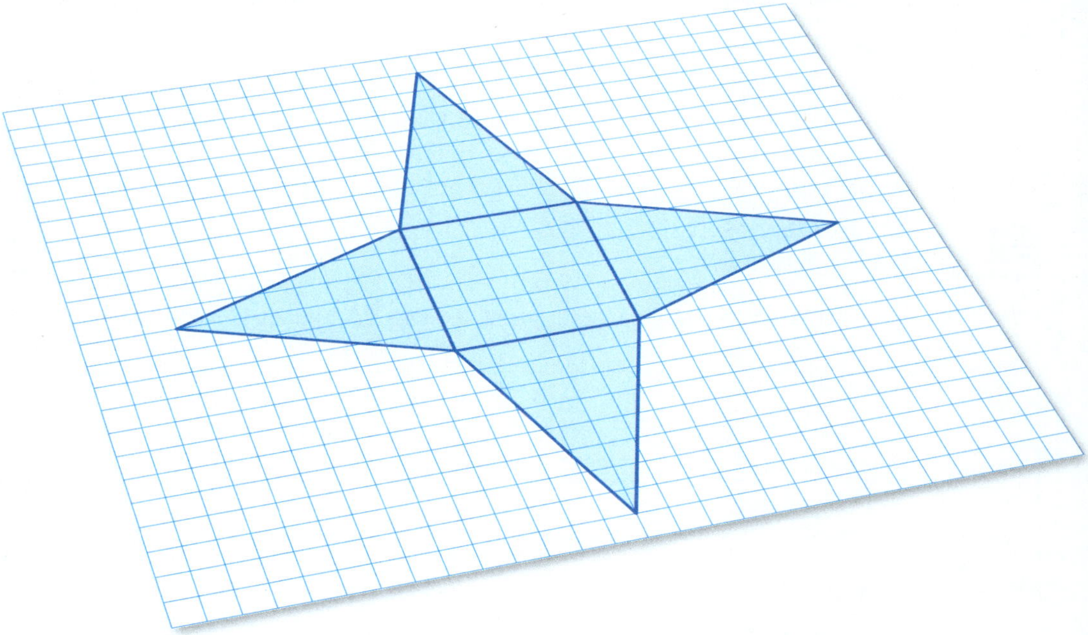

a. Cut out and fold the net to form a solid. What type of solid does the net form?

b. What is the surface area of the solid?

GO DIGITAL

Exploration 2 Finding Surface Areas of Solids

Work with a partner. Find the surface area of each solid. Explain your reasoning.

a.

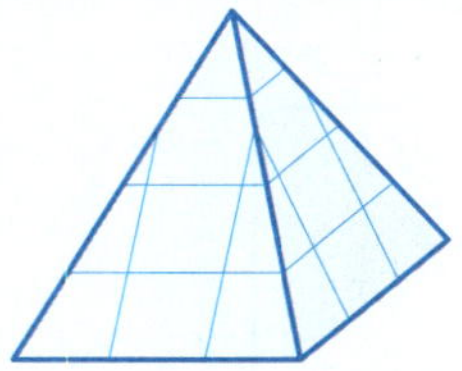

b.

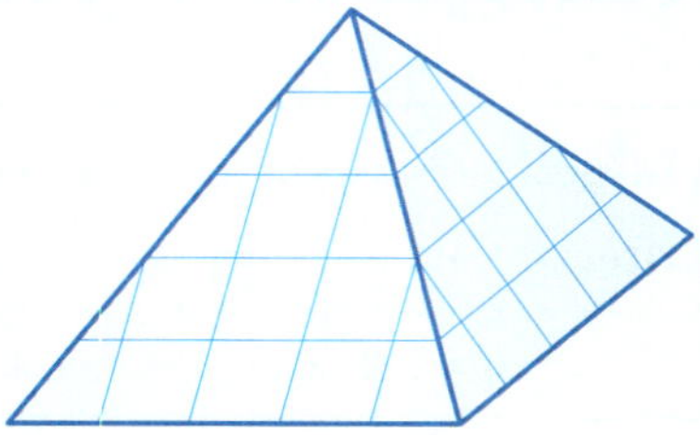

1 MTR ANALYZE A PROBLEM

What information can you determine from the diagram? How does this help you find the surface area of the solid?

10.7 Lesson

Key Idea

Net of a Rectangular Pyramid

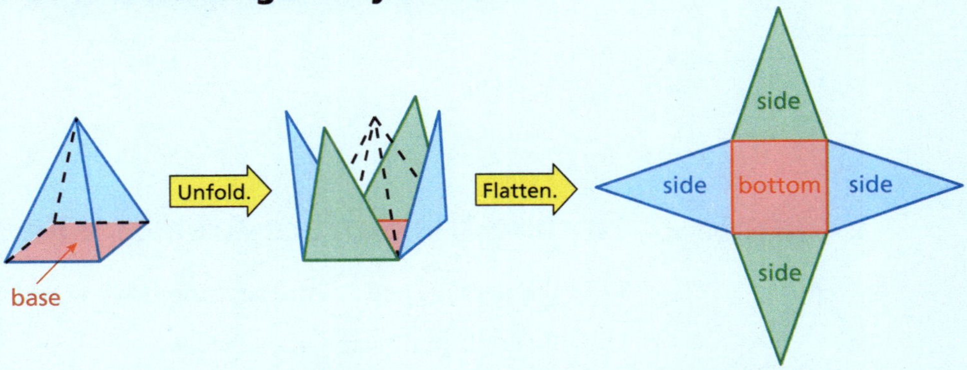

Right rectangular pyramids have four triangular lateral faces. Opposite lateral faces are identical. When the base is a square, all four lateral faces are identical.

Remember

A *rectangular pyramid* is a pyramid with a rectangular base. One type of rectangular pyramid is a *square* pyramid.

Example 1 Finding the Surface Area of a Square Pyramid

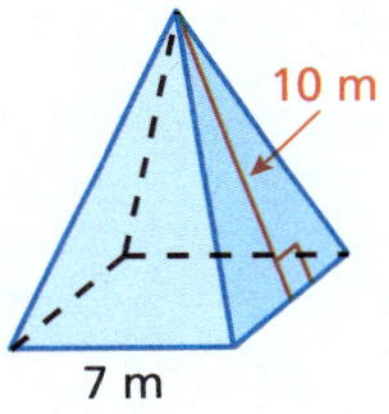

Find the surface area of the square pyramid.

Use a net to find the area of each face.

Bottom: $7 \cdot 7 = 49$

Side: $\frac{1}{2} \cdot 7 \cdot 10 = 35$

Side: $\frac{1}{2} \cdot 7 \cdot 10 = 35$

Side: $\frac{1}{2} \cdot 7 \cdot 10 = 35$

Side: $\frac{1}{2} \cdot 7 \cdot 10 = 35$

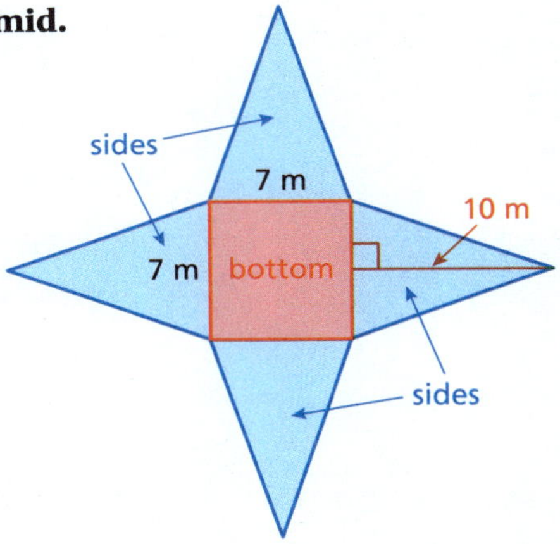

Find the sum of the areas of the faces.

Surface Area = Area of bottom + Area of a side + Area of a side + Area of a side + Area of a side

$S = 49 + 35 + 35 + 35 + 35 = 189$

▶ So, the surface area is 189 square meters.

5 MTR USE A PATTERN

How can you find the surface area of a square pyramid by calculating the area of only two of its faces?

____

____

GO DIGITAL

Try It Find the surface area of the square pyramid.

1.

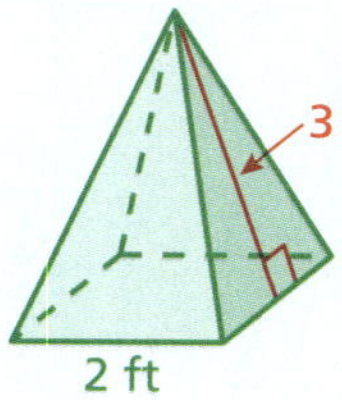

2.

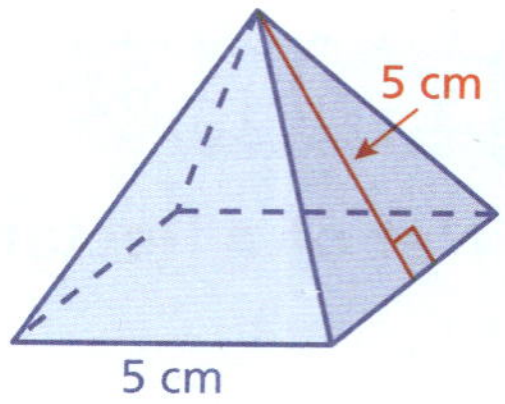

Example 2 Finding the Surface Area of a Rectangular Pyramid

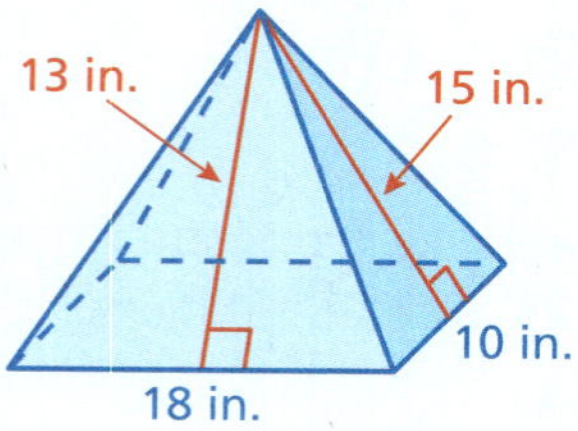

Find the surface area of the rectangular pyramid.

Use a net to find the area of each face.

Bottom: $18 \cdot 10 = 180$

Side: $\frac{1}{2} \cdot 10 \cdot 15 = 75$

Side: $\frac{1}{2} \cdot 10 \cdot 15 = 75$

Side: $\frac{1}{2} \cdot 18 \cdot 13 = 117$

Side: $\frac{1}{2} \cdot 18 \cdot 13 = 117$

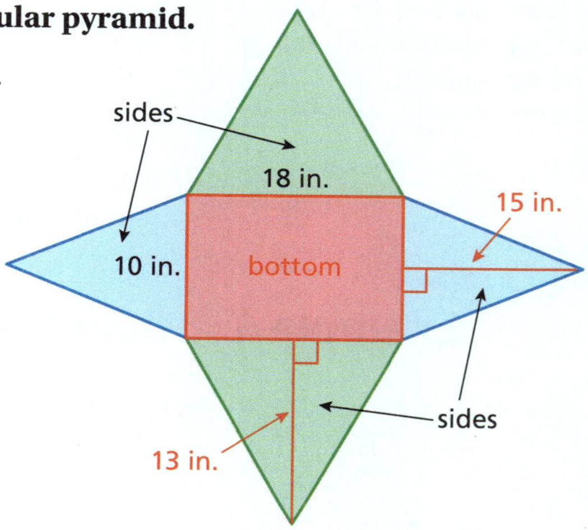

Find the sum of the areas of the faces.

Surface Area	=	Area of bottom	+	Area of a side	+	Area of a side	+	Area of a side	+	Area of a side
S	=	180	+	75	+	75	+	117	+	117
	=	564								

▶ So, the surface area is 564 square inches.

Try It Find the surface area of the rectangular pyramid.

3.

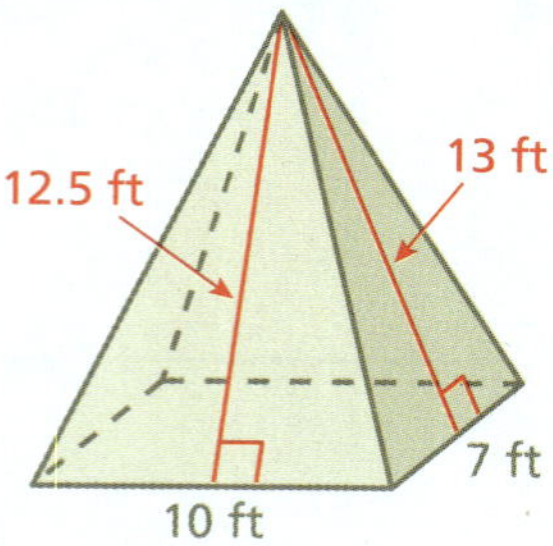

4.

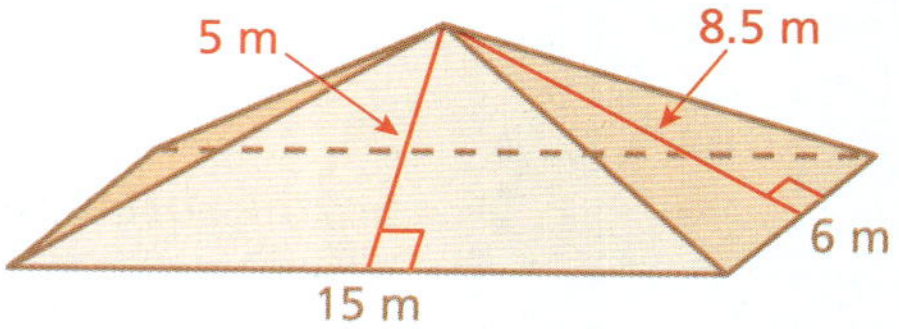

GO DIGITAL

In-Class Practice

1 I don't understand yet. **2** I can do it with help. **3** I can do it on my own. **4** I can teach someone else.

5. **WRITING** Explain how to find the surface area of a rectangular pyramid.

FINDING SURFACE AREA **Find the surface area of the rectangular pyramid.**

6.

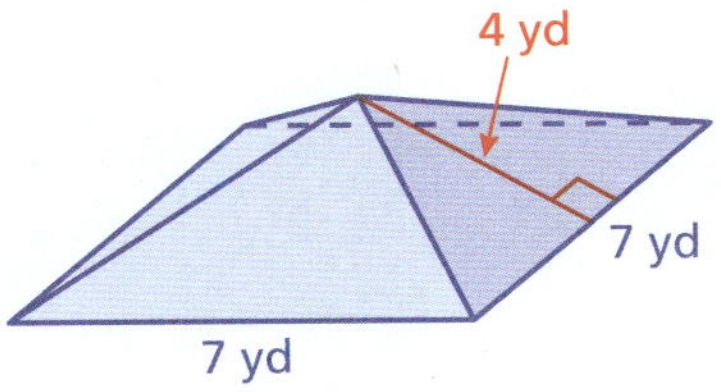

7.

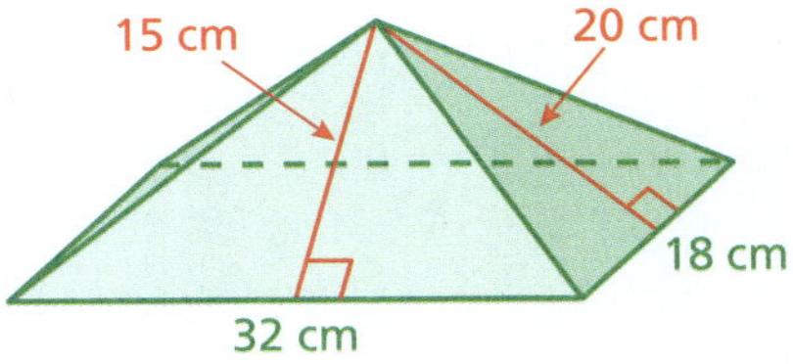

Example 3 Modeling Real Life

7 MTR

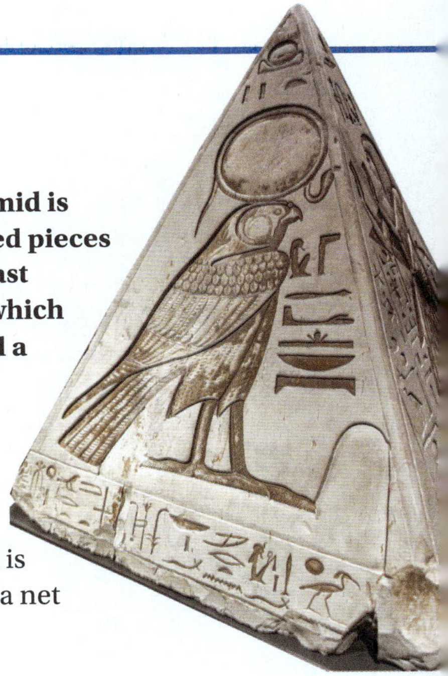

The uppermost piece of an ancient Egyptian pyramid is called a pyramidion. These square pyramid-shaped pieces were sometimes covered with gold. What is the least amount of gold needed to cover a pyramidion in which each triangular face has a height of 1.2 meters and a base of 1.5 meters?

Understand the problem.

You are given the dimensions of a pyramidion. You are asked to find the least amount of gold needed to cover the pyramidion.

Make a plan.

Draw the pyramid. The least amount of gold needed is represented by the surface area of the pyramid. Use a net to find the surface area.

Solve and check.

Bottom: $1.5 \cdot 1.5 = 2.25$

Side: $\frac{1}{2} \cdot 1.5 \cdot 1.2 = 0.9$

Side: $\frac{1}{2} \cdot 1.5 \cdot 1.2 = 0.9$

Side: $\frac{1}{2} \cdot 1.5 \cdot 1.2 = 0.9$

Side: $\frac{1}{2} \cdot 1.5 \cdot 1.2 = 0.9$

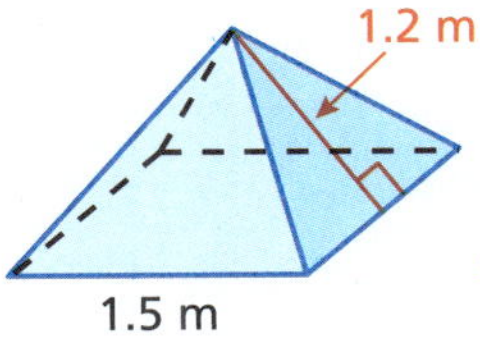

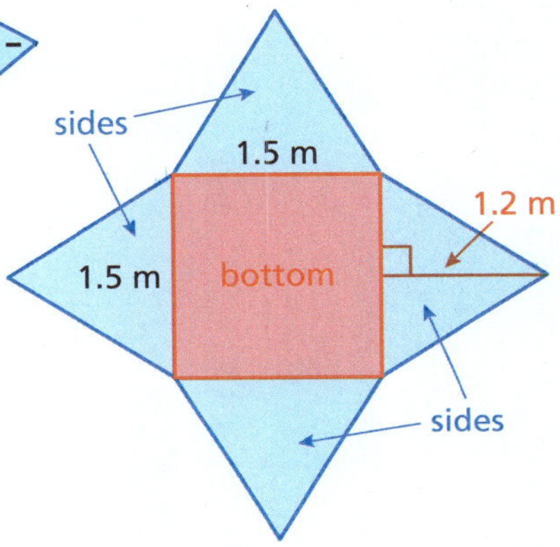

$S = 2.25 + 0.9 + 0.9 + 0.9 + 0.9 = 5.85$

▶ So, the least amount of gold needed is 5.85 square meters.

Check Reasonableness

The surface area should be less than the surface area of a cube with an edge length of 1.5 meters.

$S = 6(1.5)^2$
$= 13.5 \text{ m}^2$ ✓

GO DIGITAL

In-Class Practice

1 I don't understand yet. 2 I can do it with help. 3 I can do it on my own. 4 I can teach someone else.

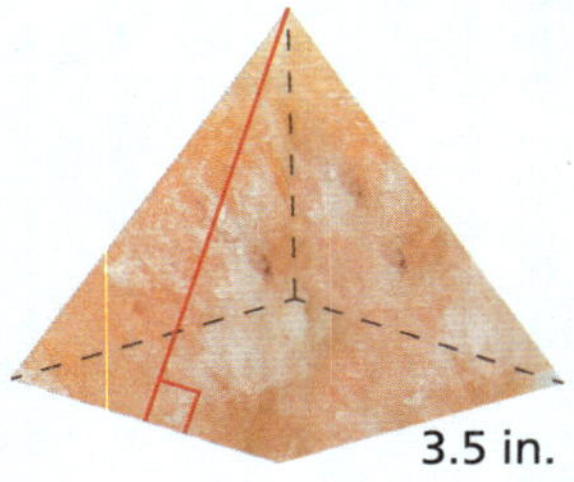

Surface Area = 40.25 in.^{2}

8. A salt lamp is shaped like a square pyramid. Find the area of each triangular face.

9. Originally, each triangular face of the Great Pyramid of Giza had a height of 612 feet and a base of 756 feet. Today, the height of each triangular face of the square pyramid is 592 feet. Find the change in the total surface area of the four triangular faces of the Great Pyramid of Giza.

10.7 Practice WITH CalcChat® AND CalcView®

Review & Refresh

Find the surface area of the prism.

1.

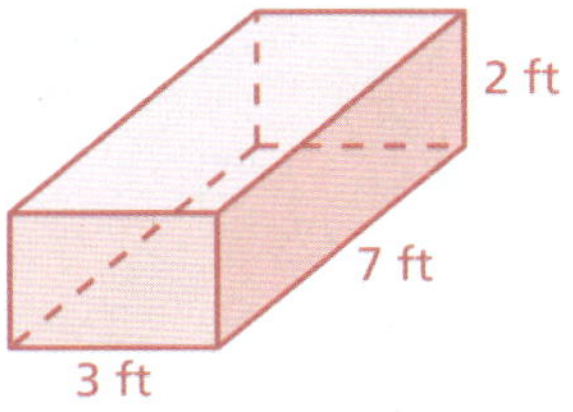

2.

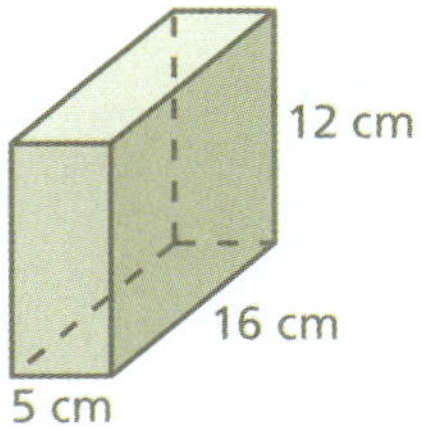

3.

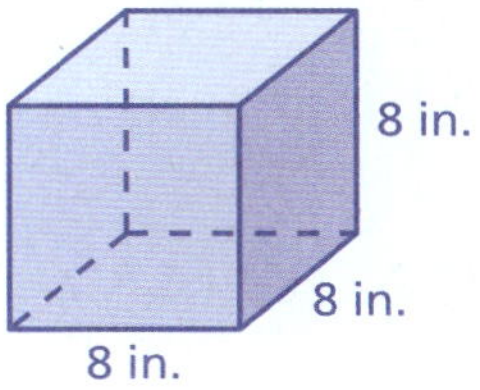

Match the expression with an equivalent expression.

4. $3(4n + 2)$ **5.** $6(2n + 3)$ **6.** $4(3n + 4)$ **7.** $12(n + 1)$

A. $2(6n + 6)$ **B.** $12n + 18$ **C.** $2(6n + 3)$ **D.** $12n + 16$

Concepts, Skills, & Problem Solving

2 MTR **MODEL A PROBLEM** **Use a net to find the surface area of the solid. Explain your reasoning.** (See Exploration 2.)

8.

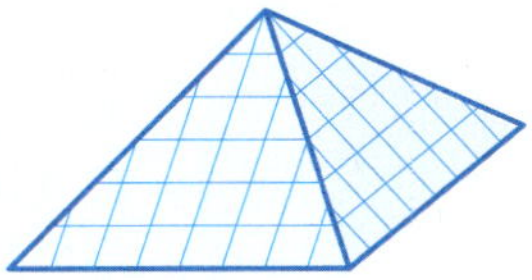

9.

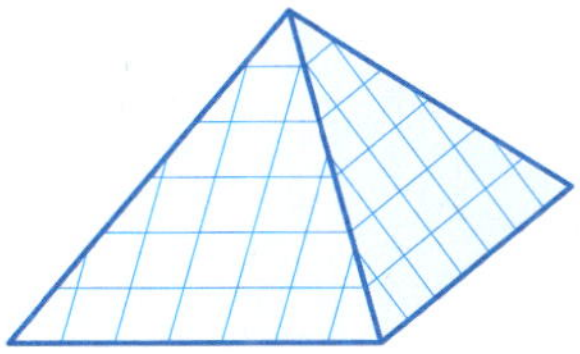

10.

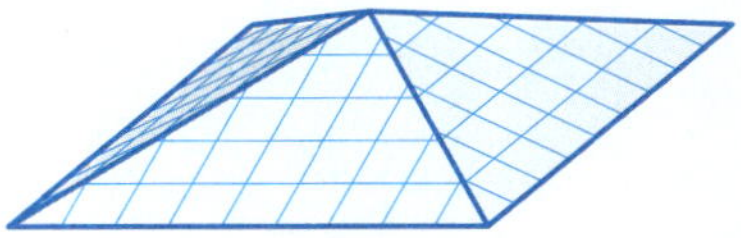

GO DIGITAL

FINDING SURFACE AREA **Find the surface area of the pyramid represented by the net.**

11.

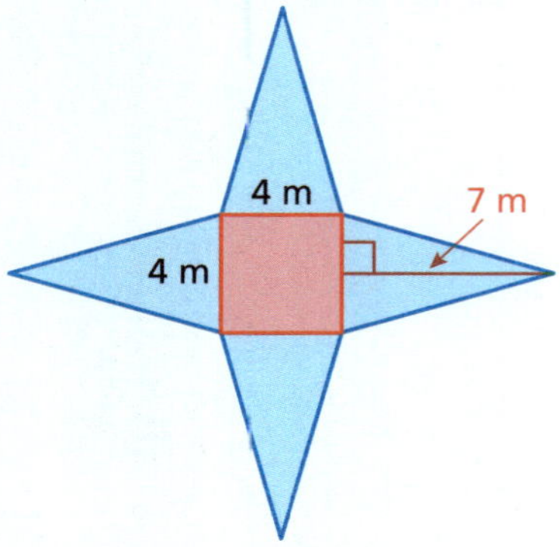

12.

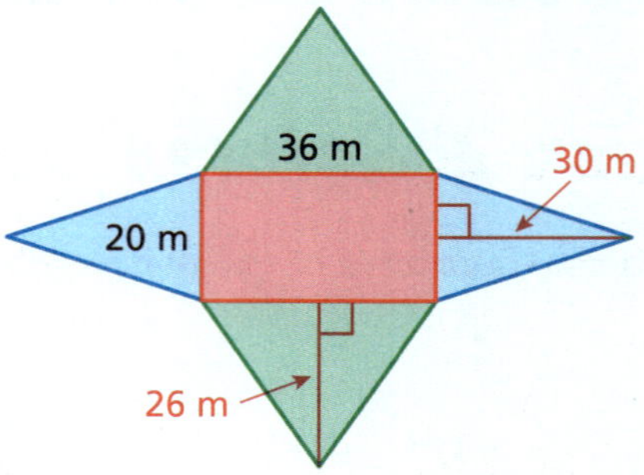

FINDING SURFACE AREA **Find the surface area of the pyramid.** (See Examples 1 and 2.)

13.

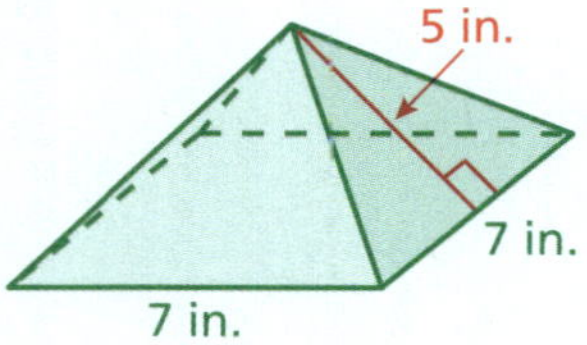

14.

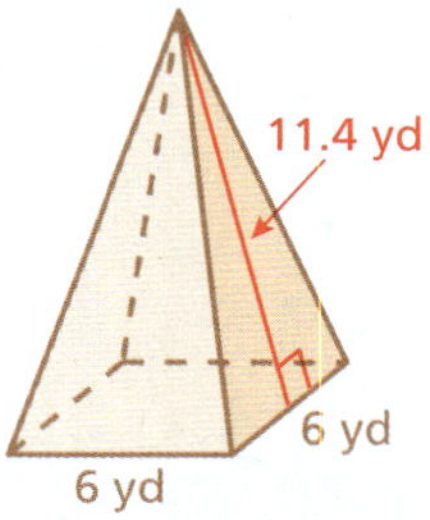

15.

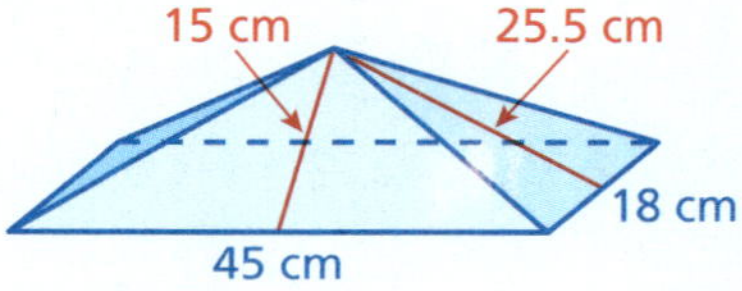

7 MTR

16. **MODELING REAL LIFE** A paperweight is shaped like a rectangular pyramid. Find the surface area of the paperweight. (See Example 3.)

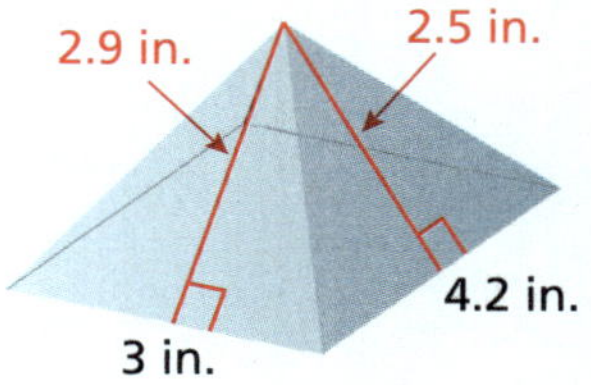

17. PROBLEM SOLVING The entrance to the Louvre Museum in Paris, France, is a square pyramid. The side length of the base is 116 feet, and the height of one of the triangular faces is 91.7 feet. Find the surface area of the four triangular faces of the entrance to the Louvre Museum.

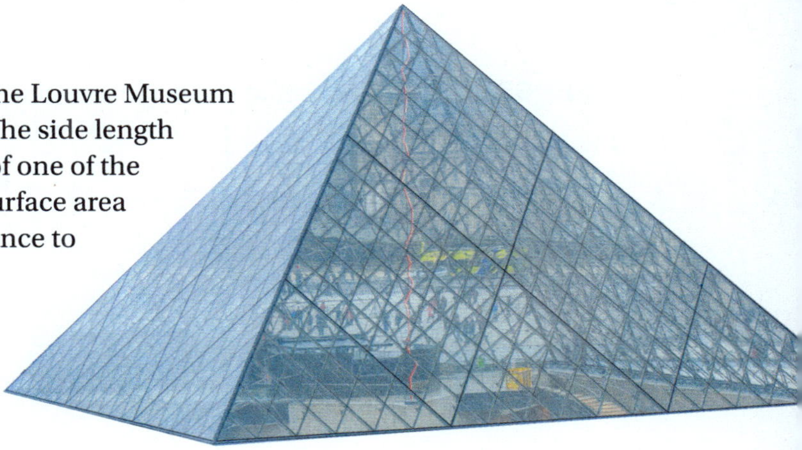

18. PROBLEM SOLVING A company builds windows in the shape of square pyramids that are rated to withstand hurricane-force winds. Each window has a base side length of 6.2 feet, and the height of each of the triangular faces is 5 feet. Find the surface area of the window. How does the surface area change if the height of each triangular face changes to 7 feet?

7 MTR

19. MODELING REAL LIFE A silicon wafer is textured to minimize light reflection. This results in a surface made up of square pyramids. Each triangular face of one of the pyramids has a base of 5 micrometers and a height of 5.6 micrometers. Find the surface area of the pyramid, including the base.

20. REASONING A hanging light cover made of glass is shaped like a square pyramid. The cover does not have a bottom. One square foot of the glass weighs 2.45 pounds. The chain can support 35 pounds. Will the chain support the light cover? Explain.

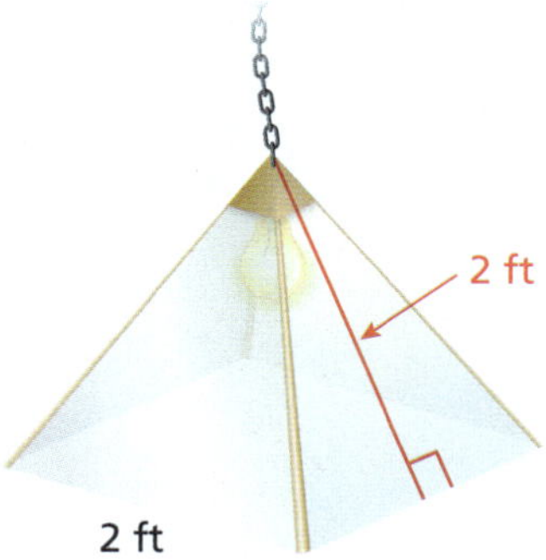

21. GEOMETRY The surface area of the square pyramid shown is 84 square inches. What is the value of x?

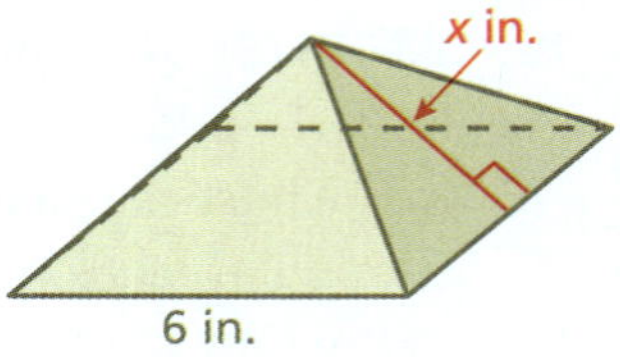

5 MTR

22. STRUCTURE A rectangular prism and a square pyramid have the same base. The height of each triangular face of the pyramid is equal to the height of the rectangular prism. Compare the surface areas. What do you notice? Explain.

4 MTR

23. MAKING AN ARGUMENT Can you form a square pyramid using a square with side lengths of 14 inches and four of the triangles shown? Justify your answer.

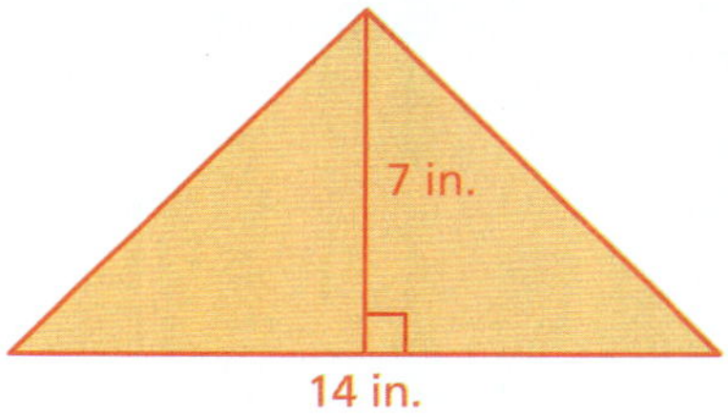

10.8 Volumes of Rectangular Prisms

Learning Target: Find volumes and missing dimensions of rectangular prisms.

Success Criteria:
- I can find the volume of a rectangular prism.
- I can solve for a missing dimension of a rectangular prism.
- I can solve real-life problems involving volumes of rectangular prisms.

Geometric Reasoning
MA.6.GR.2.3 Solve mathematical and real-world problems involving the volume of right rectangular prisms with positive rational number edge lengths using a visual model and a formula.

Recall that the **volume** of a three-dimensional figure is a measure of the amount of space that it occupies. Volume is measured in *cubic units*.

Exploration 1 Using a Unit Cube

Work with a partner. A *unit cube* is a cube with an edge length of 1 unit.

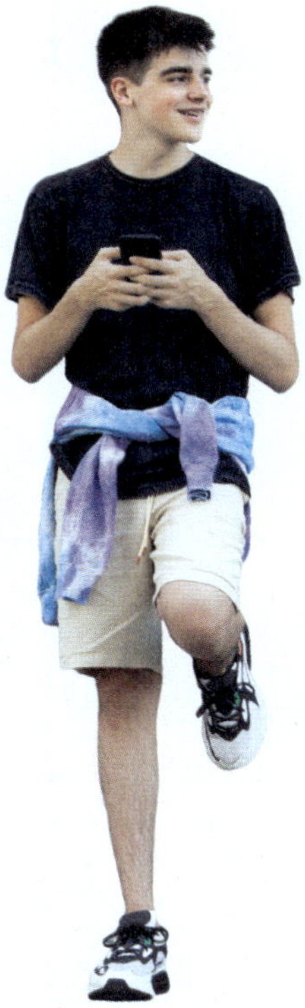

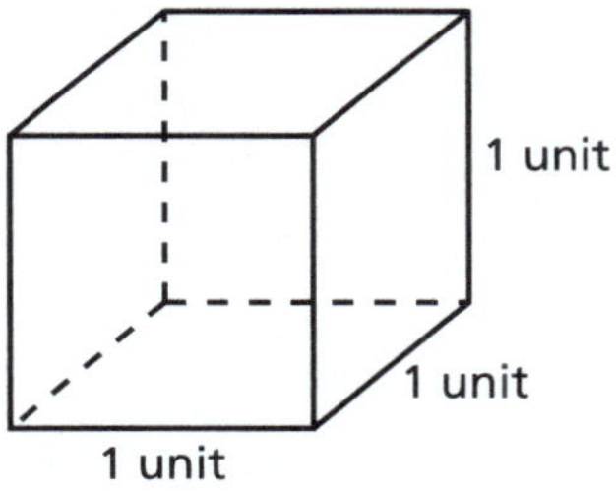

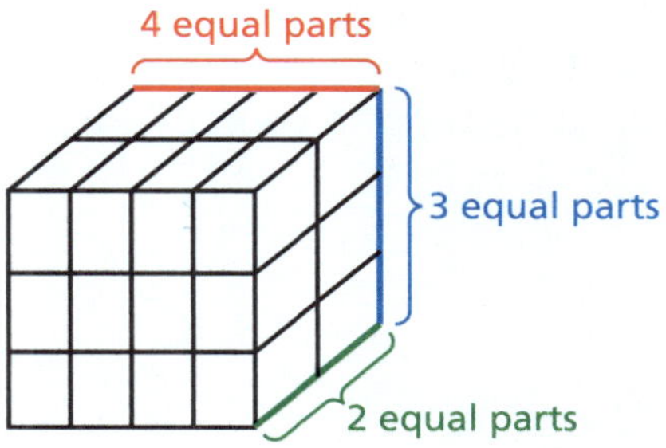

a. The parallel edges of the unit cube have been divided into 2, 3, and 4 equal parts to create smaller rectangular prisms that are identical. What can you determine about the volumes of the identical prisms? Explain.

b. Find the volume of each of the identical prisms in part (a).

GO DIGITAL

c. Use the identical prisms in part (a) to find the volume of the prism at the right. Explain.

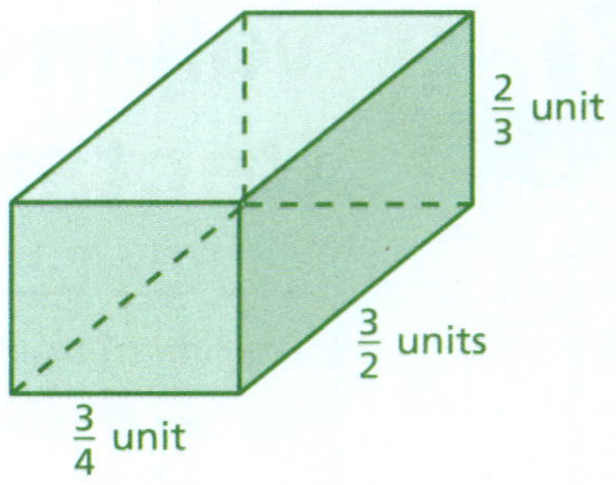

d. How can you use a unit cube to find the volume of the prism at the left? Explain.

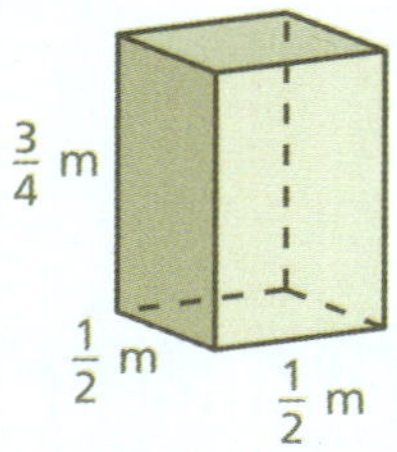

e. Do the formulas $V = Bh$ and $V = \ell wh$ work for rectangular prisms with fractional edge lengths? Give examples to support your answer.

4 MTR **COMMUNICATE CLEARLY**

In part (d), explain why you decided to divide the unit cube in the way you did.

10.8 Lesson

Key Vocabulary
volume, *p. 855*

Key Idea

Volume of a Rectangular Prism

Words The volume V of a rectangular prism is the product of the area of the base and the height of the prism.

Algebra $V = Bh$ or $V = \ell wh$

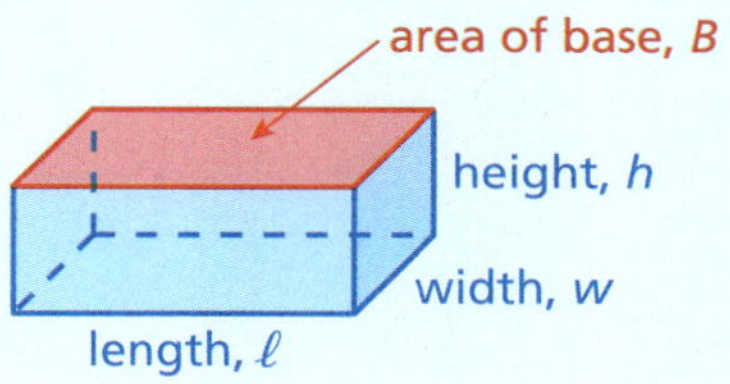

When a rectangular prism is a cube with an edge length of s, you can also use the formula $V = s^3$ to find the volume V of the cube.

Example 1 Finding Volumes of Rectangular Prisms

Find the volume of each prism.

a.

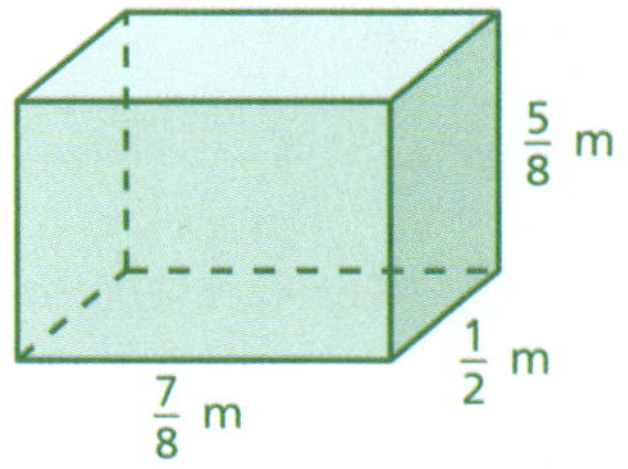

$$V = \ell wh$$

$$= \frac{7}{8}\left(\frac{1}{2}\right)\left(\frac{5}{8}\right)$$

$$= \frac{35}{128}$$

▶ So, the volume is $\frac{35}{128}$ cubic meter.

b.

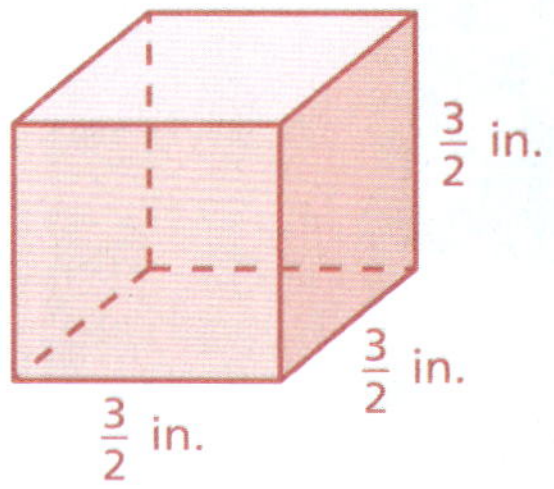

$$V = s^3$$

$$= \left(\frac{3}{2}\right)^3$$

$$= \frac{3}{2}\left(\frac{3}{2}\right)\left(\frac{3}{2}\right)$$

$$= \frac{27}{8}, \text{ or } 3\frac{3}{8}$$

▶ So, the volume is $3\frac{3}{8}$ cubic inches.

4 MTR COMPARE METHODS

In Example 1(b), do you get the same answer when you use the formula $V = \ell wh$ to find the volume? Explain why or why not.

____

____

____

____

GO DIGITAL

Try It **Find the volume of the prism.**

1. 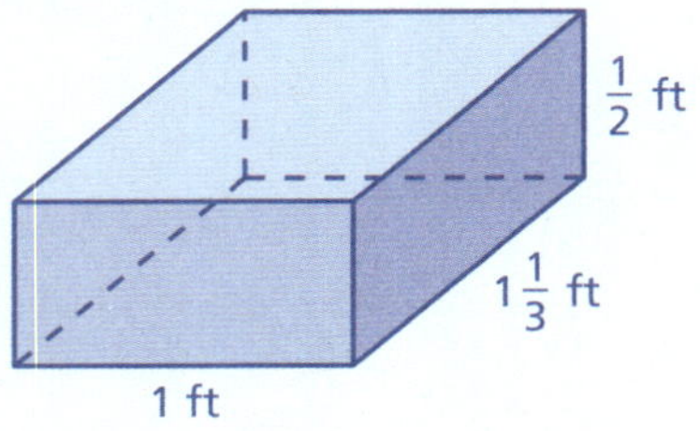

2. 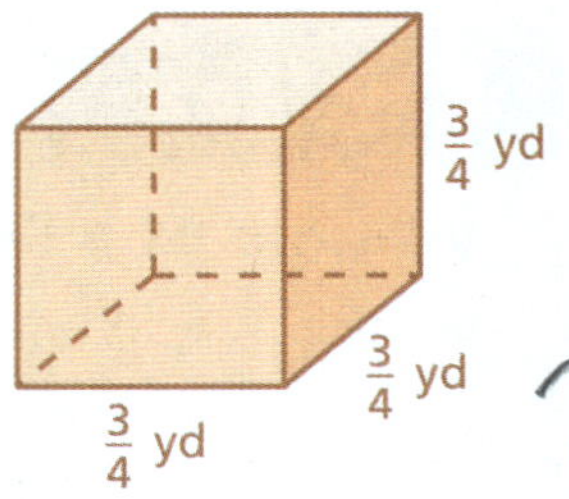

Example 2 Finding a Missing Dimension of a Rectangular Prism

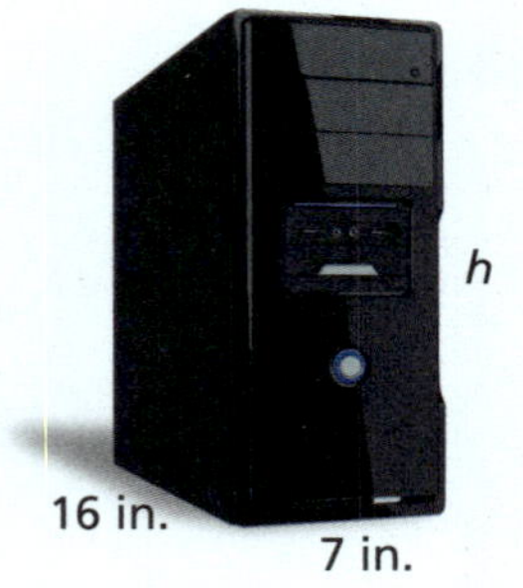

Volume = 1792 in.^{3}

Find the height of the computer tower.

$V = \ell wh$	Write formula for volume.
$1792 = 16(7)h$	Substitute values.
$1792 = 112h$	Simplify.
$\frac{1792}{112} = \frac{112h}{112}$	Division Property of Equality
$16 = h$	Simplify.

▶ So, the height of the computer tower is 16 inches.

Try It **Find the missing dimension of the prism.**

3. Volume = 72 in.^{3}

4. Volume = 1320 cm^3

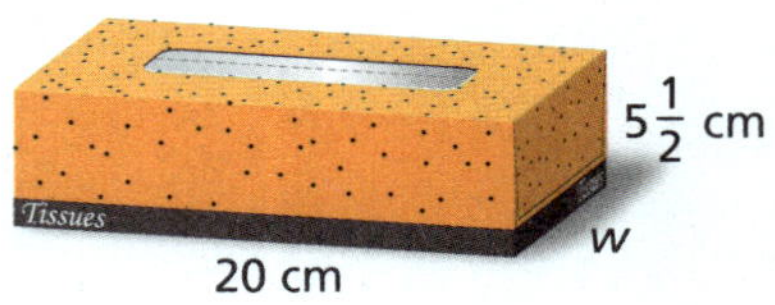

GO DIGITAL

In-Class Practice

1 I don't understand yet. 2 I can do it with help. 3 I can do it on my own. 4 I can teach someone else.

5. **VOCABULARY** Explain how volume and surface area are different.

6. **FINDING A MISSING DIMENSION** The base of a rectangular prism has an area of 24 square millimeters. The volume of the prism is 144 cubic millimeters. Make a sketch of the prism. Then find the height of the prism.

FINDING VOLUME **Find the volume of the prism.**

7.
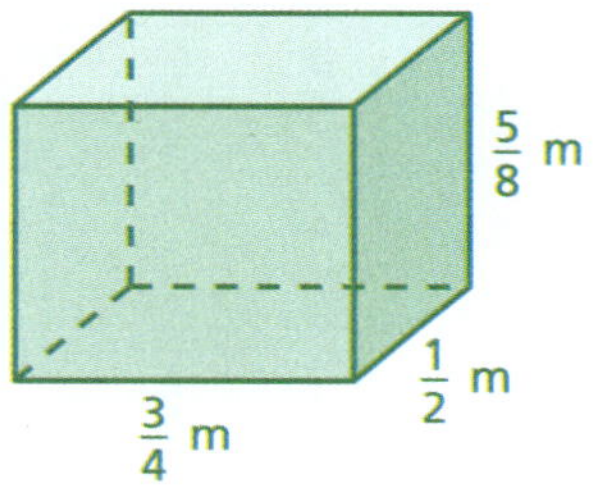

8.
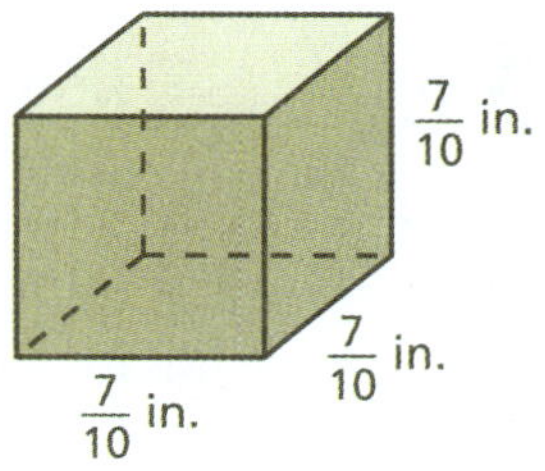

When finding volumes, you may need to convert cubic units. The diagrams show that there are 27 cubic feet per cubic yard.

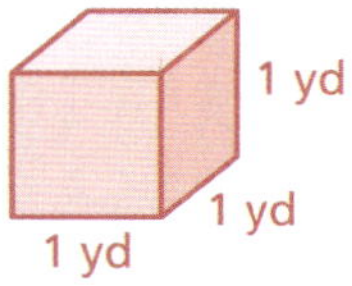

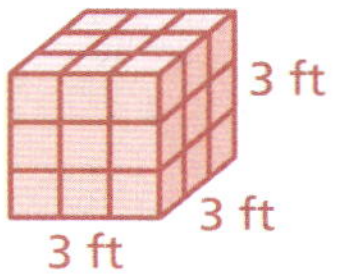

$$1 \text{ yd}^3 = (1 \text{ yd})(1 \text{ yd})(1 \text{ yd}) = (3 \text{ ft})(3 \text{ ft})(3 \text{ ft}) = 27 \text{ ft}^3$$

You can use a similar procedure to convert other cubic units.

Example 3 Modeling Real Life

7 MTR

The dump truck shown delivers dirt for $18 per cubic yard. About how much does a full load of dirt cost?

Find the volume of a full load of dirt in cubic feet.

$V = \ell wh$ Write formula for volume.

$= 17(8)\left(4\frac{3}{4}\right)$ Substitute values.

$= 646$ Multiply.

A full load of dirt is 646 cubic feet. Because $1 \text{ yd}^3 = 27 \text{ ft}^3$, convert the volume to cubic yards using a conversion factor.

$$646 \text{ ft}^3 = 646 \cancel{\text{ft}^3} \times \frac{1 \text{ yd}^3}{27 \cancel{\text{ft}^3}} \approx 24 \text{ yd}^3$$

▶ So, a full load of dirt costs about $24 \cancel{\text{yd}^3} \times \frac{\$18}{1 \cancel{\text{yd}^3}} = \$432$.

In-Class Practice

1 I don't understand yet. 2 I can do it with help. 3 I can do it on my own. 4 I can teach someone else.

9. **Dig Deeper** The shark cage is in the shape of a rectangular prism and has a volume of 315 cubic feet. Find a set of reasonable dimensions for the base of the cage. Justify your answer.

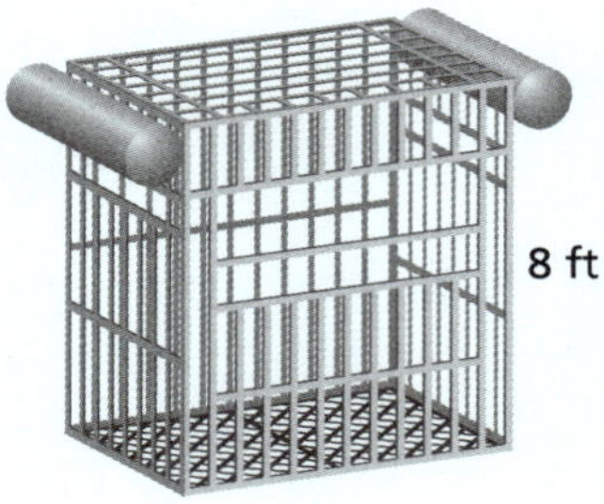

10. The hot tub is in the shape of a rectangular prism. How many pounds of water can the hot tub hold? One cubic foot of water weighs about 62.4 pounds.

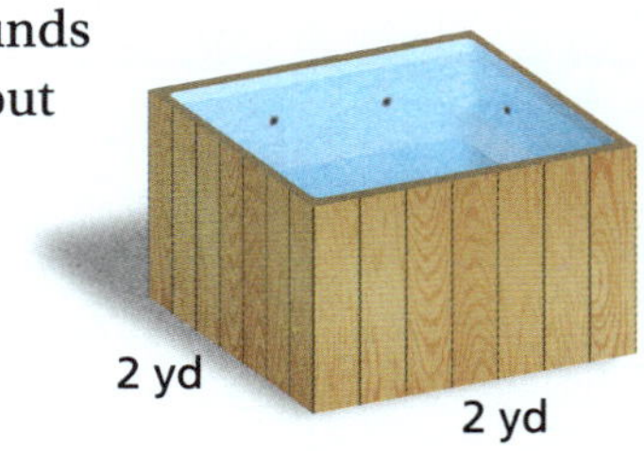

GO DIGITAL

10.8 Practice WITH CalcChat® AND CalcView®

Review & Refresh

Find the surface area of the pyramid.

1.

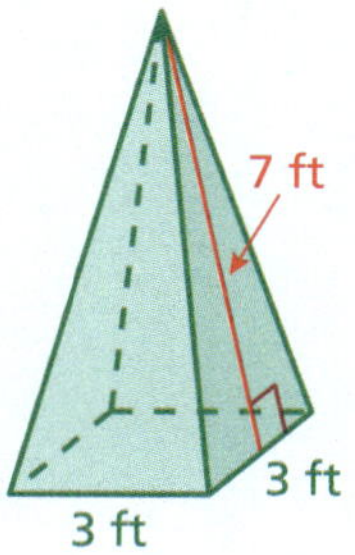

2.

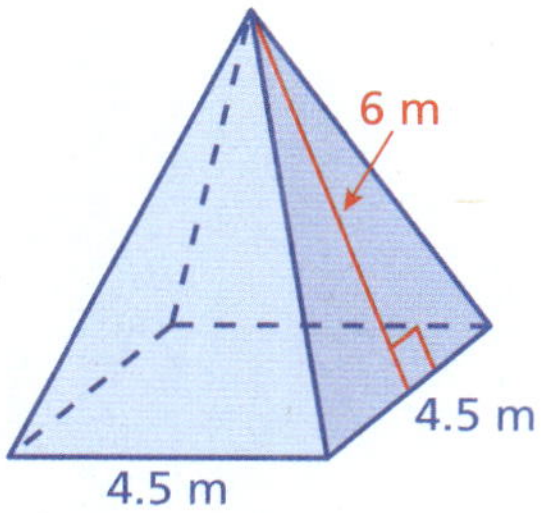

3.

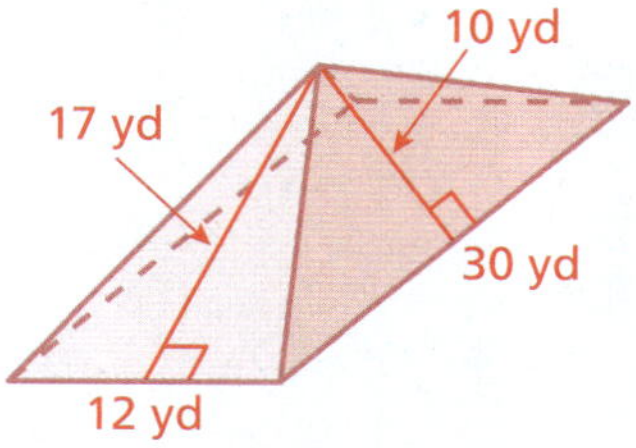

Write the phrase as an expression. Then evaluate the expression when $x = -2$ and $y = 12$.

4. 8 more than a number x

5. the difference of a number y and -9

Concepts, Skills, & Problem Solving

5 MTR **STRUCTURE** **The unit cube is divided into identical rectangular prisms. What is the volume of one of the identical prisms?** (See Exploration 1.)

6.

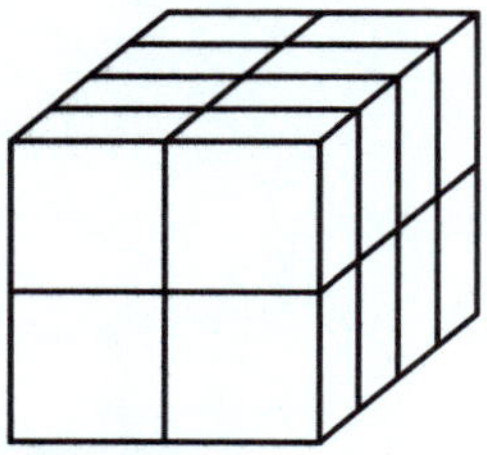

7.

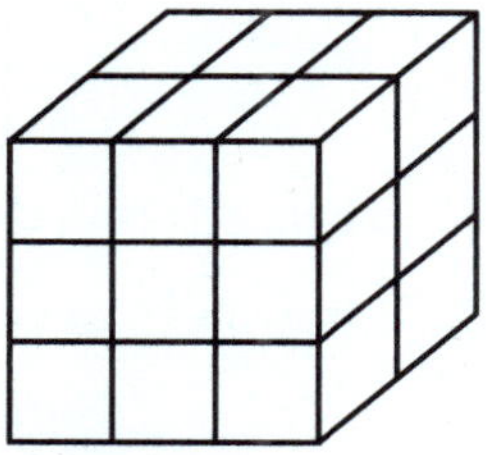

8.

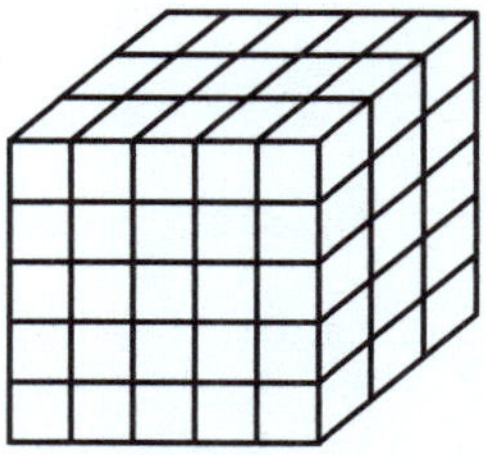

GO DIGITAL

FINDING VOLUME Find the volume of the prism. (See Example 1.)

9.

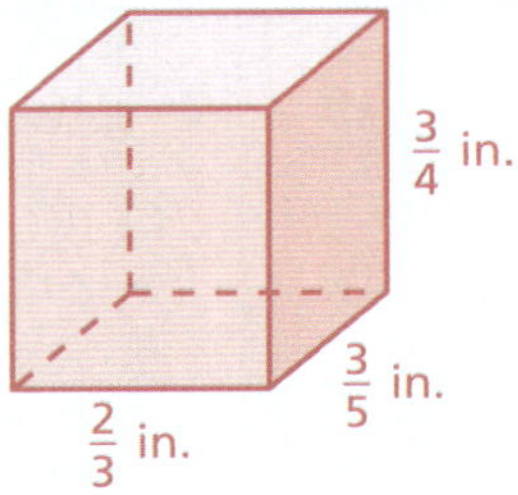

10.

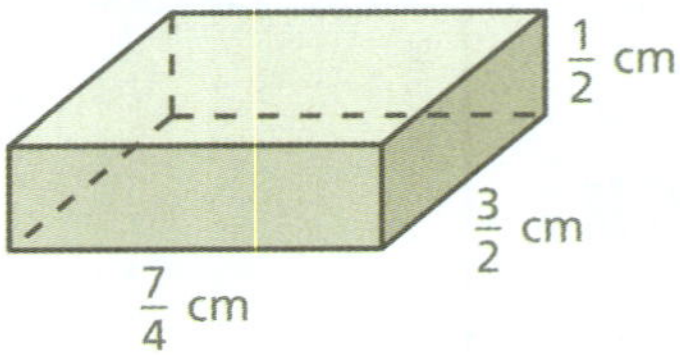

11.

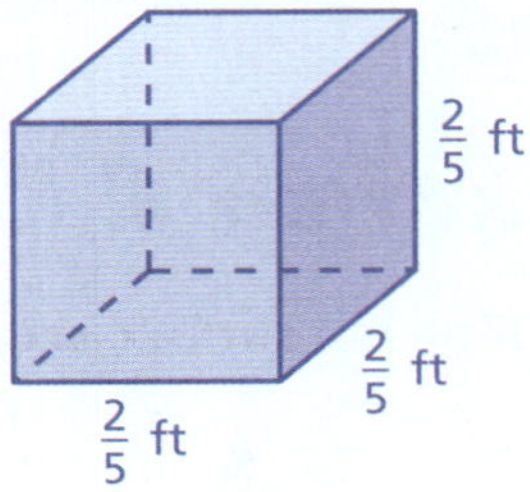

12.

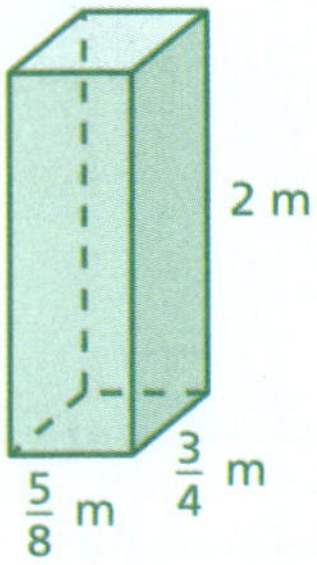

13.

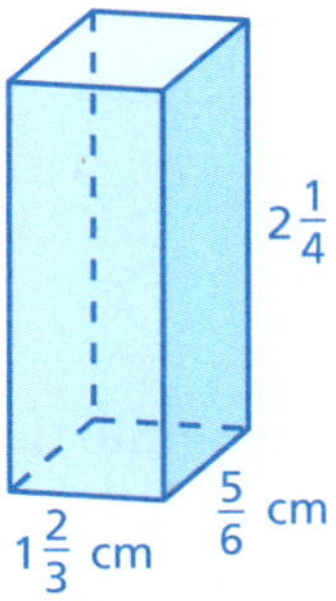

14.

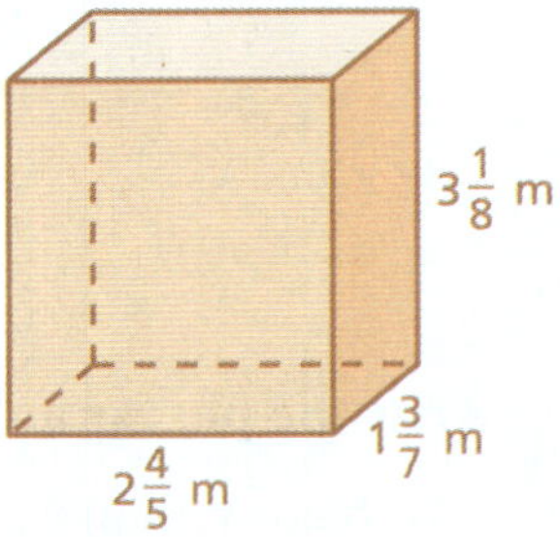

FINDING A MISSING DIMENSION Find the missing dimension of the prism. (See Example 2.)

15. Volume = 1620 cm^3

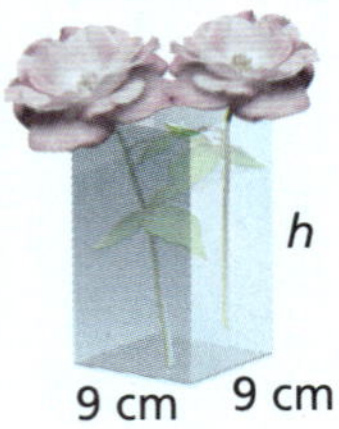

16. Volume = 245 cm^3

17. Volume = 532 $in.^3$

7 MTR

18. **MODELING REAL LIFE** An FBI agent orders a block of ballistics gel that is 6 inches wide, 20 inches long, and 6 inches tall. The gel weighs 54 pounds per cubic foot. What is the weight of the block of gel? (See Example 3.)

7 MTR

19. **MODELING REAL LIFE**

a. Estimate the amount of casserole left in the dish.

b. Will the casserole fit in the storage container? Explain your reasoning.

20. **GEOMETRY** How many $\frac{3}{4}$-meter cubes fit inside a cube with an edge length of 12 meters?

21. **REASONING** How many 1-millimeter cubes do you need to fill a cube that has an edge length of 1 centimeter? How can this result help you convert a volume from cubic millimeters to cubic centimeters? from cubic centimeters to cubic millimeters?

22. **REASONING** The container is partially filled with unit cubes. How many unit cubes fit in the container? Explain your reasoning.

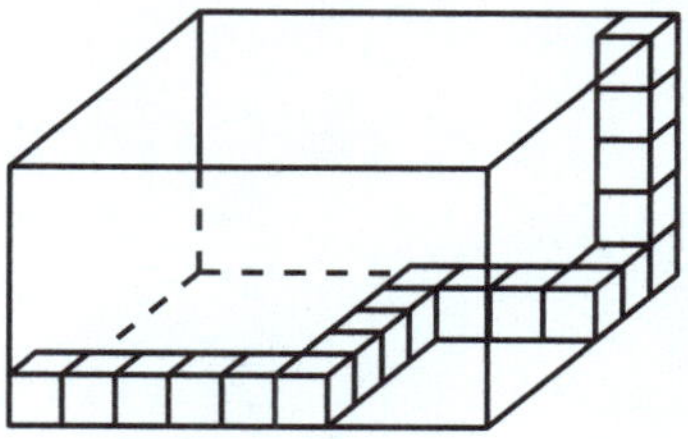

23. PROBLEM SOLVING The area of the shaded face is 96 square centimeters. What is the volume of the rectangular prism?

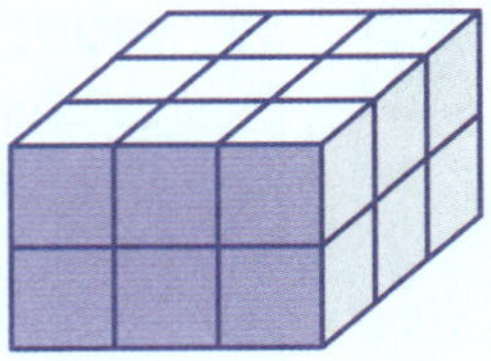

24. Dig Deeper Is the combined volume of a 4-foot cube and a 6-foot cube equal to the volume of a 10-foot cube? Use a diagram to justify your answer.

6 MTR

25. ASSESS REASONABLENESS You have 1400 square feet of boards.

a. Design a tree house that has a volume of at least 250 cubic feet. Include sketches of your tree house.

b. Are your dimensions reasonable? Explain your reasoning.

10 Connecting Concepts

5 MTR *Mathematicians who use patterns and structure to help understand and connect mathematical concepts relate previously learned concepts to new concepts.*

1 MTR **1. ANALYZE A PROBLEM** A sports complex has two swimming pools that are shaped like rectangular prisms. The amount of water in the smaller pool is what percent of the amount of water in the larger pool?

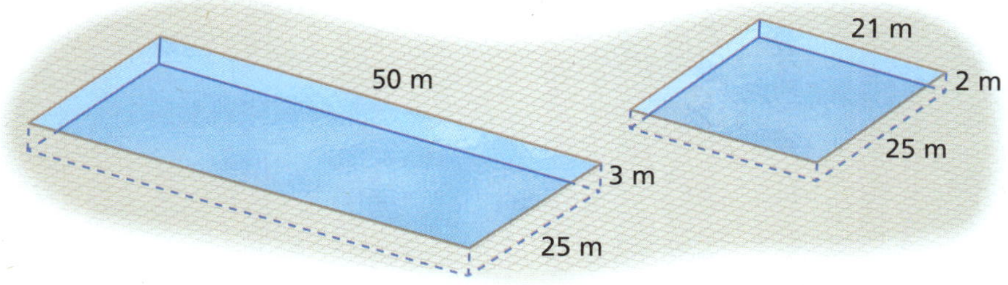

Using the Problem-Solving Plan

Understand the problem. You know the shape and the dimensions of the two swimming pools. You are asked to find the amount of water in the smaller pool as a percent of the amount of water in the larger pool.

Make a plan. First, find the volume of each pool. Then represent the amount of water in the smaller pool as a fraction of the amount of water in the larger pool. Find an equivalent fraction whose denominator is 100 to find the percent.

Solve and check. Use the plan to solve the problem. Then check your solution.

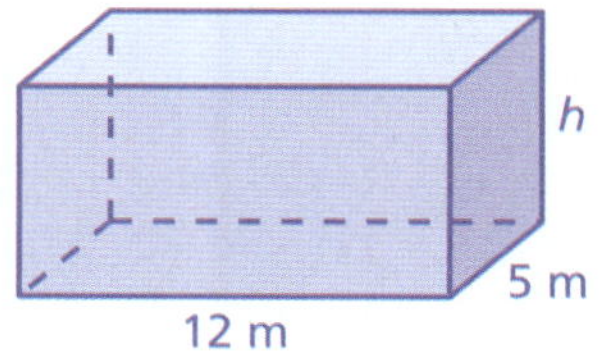

2. Find the height of the rectangular prism when the surface area of the prism is 324 square meters.

3. A toy company sells two different toy chests. The toy chests have different dimensions, but the same volume. What is the width w of Toy Chest 2?

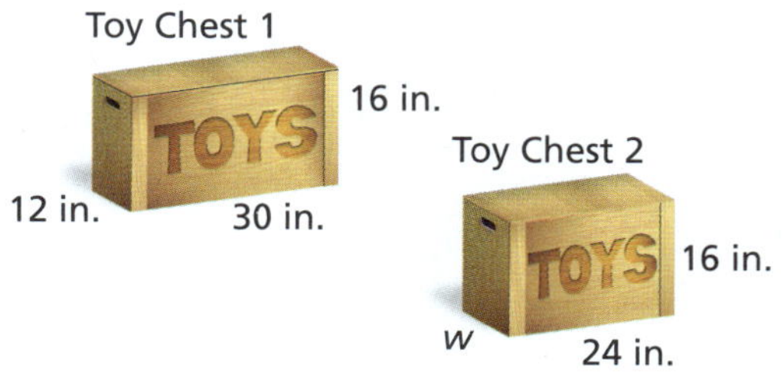

STEAM Performance Task — Maximizing the Volumes of Boxes

At the beginning of this chapter, you watched a STEAM Video called "Packaging Design." You are now ready to complete the STEAM Performance Task, where you will determine how many smaller boxes fit inside of a larger box. Be sure to use mathematical thinking and reasoning, and the problem-solving plan as you work through the performance task.

10

Chapter Review WITH CalcChat®

Review Vocabulary

Write the definition and give an example of each vocabulary term.

composite figure, *p. 795*

edge, *p. 827*

surface area, *p. 837*

solid, *p. 827*

vertex, *p. 827*

net, *p. 837*

polyhedron, *p. 827*

prism, *p. 827*

volume, *p. 855*

face, *p. 827*

pyramid, *p. 827*

GO DIGITAL

Graphic Organizers

You can use a **Four Square** to organize information about a concept. Each of the four squares can be a category, such as definition, vocabulary, example, non-example, words, algebra, table, numbers, visual, graph, or equation. Here is an example of a Four Square for the ***area of a triangle***.

Area of a triangle

Words

The area A of a triangle is one-half the product of its base b and its height h.

Algebra

$A = \frac{1}{2}bh$

Example

$A = \frac{1}{2}bh$

$= \frac{1}{2}(4)(6)$

$= 12$

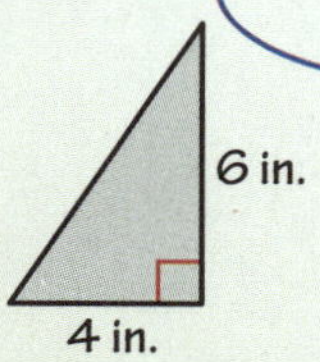

The area of the triangle is 12 square inches.

Example

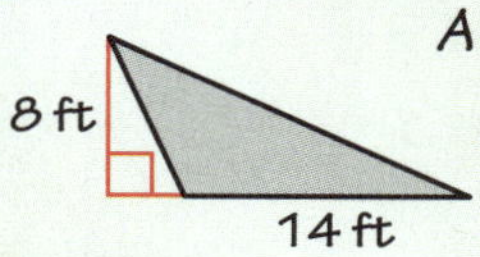

$A = \frac{1}{2}bh$

$= \frac{1}{2}(14)(8)$

$= 56$

The area of the triangle is 56 square feet.

Choose and complete a graphic organizer to help you study the concept.

1. area of a quadrilateral

2. area of a composite figure

3. polyhedron

4. surface area of a rectangular prism

5. surface area of a rectangular pyramid

6. volume of a rectangular prism

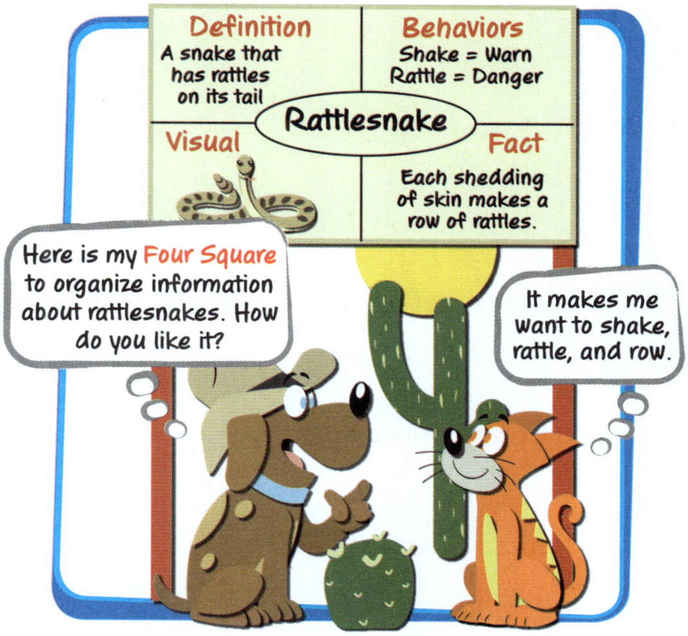

Chapter Learning Target: Understand area and volume of figures.

Chapter Success Criteria:
- ◆ I can find the area of a polygon.
- ◆ I can find the surface area of a solid.
- ■ I can find the volume of a solid.
- ■ I can apply units of measure to solve real-life problems.

◆ Surface
■ Deep

Rate your understanding after each section.

1	2	3	4
I don't understand yet.	I can do it with help.	I can do it on my own.	I can teach someone else.

10.1 Areas of Triangles *(pp. 781–790)*

Learning Target: Find areas and missing dimensions of triangles.

Find the area of the triangle.

1.

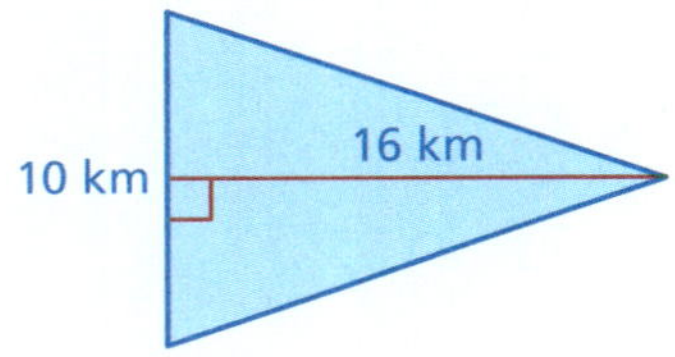

2.

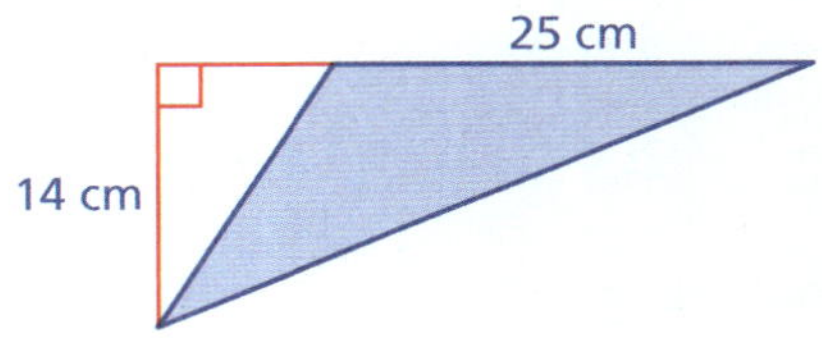

Find the missing dimension of the triangle.

3. Area = 35 mi²

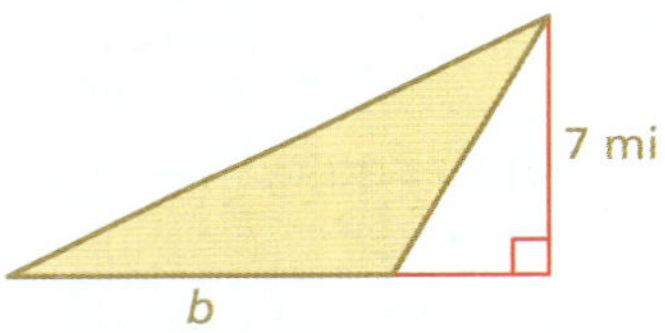

4. Area = 6 cm²

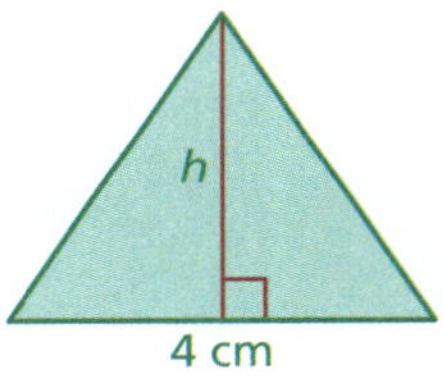

10.2 Areas of Quadrilaterals and Composite Figures *(pp. 791–800)*

Learning Target: Find areas and missing dimensions of quadrilaterals and composite figures.

Find the area of the figure.

5.

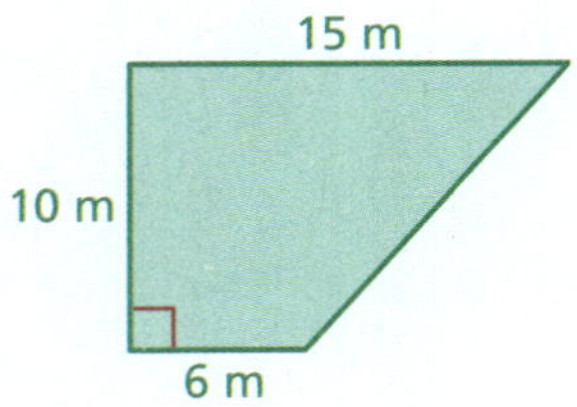

6.

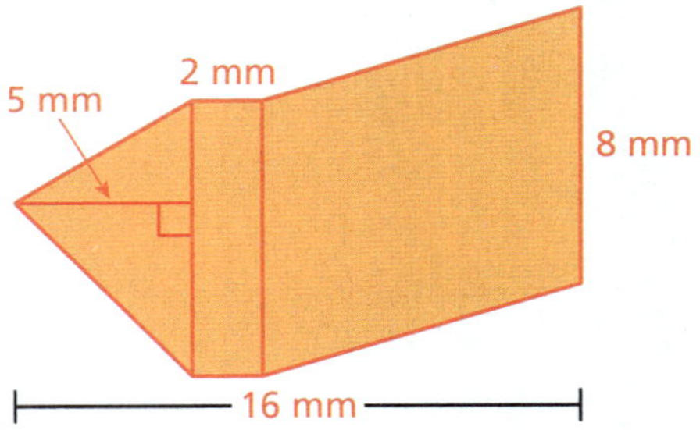

7. Draw a composite figure that has an area less than 35 square inches.

GO DIGITAL

10.3 Areas of Parallelograms *(pp. 801–810)*

Learning Target: Find areas and missing dimensions of parallelograms.

Find the area of the parallelogram.

8.

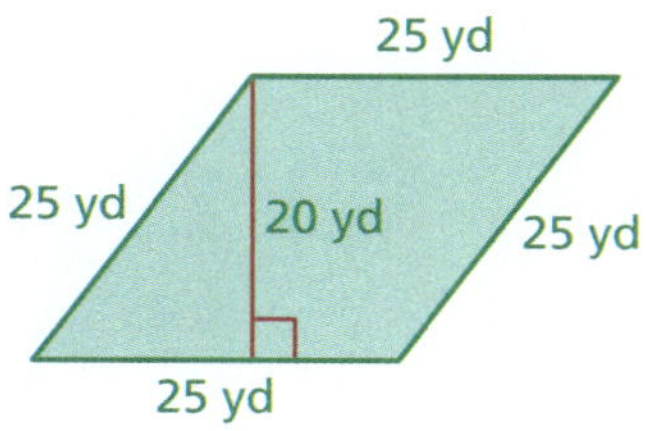

9.

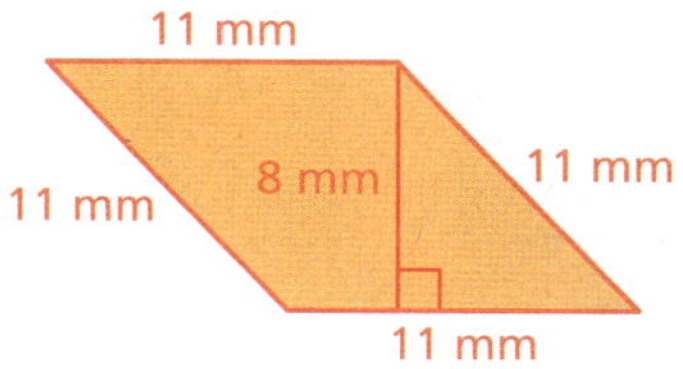

10.

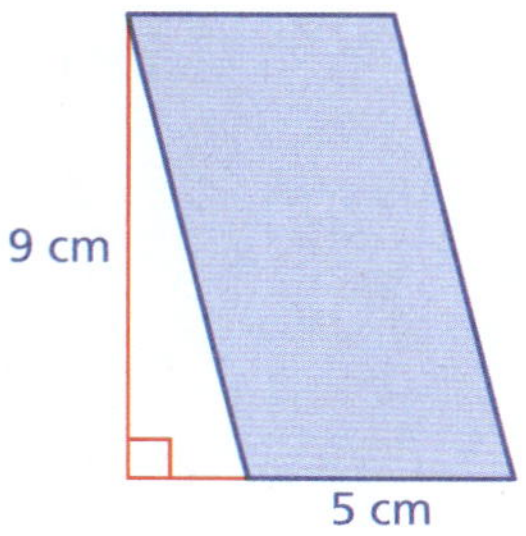

11.

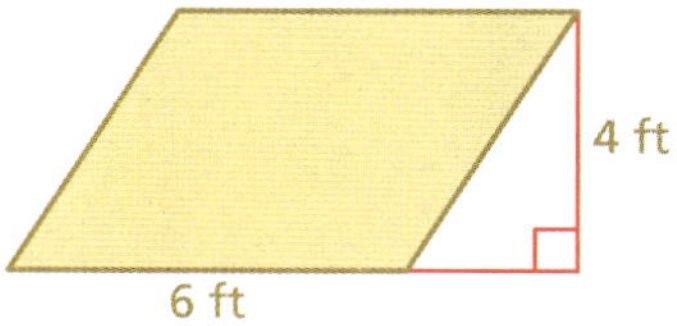

12. Find the area (in square inches) of the parallelogram.

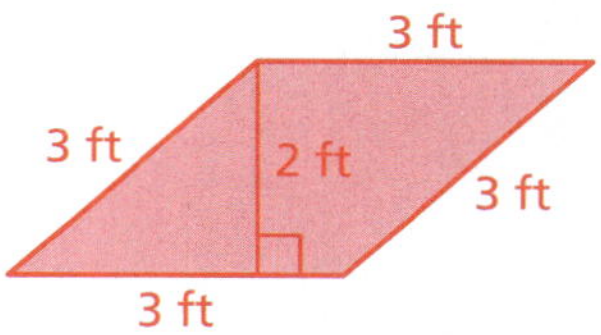

13. The billboard shown is in the shape of a parallelogram with a base of 48 feet. What is the height of the billboard?

14. The freeway noise barrier shown is made of identical parallelogram-shaped sections. The area of each section is 7.5 square meters, and the height of the barrier is 5 meters. How many meters wide is each section of the noise barrier?

15. Draw a parallelogram that has an area between 58 and 60 square centimeters.

10.4 Areas of Trapezoids and Other Polygons *(pp. 811–824)*

Learning Target: Find areas of trapezoids and other polygons.

Find the area of the trapezoid.

16.

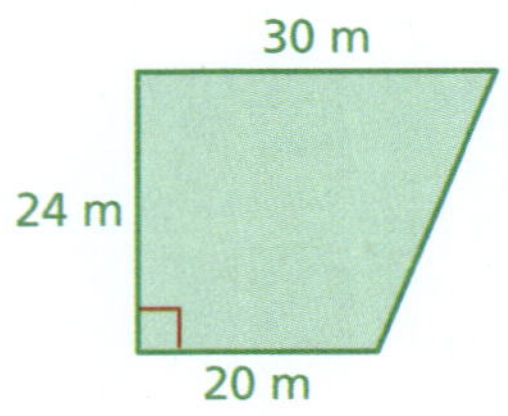

17.

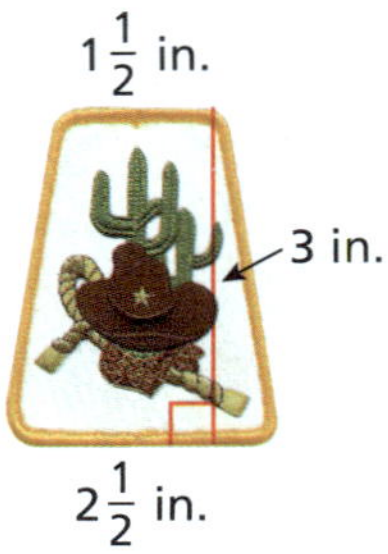

18.

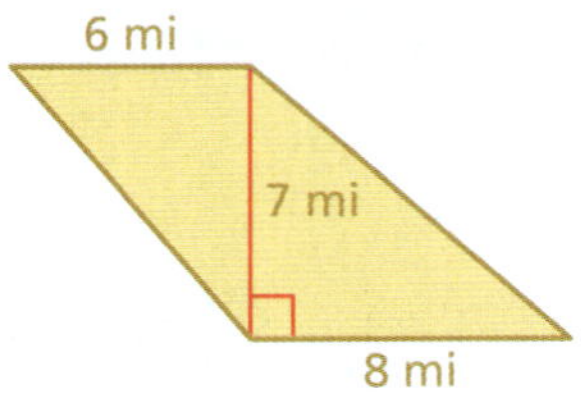

Find the area of the figure.

19.

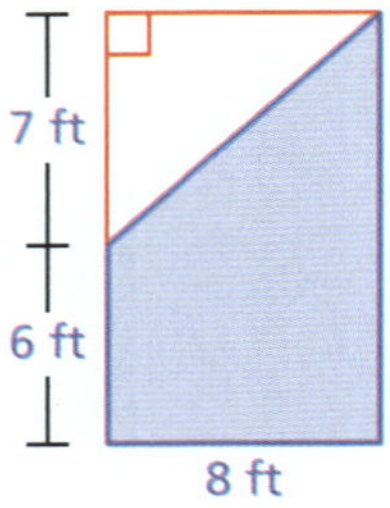

20.

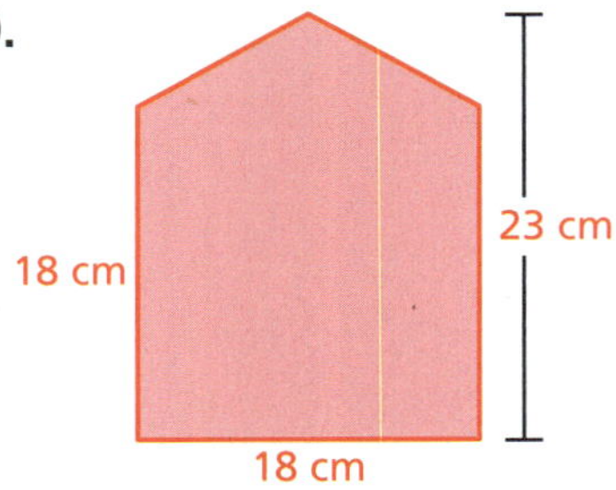

21.

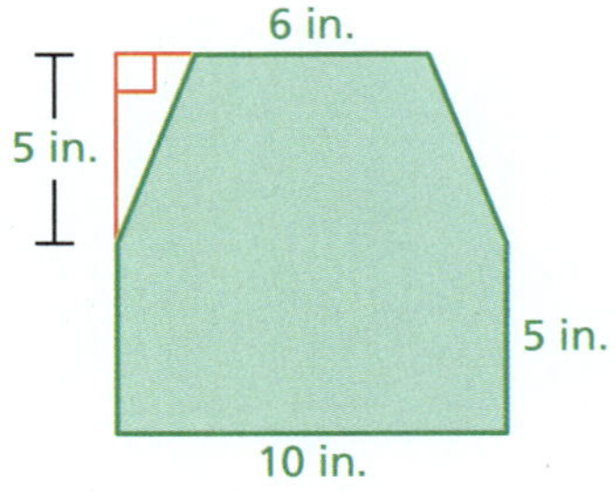

Find the height of the trapezoid.

22. Area $= 29.6 \text{ mi}^2$

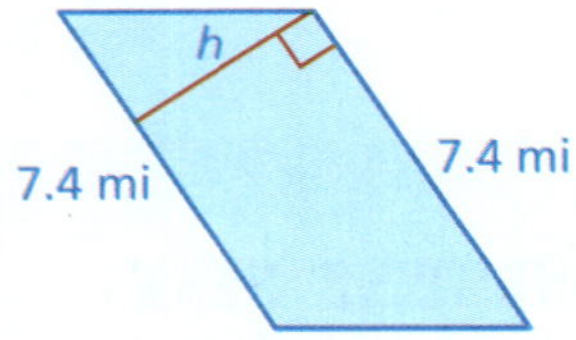

23. Area $= 91.7 \text{ mm}^2$

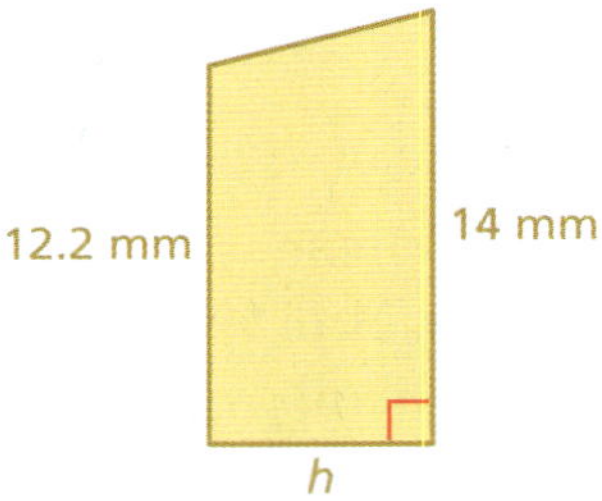

24. Area $= 882 \text{ ft}^2$

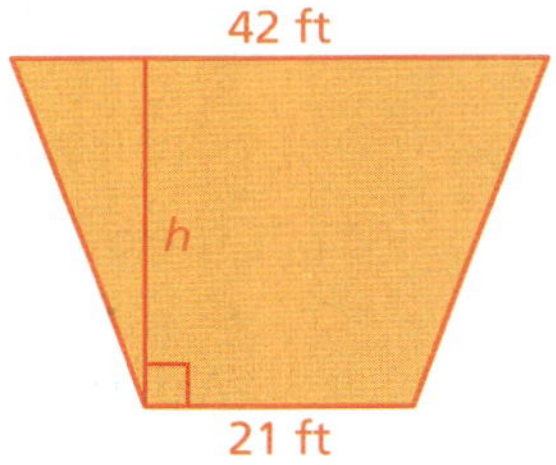

25. You are creating a design for the side of the soapbox car. How much area do you have for the design?

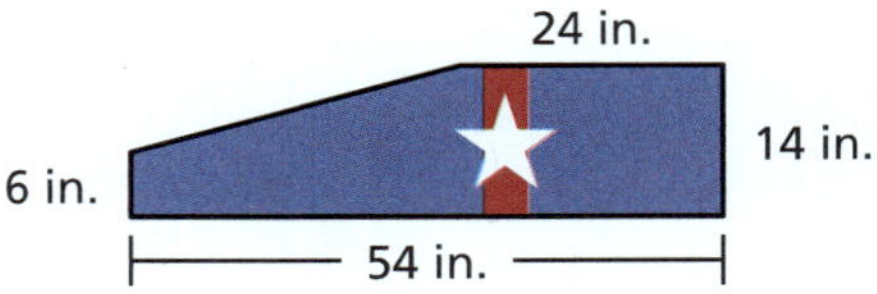

26. Find the area (in square centimeters) of a trapezoid with a height of 2 meters and base lengths of 3 meters and 5 meters.

10.5 Three-Dimensional Figures *(pp. 825–834)*

Learning Target: Describe and draw three-dimensional figures.

Find the numbers of faces, edges, and vertices of the solid.

27. 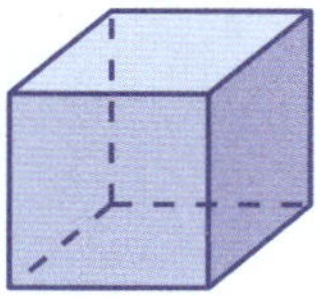

28. 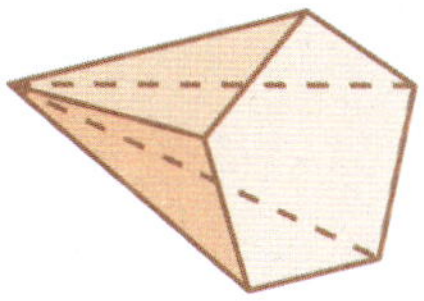

29. 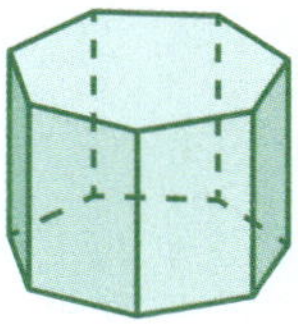

Draw the solid.

30. square pyramid

31. hexagonal prism

Draw the front, side, and top views of the solid.

32. 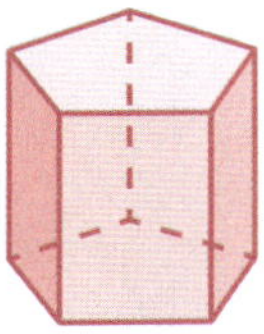

33. 

34. 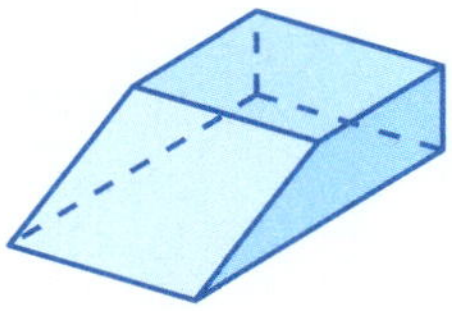

10.6 Surface Areas of Rectangular Prisms *(pp. 835–844)*

Learning Target: Find surface areas of rectangular prisms.

Find the surface area of the prism.

35.

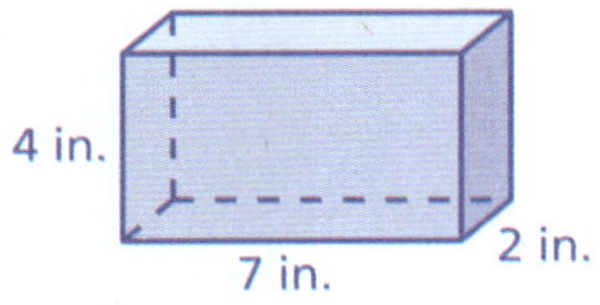

36.

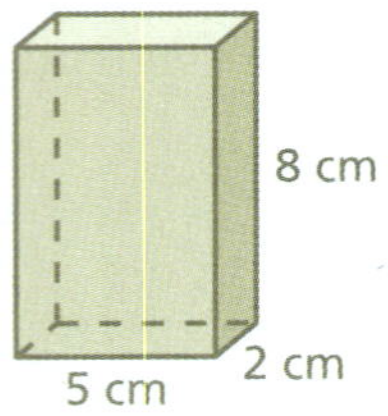

37.

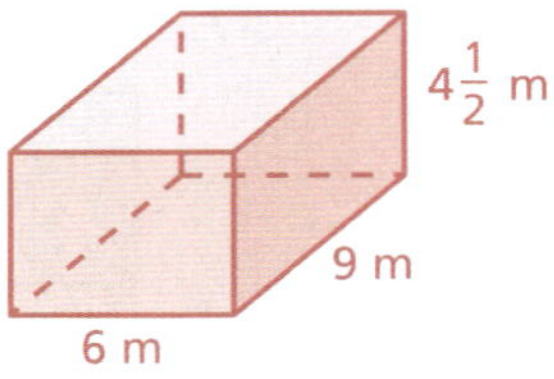

38.

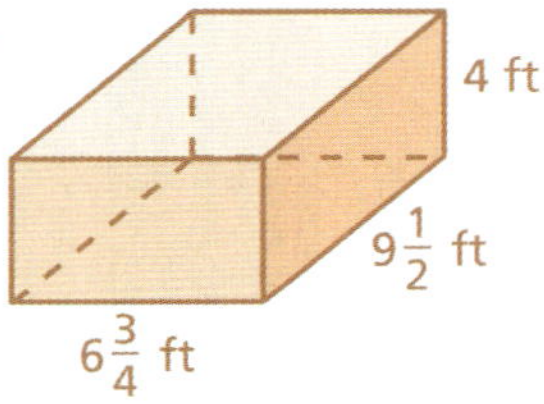

39.

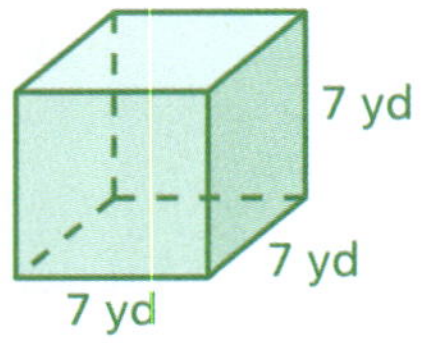

40.

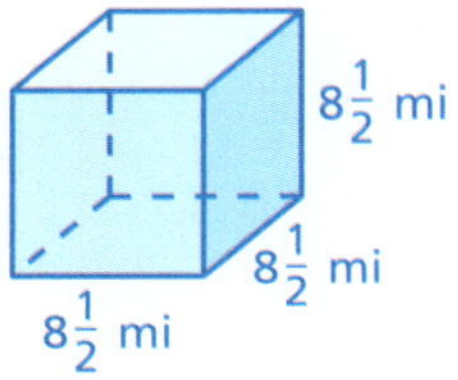

41. One quart of water-resistant paint covers 75 square feet. A swimming pool is in the shape of a rectangular prism with a length of 20 feet, a width of 10 feet, and a height of 5 feet. How many quarts should you buy to paint the swimming pool with two coats of paint?

10.7 Surface Areas of Rectangular Pyramids (pp. 845–854)

Learning Target: Find surface areas of rectangular pyramids.

Find the surface area of the pyramid.

42.

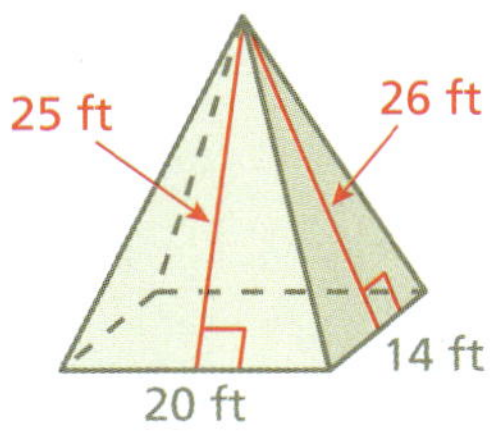

43.

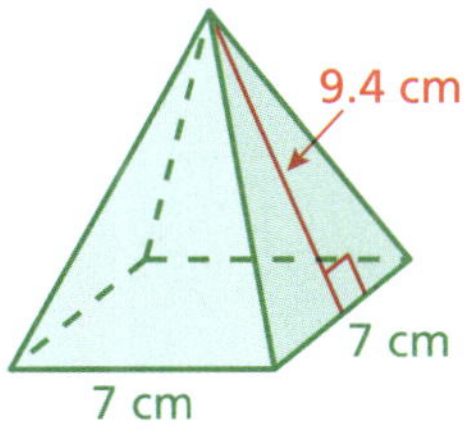

44.

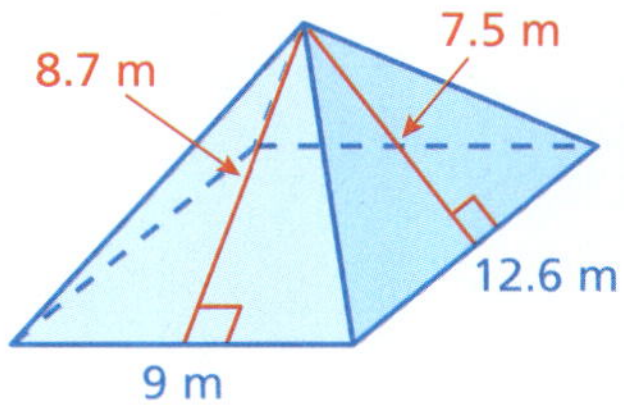

45. You make a square pyramid for a school project. Find the surface area of the pyramid.

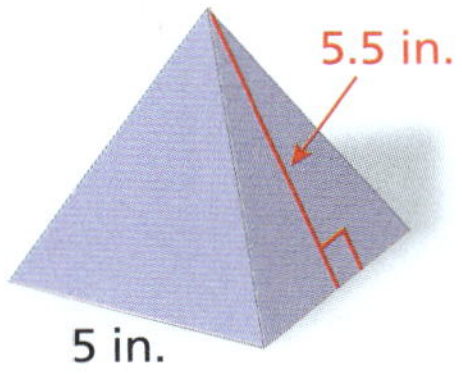

10.8 Volumes of Rectangular Prisms (pp. 855–864)

Learning Target: Find volumes and missing dimensions of rectangular prisms.

Find the volume of the prism.

46.

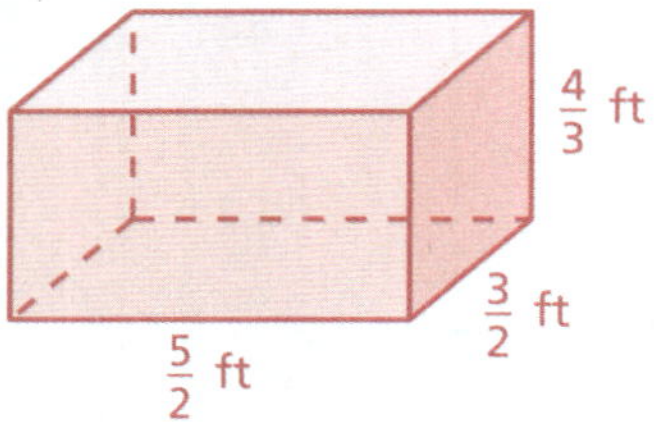

47.

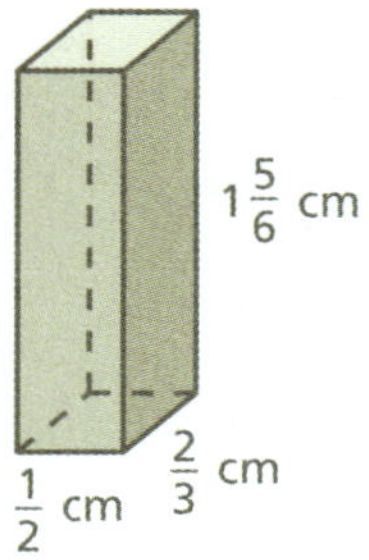

48.

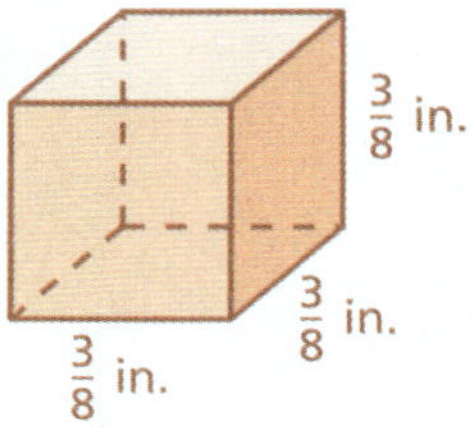

49. The prism has a volume of 150 cubic feet. Find the length of the prism.

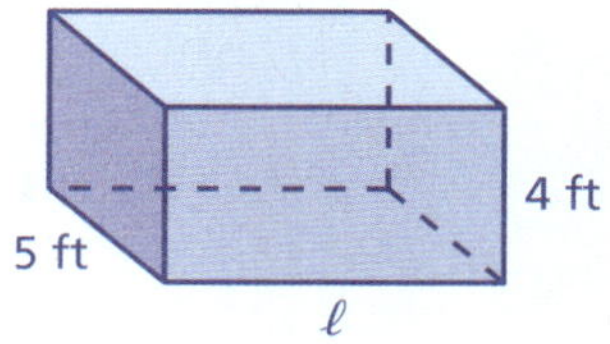

50. How many cubic inches of tissues can the box hold?

51. Draw a rectangular prism that has a volume less than 1 cubic inch.

10 Practice Test WITH CalcChat®

Find the area of the figure.

1. 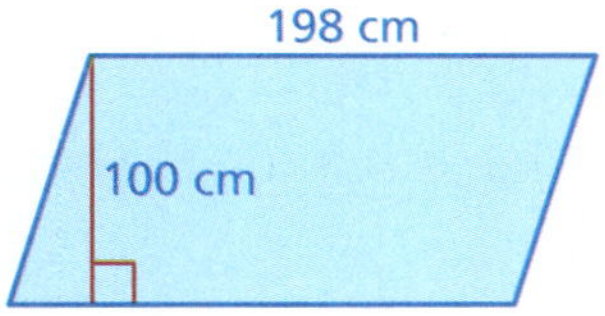

2. 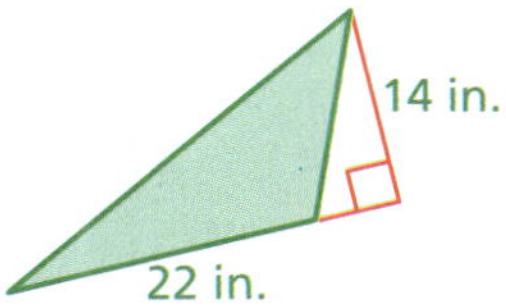

3. 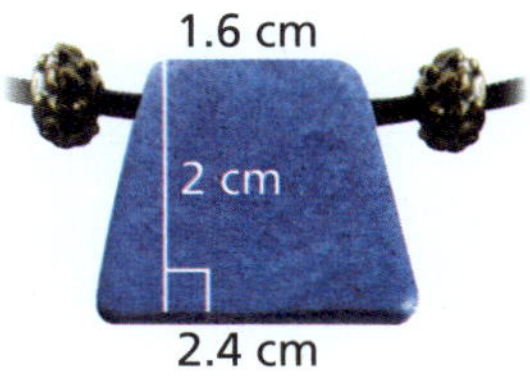

Find the surface area of the solid.

4. 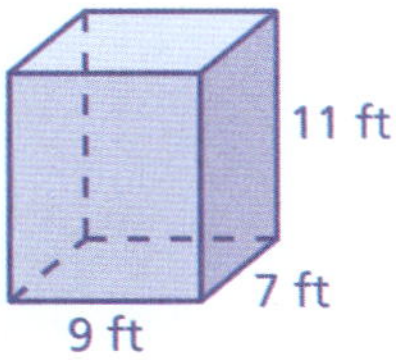

5. 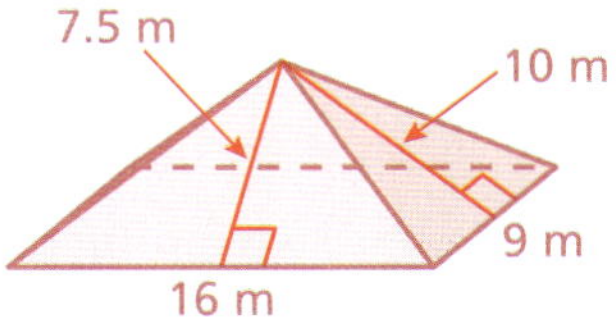

Find the volume of the prism.

6. 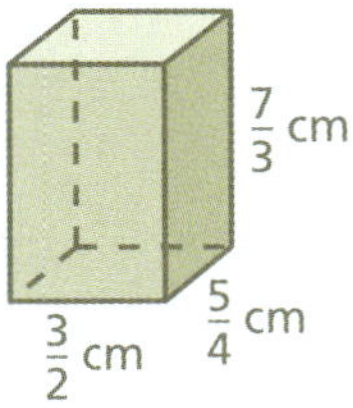

7. 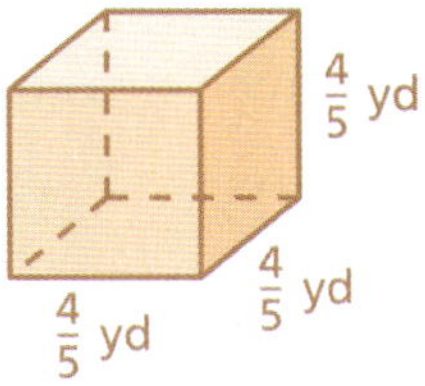

8. Draw an octagonal prism.

GO DIGITAL

9. Find the numbers of faces, edges, and vertices of the solid.

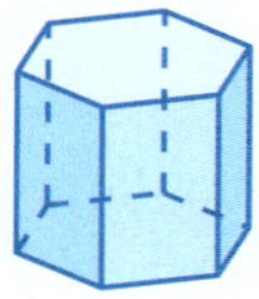

10. The area of a triangle is 154 square meters. What is the height of the triangle when the base is 22 meters?

11. A candle is shaped like a square pyramid. Find the surface area of the candle.

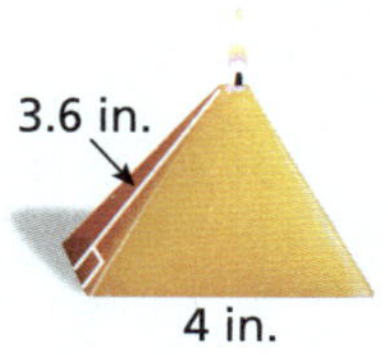

12. You are wrapping the boxed Blu-ray collection as a present. What is the least amount of wrapping paper needed to wrap the box?

13. A cube has an edge length of 4 inches. You double the edge lengths. How many times greater is the volume of the new cube?

14. The Pentagon in Arlington, Virginia, is the headquarters of the U.S. Department of Defense. The building's center contains a pentagon-shaped courtyard with an area of about 5 acres. Find the land areas (in square feet) of the courtyard and the building.

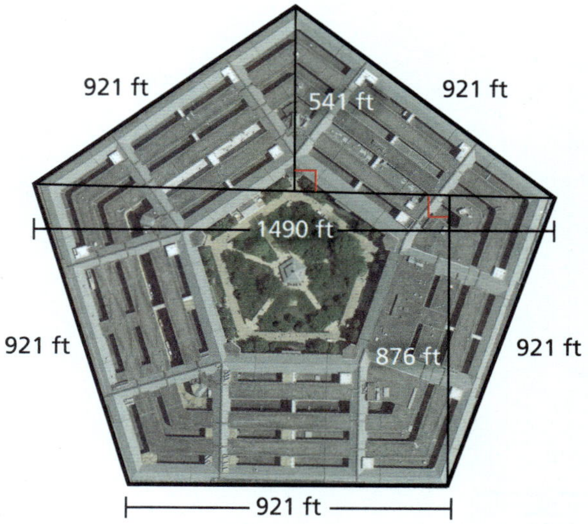

10 Review & Refresh WITH CalcChat®
while Building Fluency

Solve the equation. Check your answer.

1. $t - 7 = -31$
2. $42h = 168$
3. $\frac{4}{9} + p = \frac{2}{3}$

Simplify the expression.

4. $|6 - 11|$
5. $|-26 \div 13|$
6. $|-7 + 8|$
7. $|5 \cdot (-51)|$

8. During a reef snorkeling tour, you see 5 angelfish for every 2 starfish. How many angelfish do you see when you see 6 starfish?

9. You enter a pie baking competition. After 3 rounds of judging, 12 contestants are eliminated. There are 5 contestants remaining. How many contestants entered the contest?

GO DIGITAL

10. The pygmy shrew has the fastest heartbeat of any mammal, with a heartbeat of 1200 beats per minute.

a. How many times does the shrew's heart beat in 3 minutes?

b. How many minutes does it take for the shrew's heart to beat 6000 times?

Write the mixed number as an improper fraction.

11. $2\frac{1}{2}$

12. $7\frac{4}{7}$

13. $-3\frac{3}{5}$

Find the LCM of the numbers.

14. 3, 18, 24

15. 8, 10, 20

16. 4, 16, 24

17. How much paper do you need (in square feet) to make the poster?

18. The average growth rate of a sabal palm tree is $\frac{1}{2}$ foot each year from the time it sprouts. How tall would you expect a sabal palm to be after 7 years, when one year it grew $\frac{1}{4}$ foot less than the average?

The sabal palm tree is Florida's state tree.

10 B.E.S.T. Test Prep WITH CalcChat®
Cumulative Practice

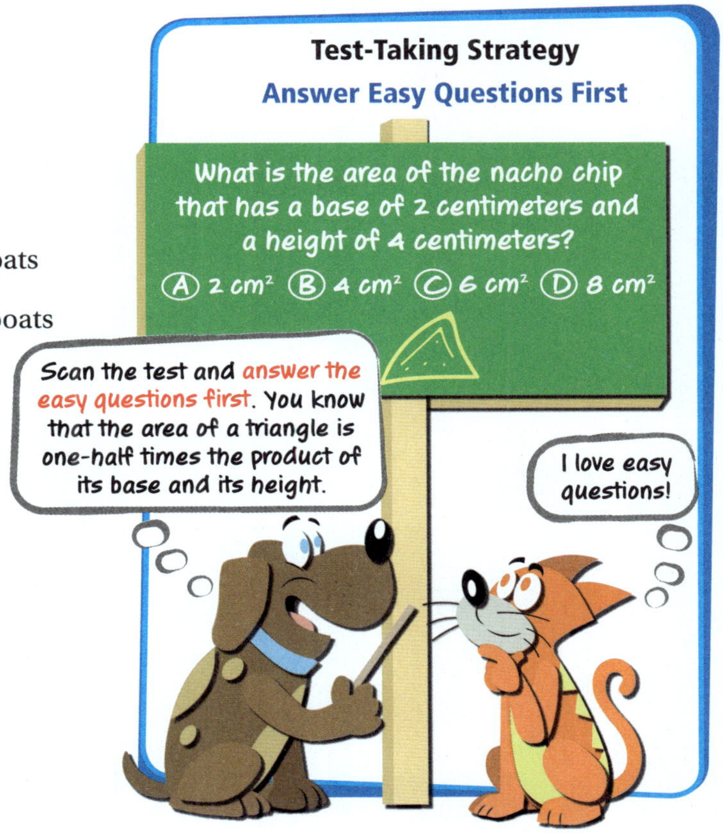

1. A cruise ship is carrying a total of 4971 people. Each lifeboat can hold a maximum of 150 people. What is the minimum number of lifeboats needed to evacuate everyone on the cruise ship?

Ⓐ 33 lifeboats

Ⓑ 34 lifeboats

Ⓒ 54 lifeboats

Ⓓ 332 lifeboats

2. What number is equivalent to the expression?

GRIDDED RESPONSE

$$3 \cdot 4^2 + 6 \div 2$$

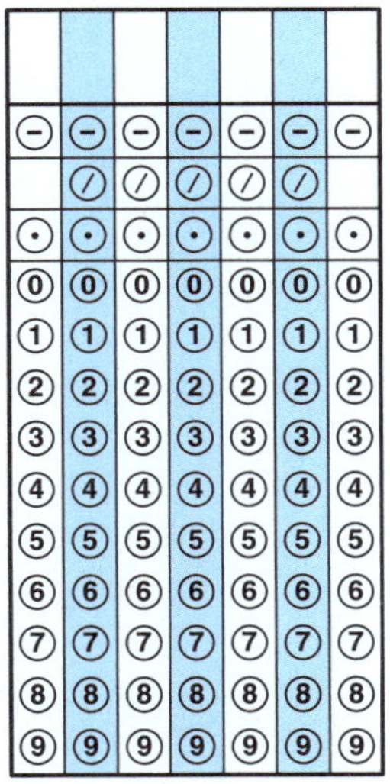

3. What is the volume of the package?

Ⓐ 240 in.^{3}

Ⓑ 376 in.^{3}

Ⓒ 480 in.^{3}

Ⓓ 960 in.^{3}

GO DIGITAL

4. What is the missing quantity in the double number line?

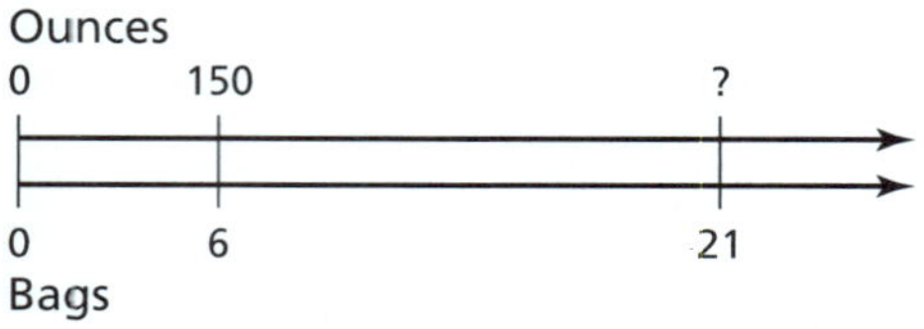

Ⓐ 25 ounces

Ⓑ 165 ounces

Ⓒ 525 ounces

Ⓓ 600 ounces

5. What is the value of m that makes the equation true?

GRIDDED RESPONSE

$$4m = 28$$

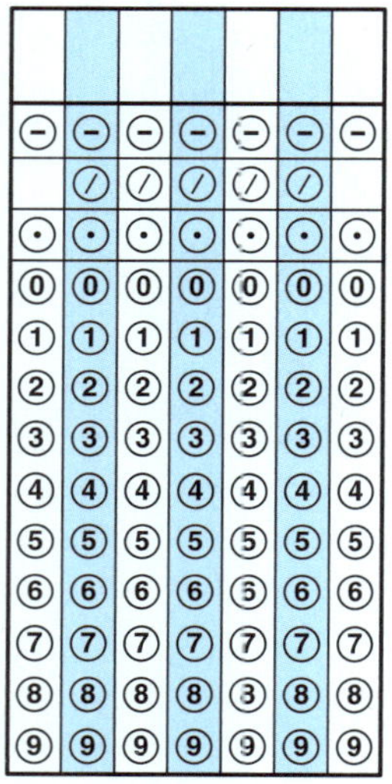

6. What is the surface area of the square pyramid?

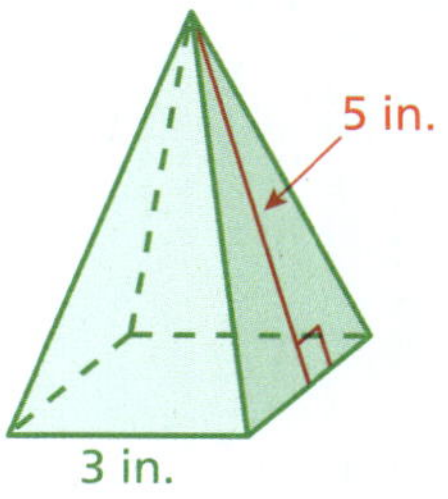

Ⓐ 30 in.^2

Ⓑ 31.5 in.^2

Ⓒ 39 in.^2

Ⓓ 69 in.^2

7. On Saturday, you earned \$35 mowing lawns. This was x dollars more than you earned on Thursday. Which expression represents the amount (in dollars) you earned mowing lawns on Thursday?

Ⓐ $35x$ Ⓒ $x - 35$

Ⓑ $x + 35$ Ⓓ $35 - x$

8. What is the area (in square yards) of the triangle?

GRIDDED RESPONSE

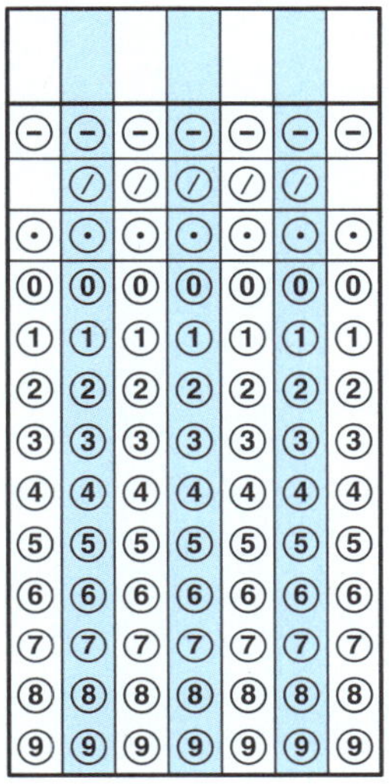

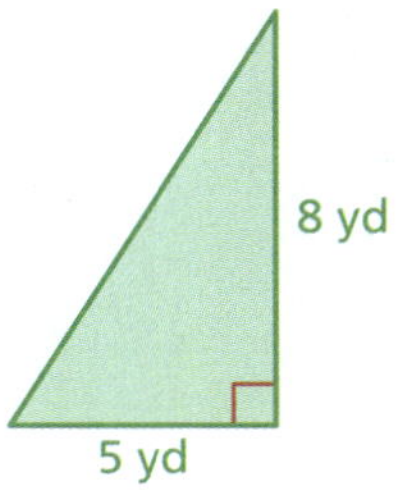

9. Which expression is equivalent to $\frac{12}{35}$?

Ⓐ $\frac{5}{6} \div \frac{2}{7}$ Ⓑ $\frac{2}{7} \div \frac{6}{5}$ Ⓒ $\frac{2}{7} \div \frac{5}{6}$ Ⓓ $\frac{5}{6} \div \frac{7}{2}$

11

Statistical Measures and Data Displays

Chapter Learning Target:
Understand statistical measures and data displays.

Chapter Success Criteria:
- ◆ I can construct a data set or data display.
- ◆ I can interpret data.
- ■ I can find and interpret the measures of center and the measures of variation for a data set.
- ■ I can compare data sets using different data displays.

◆ Surface ■ Deep

Sunrise in the City

GO DIGITAL

7 MTR *Mathematicians who apply mathematics to real-world contexts connect mathematical concepts to everyday experiences.*

STEAM Video Daylight in the Big City

Robert and Tory use data about the hours of daylight in many cities in the United States to find the average amount of daylight.

1. Why do different cities have different amounts of daylight throughout the year?

7 MTR **2.** **MODELING REAL LIFE** Robert's table includes the difference of the greatest amount of daylight and the least amount of daylight in Lagos, Nigeria, and in Moscow, Russia.

Lagos: 44 minutes **Moscow:** 633 minutes

Use these values to make a prediction about the difference of the greatest amount of daylight and the least amount of daylight in a city in Alaska.

STEAM Performance Task Which Measure of Center Is Best: Mean, Median, or Mode?

After completing this chapter, you will be able to use the concepts you learned to answer the questions in the *STEAM Performance Task*. You will be given the greatest and least amounts of daylight in the 15 cities in the United States with the greatest populations.

	Greatest	Least
New York:	905 minutes	555 minutes
Los Angeles:	865 minutes	593 minutes
Chicago:	913 minutes	548 minutes

You will determine which measure of center best represents the data.

GO DIGITAL

Getting Ready for Chapter 11 WITH CalcChat®

Chapter Exploration

1 MTR **Actively Participate in Effortful Learning Collectively**
Work with a partner to prepare for concepts in this chapter.

Write the number of letters in each of your first names on the board.

1. Write all of the numbers on a piece of paper. The collection of numbers is called *data*.

2. Talk with your partner about how you can organize the data. What conclusions can you make about the numbers of letters in the first names of the students in your class?

3. Use the grid to draw a graph of the data.

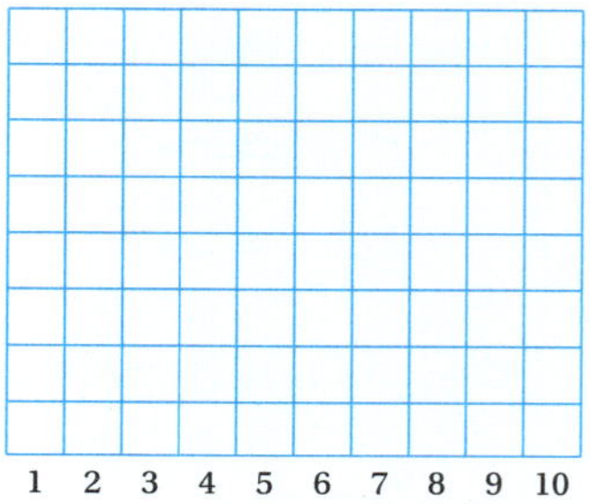

4 MTR 4. **DISCUSS MATHEMATICAL THINKING** Use the graph of the data in Exercise 3 to answer the following.

a. Is there one number that occurs more than any of the other numbers? If so, write a sentence that interprets this number in the context of your class.

b. Complete the sentence "In my class, the average number of letters in a student's first name is __________." Justify your reasoning.

c. Organize your data using a different type of graph. Describe the advantages or disadvantages of this graph.

Vocabulary

The following terms are defined in this chapter. Think about what each might mean and record your thoughts.

statistical question	median	frequency table
mean	measure of variation	box-and-whisker plot
measure of center	range	five-number summary

GO DIGITAL

11.1 Introduction to Statistics

Learning Target: Identify statistical questions and use data to answer statistical questions.

Success Criteria:
- I can recognize questions that anticipate a variety of answers.
- I can construct and interpret a dot plot.
- I can use data to answer a statistical question.

Data Analysis and Probability

MA.6.DP.1.1 Recognize and formulate a statistical question that would generate numerical data.

MA.6.DP.1.4 Given a histogram or line plot within a real-world context, qualitatively describe and interpret the spread and distribution of the data, including any symmetry, skewness, gaps, clusters, outliers and the range.

Exploration 1 Using Data to Answer a Question

Work with a partner.

a. Use your pulse to find your heart rate in beats per minute.

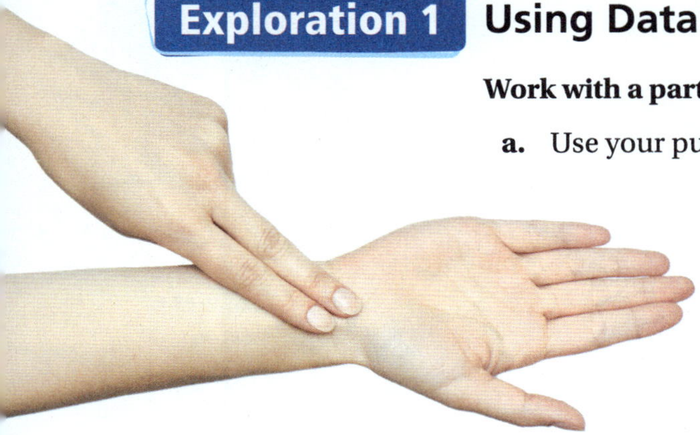

Places to check your pulse:
- on your wrist
- inside your elbow
- on the side of your neck
- on top of your foot

b. Collect the recorded heart rates of the students in your class, including yourself. How spread out are the data? Use a diagram to justify your answer.

5 MTR **c.** **MAKE A PLAN** How would you answer the following question by using only one value? Explain your reasoning.

"What is the heart rate of a sixth-grade student?"

GO DIGITAL

Exploration 2 Identifying Types of Questions

Work with a partner.

a. Answer each question on your own. Then compare your answers with your partner's answers. For which questions should your answers be the same? For which questions might your answers be different?

1. How many states are in the United States?

2. How much does a movie ticket cost?

3. What color fur do bears have?

4. How tall is your math teacher?

b. Some of the questions in part (a) are considered *statistical* questions. Which ones are they? Explain.

4 MTR COMMUNICATE CLEARLY

Explain how you determined which questions should have the same answers and which questions might have different answers.

11.1 Lesson

Key Vocabulary
statistics, *p. 887*
statistical question, *p. 887*

Statistics is the science of collecting, organizing, analyzing, and interpreting data. A **statistical question** is one for which you do not expect to get a single answer. Instead, you expect a variety of answers, and you are interested in the *distribution* and *tendency* of those answers.

Example 1 Identifying Statistical Questions

Determine whether the question is a statistical question. Explain.

a. How many countries start with the letter Z?

▶ Because there is only one answer, it is not a statistical question.

b. How much do bags of pretzels cost at the grocery store?

▶ Because you can anticipate that the prices will vary, it is a statistical question. The table at the right may represent prices of several bags of pretzels at a grocery store.

Prices	
\$0.99	\$2.99
\$1.99	\$2.99
\$1.99	\$4.29

c. How many days does your school have off for spring break this year?

▶ Because there is only one answer, it is not a statistical question.

d. What are the hair colors of students in your class?

▶ Because you can anticipate that the colors will vary, it is a statistical question. The table below may represent hair colors of several students in a class.

Hair Colors	
Brown	Red
Black	Blond
Blond	Pink
Red	Brown

GO DIGITAL

Try It **Determine whether the question is a statistical question. Explain.**

1. What types of cell phones do students in your class have?

2. How many desks are in your classroom?

3. How much do virtual-reality headsets cost?

4. How many minutes are in your lunch period?

A *dot plot* uses a number line to show the number of times each value in a data set occurs. Dot plots show the *spread* and the *distribution* of a data set.

Example 2 Using a Dot Plot

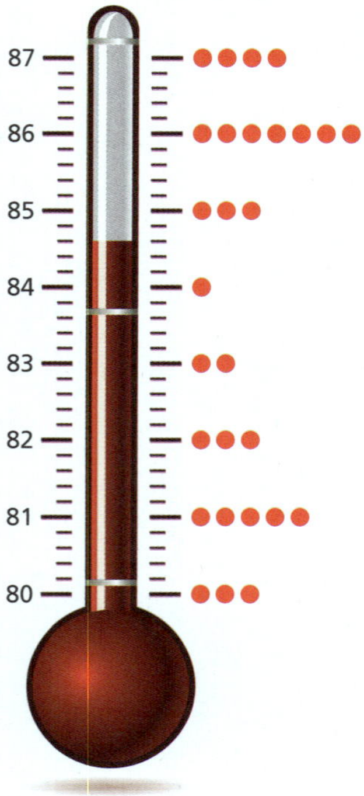

You record the high temperature every day while at summer camp in August. Then you create the vertical dot plot.

a. Find and interpret the number of data values on the dot plot.

▶ There are 28 data values on the dot plot. So, you were at camp 28 days, or 4 weeks.

b. How can you collect these data? What are the units?

▶ You can collect these data with a thermometer. The units are degrees Fahrenheit (°F).

c. Write a statistical question that you can answer using the dot plot. Then answer the question.

One possible statistical question is

"What is the daily high temperature in August?"

▶ The high temperatures are spread out with about half of the temperatures around 81°F and half of the temperatures around 86°F.

GO DIGITAL

Try It

5. Repeat parts (a)–(c) using the dot plot below that shows the times of students in a 100-meter race.

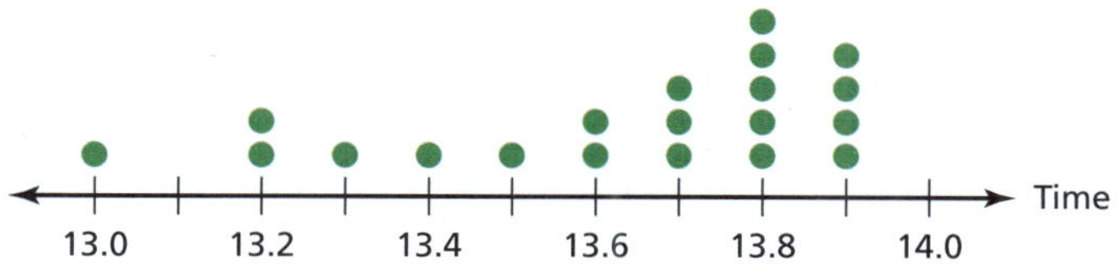

In-Class Practice

1 I don't understand yet. | 2 I can do it with help. | 3 I can do it on my own. | 4 I can teach someone else.

6. OPEN-ENDED Write and answer a statistical question using the dot plot. Then find and interpret the number of data values.

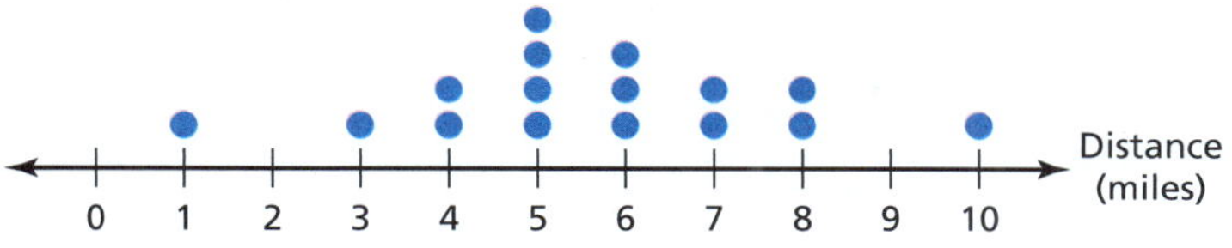

4 MTR

7. WHICH ONE DOESN'T BELONG Which question does *not* belong with the other three? Explain your reasoning.

How tall are teenagers?	What sports do students watch the most?
How many books do students own?	What is your favorite food?

Example 3 Modeling Real Life

7 MTR

Your teacher asks you, "What is the mass of a typical mouse?" You conduct a science experiment on house mice. Use the data in the table to answer the question.

Masses (grams)			
20	19	21	20
18	20	27	21
28	23	20	19
20	21	18	27
19	22	21	20

Understand the problem.

You know the masses of several mice. You are asked to use the data to answer a statistical question.

Make a plan.

Display the data in a dot plot. Identify any clusters, peaks, or gaps in the data. Then use the distribution of the data to answer the question.

Solve and check.

Draw a number line that includes the least value, 18, and the greatest value, 28. Then place a dot above the number line for each data value.

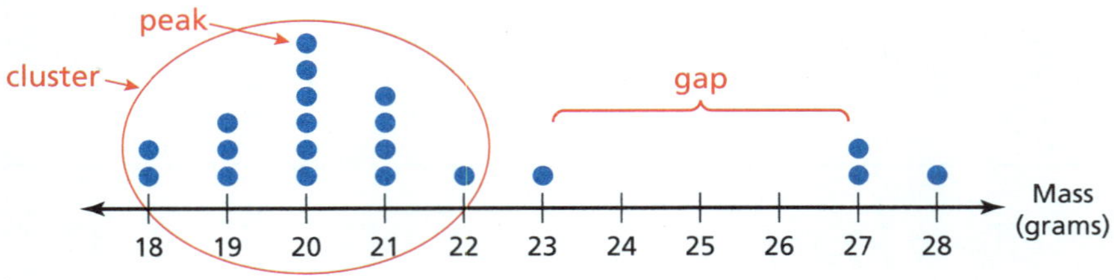

Most of the data are clustered around 20. There is a peak at 20 and a gap between 23 and 27.

▶ A typical mouse has a mass of about 20 grams.

Check Reasonableness
65% of the mice have a mass of 19, 20, or 21 grams. So, it is reasonable to say that a typical mouse has a mass of about 20 grams. ✓

In-Class Practice

1 I don't understand yet. 2 I can do it with help. 3 I can do it on my own. 4 I can teach someone else.

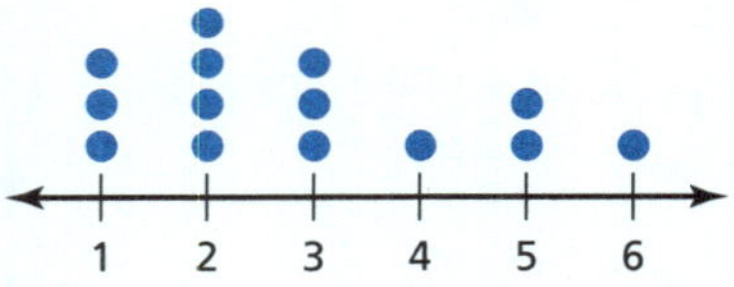

8. You record the amount of time several visitors spend at an amusement park. Then you create the dot plot.

a. Find and interpret the number of data values on the dot plot.

b. How can you collect these data? What are the units?

c. Write a statistical question that you can answer using the dot plot. Then answer the question.

Hours of Exercise				
5	1	5	3	5
4	5	2	5	4
3	4	6	5	6

9. You conduct a survey to answer, "How many hours does a typical sixth-grade student spend exercising during a week?" Use the data in the table to answer the question.

GO DIGITAL

11.1 Practice WITH CalcChat® AND CalcView®

Review & Refresh

Find the volume of the prism.

1.
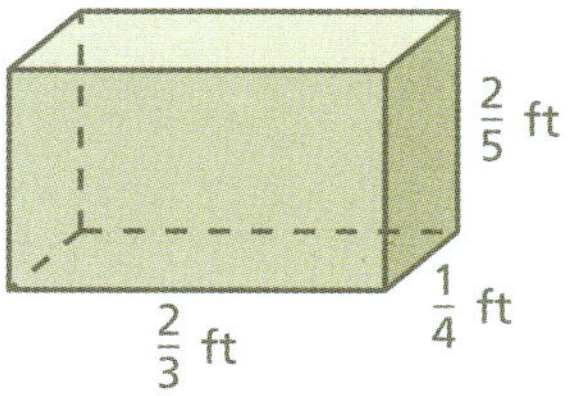

2.
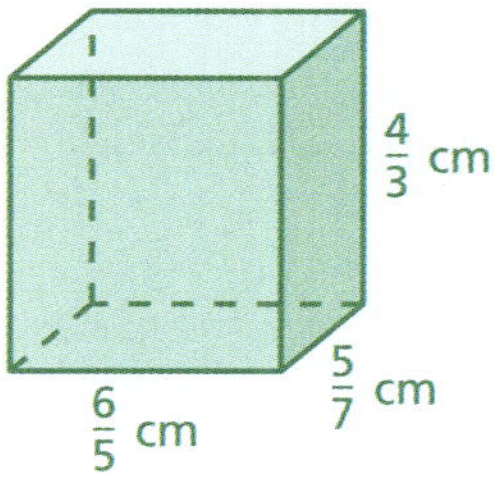

3.
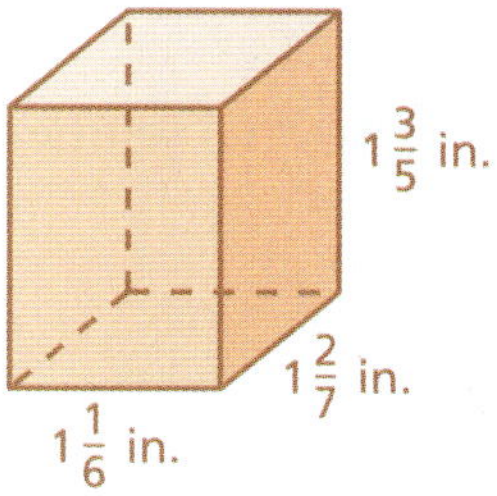

4. A point is reflected in the x-axis. The reflected point is (4, −3). Find the original point.

Order the numbers from least to greatest.

5. 24%, $\frac{1}{4}$, 0.2, $\frac{7}{20}$, 0.32

6. $\frac{7}{8}$, 85%, 0.88, $\frac{3}{4}$, 78%

Concepts, Skills, & Problem Solving

IDENTIFYING TYPES OF QUESTIONS Answer the question. Tell whether your answer should be the same as your classmates'. (See Exploration 2.)

7. How many inches are in 1 foot?

8. How many pets do you have?

9. On what day of the month were you born?

10. How many senators are in Congress?

GO DIGITAL

IDENTIFYING STATISTICAL QUESTIONS Determine whether the question is a statistical question. Explain. (See Example 1.)

11. What are the eye colors of sixth-grade students?

12. At what temperature (in degrees Fahrenheit) does water freeze?

13. How many pages are in the favorite books of students your age?

14. How many hours do sixth-grade students use the Internet each week?

7 MTR

15. **MODELING REAL LIFE** The vertical dot plot shows the heights of the players on a NBA championship team. (See Example 2.)

a. Find and interpret the number of data values on the dot plot.

b. How can you collect these data? What are the units?

c. Write a statistical question that you can answer using the dot plot. Then answer the question.

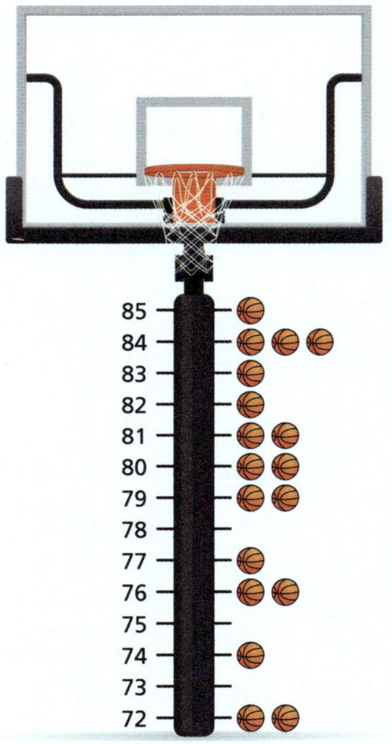

7 MTR

16. MODELING REAL LIFE The dot plot shows the lengths of earthworms.

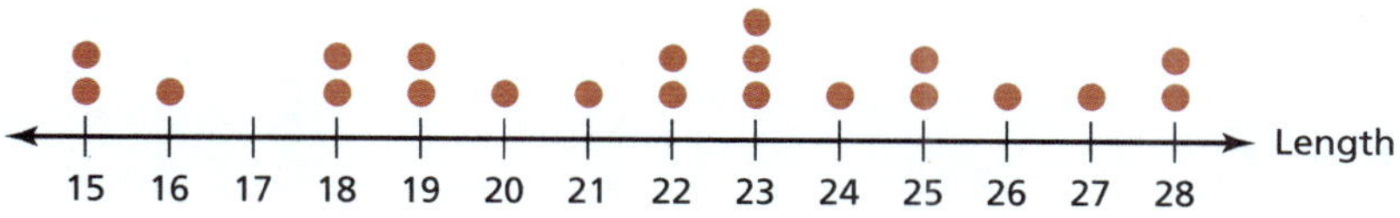

a. Find and interpret the number of data values on the dot plot.

b. How can you collect these data? What are the units?

c. Write a statistical question that you can answer using the dot plot. Then answer the question.

DESCRIBING DATA **Display the data in a dot plot. Identify any clusters, peaks, or gaps in the data.**

17.

Camper Registrations				
21	25	25	22	21
23	24	26	25	16
24	26	22	25	22

18.

Test Scores				
85	80	83	90	88
82	83	81	80	89
89	84	86	87	83

INTERPRETING DATA **The dot plot shows the speeds of cars in a traffic study. Estimate the speed limit. Explain your reasoning.**

19.

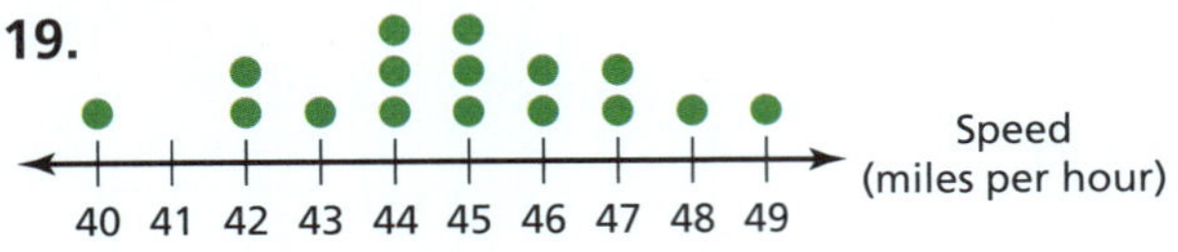

20.

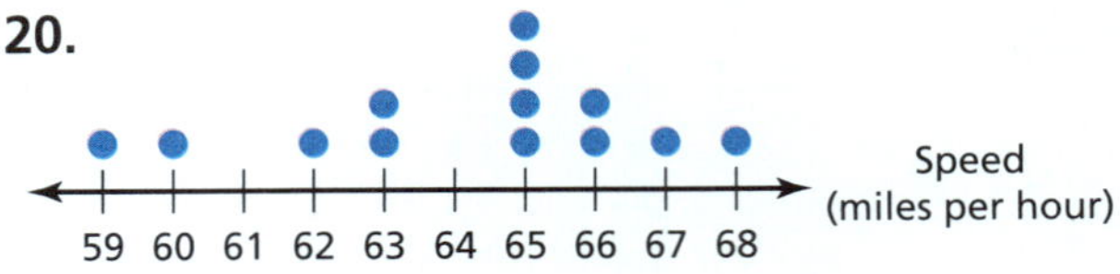

7 MTR ▶ **21. MODELING REAL LIFE** You conduct a survey to answer, "How many hours does a sixth-grade student spend on homework during a school night?" The table shows the results. (See Example 3.)

Hours of Homework			
2	4	3	2
1	2	2	1
2	3	5	2

a. Is this a statistical question? Explain.

b. Identify any clusters, peaks, or gaps in the data.

c. Use the distribution of the data to answer the question.

7 MTR **INVESTIGATE** **Use the Internet to research and identify the method of measurement and the units used when collecting data about the topic.**

22. wind speed

23. amount of rainfall

24. earthquake intensity

25. REASONING Write a question about letters in the English alphabet that is *not* a statistical question. Then write a question about letters that is a statistical question. Explain your reasoning.

26. REASONING A bar graph shows the favorite colors of 30 people. Does it make sense to describe clusters in the data? peaks? gaps? Explain.

11.2 Mean

Learning Target: Find and interpret the mean of a data set.

Success Criteria:
- I can explain how the mean summarizes a data set.
- I can find the mean of a data set.
- I can use the mean of a data set to answer a statistical question.

Data Analysis and Probability

MA.6.DP.1.2 Given a numerical data set within a real-world context, find and interpret mean, median, mode and range.

MA.6.DP.1.6 Given a real-world scenario, determine and describe how changes in data values impact measures of center and variation.

Exploration 1 Finding a Balance Point

Work with a partner. The diagrams show the numbers of tokens brought to a batting cage. Where on the number line is the data set *balanced*? Is this a good representation of the average? Explain.

a.

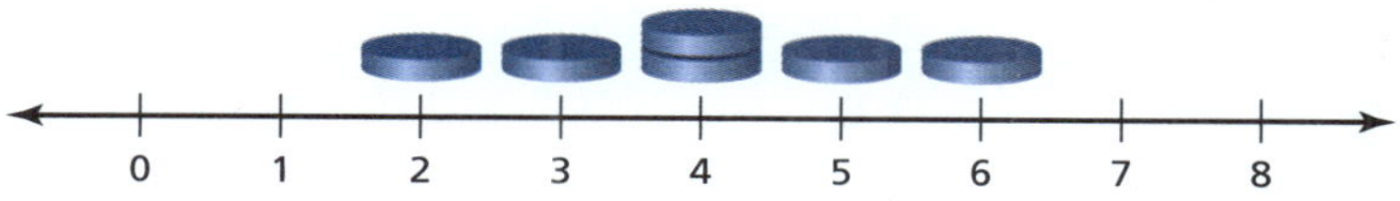

b.

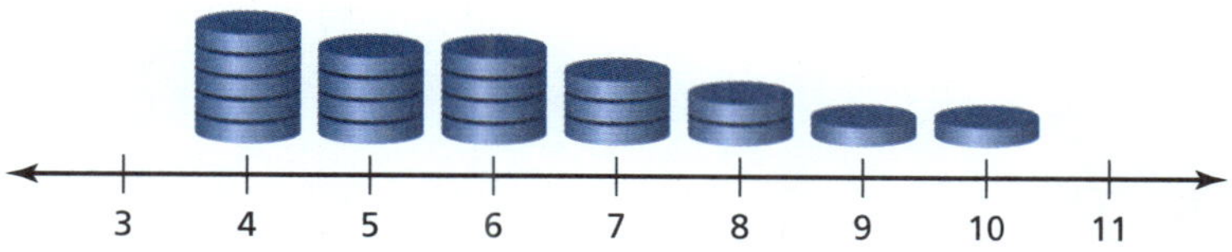

GO DIGITAL

Exploration 2 Finding a Fair Share

Work with a partner. One token lets you hit 12 baseballs in a batting cage. The table shows the numbers of tokens six friends bring to the batting cage.

Tokens					
John	Lisa	Miguel	Matt	Cheryl	Anika
6	3	4	5	2	4

a. Regroup the tokens so that everyone has the same amount. How many times can each friend use the batting cage? Explain how this represents a "fair share."

b. How can you find the answer in part (a) algebraically?

c. Write a statistical question that can be answered using the value in part (a).

7 MTR APPLY MATHEMATICS

What does it mean for data to have an average? Describe a real-life situation where it is helpful to determine an average value.

11.2 Lesson

Key Vocabulary
mean, *p. 897*
outlier, *p. 899*

You can use braces { } to represent a data set. This is called *set notation*.

Key Idea

Mean

Words The **mean** of a data set is the sum of the data divided by the number of data values. The mean is a type of average.

Numbers **Data set:** {8, 5, 6, 9} **Mean:** $\frac{8+5+6+9}{4} = \frac{28}{4} = 7$

4 data values

Example 1 B.E.S.T. Test Prep: Finding the Mean

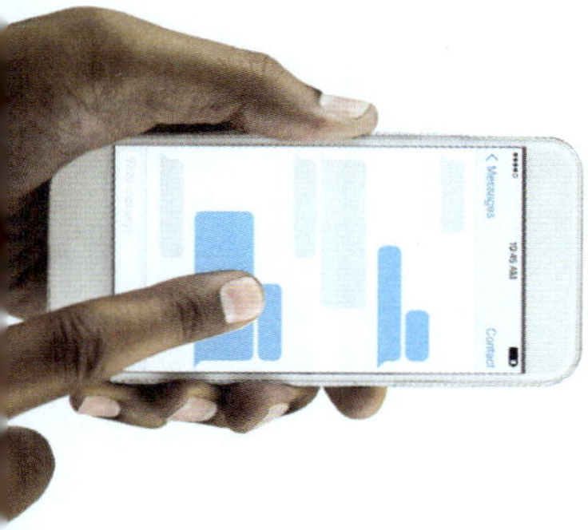

The numbers of text messages sent by several friends over 1 week are represented by the data set {120, 95, 101, 125, 82, 108, 90}. What is the mean number of messages sent?

Ⓐ 100 Ⓑ 102 Ⓒ 103 Ⓓ 104

$$\text{Mean} = \frac{120+95+101+125+82+108+90}{7}$$

sum of the data

number of values

$$= \frac{721}{7}$$ Add values in numerator.

$$= 103$$ Divide.

▶ The mean number of text messages sent is 103. The correct answer is Ⓒ.

Try It **Find the mean of the data.**

1.

Dog Weights (pounds)
Sparky: 18
Spot: 9
Rover: 60
Newton: 89
Diego: 44
Ruby: 13
Mookie: 54
Fido: 45

2.

Airbag Backpack Costs				
\$600	\$450	\$350	\$650	\$800
\$300	\$550	\$500	\$600	\$750

GO DIGITAL

Example 2 Comparing Means

The double bar graph shows the monthly rainfall amounts for two cities over a six-month period. Compare the mean monthly rainfalls.

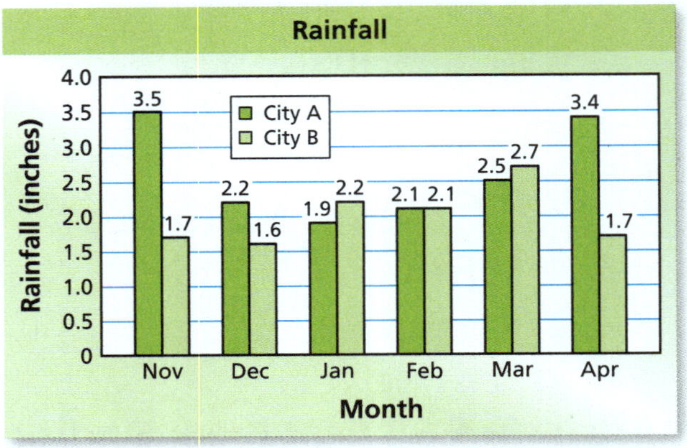

City A mean: $\frac{3.5 + 2.2 + 1.9 + 2.1 + 2.5 + 3.4}{6} = \frac{15.6}{6}$

$= 2.6$

City B mean: $\frac{1.7 + 1.6 + 2.2 + 2.1 + 2.7 + 1.7}{6} = \frac{12}{6}$

$= 2$

▶ Because 2.6 is greater than 2, City A averaged more rainfall.

Try It

3. **WHAT IF?** The monthly rainfall in May was 0.5 inch in City A and 2 inches in City B. Does this affect your answer in Example 2? Explain.

In-Class Practice

1 I don't understand yet. 2 I can do it with help. 3 I can do it on my own. 4 I can teach someone else.

4. **NUMBER SENSE** Is the mean always equal to a value in the data set? Explain.

4 MTR 5. **DISCUSS MATHEMATICAL THINKING** Explain why the mean describes a typical value in a data set.

GO DIGITAL

In-Class Practice

1 I don't understand yet. | 2 I can do it with help. | 3 I can do it on my own. | 4 I can teach someone else.

6. **NUMBER SENSE** What can you determine when the mean of one data set is greater than the mean of another data set? Explain your reasoning.

7. **COMPARING MEANS** Compare the means of the data sets.

   Data set A: {43, 32, 16, 41, 24, 19, 30, 27}

   Data set B: {44, 18, 29, 24, 36, 22, 26, 21}

   Data set C: {45, 28, 14, 32, 32, 19, 40, 37}

An **outlier** is a data value that is much greater or much less than the other values. When included in a data set, it can affect the mean.

Example 3 Modeling Real Life

7 MTR

The table shows the heights of several Shetland ponies. Describe how the outlier affects the mean. Then use the data to answer the statistical question "What is the height of a typical Shetland pony?"

Shetland Pony Heights (inches)				
40	37	39	40	42
38	38	37	28	40

Display the data in a dot plot to see the distribution of the data.

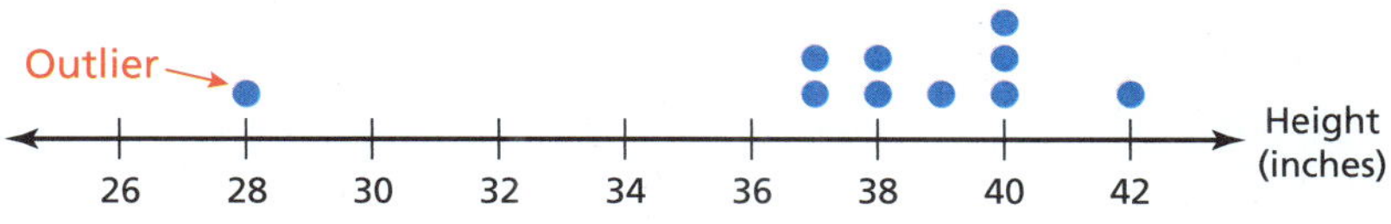

The height of 28 inches is much less than the other heights. So, it is an outlier. Find the mean with and without the outlier.

Mean with outlier:

$$\frac{40 + 37 + 39 + 40 + 42 + 38 + 38 + 37 + 28 + 40}{10} = \frac{379}{10} = 37.9$$

Mean without outlier:

$$\frac{40 + 37 + 39 + 40 + 42 + 38 + 38 + 37 + 40}{9} = \frac{351}{9} = 39$$

▶ With the outlier, the mean is less than all but three of the heights. Without the outlier, the mean better represents the heights. So, the height of a typical Shetland pony is about 39 inches.

GO DIGITAL

In-Class Practice

1 I don't understand yet. | 2 I can do it with help. | 3 I can do it on my own. | 4 I can teach someone else.

8. The monthly numbers of customers at a store in the first half of a year are represented by the data set {282, 270, 320, 351, 319, 252}. The monthly numbers of customers in the second half of the year are represented by the data set {211, 185, 192, 216, 168, 144}. Compare the mean monthly customers in the first half of the year with the mean monthly customers in the second half of the year.

9. The table shows tournament finishes for a golfer. What place does the golfer typically finish in tournaments? Explain how you found your answer.

Tournament Finishes			
5	1	2	5
21	2	8	3
1	5	9	10
3	4	8	9

11.2 Practice WITH CalcChat® AND CalcView®

Review & Refresh

Determine whether the question is a statistical question. Explain.

1. How tall are sixth-grade students?

2. How many minutes are in 1 year?

3. How many counties are in Florida?

4. What is a student's favorite sport?

Write the percent as a fraction or mixed number in simplest form.

5. 84%

6. 71%

7. 353%

8. 0.2%

Divide. Check your answer.

9. $11.7 \div 9$

10. $5\overline{)72.8}$

11. $6.8\overline{)28.56}$

12. $93 \div 3.75$

Concepts, Skills, & Problem Solving

FINDING A FAIR SHARE **Regroup the amounts so that each person has the same amount. What is the amount?** (See Exploration 2.)

13. Dollars brought by friends to a fair: 11, 12, 12, 12, 12, 12, 13

14. Tickets earned by friends playing an arcade game: 0, 0, 0, 1, 1, 2, 3

GO DIGITAL

FINDING THE MEAN Find the mean of the data. (See Example 1.)

15. 16, 14, 13, 15, 15, 17

16. 56, 54, 61, 63, 68, 58

17.

Pets Owned	
Brandon	I
Jill	III
Mark	II
Nicole	IIII
Steve	0

18.

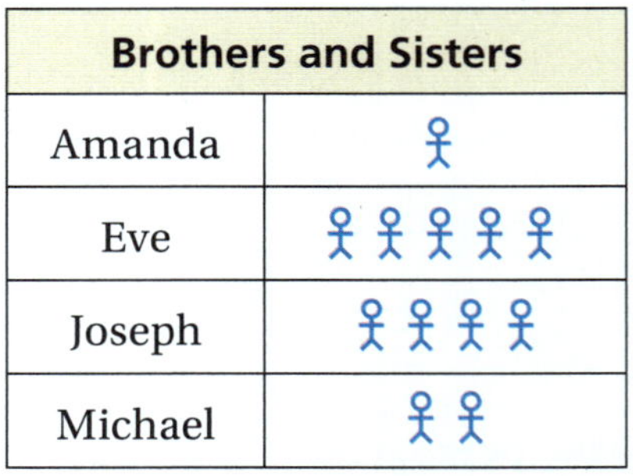

19.

Sit-Ups		
108	85	94
103	112	115
98	119	126
105	82	89

20.

21. **B.E.S.T. Test Prep** What is the mean of the data set {4.2, 3.5, 4.55, 2.75, 2.25}?

Ⓐ 3.4 Ⓑ 3.45 Ⓒ 3.55 Ⓓ 3.5

7 MTR

22. MODELING REAL LIFE The table shows the monthly rainfall amounts at a measuring station. (See Example 2.)

Month	Jan	Feb	Mar	Apr	May	Jun	Jul	Aug	Sep	Oct	Nov	Dec
Rainfall (inches)	2.22	1.51	1.86	2.06	3.48	4.47	3.37	5.40	5.45	4.34	2.64	2.14

a. What is the mean monthly rainfall?

b. Compare the mean monthly rainfall for the first half of the year with the mean monthly rainfall for the second half of the year.

23. OPEN-ENDED Create two different data sets that have six values and a mean of 21.

7 MTR

24. MODELING REAL LIFE The bar graph shows your cell phone data usage for five months. Describe how the outlier affects the mean. Then use the data to answer the statistical question "How much cell phone data do you use in a month?" (See Example 3.)

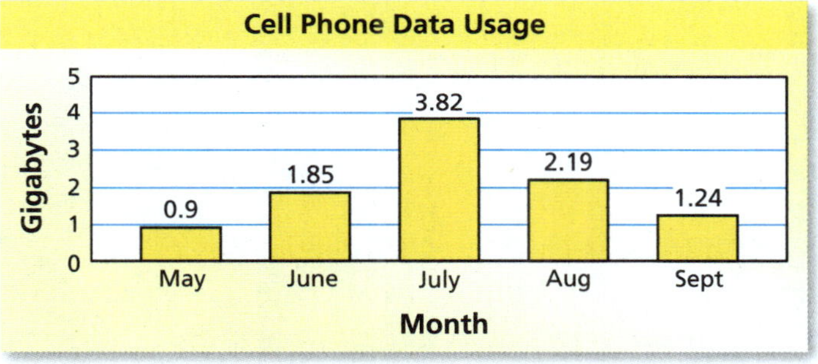

7 MTR

▶ **25. MODELING REAL LIFE** The heights of the volleyball players on two teams are represented by the following data sets.

Dolphins: {59, 65, 53, 56, 58, 61, 64, 68, 51, 56, 54, 57}
Tigers: {63, 68, 66, 58, 54, 55, 61, 62, 53, 70, 64, 64}

Compare the mean heights of the two teams. Do outliers affect either mean? Explain.

26. **REASONING** Use a dot plot to explain why the mean of the data below is the point where the data is balanced.

11, 13, 17, 15, 12, 18, 12

27. **Dig Deeper** In your class, 7 students do not receive a weekly allowance, 5 students receive \$3, 7 students receive \$5, 3 students receive \$6, and 2 students receive \$8.

a. What is the mean weekly allowance? Explain how you found your answer.

b. A new student who joins your class receives a weekly allowance of \$3.50. Without calculating, explain how this affects the mean.

28. **MAINTAIN ACCURACY** A collection of 8 geodes has a mean weight of 14 ounces. A different collection of 12 geodes has a mean weight of 9 ounces. What is the mean weight of the 20 geodes? Explain how you found your answer.

11.3 Measures of Center

Learning Target: Find and interpret the median and mode of a data set.

Success Criteria:
- I can explain how the median and mode summarize a data set.
- I can find the median and mode of a data set.
- I can explain how changes to a data set affect the measures of center.
- I can use a measure of center to answer a statistical question.

Data Analysis and Probability

MA.6.DP.1.2 Given a numerical data set within a real-world context, find and interpret mean, median, mode and range.

MA.6.DP.1.6 Given a real-world scenario, determine and describe how changes in data values impact measures of center and variation.

Exploration 1 Finding the Median

Work with a partner.

a. Write the total numbers of letters in the first and last names of 15 celebrities, historical figures, or people you know. One person is already listed for you.

Person	Number of Letters in First and Last Name
Rosa Parks	9

Rosa Parks is honored as "the first lady of civil rights" and "the mother of the freedom movement" for her role in the civil rights movement of the 1950s and 1960s.

GO DIGITAL

b. Order the values in your data set from least to greatest. Then write the data on the strip of grid paper.

c. The *middle value* of the data set is called the *median*. The value (or values) that occur most often is called the *mode*. Find the median and the mode of your data set. Explain how you found your answers.

d. Why are the median and the mode considered averages of a data set?

2 MTR **MODEL A PROBLEM**

How can you use a dot plot to find the mode?

11.3 Lesson

Key Vocabulary
measure of center, *p. 907*
median, *p. 907*
mode, *p. 907*

A **measure of center** is a measure that describes the typical value of a data set. The mean is one type of measure of center. Here are two others.

Key Ideas

Median

Words Order the data. For a set with an odd number of values, the **median** is the middle value. For a set with an even number of values, the **median** is the mean of the two middle values.

Numbers **Data set:** {5, 8, 9, 12, 14} The median is 9.

Data set: {2, 3, 5, 7, 10, 11} The median is $\frac{5+7}{2} = 6$.

Mode

Words The **mode** of a data set is the value or values that occur most often. Data can have one mode, more than one mode, or no mode. When all values occur only once, there is no mode.

Numbers **Data set:** {11, 13, 15, 15, 18, 21, 24, 24}

The modes are 15 and 24.

The mode is the only measure of center that you can use to describe a set of data that is *not* made up of numbers.

Example 1 Finding the Median and Mode

Find the median and mode of the bowling scores.

Bowling Scores				
120	135	160	125	90
205	160	175	105	145

90, 105, 120, 125, 135, 145, 160, 160, 175, 205 Order the data.

Median: $\frac{135 + 145}{2} = \frac{280}{2} = 140$ Add the two middle values and divide by 2.

Mode: 90, 105, 120, 125, 135, 145, 160, 160, 175, 205.

The value 160 occurs most often.

▶ The median is 140. The mode is 160.

GO DIGITAL

Try It Find the median and mode of the data.

1. 20, 4, 17, 8, 12, 9, 5, 20, 13

2. 100, 75, 90, 80, 110, 102

Example 2 Finding the Mode

Favorite Types of Movies

Comedy	Drama	Horror
Horror	Drama	Horror
Comedy	Comedy	Action
Action	Comedy	Action
Horror	Drama	Comedy
Comedy	Comedy	Horror
Horror	Comedy	Action
Horror	Action	Drama

The list shows the favorite types of movies for students in a class. Organize the data in a table. Then find the mode.

Type	Number of Students
Action	5
Comedy	8
Drama	4
Horror	7

Comedy received the most votes.

▶ So, the mode is comedy.

Try It

3. One member of the class was absent and votes for horror. Does this change the mode? Explain.

Example 3 Removing an Outlier

The numbers of bugs eaten by seven competitors in a bug-eating competition are represented by the data set {26, 33, 34, 2, 32, 34, 42}. Find the mean, median, and mode of the data with and without the outlier. Which measure does the outlier affect the most?

The competitor with 2 bugs ate much less than any other competitor. So, the outlier is 2.

	Mean	Median	Mode
With Outlier	29	33	34
Without Outlier	33.5	33.5	34

Removing the outlier increases the mean more than the median. The mode is not affected.

▶ So, the mean is affected the most by the outlier.

GO DIGITAL

Try It

4. The times (in minutes) it takes six students to travel to school are represented by the data set {8, 10, 10, 15, 20, 45}. Find the mean, median, and mode of the data with and without the outlier. Which measure does the outlier affect the most?

Example 4 Changing the Values of a Data Set

The prices of six video games at an online store are shown in the table. The price of each game increases by $4.98 when a shipping charge is included. How does this increase affect the mean, median, and mode?

Video Game Prices	
$53.42	$35.69
$18.99	$25.13
$27.97	$53.42

Make a new table by adding $4.98 to each price. Then find the mean, median, and mode of both data sets.

Video Game Prices with Shipping Charge	
$58.40	$40.67
$23.97	$30.11
$32.95	$58.40

	Mean	Median	Mode
Original Price	35.77	31.83	53.42
Price with Shipping Charge	40.75	36.81	58.40

Compare the measures of center of both data sets.

Mean: $40.75 - 35.77 = 4.98$

Median: $36.81 - 31.83 = 4.98$

Mode: $58.4 - 53.42 = 4.98$

▶ By increasing each video game price by $4.98 for shipping, the mean, median, and mode all increase by $4.98.

Try It

5. **WHAT IF?** The store decreases the price of each video game by $3. How does this decrease affect the mean, median, and mode?

GO DIGITAL

In-Class Practice

1 I don't understand yet. | 2 I can do it with help. | 3 I can do it on my own. | 4 I can teach someone else.

6. FINDING MEASURES OF CENTER Consider the data set below.

$$\{15, 18, 13, 11, 12, 21, 9, 11\}$$

a. Find the mean, median, and mode of the data.

b. Each value in the data set is decreased by 7. How does this change affect the mean, median, and mode?

7. WRITING Explain why a typical value in a data set can be described by the median or the mode.

Example 5 Modeling Real Life

Use the data to answer the statistical question "How much do climbing shoes cost?"

Find the mean, median, and mode of the data. Then answer the question by creating a dot plot to determine which measure best represents the data.

Mean: $\dfrac{40 + 51 + 142 + 68 + 57 + 40 + 65 + 85}{8} = \dfrac{548}{8} = 68.5$

Median: 40, 40, 51, 57, 65, 68, 85, 142   Order the data.

$$\frac{57 + 65}{2} = \frac{122}{2} = 61$$

Mode: 40, 40, 51, 57, 65, 68, 85, 142   The value 40 occurs most often.

▶ The mode is less than most of the data, and the mean is greater than most of the data. The median best represents the data. So, climbing shoes cost about $61.

Check Reasonableness
It is reasonable to expect some styles or brands of climbing shoes to cost much more than others, which affects the mean price. So, it makes sense that the median is a better representation of the data than the mean in this scenario. ✓

GO DIGITAL

In-Class Practice

1 I don't understand yet. 2 I can do it with help. 3 I can do it on my own. 4 I can teach someone else.

8. How does removing the outlier affect your answer in Example 5?

9. The numbers of seconds that it takes 10 contestants on a television show to cross a canyon on a zipline are represented by the data set {43, 41, 62, 40, 44, 43, 44, 46, 45, 41}. Find the mean, median, and mode of the data with and without the outlier. Which measure does the outlier affect the most?

10. The table shows the weights of several great white sharks. Use the data to answer the statistical question "What is the weight of a great white shark?"

Great White Shark Weights (pounds)					
1700	1500	1700	2100	1700	2200
1600	1700	1900	3900	1800	1600

11.3 Practice WITH CalcChat® AND CalcView®

Review & Refresh

Find the mean of the data.

1. 1, 5, 8, 4, 5, 7, 6, 6, 2, 3

2. 9, 12, 11, 11, 10, 7, 4, 8

3. 26, 42, 31, 50, 29, 37, 44, 31

4. 53, 45, 43, 55, 28, 21, 61, 29, 24, 40, 27, 42

5. A shelf in your room can hold at most 30 pounds. There are 12 pounds of books on the shelf. Write an inequality that represents the numbers of pounds you can add to the shelf.

Find the missing values in the ratio table. Then write the equivalent ratios.

6.

Turkey (pounds)	1	2	3
Price (dollars)	5.50		

7.

Guests	5	30	6
Cost (dollars)	100		

Find the surface area of the prism.

8.

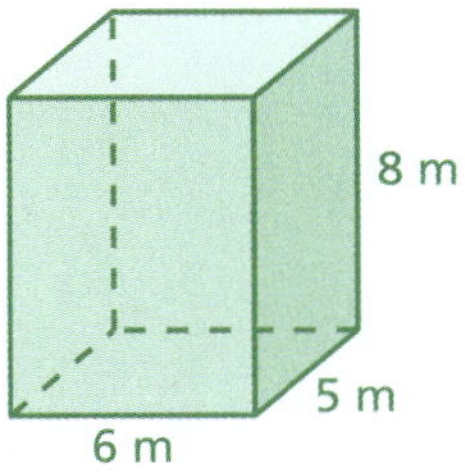

9.

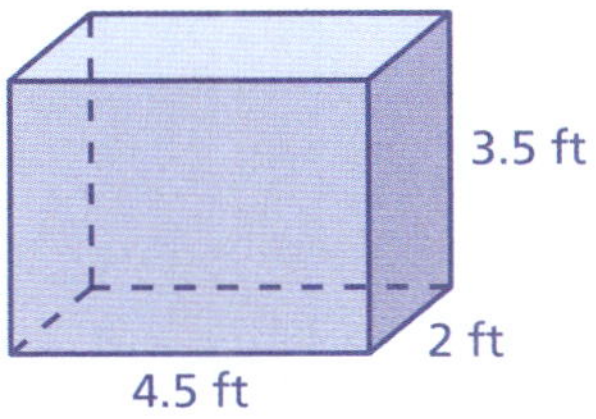

10.

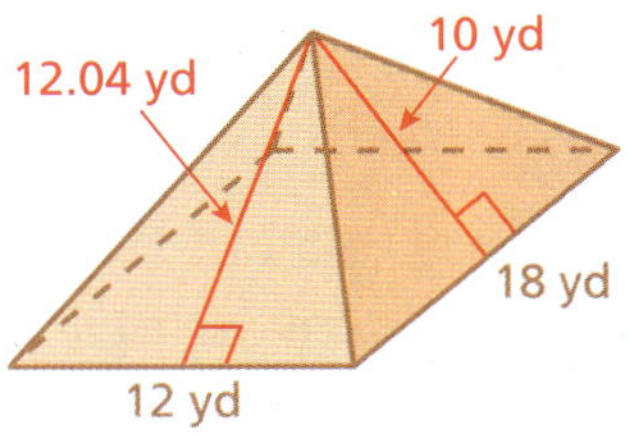

GO DIGITAL

Concepts, Skills, & Problem Solving

FINDING THE MEDIAN Use grid to find the median of the data. (See Exploration 1.)

11. 9, 7, 2, 4, 3, 5, 9, 6, 8, 0, 3, 8

12. 16, 24, 13, 36, 22, 26, 22, 28, 25

FINDING THE MEDIAN AND MODE Find the median and mode of the data. (See Example 1.)

▶**13.** 3, 5, 7, 9, 11, 3, 8

14. 14, 19, 16, 13, 16, 14

15. 93, 81, 94, 71, 89, 92, 94, 99

16. 44, 13, 36, 52, 19, 27, 33

17. 12, 33, 18, 28, 29, 12, 17, 4, 2

18. 55, 44, 40, 55, 48, 44, 58, 67

4 MTR

19. YOU BE THE TEACHER Your friend finds the median of the data. Is your friend correct? Explain your reasoning.

The median is 58.

63, 55, 49, 58, 50, 59, 51

FINDING THE MODE **Find the mode of the data.** (See Example 2.)

20.

Shirt Colors

Black	Blue	Red
Pink	Black	Black
Gray	Green	Blue
Blue	Blue	Red
Yellow	Blue	Blue
Black	Orange	Black
Black		

▶**21.**

Talent Show Acts

Singing	Dancing	Comedy
Singing	Singing	Dancing
Juggling	Dancing	Singing
Singing	Poetry	Dancing
Comedy	Magic	Dancing
Poetry	Singing	Singing

4 MTR **22. DISCUSS MATHEMATICAL THINKING** In Exercises 20 and 21, can you find the mean and median of the data? Explain.

FINDING MEASURES OF CENTER **Find the mean, median, and mode of the data.**

23. 4.7, 8.51, 6.5, 7.42, 9.64, 7.2, 9.3

24. $8\frac{1}{2}, 6\frac{5}{8}, 3\frac{1}{8}, 5\frac{3}{4}, 6\frac{5}{8}, 5\frac{1}{4}, 10\frac{5}{8}, 4\frac{1}{2}$

7 MTR **25. MODELING REAL LIFE** The lengths (in inches) of several green anoles are shown in the table. Find the mean, median, and mode of the lengths.

Green Anole Lengths (inches)		
5.2	6.5	3.2
6.4	4.8	2.9
4.6	5.0	5.5

REMOVING AN OUTLIER Find the mean, median, and mode of the data with and without the outlier. Which measure does the outlier affect the most? (See Example 3.)

26. 45, 52, 17, 63, 57, 42, 54, 58

27. 85, 77, 211, 88, 91, 84, 85

28. 23, 73, 45, 27, 23, 25, 43, 45

29. 101, 110, 99, 100, 64, 112, 110, 111, 102

7 MTR

30. MODELING REAL LIFE The table shows the monthly salaries for employees at a company. (See Example 4.)

Monthly Salaries (dollars)				
1940	1660	1860	2100	1720
1540	1760	1940	1820	1600

a. Find the mean, median, and mode of the data.

b. Each employee receives a 5% raise. Find the mean, median, and mode of the data with the raise. How does this increase affect the mean, median, and mode of the data?

c. How are the mean, median, and mode of the monthly salaries related to the mean, median, and mode of the annual salaries?

CHOOSING A MEASURE OF CENTER Find the mean, median, and mode of the data. Choose the measure that best represents the data. Explain your reasoning.

31. 48, 12, 11, 45, 48, 48, 43, 32

32. 12, 13, 40, 95, 88, 7, 95

33. 2, 8, 10, 12, 56, 9, 5, 2, 4

34. 126, 62, 144, 81, 144, 103

7 MTR ▶ **35. MODELING REAL LIFE** Use the data to answer the statistical question "What is the typical high temperature during the week?" (See Example 5.)

	Sun	Mon	Tue	Wed	Thu	Fri	Sat

High	90°F	91°F	89°F	97°F	101°F	99°F	91°F
Low	74°F	78°F	77°F	77°F	83°F	78°F	72°F

6 MTR **36. ASSESS REASONABLENESS** Is the mean or median a better representation of the data in Exercise 35? Explain your reasoning.

7 MTR **37. INVESTIGATE** Find the costs of 10 different boxes of cereal. Use the data to answer the statistical question "How much does a box of cereal cost?"

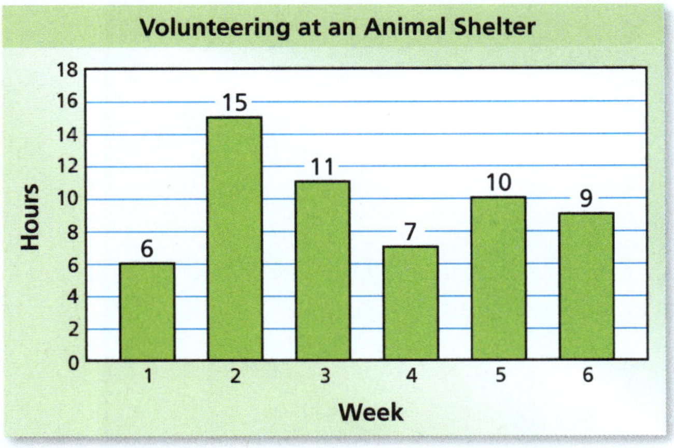

38. PROBLEM SOLVING The bar graph shows the numbers of hours you volunteered at an animal shelter. What is the minimum number of hours you need to volunteer in the seventh week to justify that you volunteered an average of 10 hours per week for the 7 weeks? Explain your answer using measures of center.

39. REASONING Why is the mode the least frequently used measure of center to describe a data set? Explain.

40. **Dig Deeper** The data are the prices of several fitness wristbands at a store.

{\$130 \$170 \$230 \$130 \$250 \$275 \$130 \$185}

a. Does the price shown in the advertisement represent the prices well? Explain.

b. Why might the store use this advertisement?

c. In this situation, why might a person want to know the mean? the median? the mode? Explain.

41. NUMBER SENSE The expressions below form a data set. Assume $x > 0$.

$\{3x, 9x, 4x, 23x, 6x, 3x\}$

a. Find the mean, median, and mode of the data.

b. Is there an outlier? If so, what is it?

11.4 Measures of Variation

Learning Target: Find and interpret the range and interquartile range of a data set.

Success Criteria:
- I can find the range and interquartile range of a data set.
- I can explain how the range and interquartile range describe the variability of a data set.
- I can explain how changes to a data set affect the measures of variation.
- I can use the interquartile range to identify outliers.

Data Analysis and Probability

MA.6.DP.1.2 Given a numerical data set within a real-world context, find and interpret mean, median, mode and range.

MA.6.DP.1.3 Given a box plot within a real-world context, determine the minimum, the lower quartile, the median, the upper quartile and the maximum. Use this summary of the data to describe the spread and distribution of the data.

Also MA.6.DP.1.4, MA.6.DP.1.6

Exploration 1 Interpreting Statements

Work with a partner. There are 24 students in your class. Your teacher makes the following statements.

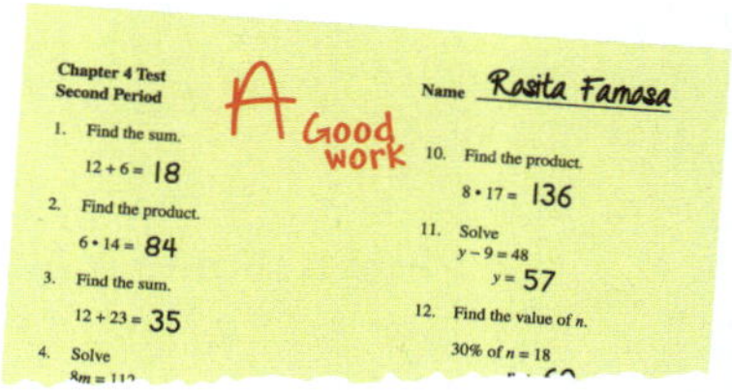

- ***"The exam scores range from 75% to 96%."***
- ***"Most of the students received high scores."***

a. What does each statement mean? Explain.

b. Use your teacher's statements to make a dot plot that can represent the distribution of the exam scores of the class.

c. Compare your dot plot with other groups'. How are they alike? different?

1 MTR ANALYZE A PROBLEM

How can you use the given information to determine how spread out the data are?

GO DIGITAL

Exploration 2 Grouping Data

Work with a partner. The numbers of U.S. states visited by students in a sixth-grade class are shown.

Number of States Visited				
1	7	5	2	4
18	1	6	11	6
3	20	2	7	1
8	10	2	12	5
3	21			

a. Represent the data using a dot plot. Between what values do the data range?

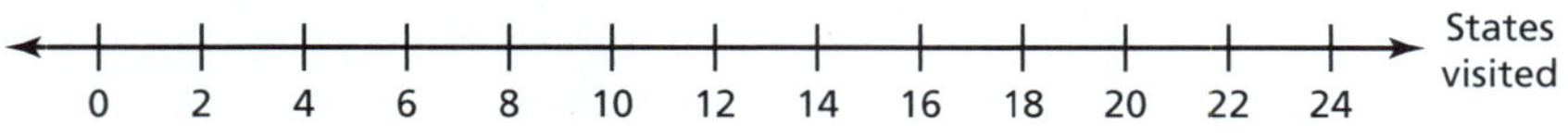

b. Use the dot plot to make observations about the data.

c. How can you describe the *middle half* of the data?

11.4 Lesson

A **measure of variation** is a measure that describes the distribution of a data set. A simple measure of variation to find is the *range*. The **range** of a data set is the difference of the greatest value and the least value.

Key Vocabulary

measure of variation, *p. 921*
range, *p. 921*
quartiles, *p. 921*
first quartile, *p. 921*
third quartile, *p. 921*
interquartile range, *p. 921*

Example 1 Finding the Range

The table shows the lengths of several Burmese pythons captured for a study. Find and interpret the range of the lengths.

Lengths (feet)	
18.5	8
11	10
14	15.5
12.5	6.25
16.25	5

To find the least and the greatest values, order the lengths from least to greatest.

5, 6.25, 8, 10, 11, 12.5, 14, 15.5, 16.25, 18.5

The least value is 5. The greatest value is 18.5.

▶ So, the range of the lengths is $18.5 - 5 = 13.5$ feet. This means that the lengths vary by no more than 13.5 feet.

Try It

1. The ages of people in line for a roller coaster are represented by the data set {15, 17, 21, 32, 41, 30, 25, 52, 16, 39, 11, 24}. Find and interpret the range of the ages.

Key Ideas

Quartiles

The **quartiles** of a data set divide the data into four equal parts. Recall that the median (second quartile) divides the data set into two halves.

Median = 29

{18 21 **22** 24 28 30 31 **32** 36 37}

The median of the lower half is the **first quartile**, Q_1.

The median of the upper half is the **third quartile**, Q_3.

Reading

The first quartile can also be called the *lower quartile*. The third quartile can also be called the *upper quartile*.

Interquartile Range (IQR)

The difference of the third quartile and the first quartile is called the **interquartile range**. The IQR represents the range of the middle half of the data and is another measure of variation.

{18 21 **22** 24 28 30 31 **32** 36 37}

$$\text{IQR} = Q_3 - Q_1$$
$$= 32 - 22$$
$$= 10$$

GO DIGITAL

Example 2 Finding the Interquartile Range

The dot plot shows the top speeds of 12 models of sports cars. Find and interpret the interquartile range of the data.

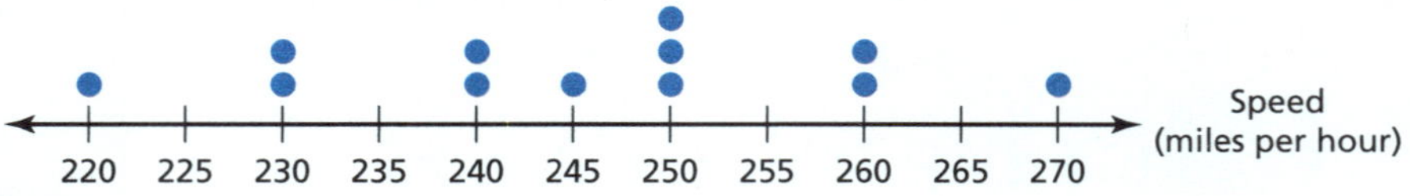

Order the speeds from slowest to fastest. Find the quartiles.

Median: $\frac{245 + 250}{2} = 247.5$

220 230 230 240 240 245 | 250 250 250 260 260 270

Q_1: $\frac{230 + 240}{2} = 235$

Q_3: $\frac{250 + 260}{2} = 255$

▶ So, the interquartile range is $255 - 235 = 20$. This means that the middle half of the speeds vary by no more than 20 miles per hour.

Try It

Pages in Novels		
356	364	390
468	400	382
376	396	350

2. The table shows the number of pages in each of an author's novels. Find and interpret the interquartile range of the data.

Example 3 Changing the Values of a Data Set

Next year, the top speed of each model of car in Example 2 increases by 5 miles per hour. How does this increase affect the range and IQR?

Adding 5 miles per hour to each speed in Example 2 results in the data set

$$\{225, 235, 235, 245, 245, 250, 255, 255, 255, 265, 265, 275\}.$$

Find the range and IQR of both data sets.

	Range	IQR
Original Top Speed	50	20
Next Year's Top Speed	$275 - 225 = 50$	$260 - 240 = 20$

▶ After increasing each speed by 5 miles per hour, the range and IQR remain the same.

GO DIGITAL

Try It

3. WHAT IF? The top speed of each model of car decreases by 10 miles per hour. How does this decrease affect the range and the IQR?

You can use the quartiles and the interquartile range to check for outliers.

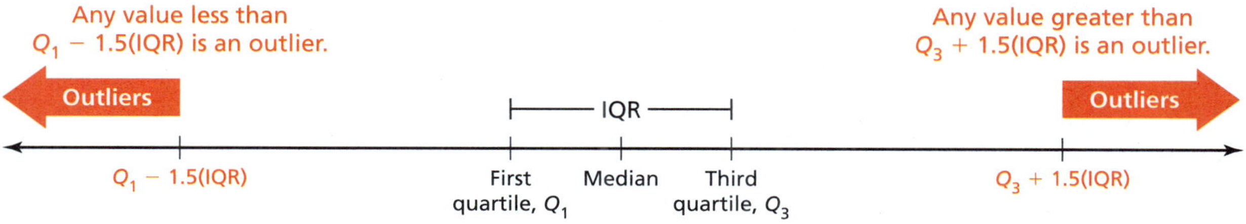

Example 4 Identifying Outliers

Bugs Eaten			
26	34	32	42
33	2	34	

Section 11.3 Example 3 identifies 2 as an outlier of the data in the table shown. Use the IQR to determine whether it is the only outlier.

Order the data values from least to greatest. Find the quartiles.

2 26 32 33 34 34 42

$Q_1 = 26$ Median $= 33$ $Q_3 = 34$

The IQR is $34 - 26 = 8$. Use the IQR to find the outlier boundaries.

$$Q_1 - 1.5(\text{IQR}) = 26 - 1.5(8) = 14 \qquad Q_3 + 1.5(\text{IQR}) = 34 + 1.5(8) = 46$$

▶ The only data value less than 14 is 2. There are no data values greater than 46. So, the only outlier is 2.

46 8
350
118

Try It

4. WHAT IF? An eighth competitor eats 50 bugs. Is this value an outlier? Explain your reasoning.

GO DIGITAL

In-Class Practice

1 I don't understand yet. 2 I can do it with help. 3 I can do it on my own. 4 I can teach someone else.

5. **WRITING** Explain why the variability of a data set can be described by the range or the interquartile range.

6. **IDENTIFYING OUTLIERS** Identify any outliers in the data set.

$\{2, 9, 10, 10, 12, 14, 15, 15, 24\}$

4 MTR

7. **DIFFERENT WORDS, SAME QUESTION** Which is different? Find "both" answers.

53, 47, 60, 45, 62, 59, 65, 50, 56, 48

What is the range of the data?

What is $Q_3 - Q_1$ for the data?

What is the range of the middle half of the data?

What is the IQR of the data?

Example 5 Modeling Real Life

7 MTR

County	Population (millions)
Miami-Dade	2.72
Broward	1.91
Palm Beach	1.45
Hillsborough	1.38
Orange	1.32
Pinellas	0.96
Duval	0.92
Lee	0.72
Polk	0.67
Brevard	0.58

The table shows the approximate populations of the 10 most populous counties in Florida. Describe the variability of the data. Which measure of variation better represents the data?

Find the range and the IQR of the data. Then use a dot plot to determine which measure better represents the data.

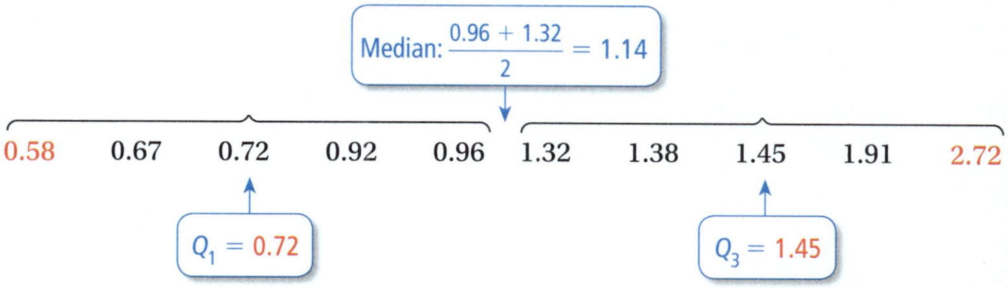

Range: $2.72 - 0.58 = 2.14$

IQR: $1.45 - 0.72 = 0.73$

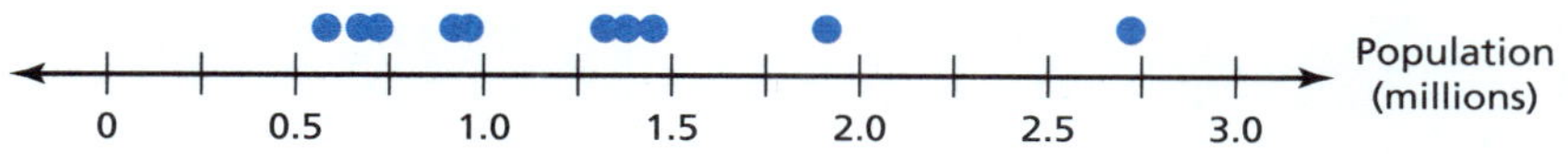

▶ The range is 2.14 million and the IQR is 0.73 million. Because most of the populations are between 0.5 million and 1.5 million, the IQR better represents the data.

Check Reasonableness
It is reasonable to expect some counties to have much greater populations than others, which affects the range. So, it makes sense that the IQR is a better representation of the data than the range in this scenario. ✓

5 MTR **RELATE CONCEPTS**

Use the IQR to identify the outlier in the data set. Does the outlier have a greater effect on the range or on the interquartile range?

GO DIGITAL

In-Class Practice

1 I don't understand yet.	2 I can do it with help.	3 I can do it on my own.	4 I can teach someone else.

Distances (feet)			
$13\frac{1}{2}$	$21\frac{1}{2}$	21	$16\frac{3}{4}$
$10\frac{1}{4}$	19	$44\frac{1}{2}$	$26\frac{1}{2}$
29	$16\frac{1}{4}$	$28\frac{1}{2}$	$18\frac{1}{2}$

8. The table shows the distances traveled by a paper airplane. Describe the variability of the data. Which measure of variation better represents the data?

9. The table shows the years of teaching experience of math teachers at a school. How do the outlier or outliers affect the variability of the data?

Teaching Experience (years)	5	10	7	8	10
	11	22	8	6	35

11.4 Practice WITH CalcChat® AND CalcView®

Review & Refresh

Find the mean, median, and mode of the data.

1. 4, 8, 11, 6, 4, 5, 9, 10, 10, 4

2. 74, 78, 86, 67, 80

3. 15, 18, 17, 17, 15, 16, 14

4. 31, 14, 18, 26, 17, 32

Complete the statement using < or >.

5. $6 \; \square \; -7$

6. $-3 \; \square \; 0$

7. $14 \; \square \; -14$

8. $8 \; \square \; -10$

Find the surface area of the pyramid.

9.

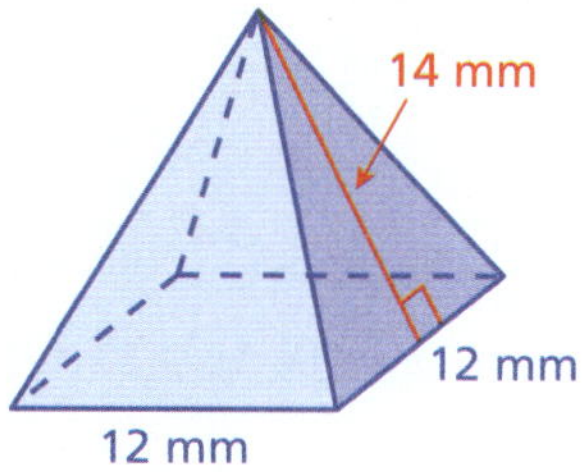

10.

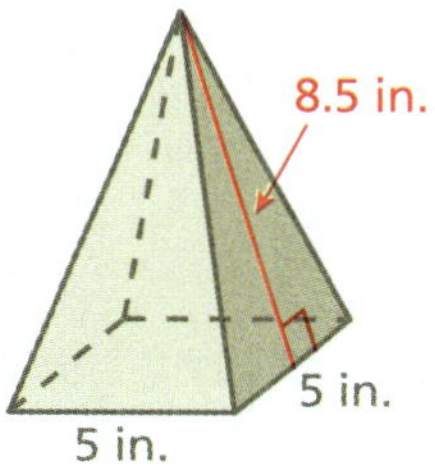

11.

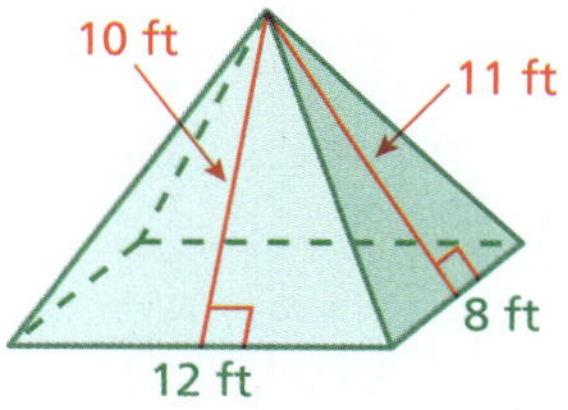

Concepts, Skills, & Problem Solving

INTERPRETING STATEMENTS There are 20 students in your class. Your teacher makes the two statements shown. Use your teacher's statements to make a dot plot that can represent the distribution of the scores of the class. (See Exploration 1.)

12. *"The quiz scores range from 65% to 95%."*

"The scores were evenly spread out."

13. *"The project scores range from 78% to 93%."*

"Most of the students received low scores."

GO DIGITAL

FINDING THE RANGE **Find and interpret the range of the data.** (See Example 1.)

14. 4, 8, 2, 9, 5, 3

15. 28, 42, 36, 23, 14, 47, 40

16. 26, 21, 27, 33, 24, 29

17. 52, 40, 49, 48, 62, 54, 44, 58, 39

18. 133, 117, 152, 127, 168, 146, 174

19. 4.8, 5.5, 4.2, 8.9, 3.4, 7.5, 1.6, 3.8

4 MTR

20. YOU BE THE TEACHER Your friend finds the range of the data. Is your friend correct? Explain your reasoning.

29, 48, 51, 28, 35, 44, 59

The range is 59 − 29 = 30.

FINDING THE INTERQUARTILE RANGE **Find and interpret the interquartile range of the data.** (See Example 2.)

21. 4, 6, 4, 2, 9, 1, 12, 7

22. 18, 22, 15, 16, 15, 13, 19, 18

23. 40, 33, 37, 54, 41, 34, 27, 39, 35

24. 84, 75, 90, 87, 99, 91, 85, 88, 76, 92, 94

25. 132, 127, 106, 140, 158, 135, 129, 138

26. 38, 55, 61, 56, 46, 67, 59, 75, 65, 58

CHANGING THE VALUES OF A DATA SET **Each value in the data set is changed as described. How does the change affect the range and IQR?** (See Example 3.)

27. {4, 6, 4, 2, 9, 1, 12, 7}, increase of 4

28. {40, 33, 37, 54, 41, 34, 27, 39, 35}, decrease of 15

29. {3, 12, 7, 9, 11, 5, 15, 20}, multiplied by 2

30. {132, 147, 105, 150, 168, 135, 129, 138}, multiplied by $\frac{1}{3}$

7 MTR

31. **MODELING REAL LIFE** You are planning a surf trip to Cocoa Beach. You want to rent surf equipment for 8 hours. The table shows the cost (in dollars) to rent surf equipment from different shops.

Cost of Surf Equipment					
20	25	44	35	25	40
52	33	45	24	50	38

a. Find the range and IQR of the cost to rent surf equipment.

b. Each shop offers a 5% discount. How does this affect the range and the IQR of the data?

IDENTIFYING OUTLIERS Identify any outliers in the data set. (See Example 4.)

32. $\{3, 6, 10, 6, 17, 5\}$

33. $\{18, 28, 21, 19, 14, 24, 35, 18\}$

34. $\{40, 33, 37, 15, 41, 34, 27, 39, 35\}$

35. $\{55, 75, 90, 87, 99, 91, 85, 88, 76, 92, 94, 89, 125\}$

7 MTR

36. MODELING REAL LIFE The table shows the numbers of tornadoes in Florida each year for several years. Describe the variability of the data. Which measure better represents the data? (See Example 5.)

Numbers of Tornadoes			
29	46	57	45
30	30	38	48
39	21	38	56

37. REASONING Use the data in Exercise 36. How are the range and interquartile range affected if there are half as many tornadoes each year? Round to the nearest integer, if necessary.

38. NUMBER SENSE Is it possible for the range of a data set to be equal to the interquartile range? Explain your reasoning.

39. REASONING How does an outlier affect the range of a data set? Explain.

7 MTR

40. MODELING REAL LIFE The table shows the numbers of points scored by players on a sixth-grade basketball team in a season.

Points Scored					
21	53	74	82	84	93
103	108	116	122	193	

a. Find the range and interquartile range of the data.

b. Identify the outlier(s) in the data set. Find the range and interquartile range of the data set without the outlier(s). Which measure does the outlier or outliers affect more?

c. Which measure better represents the data?

41. REASONING The data sets show the years of the coins in two collections.

Your collection: {1950, 1952, 1908, 1902, 1955, 1954, 1901, 1910}

Your friend's collection: {1929, 1935, 1928, 1930, 1925, 1932, 1933, 1920}

Compare the measures of center and the measures of variation for each data set. What can you conclude?

42. **Dig Deeper** Two data sets have the same range. Can you assume that the interquartile ranges of the two data sets are about the same? Give an example to justify your answer.

7 MTR

43. **MODELING REAL LIFE** The tables show the ages of the finalists for two reality singing competitions.

a. Find the mean, median, range, and interquartile range of the ages for each show. Compare the results.

Ages for Show A	
18	17
15	21
22	16
18	28
24	21

Ages for Show B	
21	20
23	13
15	18
17	22
36	25

b. Which measures best represent the data for each show?

c. A 21-year-old is voted off Show A, and the 36-year-old is voted off Show B. How do these changes affect the measures in part (a)? Explain.

44. **OPEN-ENDED** Create a set of data with 7 values that has a mean of 30, a median of 26, a range of 50, and an interquartile range of 36.

11.5 Histograms

Learning Target: Display and interpret data in histograms.

Success Criteria:
- I can explain how to display data in a histogram.
- I can make and interpret a histogram.
- I can determine whether a question can be answered using a histogram.

Data Analysis and Probability

MA.6.DP.1.4 Given a histogram or line plot within a real-world context, qualitatively describe and interpret the spread and distribution of the data, including any symmetry, skewness, gaps, clusters, outliers and the range.

MA.6.DP.1.5 Create box plots and histograms to represent sets of numerical data within real-world contexts.

Exploration 1 Performing an Experiment

Work with a partner.

a. Make the airplane shown from a single sheet of $8\frac{1}{2}$-by-11-inch paper. Then design and make your own paper airplane.

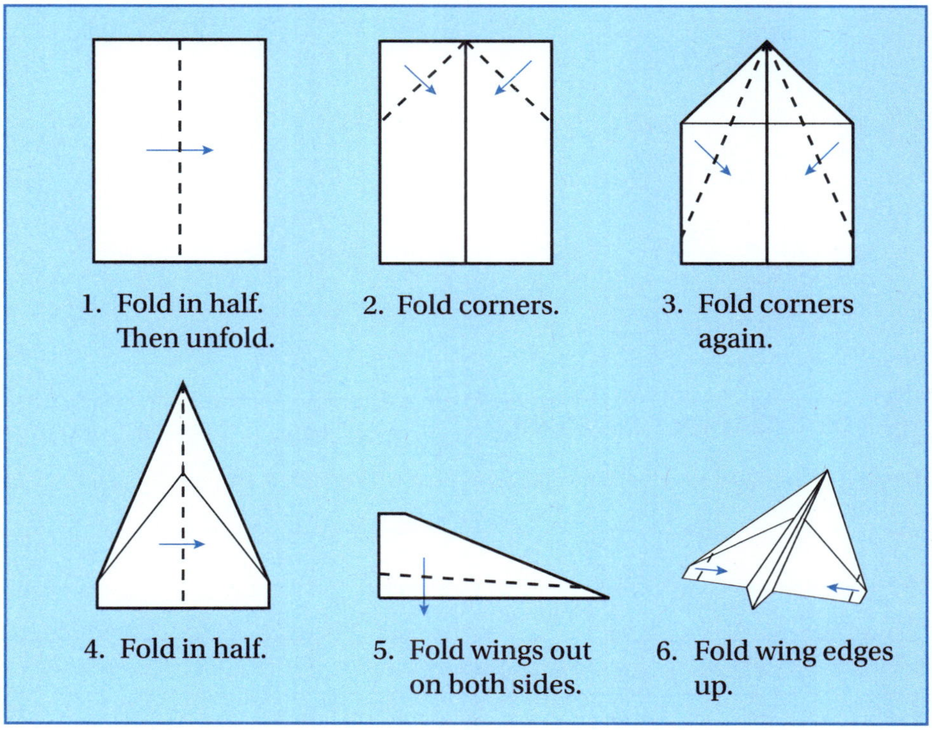

b. Fly each airplane 20 times. Keep track of the distance flown each time.

GO DIGITAL

c. A **frequency table** groups data values into intervals. The **frequency** is the number of values in an interval. Use a frequency table to organize the results for each airplane.

2 MTR d. **MODEL A PROBLEM** Represent the data in the frequency tables graphically. Which airplane flies farther? Explain your reasoning.

4 MTR **CONSTRUCT AN ARGUMENT**

Why is it beneficial to represent data graphically? Use your answer in part (d) to support your argument.

11.5 Lesson

Key Vocabulary
frequency table, *p. 934*
frequency, *p. 934*
histogram, *p. 935*

Key Idea

Histograms

A **histogram** is a bar graph that shows the frequencies of data values in intervals of the same size.

The height of a bar represents the frequency of the values in the interval.

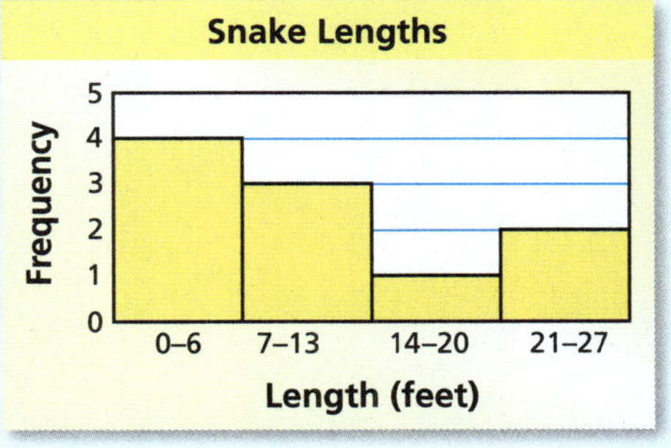

Example 1 Making a Histogram

5 MTR MAKE A PLAN

In Example 1, it may be difficult or impossible to record the numbers of laps completed by every person at the pool. What factors might affect the accuracy of your data? How can you ensure that you collect truthful data?

The frequency table shows the numbers of laps that several people at a pool completed today. Display the data in a histogram. Identify any gaps, clusters, and outliers.

Number of Laps	Frequency
1–3	9
4–6	10
7–9	4
10–12	0
13–15	1

Step 1: Draw and label the axes.

Step 2: Draw a bar to represent the frequency of each interval.

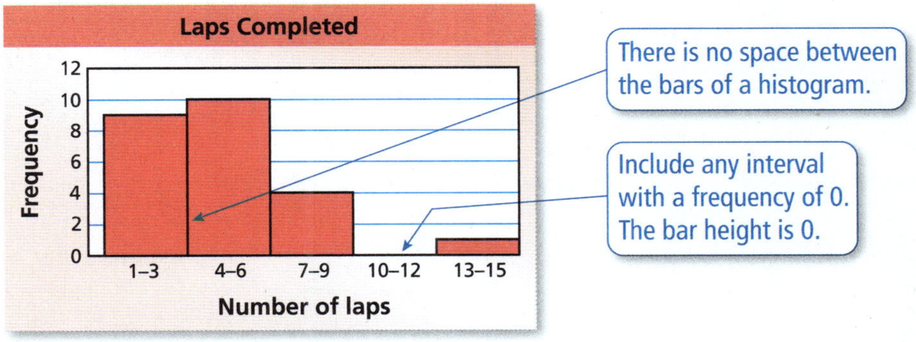

▶ There is a gap from 9–13 and a cluster from 1–6. There appears to be an outlier from 13–15.

GO DIGITAL

Try It

1. The frequency table shows the ages of people riding a roller coaster. Display the data in a histogram. Identify any gaps, clusters, and outliers.

Age	10–19	20–29	30–39	40–49	50–59
Frequency	16	11	5	2	4

Example 2 Using a Histogram

The histogram shows winning speeds at the Daytona 500.

a. Which interval contains the most data values?

b. How many of the winning speeds are less than 140 miles per hour?

c. How many of the winning speeds are at least 160 miles per hour?

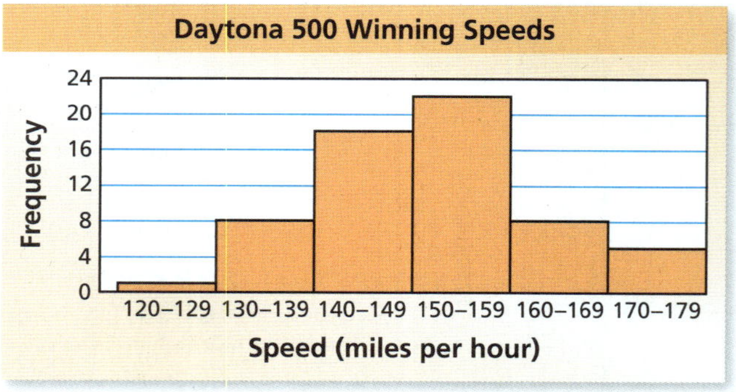

a. The interval with the tallest bar contains the most data values.

▶ So, the 150–159 miles per hour interval contains the most data values.

b. One winning speed is in the 120–129 miles per hour interval, and eight winning speeds are in the 130–139 miles per hour interval.

▶ So, $1 + 8 = 9$ winning speeds are less than 140 miles per hour.

c. Eight winning speeds are in the 160–169 miles per hour interval, and five winning speeds are in the 170–179 miles per hour interval.

▶ So, $8 + 5 = 13$ winning speeds are at least 160 miles per hour.

GO DIGITAL

Try It

2. The histogram shows the numbers of hours that students in a class slept last night.

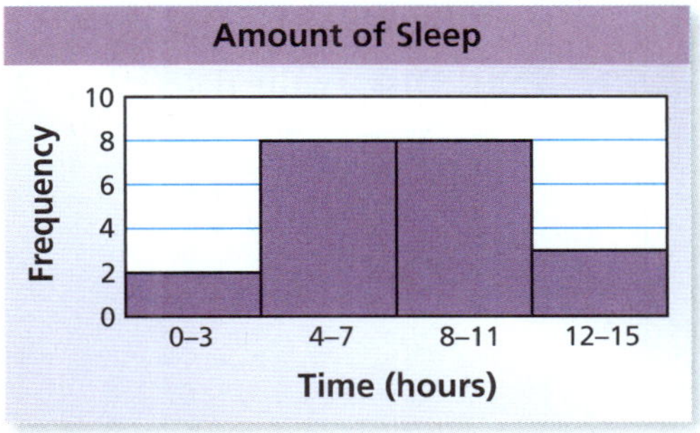

a. How many students slept at least 8 hours?

b. How many students slept less than 12 hours?

Example 3 Comparing Data Displays

The data displays show how many push-ups students in a class completed for a physical fitness test. Which data display can you use to find how many students are in the class? Explain.

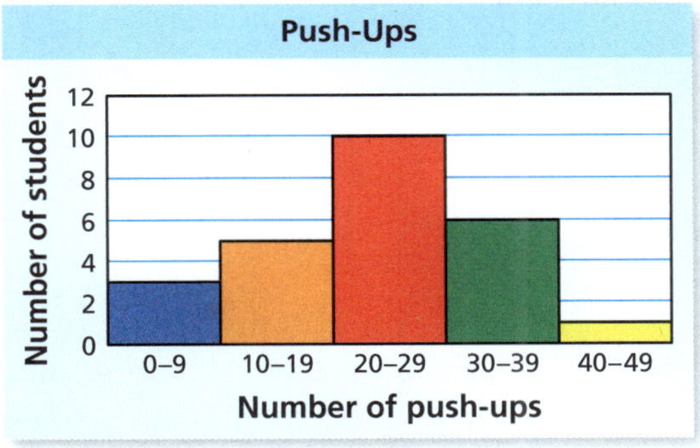

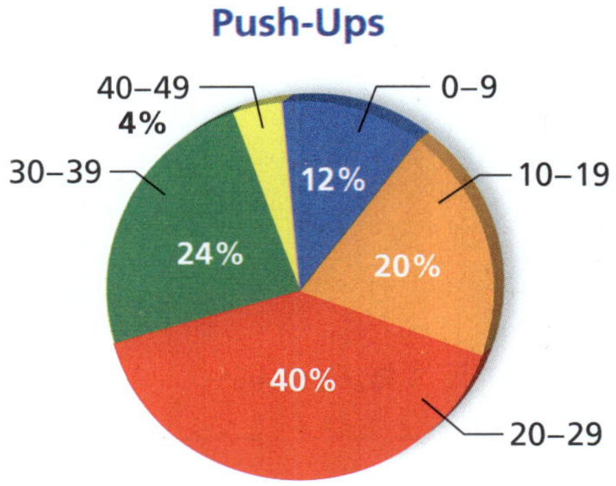

▶ You can use the histogram because it shows the number of students in each interval. The sum of these values represents the number of students in the class. You cannot use the circle graph, because it does not show the number of students in each interval.

Try It

3. Which data display should you use to describe the portion of the entire class that completed 30–39 push-ups? Explain.

GO DIGITAL

In-Class Practice

1 I don't understand yet. | 2 I can do it with help. | 3 I can do it on my own. | 4 I can teach someone else.

Numbers of Siblings			
0	2	3	6
3	2	1	2
1	4	0	3
7	3	1	5
4	3	2	0

4. **MAKING A HISTOGRAM** The table shows the numbers of siblings of students in a class.

   a. Display the data in a histogram.

   b. Explain how you chose reasonable intervals for your histogram in part (a).

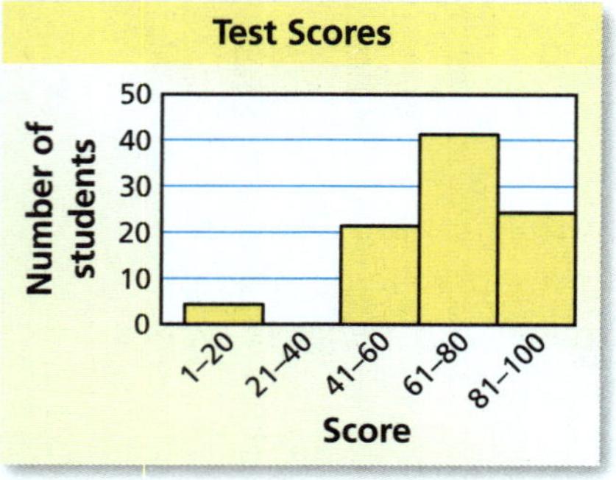

5. **NUMBER SENSE** Can you find the range and the interquartile range of the data in the histogram? If so, find them. If you cannot find them, explain why not.

Example 4 B.E.S.T. Test Prep: Modeling Real Life

7 MTR

Which statement *cannot* be made using the data displays in Example 3?

Ⓐ The range of the number of push-ups completed is at least 31 and at most 49.

Ⓑ Five students completed at least 10 and at most 19 push-ups.

Ⓒ At least one student completed more than 39 push-ups.

Ⓓ Less than $\frac{1}{4}$ of the class completed 30 or more push-ups.

The histogram shows that the least number of push-ups completed is from 0 to 9 and the greatest number is from 40 to 49. So, the least possible value for the range is $40 - 9 = 31$, and the greatest possible value for the range is $49 - 0 = 49$. So, Statement A can be made.

In the histogram, the bar height for the 10–19 interval is 5, and the bar height for the 40–49 interval is 1. So, Statements B and C can be made.

The circle graph shows that 24% completed 30–39 push-ups, and 4% completed 40–49 push-ups. So, $24\% + 4\% = 28\%$ completed 30 or more push-ups. Because $\frac{1}{4} = 25\%$ and $28\% > 25\%$, Statement D cannot be made.

▶ The correct answer is Ⓓ.

GO DIGITAL

In-Class Practice

1 I don't understand yet. 2 I can do it with help. 3 I can do it on my own. 4 I can teach someone else.

6. The histogram shows the numbers of rebounds per game for a middle school basketball player in a season.

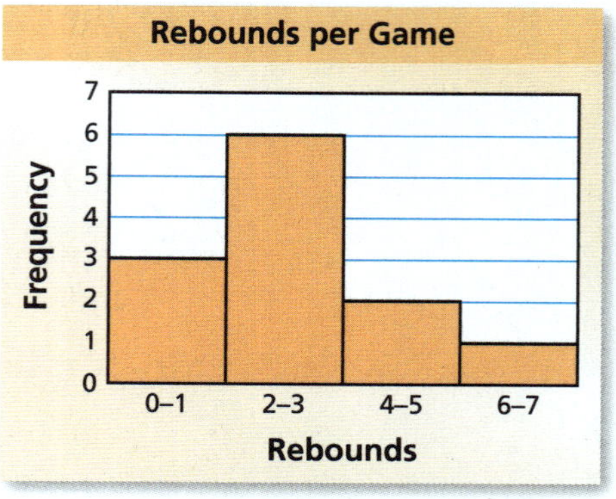

a. Which interval contains the most data values?

b. How many games did the player play during the season?

c. In what percent of the games did the player have 4 or more rebounds?

7. Determine whether you can make each statement by using the histogram in the previous exercise. Explain.

a. The basketball player had 2 rebounds in 6 different games.

b. The basketball player had more than 1 rebound in 9 different games.

11.5 Practice WITH CalcChat® AND CalcView®

Review & Refresh

Find and interpret the interquartile range of the data.

1.

Blog Posts			
7	5	12	4
20	11	9	15
4	8	6	12

2.

Social Media Comments			
4	18	1	32
10	36	16	7
44	3	7	15

Find the percent of the number.

3. 25% of 180

4. 30% of 90

5. 16% of 140

6. 64% of 80

7. Find the least common multiple of 7 and 12.

Concepts, Skills, & Problem Solving

MAKING A FREQUENCY TABLE **Organize the data using a frequency table.** (See Exploration 1.)

8.

Members of Book Clubs			
6	17	13	19
13	9	18	24
11	15	21	14

9.

Points Scored				
42	45	57	39	55
38	48	36	48	46
51	29	45	54	42

GO DIGITAL

MAKING A HISTOGRAM **Display the data in a histogram. Identify any gaps, clusters, and outliers.** (See Example 1.)

10.

States Visited	
States	**Frequency**
1–5	12
6–10	14
11–15	0
16–20	1

11.

Chess Team	
Wins	**Frequency**
10–13	3
14–17	4
18–21	4
22–25	2

12.

Movies Watched	
Movies	**Frequency**
0–1	5
2–3	11
4–5	0
6–7	1

13.

Ages of Celebrities	
Ages	**Frequency**
10–19	2
20–29	6
30–39	8
40–49	4

14.

Shoes Owned	
Pairs of Shoes	**Frequency**
1–3	3
4–6	8
7–9	10
10–12	0
13–15	1

▶15.

Steps Taken	
Steps	**Frequency**
0–1999	1
2000–3999	0
4000–5999	9
6000–7999	12
8000–9999	11

4 MTR **16.** **YOU BE THE TEACHER** Your friend displays the data in a histogram. Is your friend correct? Explain your reasoning.

Snow Days per School	
Snow Days	**Frequency**
0–1	9
2–3	6
4–5	2
6–7	1

7 MTR ▶ **17.** **MODELING REAL LIFE** The histogram shows the numbers of magazines read last month by the students in a class. (See Example 2.)

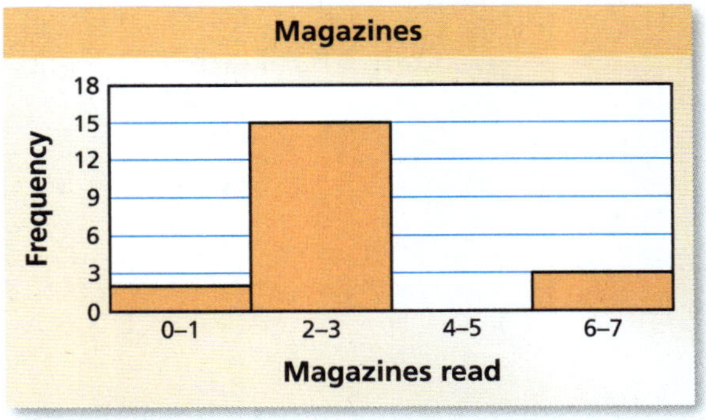

a. Which interval contains the fewest data values?

b. How many students are in the class?

c. What percent of the students read fewer than six magazines?

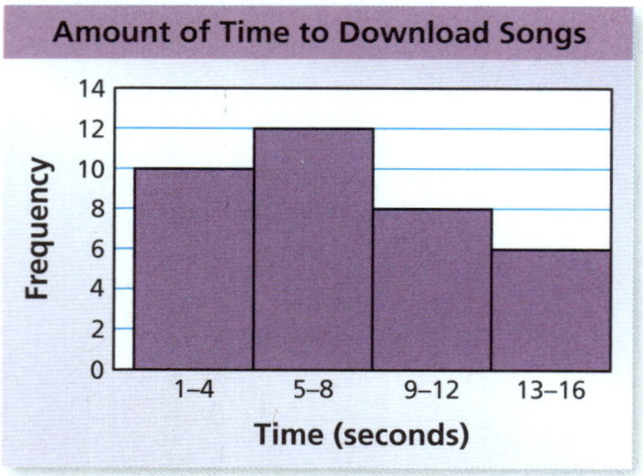

18. **B.E.S.T. Test Prep** The histogram shows the amount of time it takes to download songs. Which value describes how many songs took 5–8 seconds to download?

Ⓐ 12%

Ⓑ 10%

Ⓒ 10

Ⓓ 12

19. **REASONING** The histogram shows the percent of the voting-age population in each state who voted in a presidential election. Explain whether the graph supports each statement.

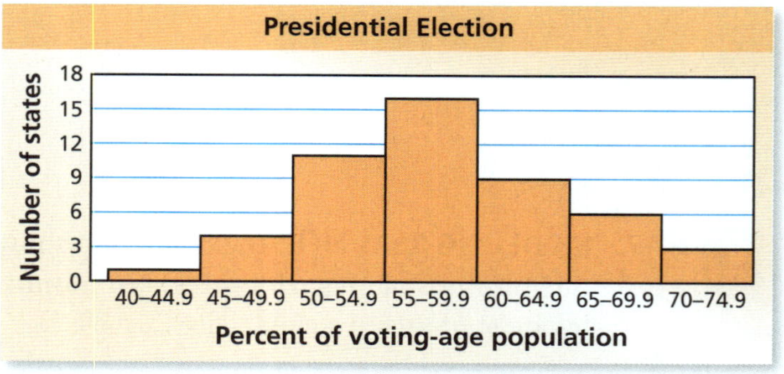

a. Only 40% of one state voted.

b. In most states, between 50% and 64.9% voted.

c. The mode of the data is between 55% and 59.9%.

20. PROBLEM SOLVING The histograms show the areas of counties in Florida and Indiana. Which state do you think has the greater area? Explain.

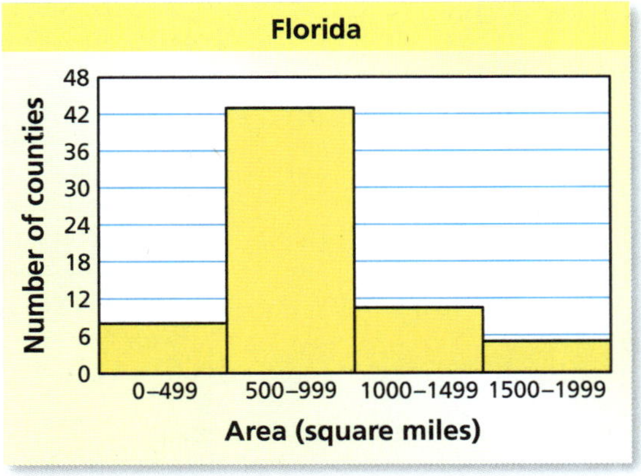

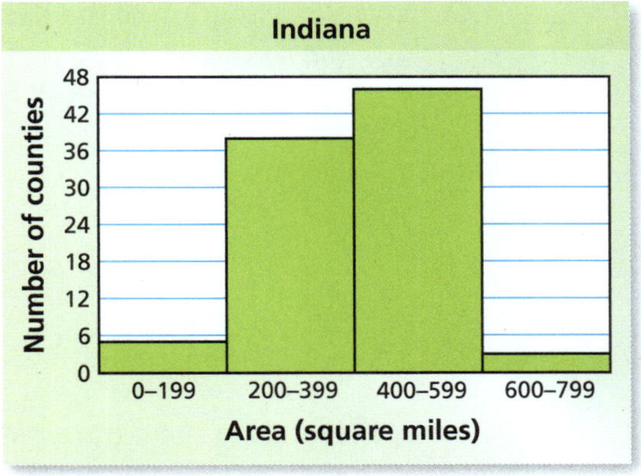

7 MTR ▶ **21. MODELING REAL LIFE** The data displays show how many pounds of garbage apartment residents produced in 1 week. Which data display can you use to find how many residents produced more than 25 pounds of garbage? Explain. (See Example 3.)

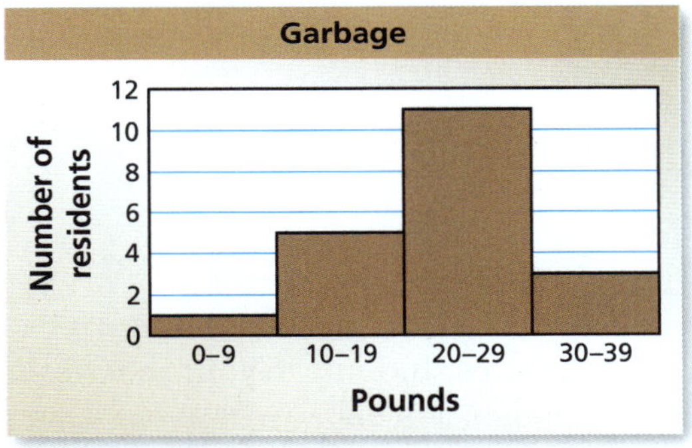

Garbage

Stem	Leaf
0	9
1	0 5 8 8 9
2	1 2 5 5 6 7 7 7 9 9 9
3	2 3 3

Key: 1 | 5 = 15 pounds

7 MTR **22. MODELING REAL LIFE** Determine whether you can make each statement by using the data displays in Exercise 21. Explain your reasoning. (See Example 4.)

a. One resident produced 10 pounds of garbage.

b. Twelve residents produced 20–29 pounds of garbage.

23. **Dig Deeper** The table shows the lengths of whales in a marine sanctuary.

Lengths (feet)				
81	88	57	82	70
71	51	82	77	79
83	80	54	80	81
59	84	75	76	68
83	78	55	67	85
85	77	73	78	79

a. Make a histogram of the data, starting with the interval 51–55.

b. Make another histogram of the data using a different-sized interval.

c. Compare and contrast the two histograms.

7 MTR

24. **INVESTIGATE** Collect data on the number of minutes students in your school spend exercising each day. Display and interpret the data in a histogram. Discuss ways to know that the data you collect is truthful data.

25. **REASONING** Can you find the mean or the median of the data in Exercise 17? Explain.

11.6 Shapes of Distributions

Learning Target: Describe and compare shapes of distributions.

Success Criteria:
- I can explain what skewed left, skewed right, and symmetric mean.
- I can use data displays to describe shapes of distributions.
- I can use shapes of distributions to compare data sets.

Data Analysis and Probability

MA.6.DP.1.4 Given a histogram or line plot within a real-world context, qualitatively describe and interpret the spread and distribution of the data, including any symmetry, skewness, gaps, clusters, outliers and the range.

MA.6.DP.1.5 Create box plots and histograms to represent sets of numerical data within real-world contexts.

The Meaning of a Word ▶ Skewed

When something is **skewed**, it has a slanted direction or position.

7 MTR APPLY MATHEMATICS

How can the word *skewed* be applied in mathematics?

GO DIGITAL

Exploration 1 Describing Shapes of Distributions

Work with a partner. The lists show the first three digits and last four digits of several phone numbers in the contact list of a cell phone.

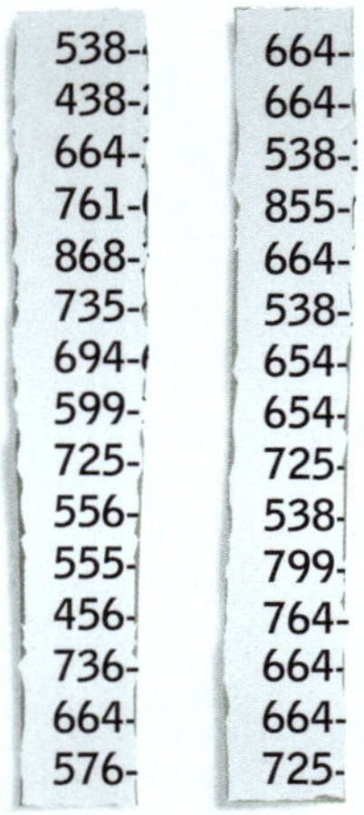

a. Compare and contrast the distribution of the last digit of each phone number to the distribution of the first digit of each phone number. Describe the shapes of the distributions.

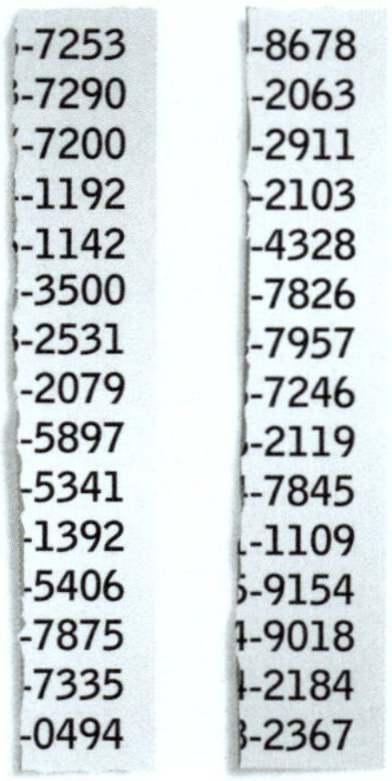

b. Describe the shape of the distribution of the data in the table below. Compare it to the distributions in part (a).

Ages of Cell Phones (years)					
0	1	0	6	4	0
2	3	5	1	1	2
0	1	2	3	1	0
0	0	1	1	1	1
7	1	4	2	2	2

11.6 Lesson

You can use dot plots and histograms to identify shapes of distributions.

Key Ideas

Symmetric and Skewed Distributions

If all the dots of a dot plot or bars of a histogram are about the same height, then the distribution is a *flat*, or *uniform*, distribution. A uniform distribution is also symmetric.

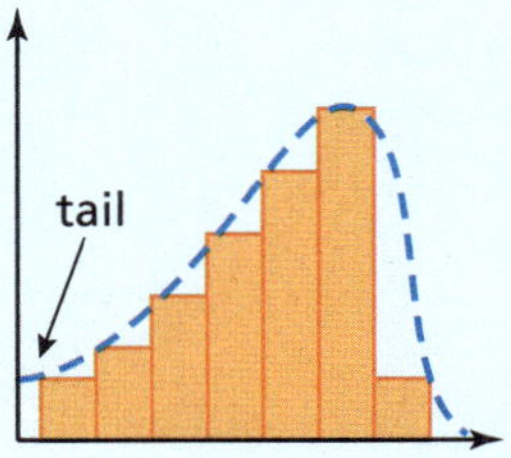

Skewed left

- The "tail" of the graph extends to the left.
- Most data are on the right.

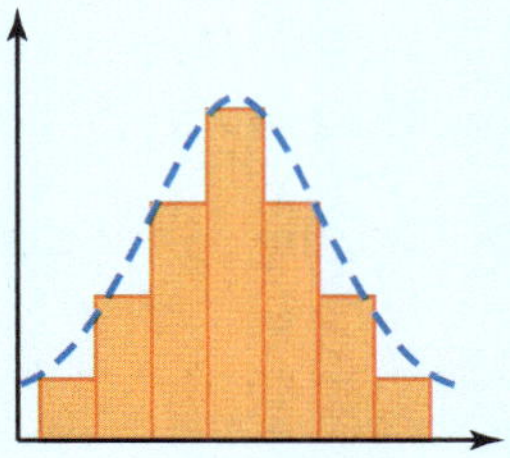

Symmetric

- The left side of the graph is a mirror image of the right side of the graph.

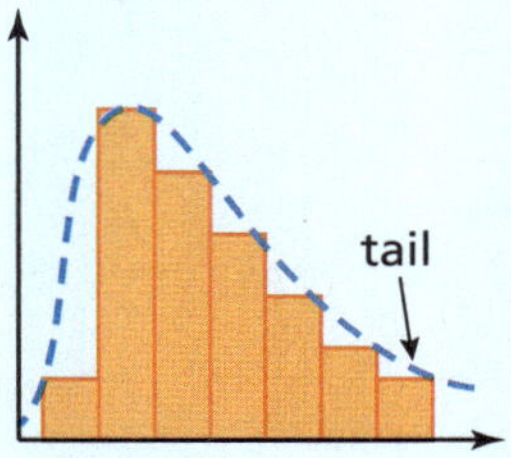

Skewed right

- The "tail" of the graph extends to the right.
- Most data are on the left.

Example 1 Describing Shapes of Distributions

Describe the shape of each distribution.

a.

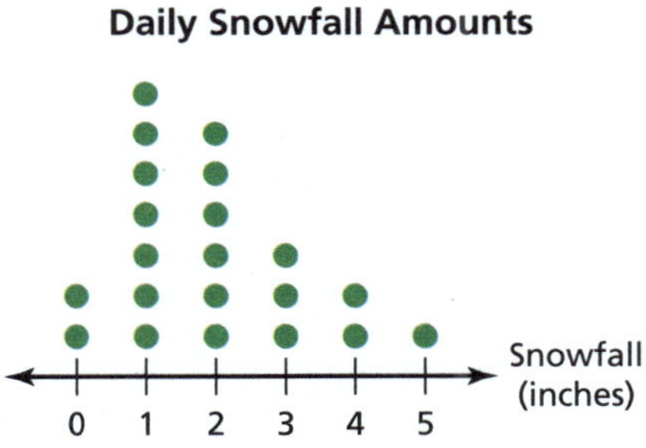

Most of the data are on the left, and the tail extends to the right.

▶ So, the distribution is skewed right.

b.

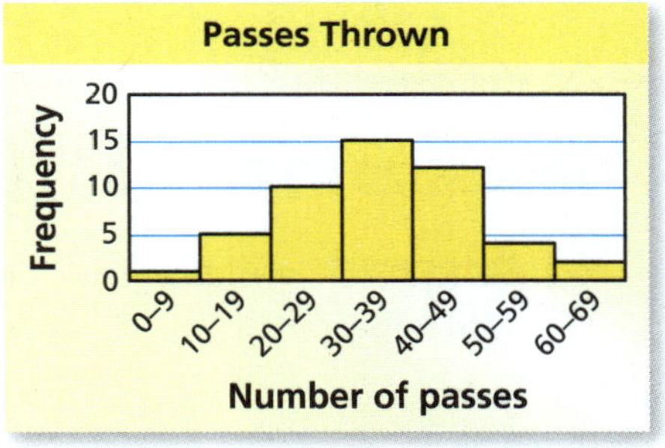

The left side of the graph is approximately a mirror image of the right side of the graph.

▶ So, the distribution is symmetric.

Try It

1. Describe the shape of the distribution.

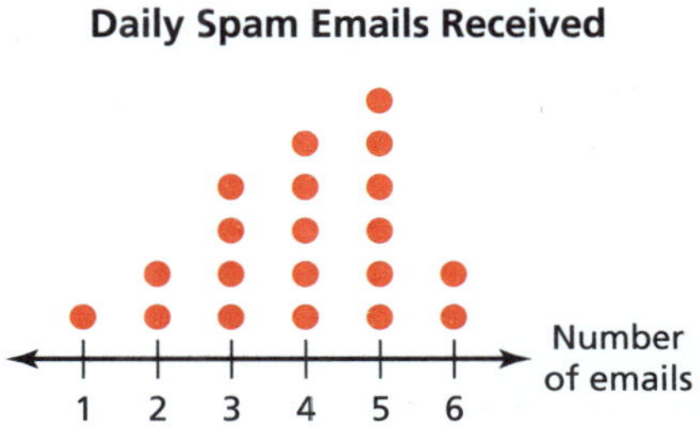

GO DIGITAL

Example 2 Describing the Shape of a Distribution

Ages	Frequency
10–13	1
14–17	3
18–21	7
22–25	12
26–29	20
30–33	18
34–37	3

The frequency table shows the ages of people watching a comedy in a theater. Display the data in a histogram. Then describe the shape of the distribution.

Draw and label the axes. Then draw a bar to represent the frequency of each interval.

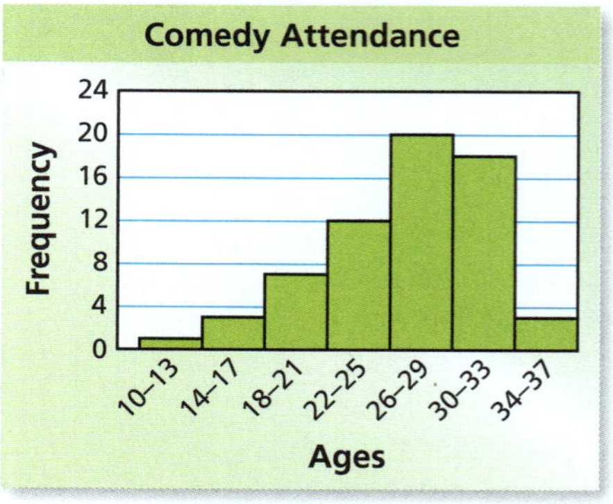

Most of the data are on the right, and the tail extends to the left.

▶ So, the distribution is skewed left.

Try It

2. The frequency table shows the ages of people watching a historical movie in a theater. Display the data in a histogram. Describe the shape of the distribution.

Ages	10–19	20–29	30–39	40–49	50–59	60–69
Frequency	3	18	36	40	14	5

In-Class Practice

1 I don't understand yet. 2 I can do it with help. 3 I can do it on my own. 4 I can teach someone else.

3. WRITING Explain in your own words what it means for a distribution to be (a) skewed left, (b) symmetric, and (c) skewed right.

4. DESCRIBING A DISTRIBUTION Display the data shown in a histogram. Describe the shape of the distribution.

Calories	Frequency
1–100	2
101–200	8
201–300	10
301–400	5
401–500	3

GO DIGITAL

In-Class Practice

1 I don't understand yet. 2 I can do it with help. 3 I can do it on my own. 4 I can teach someone else.

4 MTR

5. WHICH ONE DOESN'T BELONG? Which histogram does not belong with the other three? Explain your reasoning.

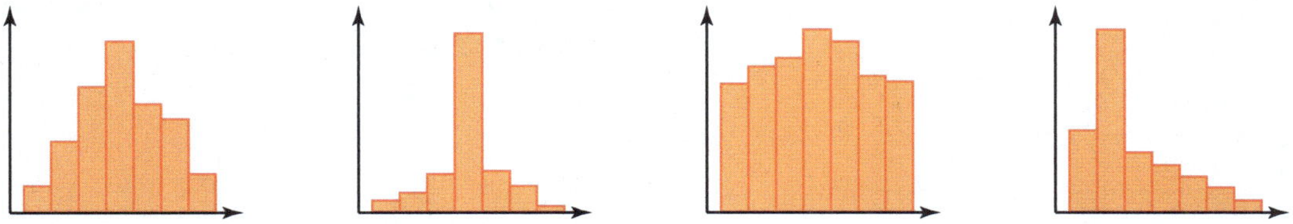

Example 3 Modeling Real Life

7 MTR

The histogram shows the ages of people watching an animated movie in the same theater as in Example 2. Which movie has an older audience?

Understand the problem.

You are given histograms that display the ages of people watching two movies. You are asked to determine which movie has an older audience.

Make a plan.

Use the intervals and distributions of the data to determine which movie has an older audience.

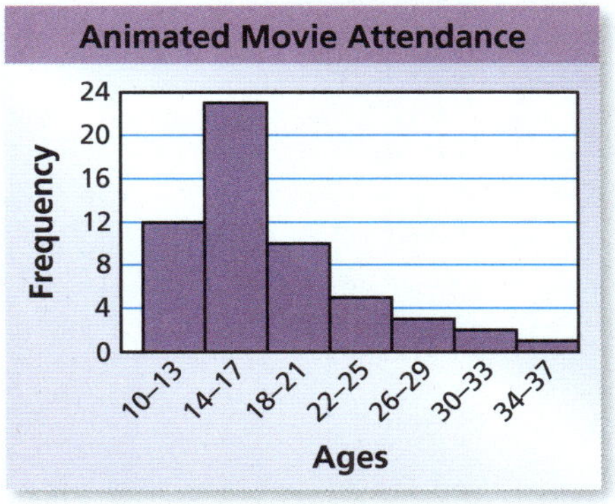

Solve and check.

The intervals in the histograms are the same. Most of the data for the animated movie are on the left, and most of the data for the comedy are on the right. This means that the people watching the comedy are generally older than the people watching the animated movie.

▶ So, the comedy has an older audience.

Check Reasonableness

The movies have similar attendance. However, only 4 people watching the comedy are 17 or under. A total of 35 people watching the animated movie are 17 or under. So, it is reasonable to conclude that the comedy has an older audience. ✓

GO DIGITAL

In-Class Practice

1 I don't understand yet. 2 I can do it with help. 3 I can do it on my own. 4 I can teach someone else.

Visitors	Tallahassee	Jacksonville
1–20	3	6
21–40	5	11
41–60	6	7
61–80	10	4
81–100	7	3

6. The frequency table shows the numbers of visitors each day to a park in Tallahassee and a park in Jacksonville in one month. Which park generally has more daily visitors? Justify your answer.

Ages	Frequency
18–24	26
25–31	26
32–38	22
39–45	14
46–52	8

Ages	Frequency
18–22	16
23–27	22
28–32	26
33–37	20
38–42	12

7. **Dig Deeper** The frequency tables below show the ages of guests on two cruises. Can you make accurate comparisons of the ages of the guests? Explain your reasoning.

11.6 Practice WITH CalcChat® AND CalcView®

Review & Refresh

Display the data in a histogram.

1.

Goals per Game	
Goals	**Frequency**
0–1	5
2–3	4
4–5	0
6–7	1

2.

Minutes Practiced	
Minutes	**Frequency**
0–19	8
20–39	10
40–59	11
60–79	2

3.

Poems Written for Class	
Poems	**Frequency**
0–4	6
5–9	16
10–14	4
15–19	2

Write a unit rate for the situation.

4. \$200 per 8 days

5. 60 kilometers for every 1.5 hours

Concepts, Skills, & Problem Solving

DESCRIBING SHAPES OF DISTRIBUTIONS Describe the shape of the distribution of the data in the table. (See Exploration 1.)

6.

Miles Run per Day										
1	4	2	0	3	2	1	2	4	2	3
2	1	6	3	2	4	0	5	3	1	5

7.

Raffle Tickets Sold							
15	12	16	15	13	14	16	13
13	16	14	12	15	12	14	

GO DIGITAL

DESCRIBING SHAPES OF DISTRIBUTIONS **Describe the shape of the distribution.** (See Examples 1 and 2.)

8.

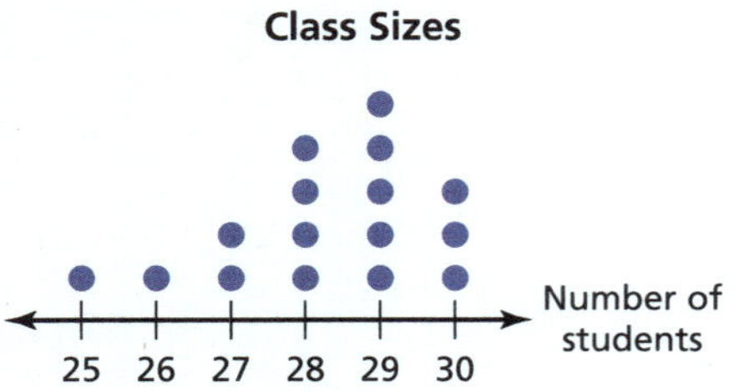

▶9.

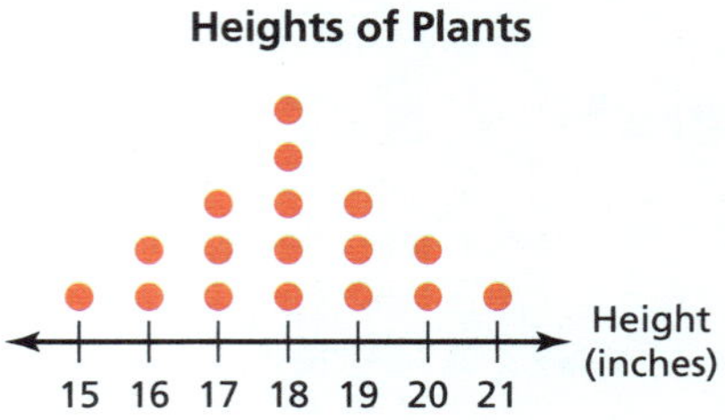

10.

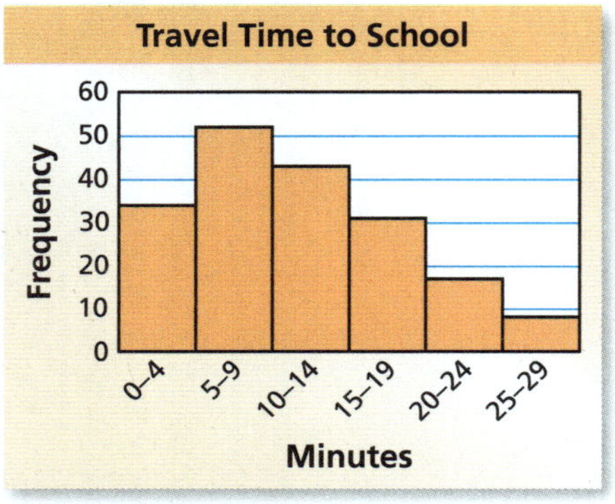

▶11.

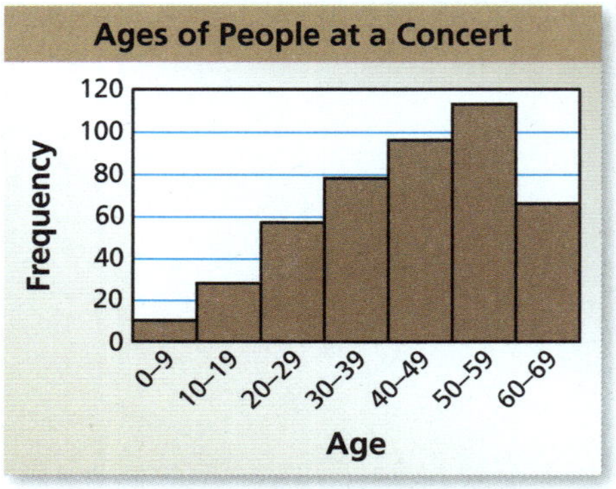

7 MTR

12. **MODELING REAL LIFE** The frequency table shows the years of experience for the medical staffs in Broward County and Polk County. Display the data for each county in a histogram. Which county's medical staff has less experience? Explain. (See Example 3.)

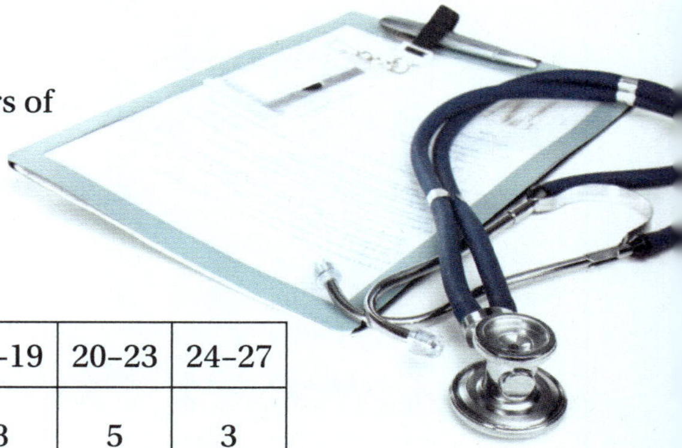

Years of Experience	0–3	4–7	8–11	12–15	16–19	20–23	24–27
Frequency for Broward County	7	15	17	12	8	5	3
Frequency for Polk County	3	5	9	14	10	6	2

13. REASONING What is the shape of the distribution of the restaurant waiting times? Explain your reasoning.

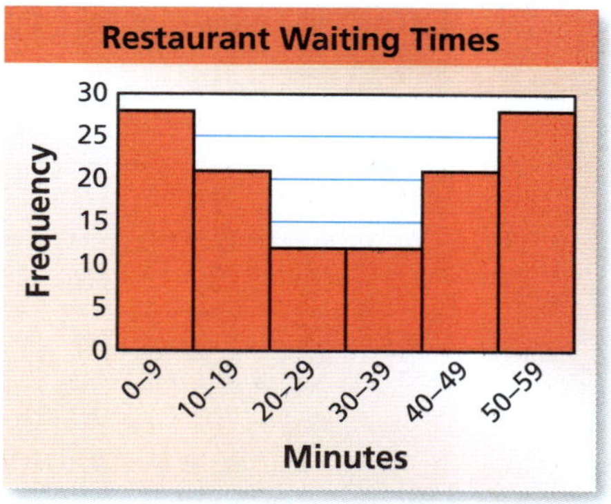

4 MTR

14. DISCUSS MATHEMATICAL THINKING Are all distributions either approximately symmetric or skewed? Explain. If not, give an example.

7 MTR

15. INVESTIGATE Research last year's average daily temperatures where you live. Make a histogram of the data and describe the shape of the distribution. Is the shape what you expected? Explain.

16. **Dig Deeper** The table shows the donation amounts received by a charity in one day.

Donations (dollars)												
20	15	40	70	20	5	25	50	47	20	62	55	40
10	50	18	20	100	40	80	60	20	80	3	30	50
25	30	10	33	20	50	7	35	40	25	70		

a. Make a histogram of the data, starting with the interval 0–14. Describe the shape of the distribution.

b. A company adds $5 to each donation. Make another histogram, starting with the same interval as in part (a). Compare the shape of this distribution with the distribution in part (a). Explain any differences in the distributions.

2 MTR

17. **MAKE A CONNECTION** Describe the shape of the distribution of each bar graph. Match the letters A, B, and C with the mean, the median, and the mode of each data set. Explain your reasoning.

a.

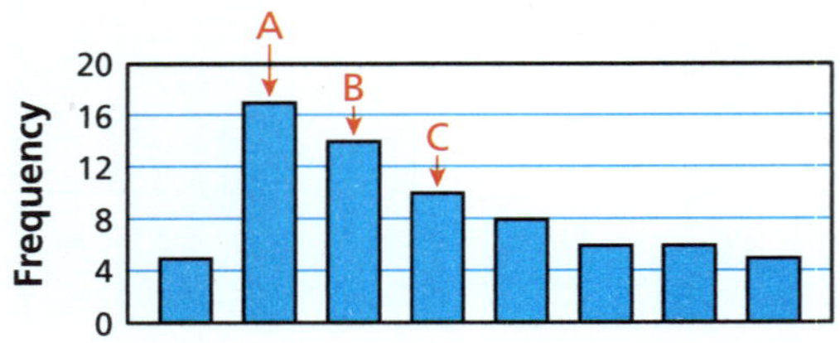

b.

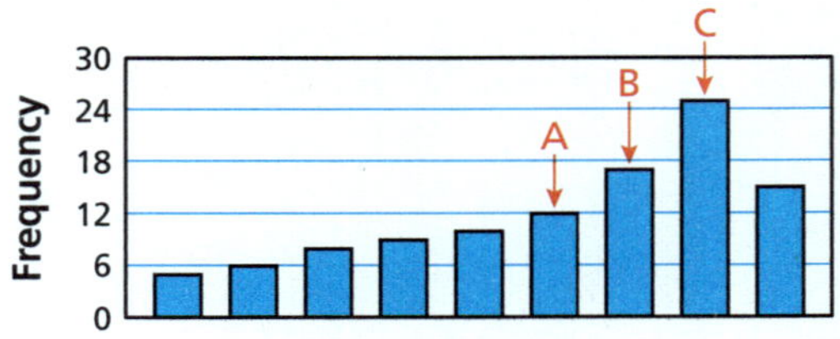

11.7 Box-and-Whisker Plots

Learning Target: Display and interpret data in box-and-whisker plots.

Success Criteria:
- I can find the five-number summary of a data set.
- I can make a box-and-whisker plot.
- I can explain what the box and the whiskers of a box-and-whisker plot represent.
- I can compare data sets represented by box-and-whisker plots.

Data Analysis and Probability

MA.6.DP.1.3 Given a box plot within a real-world context, determine the minimum, the lower quartile, the median, the upper quartile and the maximum. Use this summary of the data to describe the spread and distribution of the data.

MA.6.DP.1.5 Create box plots and histograms to represent sets of numerical data within real-world contexts.

Also MA.6.DP.1.2

Exploration 1 Drawing a Box-and-Whisker Plot

Work with a partner. Each student in a sixth-grade class is asked to choose a number from 1 to 20. The results are shown below.

Numbers Chosen

4	5	14	16
5	16	17	8
18	13	17	18
17	14	19	11
15	8	2	18
13	19	8	7

a. The *box-and-whisker plot* below represents the data set. Which part represents the *box*? the *whiskers*? Explain.

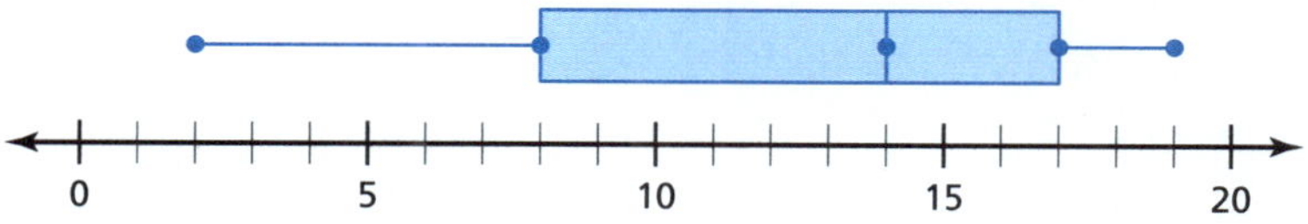

2 MTR MAKE A CONNECTION

What do the different components of a box-and-whisker plot represent?

GO DIGITAL

b. What does each of the five plotted points represent?

c. In your own words, describe what a box-and-whisker plot is and what it tells you about a data set.

d. Conduct a survey in your class. Have each student write a number from 1 to 20 on a piece of paper. Collect all of the data and draw a box-and-whisker plot that represents the data. Compare the data with the box-and-whisker plot in part (a).

11.7 Lesson

Key Vocabulary

box-and-whisker plot, *p. 959*

five-number summary, *p. 959*

Key Idea

Box-and-Whisker Plot

A **box-and-whisker plot** represents a data set along a number line by using the least value, the greatest value, and the quartiles of the data. A box-and-whisker plot shows the *variability* of a data set.

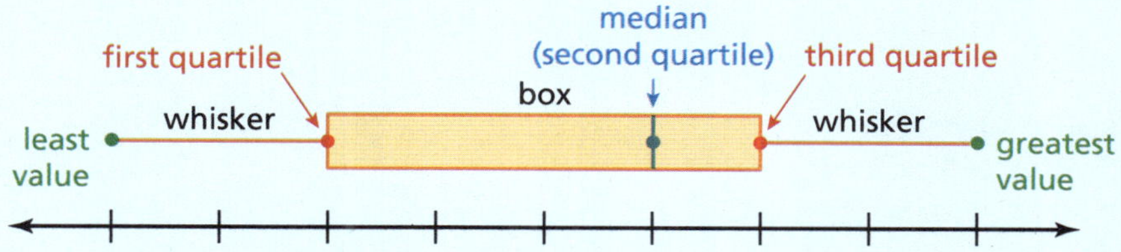

The five numbers that make up the box-and-whisker plot are called the **five-number summary** of the data set.

Example 1 Making a Box-and-Whisker Plot

Make a box-and-whisker plot for the ages (in years) of the spider monkeys at a zoo.

15, 20, 14, 38, 30, 36, 30, 30, 27, 26, 33, 35

Step 1: Order the data. Find the quartiles.

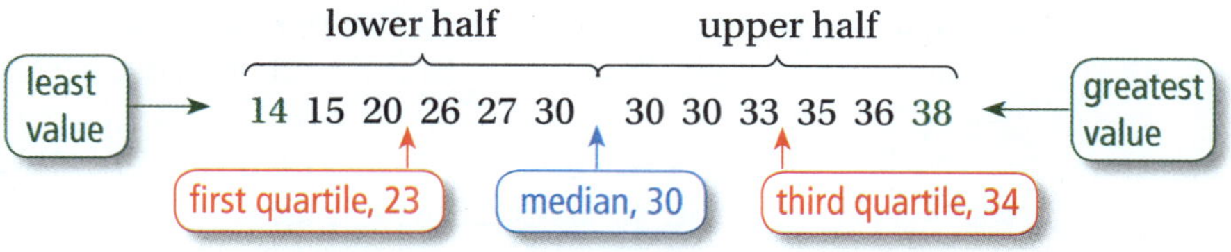

Step 2: Draw a number line that includes the least and greatest values. Graph points above the number line that represent the five-number summary.

Step 3: Draw a box using the quartiles. Draw a line through the median. Draw whiskers from the box to the least and the greatest values.

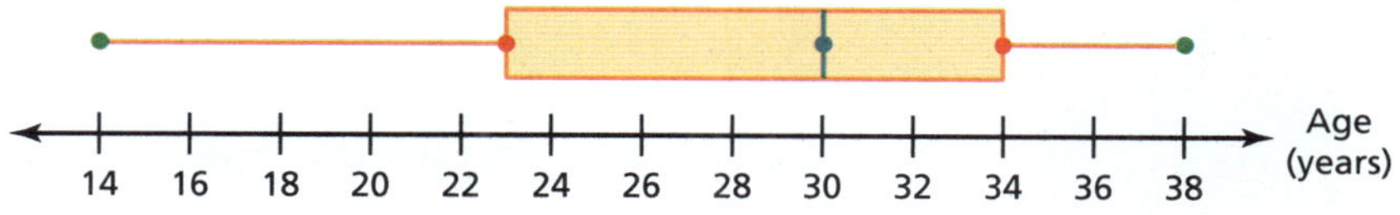

GO DIGITAL

Try It

1. The data set {1, 0, 2, 3, 4, 3, 6, 1, 0, 1, 2, 2} represents the amounts of time (in hours) that a group of friends spent online last night. Make a box-and-whisker plot for the data.

The figure shows how data are distributed in a box-and-whisker plot.

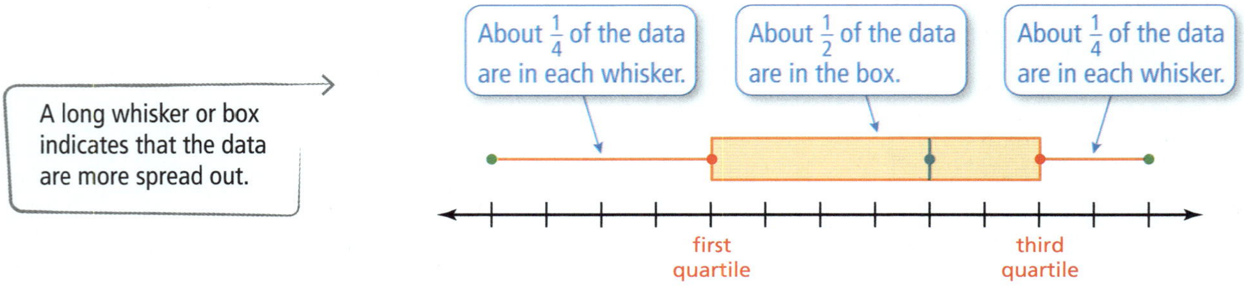

Example 2 Analyzing a Box-and-Whisker Plot

The box-and-whisker plot shows the body mass indexes (BMI) of a sixth-grade class.

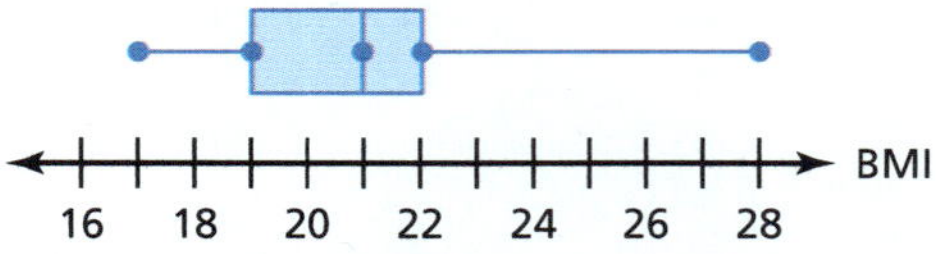

a. What fraction of the students have a BMI of at least 22?

Because the third quartile is 22, the right whisker represents students who have a BMI of at least 22.

▶ So, about $\frac{1}{4}$ of the students have a BMI of at least 22.

b. Are the data more spread out below the first quartile or above the third quartile? Explain.

length of left whisker = first quartile − least value

$= 19 - 17 = 2$

length of right whisker = greatest value − third quartile

$= 28 - 22 = 6$

▶ The right whisker is longer than the left whisker. So, the data are more spread out above the third quartile than below the first quartile.

c. Find and interpret the interquartile range of the data.

interquartile range = third quartile − first quartile

$= 22 - 19 = 3$

▶ So, the middle half of the students' BMIs varies by no more than 3.

GO DIGITAL

Try It

2. The box-and-whisker plot shows the heights of the roller coasters at an amusement park. (a) What fraction of the heights are between 120 feet and 220 feet? (b) Are the data more spread out below or above the median? Explain. (c) Find and interpret the interquartile range of the data.

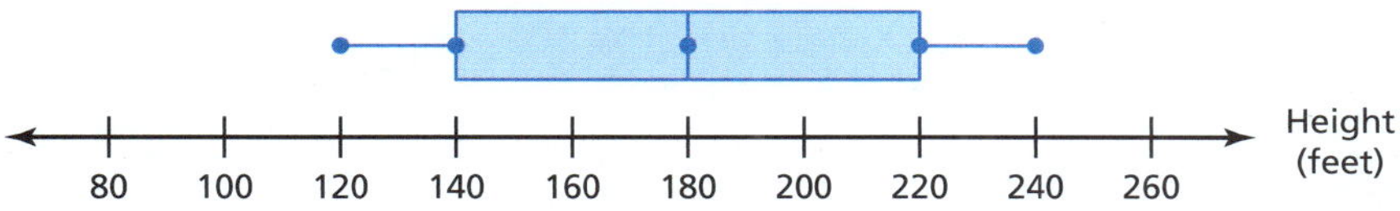

A box-and-whisker plot also shows the shape of a distribution.

If you can draw a line through the median of a box-and-whisker plot, and each side is a mirror image of the other, then the distribution is symmetric.

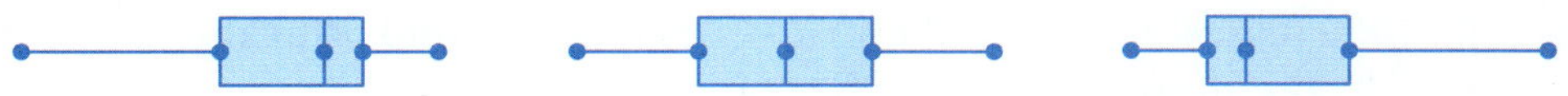

Skewed left

- The left whisker is longer than the right whisker.
- Most data are on the right.

Symmetric

- The whiskers are about the same length.
- The median is in the middle of the box.

Skewed right

- The right whisker is longer than the left whisker.
- Most data are on the left.

Example 3 Identifying Shapes of Distributions

The double box-and-whisker plot represents the life spans of crocodiles and alligators at a zoo. Identify the shape of the distribution of the life spans of alligators.

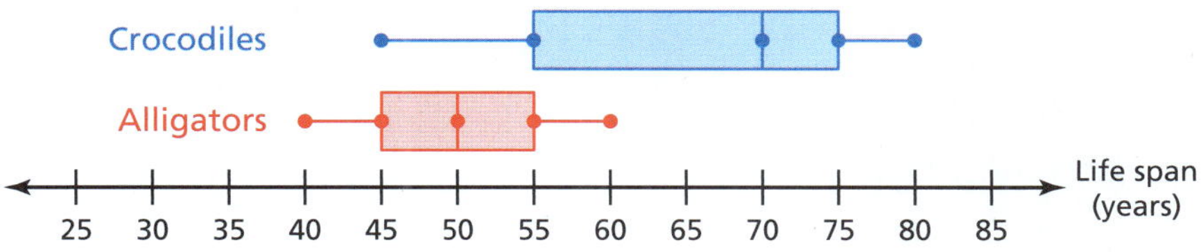

For alligator life spans, the whisker lengths are equal. The median, 50, is in the middle of the box. The left side of the box-and-whisker plot is a mirror image of the right side of the box-and-whisker plot.

▶ So, the distribution is symmetric.

GO DIGITAL

Try It

3. Identify the shape of the distribution of the life spans of crocodiles.

In-Class Practice

1	I don't understand yet.	2	I can do it with help.	3	I can do it on my own.	4	I can teach someone else.

4. VOCABULARY Explain how to find the five-number summary of a data set.

MAKING A BOX-AND-WHISKER PLOT Make a box-and-whisker plot for the data. Identify the shape of the distribution.

5. Ticket prices (dollars): 39, 42, 40, 47, 38, 39, 44, 55, 44, 58, 45

6. Number of sit-ups: 20, 20, 23, 25, 25, 26, 27, 29, 30, 30, 32, 34, 37, 38

7. NUMBER SENSE In a box-and-whisker plot, what fraction of the data is greater than the first quartile?

Example 4 Modeling Real Life 7 MTR

The double box-and-whisker plot represents the prices of skateboards at two stores.

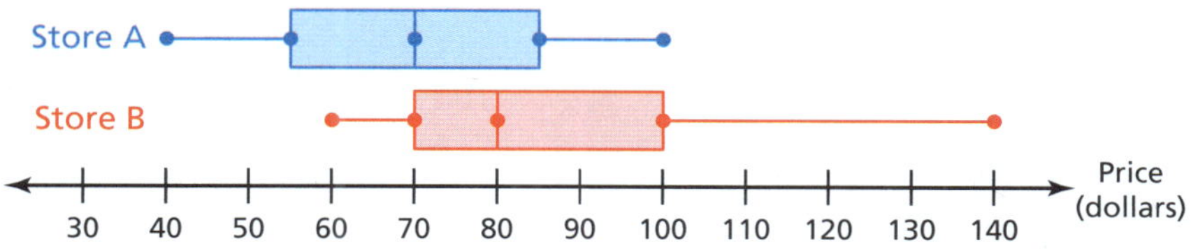

a. Which store's prices are more spread out? Explain.

Both boxes appear to be the same length. So, the interquartile range of each data set is equal. The range of the prices in Store B, however, is greater than the range of the prices in Store A.

▶ So, the prices in Store B are more spread out.

b. Which store's prices are generally higher? Explain.

For Store A, the distribution is symmetric with about one-half of the prices above \$70.

For Store B, the distribution is skewed right with about three-fourths of the prices above \$70.

▶ So, the prices in Store B are generally higher.

6 MTR **ASSESS REASONABLENESS**

Use the five-number summary for each data set to explain why the answers in parts (a) and (b) are reasonable.

GO DIGITAL

In-Class Practice

1 I don't understand yet. | 2 I can do it with help. | 3 I can do it on my own. | 4 I can teach someone else.

8. The tables show the test scores of two sixth-grade achievement tests. The same group of students took both tests. The students took one test in the fall and the other in the spring.

Test Scores				
75	65	64	79	100
75	94	52	73	80

Test Scores				
56	70	47	100	83
44	45	58	54	30

a. Analyze each distribution. Then compare and contrast the test results.

b. Which table likely represents the results of which test? Explain your reasoning.

9. Make a box-and-whisker plot that represents the heights of the boys in your class. Then make a box-and-whisker plot that represents the heights of the girls in your class. Compare and contrast the distributions.

11.7 Practice WITH CalcChat® AND CalcView®

Review & Refresh

Describe the shape of the distribution.

1.

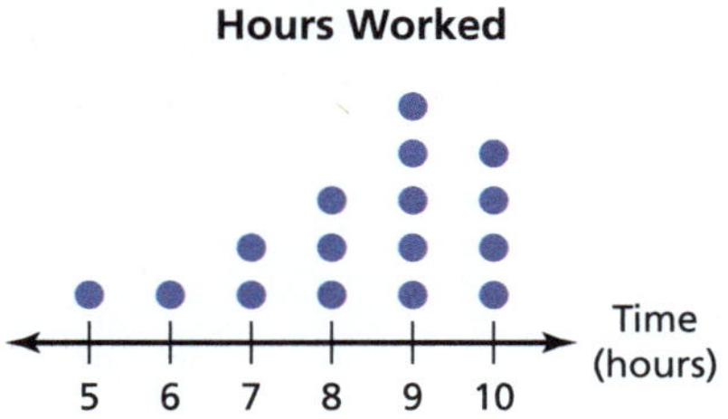

2.

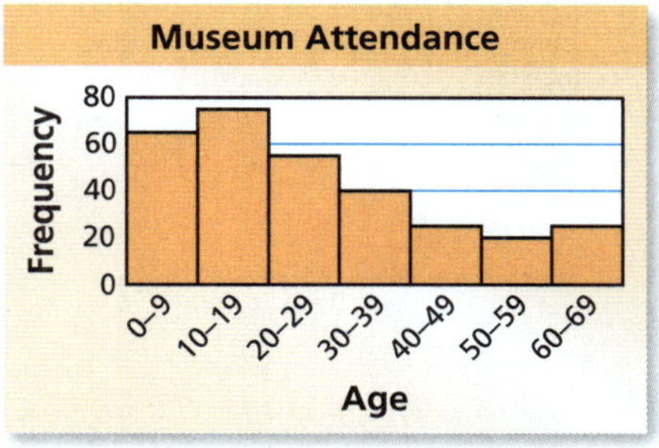

Complete the statement using < or >.

3. $-\frac{2}{3} \; \square \; -\frac{3}{4}$ **4.** $-2\frac{1}{5} \; \square \; -2\frac{1}{6}$ **5.** $-5.3 \; \square \; -5.5$

Factor the expression using the GCF.

6. $42 + 14$ **7.** $12x - 18$ **8.** $28n + 20$ **9.** $60g - 25h$

Concepts, Skills, & Problem Solving

COMPARING DATA Compare the data in the box-and-whisker plots. (See Exploration 1.)

10.

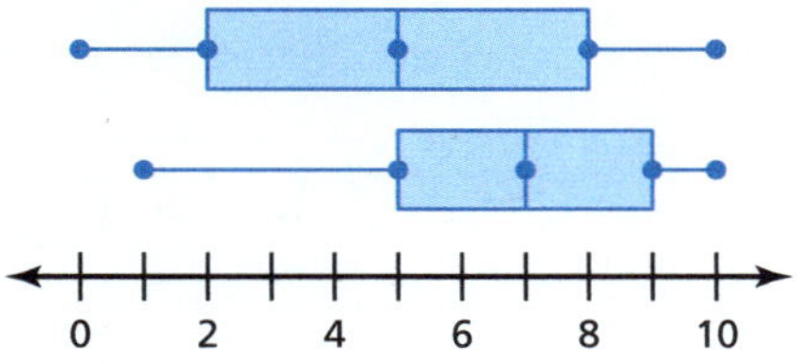

11.

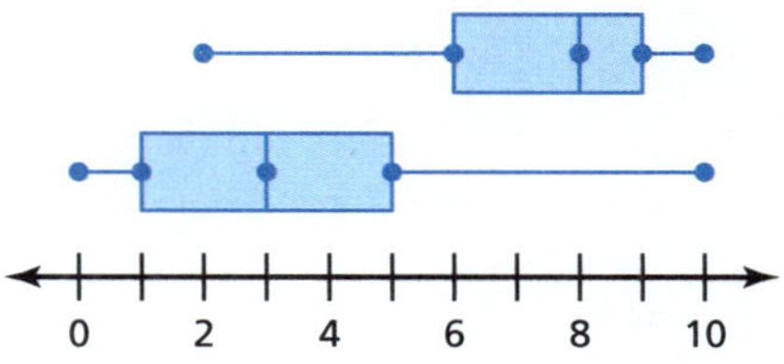

GO DIGITAL

MAKING A BOX-AND-WHISKER PLOT **Make a box-and-whisker plot for the data.** (See Example 1.)

12. Ages of teachers (in years): 30, 62, 26, 35, 45, 22, 49, 32, 28, 50, 42, 35

13. Quiz scores: $8\frac{1}{2}$, 12, 9, 10, $12\frac{1}{2}$, 8, 5, 9, $7\frac{1}{2}$, $10\frac{1}{2}$, 8, 9, 11

14. Donations (in dollars): 10, 30, 5, 15, 50, 25, 5, 20, 15, 35, 10, 30, 20

15. Science test scores: 85, 76, 99, 84, 92, 95, 68, 100, 93, 88, 87, 85

16. Shoe sizes: 12, 8.5, 9, 10, 9, 11, 11.5, 9, 9, 10, 10, 10.5, 8

17. Ski lengths (in centimeters): 180, 175, 205, 160, 210, 175, 190, 205, 190, 160, 165, 195

4 MTR

18. **YOU BE THE TEACHER** Your friend makes a box-and-whisker plot for the data shown. Is your friend correct? Explain your reasoning.

2, 6, 4, 3, 7, 4, 6, 9, 6, 8, 5, 7

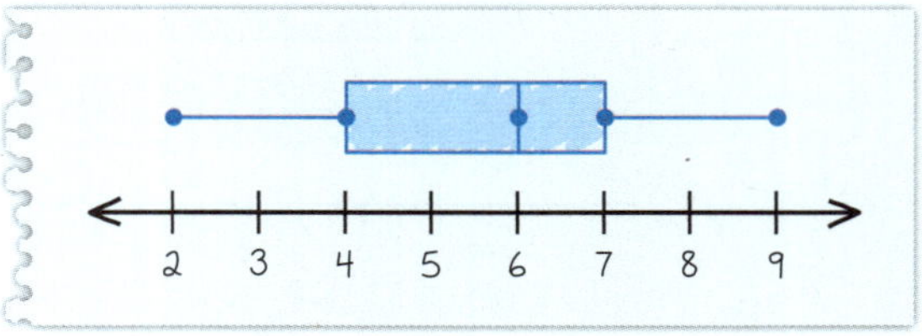

7 MTR

19. **MODELING REAL LIFE** The data set {6, 2, 0, 10, 3, 6, 6, 4, 12, 0, 6, 2} represents the numbers of days 12 friends went camping during the summer. Make a box-and-whisker plot for the data. What is the range of the data?

FIVE-NUMBER SUMMARY **Identify the five-number summary for each box-and-whisker plot.**

20.

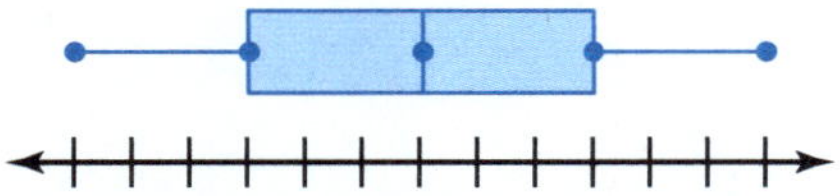

21.

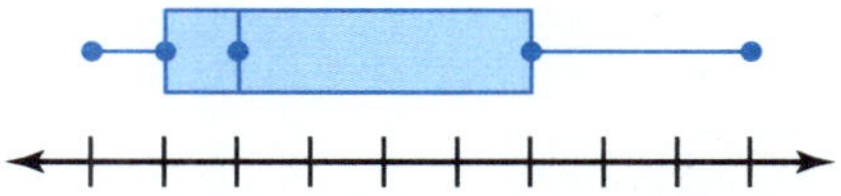

22.

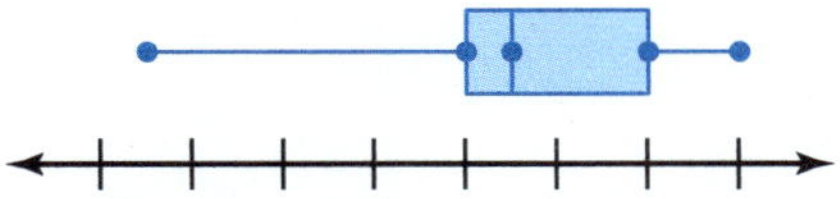

23.

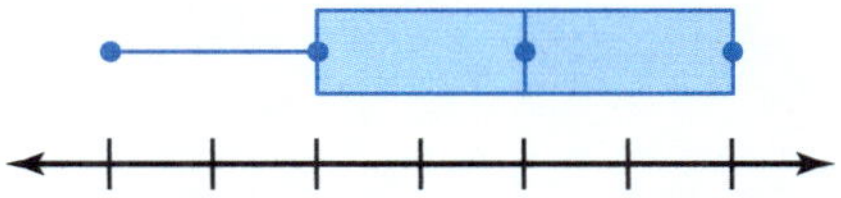

24. **ANALYZING A BOX-AND-WHISKER PLOT** The box-and-whisker plot represents the numbers of gallons of water needed to fill different types of dunk tanks offered by a company. (See Example 2.)

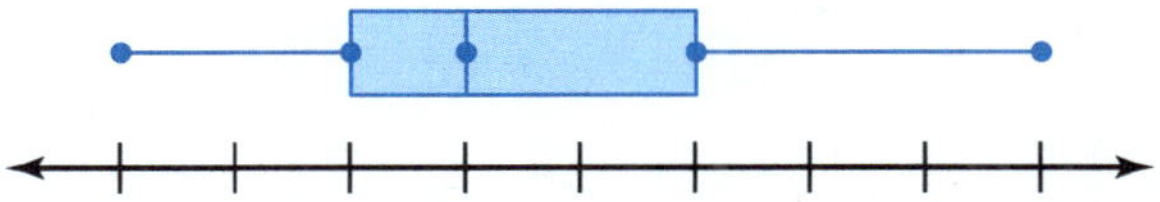

a. What fraction of the dunk tanks requires at least 500 gallons of water?

b. Are the data more spread out below the first quartile or above the third quartile? Explain.

c. Find and interpret the interquartile range of the data.

25. MULTIPLE REPRESENTATIONS The numbers of spots on several frogs in a forest are shown in the dot plot.

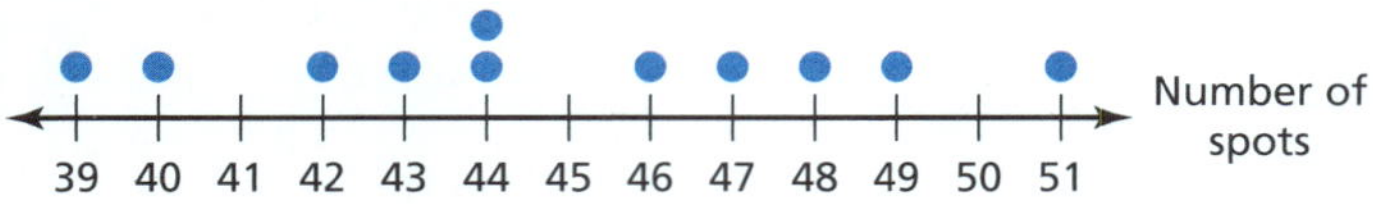

a. Make a box-and-whisker plot for the data.

b. Compare the dot plot and the box-and-whisker plot. Describe the advantages and disadvantages of each data display.

SHAPES OF BOX-AND-WHISKER PLOTS **Identify the shape of the distribution. Explain.**

26. 

27. 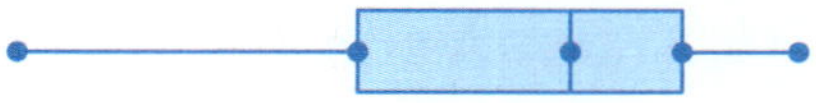

28. 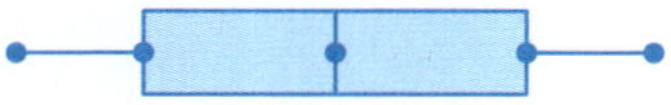

29. 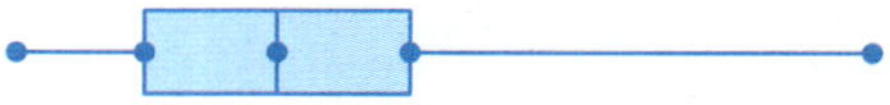

7 MTR

30. MODELING REAL LIFE The double box-and-whisker plot represents the start times of recess for classes at two schools. (See Examples 3 and 4.)

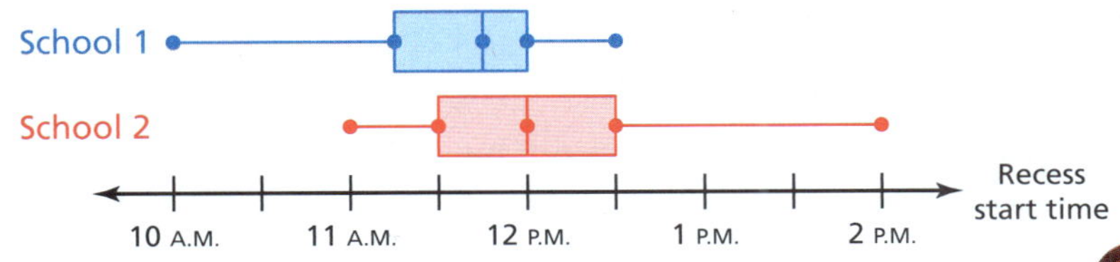

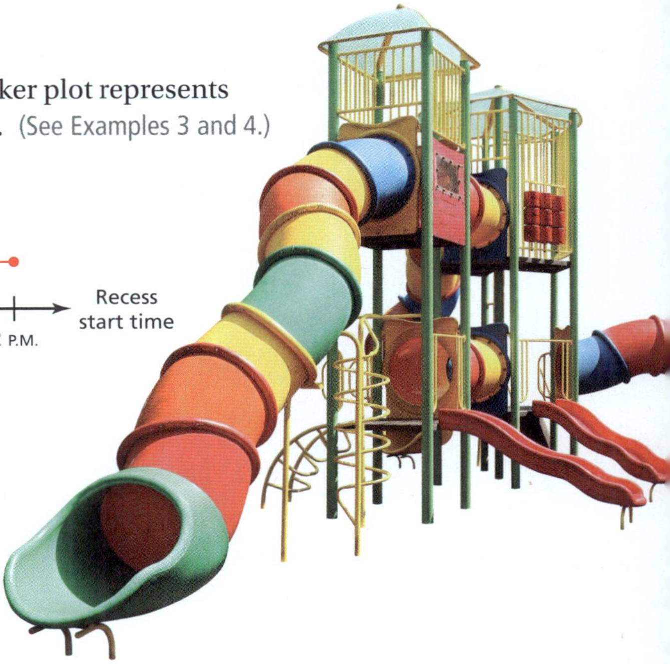

a. Identify the shape of each distribution.

b. Which school's start times for recess are more spread out? Explain.

c. You randomly pick one class from each school. Which class is more likely to have recess before lunch? Explain.

MAKING A BOX-AND-WHISKER PLOT **Make a box-and-whisker plot for the data.**

31. Temperatures (in °C): 15, 11.2, 14, 10.6, 19.8, 10, 2.7, 15, 12, 14.3, 9, 20.4, 17, 5.6

32. Checking account balances (in dollars): 30, 0, 50, 20.50, 90.50, 15, 40, 100, 45.50, 20, 70, 0

33. REASONING The data set in Exercise 31 has an outlier. Describe how removing the outlier affects the box-and-whisker plot.

34. OPEN-ENDED Write a data set with 12 values that has a symmetric box-and-whisker plot.

35. REASONING When does a box-and-whisker plot *not* have one or both whiskers?

5 MTR

36. STRUCTURE Draw a histogram that could represent the distribution shown in Exercise 28.

37. Dig Deeper The double box-and-whisker plot represents the goals scored per game by two lacrosse teams during a 16-game season.

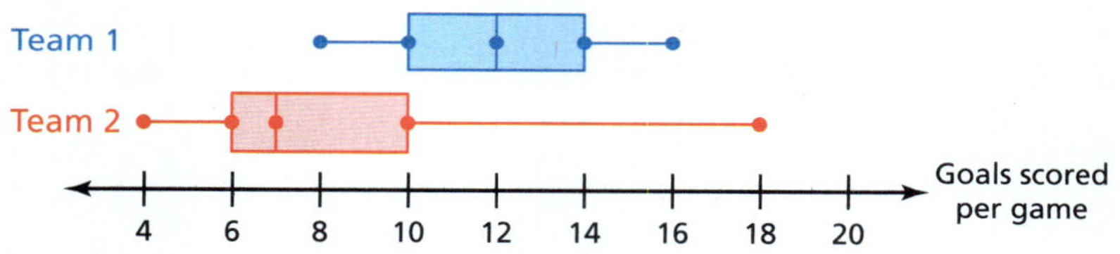

a. Which team is more consistent? Explain.

b. Team 1 played Team 2 once during the season. Which team do you think won? Explain.

c. Can you determine the number of games in which Team 2 scored 10 goals or less? Explain your reasoning.

11.8 Choosing Appropriate Measures

Learning Target: Determine which measures of center and variation are appropriate to describe a data set.

Success Criteria:
- I can determine whether a data set contains outliers.
- I can explain whether a measure of center is appropriate to describe a data set.
- I can explain whether a measure of variation is appropriate to describe a data set.

Data Analysis and Probability
MA.7.DP.1.1 Determine an appropriate measure of center or measure of variation to summarize numerical data, represented numerically or graphically, taking into consideration the context and any outliers.

Exploration 1 Choosing Appropriate Measures

Work with a partner. The first three digits of several cell phone numbers and the ages of the cell phones are shown.

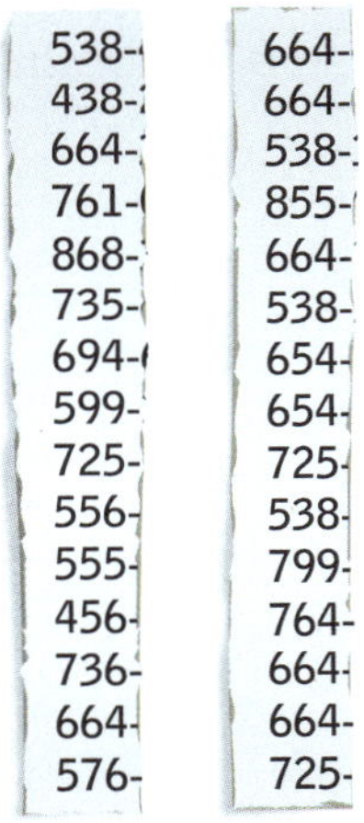

Ages of Cell Phones (years)					
1	10	1	3	5	9
9	1	1	2	1	1
1	0	9	1	0	10
2	3	1	7	2	1
4	1	0	1	1	3

a. Find the mean, median, range, and IQR of the first digit of each phone number. Then find the mean, median, range, and IQR of the cell phone ages.

GO DIGITAL

b. For each data set, compare the measures of center. Then compare the measures of variation. Does one of the measures of center or variation describe the data better than the other? Explain.

c. Does either data set contain outliers? How are outliers related to your answers in part (b)?

d. How can you choose appropriate measures to describe the center and variation of a data set?

7 MTR APPLY MATHEMATICS

Why might a cell phone manufacturer use the range to describe the variation of the ages? Why might the manufacturer choose the interquartile range?

11.8 Lesson

You can use measures of center and measures of variation to describe data sets. The context and any outliers can help you choose appropriate measures.

Regardless of whether a data set contains outliers, the IQR is considered a more accurate measure of variation than the range. However, the range can still be an appropriate choice, depending on the context.

Key Idea

Choosing Appropriate Measures

In general, the median and the interquartile range (IQR) are less affected by outliers than the mean and the range. So, when a data set contains outliers,

- use the median to describe the center and
- use the IQR to describe the variation.

When a data set does not contain outliers, you can use either the mean or the median to describe the center, and either the range or the IQR to describe the variation.

Example 1 Choosing Appropriate Measures

The box-and-whisker plot represents the number of states that border each state in the United States. For example, the states of Alabama and Georgia border the state of Florida. So there are two states that border Florida. Choose appropriate measures to describe the center and the variation.

Notice that you cannot use a box-and-whisker plot to calculate the mean. When you are given a box-and-whisker plot without the original data, you can generally use the median to describe the center and the IQR to describe the variation.

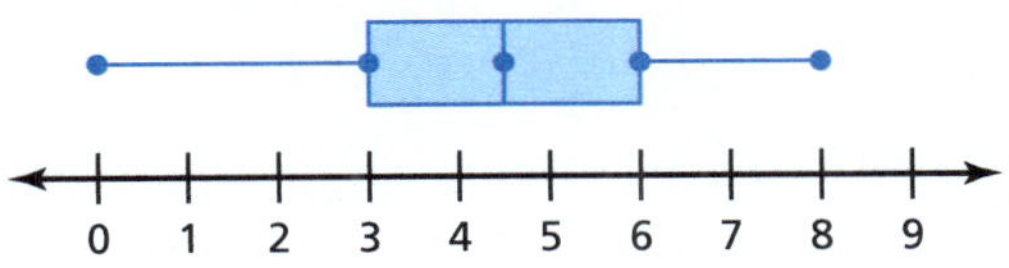

Because $Q_1 = 3$ and $Q_3 = 6$, the IQR is $6 - 3 = 3$. So, any value less than $3 - 1.5(3) = -1.5$ or greater than $6 + 1.5(3) = 10.5$ is an outlier. Because the least value in the data set is 0 and the greatest value is 8, there are no outliers.

▶ So, both the mean and the median are appropriate measures of center. Both the range and the IQR are appropriate measures of variation.

GO DIGITAL

Try It

1. The stem-and-leaf plot shows the gas mileages of several motorcycles made by a company. Choose appropriate measures to describe the center and the variation.

Stem	Leaf
4	0 9
5	3 4 4 4 5 5 5 6 6 6 7 8
6	0 0 0 1 3 3 3 4 5 5 7

Key: 5 | 6 = 56 miles per gallon

Example 2 Describing a Data Set

The dot plot shows the numbers of hours students in a class slept last night. Describe the center and the variation of the data set.

Amount of Sleep

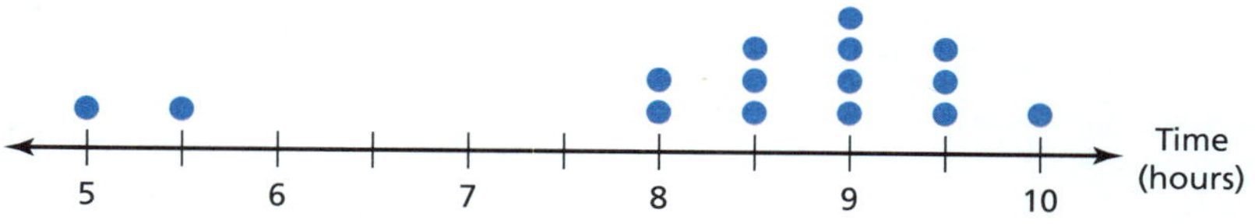

Determine whether the data set contains any outliers.

{5 5.5 8 8 8.5 8.5 8.5 9 9 9 9 9.5 9.5 9.5 10}

$Q_1 = 8$ median = 9 $Q_3 = 9.5$

The IQR is $9.5 - 8 = 1.5$. So, any value less than $8 - 1.5(1.5) = 5.75$ or greater than $9.5 + 1.5(1.5) = 11.75$ is an outlier. Because 5 and 5.5 are less than 5.75, they are outliers.

Because the data set contains outliers, use the median and the IQR to describe the center and the variation.

▶ The data are centered around 9 hours, so a typical student slept for about 9 hours. The middle half of the data varies by no more than 1.5 hours.

GO DIGITAL

Try It

2. The numbers of hours people spent at the gym last week are represented by the data set {6, 2, 5, 3, 8, 3, 7, 5, 5, 4, 4, 6, 6, 5, 7, 5, 4}. Describe the center and the variation of the data set.

In-Class Practice

1 I don't understand yet. **2** I can do it with help. **3** I can do it on my own. **4** I can teach someone else.

CHOOSING APPROPRIATE MEASURES **Choose appropriate measures to describe the center and the variation. Explain your reasoning. Then find the measures you chose.**

3.

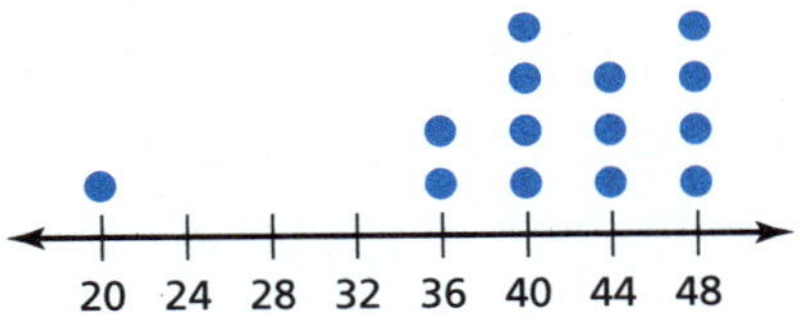

4.

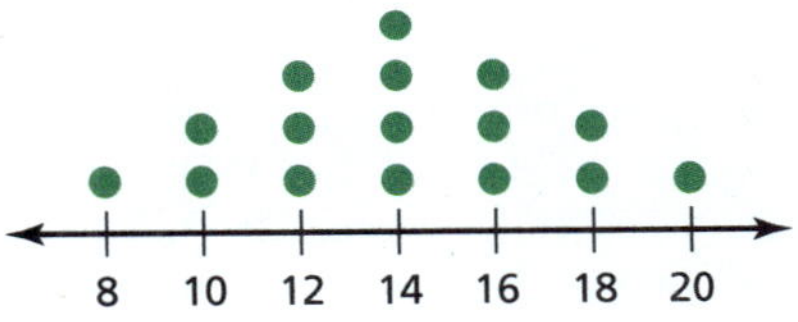

5. **WRITING** Explain why the median and the IQR are appropriate measures of center and variation when a data set contains outliers.

Example 3 Modeling Real Life

7 MTR

Basket A	
\$70	\$40
\$60	\$90
\$10	\$0
\$70	\$40
\$30	\$100
\$50	\$60
\$40	\$50
\$60	\$30

Basket B	
\$45	\$30
\$45	\$55
\$55	\$40
\$50	\$60
\$45	\$50
\$55	\$55
\$60	\$45
\$40	\$70

Two baskets each have 16 envelopes with money inside, as shown in the tables. How much does a typical envelope in each basket contain? Why might a person want to pick from Basket A instead of Basket B?

To answer each question, determine whether there are any outliers in the data sets.

Basket A: {0, 10, 30, 30, 40, 40, 40, 50, 50, 60, 60, 60, 70, 70, 90, 100}

$Q_1 = 35$ median = 50 $Q_3 = 65$

$Q_1 - 1.5(\text{IQR}) = -10$

$Q_3 + 1.5(\text{IQR}) = 110$

Basket B: {30, 40, 40, 45, 45, 45, 45, 50, 50, 55, 55, 55, 55, 60, 60, 70}

$Q_1 = 45$ median = 50 $Q_3 = 55$

$Q_1 - 1.5(\text{IQR}) = 30$

$Q_3 + 1.5(\text{IQR}) = 70$

Neither data set contains an outlier. So for each data set, both the mean and the median are appropriate measures of center, and both the range and the IQR are appropriate measures of variation.

The mean of each data set is $\frac{800}{16} = \$50$, and the median of each data set is \$50. The range of Basket A is \$100, and the IQR is \$30. The range of Basket B is \$40, and the IQR is \$10. So, Basket A has more variability.

▶ A typical envelope in each basket contains about \$50. The envelopes in Basket A range from \$0 to \$100, but the envelopes in Basket B range from only \$30 to \$70. So, a person may choose from Basket A because there is a chance to get a greater amount from Basket A than from Basket B.

4 MTR CONSTRUCT AN ARGUMENT

Explain why a person might want to pick from Basket B instead of Basket A.

____

____

____

____

____

In-Class Practice

1 I don't understand yet. 2 I can do it with help. 3 I can do it on my own. 4 I can teach someone else.

Room A	
5	25
30	45
20	15
50	20
25	30
10	25

Room B	
25	15
30	20
5	10
20	50
45	5
25	50

6. In a video game, two rooms each have 12 treasure chests containing gold coins. The tables show the numbers of coins in each chest. You pick one chest and are rewarded with the coins inside. From which room would you choose? Explain your reasoning.

7. Write the numbers of pets that students in your class own. Describe the center and the variation of the data set.

GO DIGITAL

11.8 Practice WITH CalcChat® AND CalcView®

Review & Refresh

Make a box-and-whisker plot for the data.

1. 68, 74, 67, 72, 63, 70, 78, 64, 76

2. 39, 48, 33, 24, 30, 44, 36, 41, 39, 28, 53

Convert. Round to the nearest hundredth, if necessary.

3. 7 gallons to quarts

4. 1532 kilometers to meters

The table shows the ages of people watching a movie at a theater. Display the data in a histogram.

5.

Ages	0–19	20–39	40–59	60–79
Frequency	25	31	26	18

6.

Ages	0–19	20–39	40–59	60–79
Frequency	21	20	24	35

GO DIGITAL

Concepts, Skills, & Problem Solving

CHOOSING APPROPRIATE MEASURES **Find the mean, median, range, and IQR of the data set. Which measures of center and variation best describe the data set? Explain your reasoning.** (See Exploration 1.)

7. {9, 3, 7, 7, 9, 2, 8, 9, 6, 7, 8, 9}

8. {24, 25, 27, 27, 23, 29, 26, 26, 26, 25, 28}

CHOOSING APPROPRIATE MEASURES **Choose the most appropriate measures to describe the center and the variation.** (See Example 1.)

9. **Prices of Jeans**

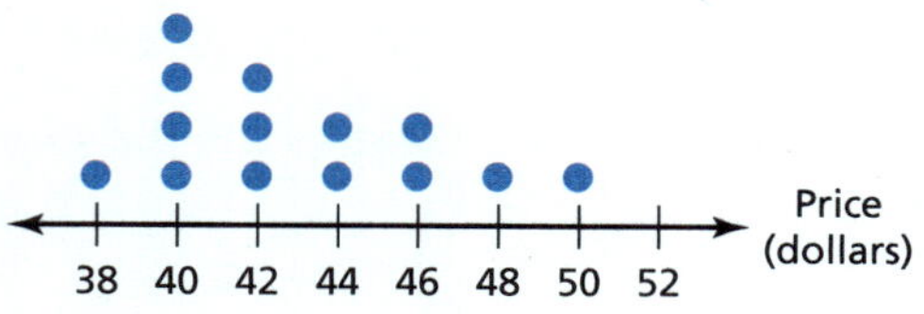

10. **Time Volunteering**

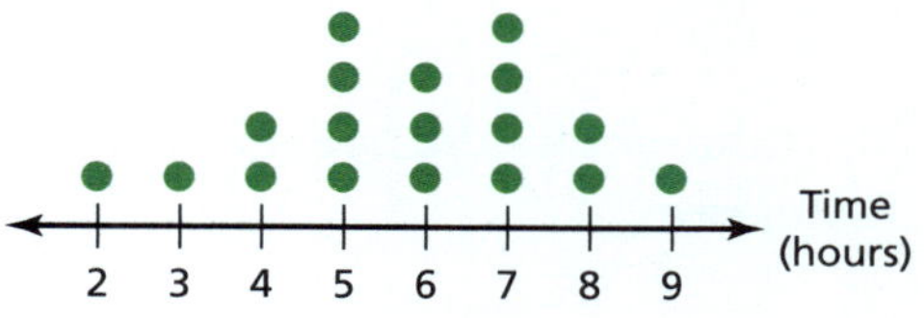

11.

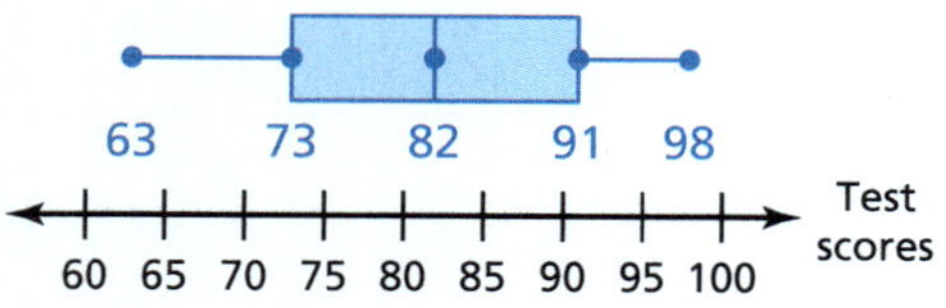

12. **Numbers of Exercises Completed**

Stem	Leaf
0	3 5 5 5 7 8 9
1	0 2 3 3 8 8 8 9
2	0 1 2 3 3 3 4 7 7 7

Key: 1 | 1 = 11

▶ **13. DESCRIBING DATA SETS** Describe the centers and the variations of the data sets in (a) Exercise 10 and (b) Exercise 11. (See Example 2.)

14. MODELING REAL LIFE The box-and-whisker plot represents the number of books each class in a school reads during the year. What are the most appropriate measures to describe the center and the variation? Explain your reasoning.

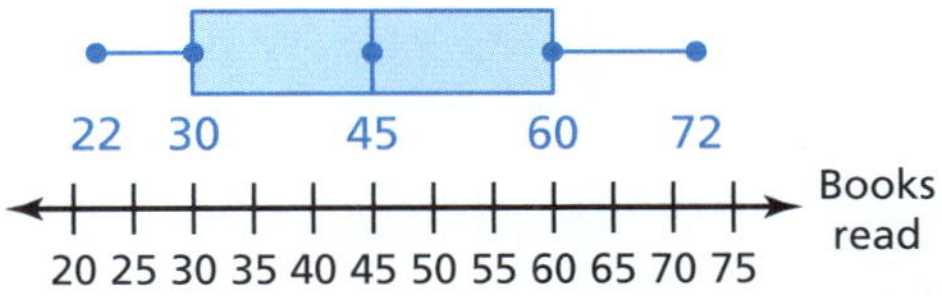

15. OPEN-ENDED Describe a real-life situation where the median and the interquartile range are likely the best measures of center and variation to describe the data. Explain your reasoning.

16. MODELING REAL LIFE The stem-and-leaf plot shows the vertical jump heights (in inches) of several professional athletes. Describe the center and the variation of the data set.

Vertical Jump Heights

Stem	Leaf
2	4 6 7 8
	8 9 9 9
3	0 0 0 0 1
	1 1 2 2 3
	4 4 5 7

Key: 2|6 = 26 inches

Pile A	
5	7
4	3
6	4
5	6
4	6
3	5
5	7

Pile B	
7	5
5	2
3	8
5	6
5	4
0	8
2	10

17. PROBLEM SOLVING You play a board game in which you draw from one of two piles of cards. Each card has a number that says how many spaces you will move your piece forward on the game board. The tables show the numbers on the cards in each pile. From which pile would you choose? Explain your reasoning. (See Example 3.)

18. REASONING You are monitoring sale prices of a bicycle at different stores, trying to find the best price. Is it better to use the range or IQR of prices? Explain your reasoning.

4 MTR

19. MAKING AN ARGUMENT The histogram shows the numbers of words that several students can form in 1 minute using the letters P, S, E, D, A. Your friend says that the median and IQR are appropriate measures of center and variation because the value in the 19–21 category must be an outlier. Is your friend correct? Explain.

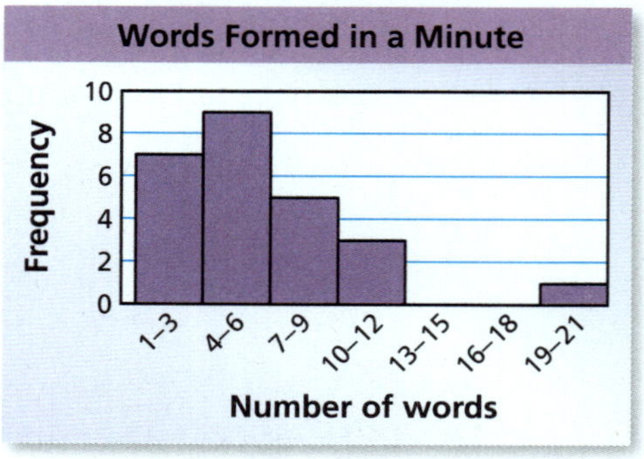

20. REASONING A bag contains 20 vouchers that can be redeemed for different numbers of tokens at an arcade, as shown in the dot plot.

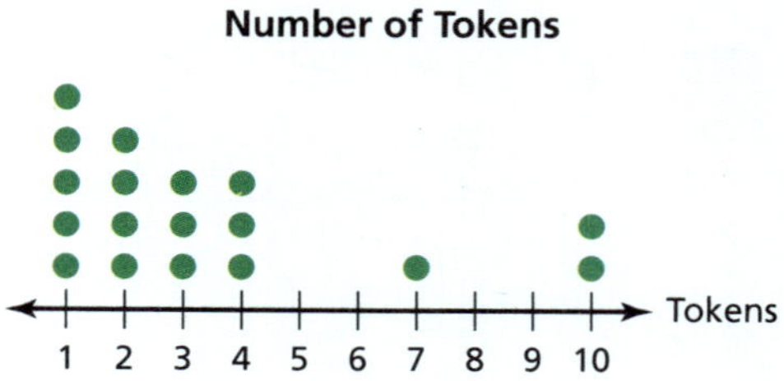

a. Find the most appropriate measure to describe the center of the data set.

b. You randomly select a voucher from the bag. How many tokens are you most likely to receive? Explain.

c. Are your answers in parts (a) and (b) the same? Explain why or why not.

11 Connecting Concepts

5 MTR *Mathematicians who use patterns and structure to help understand and connect mathematical concepts relate previously learned concepts to new concepts.*

1 MTR **1. ANALYZE A PROBLEM** Six friends play a carnival game in which a person throws darts at balloons. Each person throws the same number of darts and then records the portion of the balloons that pop. Find and interpret the mean and median of the data.

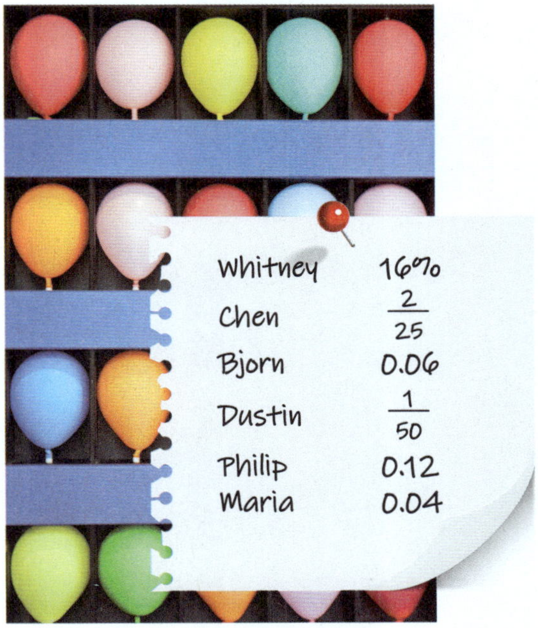

Using the Problem-Solving Plan

Understand the problem. You know that each person throws the same number of darts. You are given the portion of balloons popped by each person as a fraction, a decimal, or a percent.

Make a plan. First, write each fraction and each decimal as a percent. Next, order the percents from least to greatest. Then find and interpret the mean and median of the data.

Solve and check. Use the plan to solve the problem. Then check your solution.

2. A set of 20 data values is described below. Sketch a histogram that could represent the data set. Explain.

- least value: 10
- third quartile: 34
- first quartile: 25
- greatest value: 48
- mean: 29

SHIPPING RATES

Dimensions	Price
5 × 5 × 4	$6.80
8 × 5 × 3	$8.30
9 × 8 × 5	$9.75
10 × 6 × 6	$10.75
10 × 10 × 4	$10.75
8 × 7 × 5	$10.75
12 × 10 × 3	$11.25
15 × 10 × 3	$12.25
12 × 12 × 5	$17.40

2 MTR **3. MAKE A CONNECTION** The chart shows the dimensions (in inches) of several flat-rate shipping boxes. Each box is in the shape of a rectangular prism. Describe the distribution of the volumes of the boxes. Then find the most appropriate measure to describe the center of the volumes.

STEAM Performance Task

Which Measure of Center Is Best: Mean, Median, or Mode?

GO DIGITAL

At the beginning of this chapter, you watched a STEAM Video called "Daylight in the Big City." You are now ready to complete the STEAM Performance Task, where you will decide which measure of center best represents the amount of daylight in different cities. Be sure to use mathematical thinking and reasoning, and the problem-solving plan as you work through the performance task.

11

Chapter Review WITH CalcChat®

Review Vocabulary

Write the definition and give an example of each vocabulary term.

statistics, *p. 887*	mode, *p. 907*	interquartile range, *p. 921*
statistical question, *p. 887*	measure of variation, *p. 921*	frequency table, *p. 934*
mean, *p. 897*	range, *p. 921*	frequency, *p. 934*
outlier, *p. 899*	quartiles, *p. 921*	histogram, *p. 935*
measure of center, *p. 907*	first quartile, *p. 921*	box-and-whisker plot, *p. 959*
median, *p. 907*	third quartile, *p. 921*	five-number summary, *p. 959*

GO DIGITAL

Graphic Organizers

You can use a **Definition and Example Chart** to organize information about a concept. Here is an example of a Definition and Example Chart for the vocabulary term ***statistical question***.

Statistical Question: a question for which you do not expect to get a single answer

Example

What are the heights of sixth-grade students?

Example

What are the ages of people in the auditorium?

Example

What are the numbers of letters in the first names of sixth-grade students?

Choose and complete a graphic organizer to help you study the concept.

1. mean
2. median
3. mode
4. range
5. quartiles
6. frequency table
7. histogram
8. five-number summary

Chapter Learning Target: Understand statistical measures and data displays.

Chapter Success Criteria:
- ◆ I can construct a data set or data display.
- ◆ I can interpret data.
- ■ I can find and interpret the measures of center and the measures of variation for a data set.
- ■ I can compare data sets using different data displays.

◆ Surface
■ Deep

Rate your understanding after each section.

1	2	3	4
I don't understand yet.	I can do it with help.	I can do it on my own.	I can teach someone else.

11.1 Introduction to Statistics *(pp. 885–894)*

Learning Target: Identify statistical questions and use data to answer statistical questions.

Determine whether the question is a statistical question. Explain.

1. How many positive integers are less than 20?

2. In what month were the students in a sixth-grade class born?

3. The dot plot shows the number of televisions owned by each family on a city block.

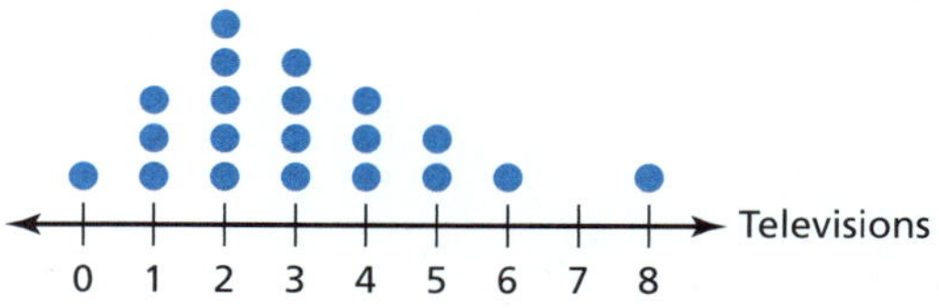

a. Find and interpret the number of data values on the dot plot.

b. Write a statistical question that you can answer using the dot plot. Then answer the question.

GO DIGITAL

Display the data in a dot plot. Identify any clusters, peaks, or gaps in the data.

4.

Distances (feet)			
56	55	56	57
58	54	51	55
51	56	49	56

5.

Weights (pounds)				
83	88	89	90	89
91	89	84	90	92
90	88	89	83	88

6. You conduct a survey to answer, "What is the heart rate of a typical sixth-grade student?" The table shows the results. Use the distribution of the data to answer the question.

Heart Rates (beats per minute)					
68	69	74	68	64	67
66	70	67	68	74	69
68	69	74	70	68	70
70	68	68	69	67	64
69	65	66	68	69	67

11.2 Mean (pp. 895–904)

Learning Target: Find and interpret the mean of a data set.

7. Find the mean of the data.

Boiling Points (°F)	Arsenic	Cadmium	Mercury	Phosphorus	Potassium	Sodium
	1117	1409	675	536	1398	1621

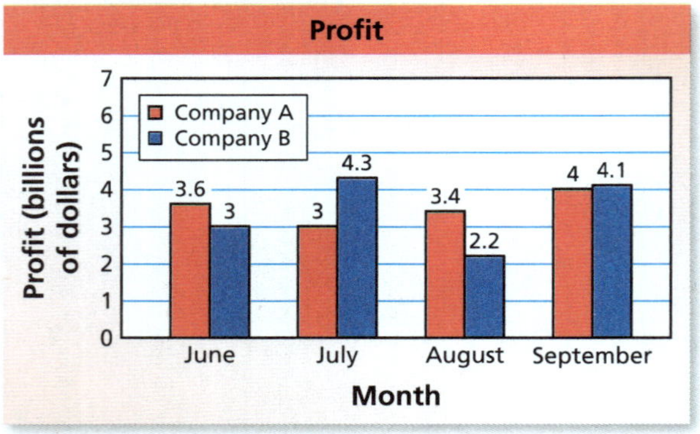

8. The double bar graph shows the monthly profit for two toy companies over a four-month period. Compare the mean monthly profits.

9. The table shows the test scores for a class of sixth-grade students. Describe how the outlier affects the mean. Then use the data to answer the statistical question "What is the typical test score for a student in the class?"

Test Scores		
91	81	100
72	83	70
85	97	75
80	90	36

11.3 Measures of Center (pp. 905–918)

Learning Target: Find and interpret the median and mode of a data set.

Find the median and mode of the data.

10. 8, 8, 6, 8, 4, 5, 6

11. 24, 74, 61, 29, 38, 27, 68, 54

12. 6.7, 12.31, 7.83, 1.36, 6.71, 11.29

13. $\frac{2}{9}, \frac{3}{4}, \frac{5}{7}, \frac{3}{8}, \frac{2}{11}, \frac{3}{4}, \frac{5}{6}$

14. Find the mean, median, and mode of the data set {67, 52, 50, 99, 66, 50, 57} with and without the outlier. Which measure does the outlier affect the most?

15. The table shows the lengths of several movies. Which measure of center best represents the data? Explain your reasoning.

Movie Lengths (minutes)		
91	112	126
142	113	112
92	144	148

16. Give an example of a data set that does not have a median. Explain why the data set does not have a median.

11.4 Measures of Variation (pp. 919–932)

Learning Target: Find and interpret the range and interquartile range of a data set.

Find the range of the data.

17. 45, 76, 98, 21, 52, 39

18. 95, 63, 52, 8, 93, 16, 42, 37, 62

19. 67, 45, 87, 23, 90, 12, 8, 16

20. 19, 6, 82, 8, 14, 84, 43

Find the interquartile range of the data.

21. 28, 46, 25, 76, 18, 25, 47, 83, 44

22. 14, 25, 97, 55, 66, 28, 92, 38, 94

23. The table shows the weights of several adult emperor penguins. Find and interpret the range and interquartile range of the data. Then determine whether there are any outliers.

Weights (kilograms)	25	36	33.5	30.75	24
	27	23.5	31.25	32	29.25

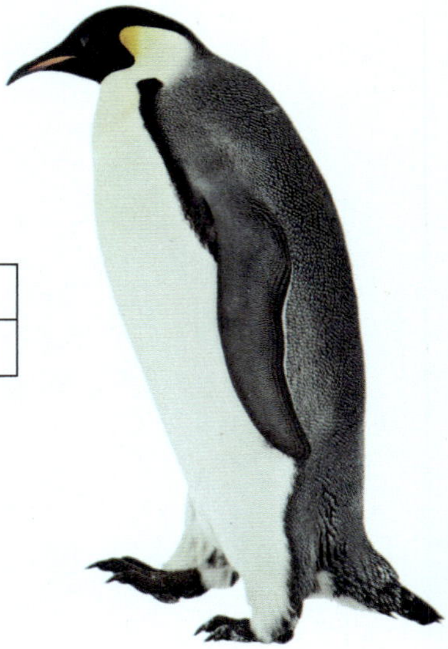

24. Two data sets have the same interquartile range. Can you assume that the ranges of the two data sets are about the same? Give an example to justify your answer.

Each value in the data set is changed as described. How does the change affect the range and IQR?

25. {9, 12, 16, 14, 9, 11, 10, 12}, decrease of 6

26. {45, 24, 28, 49, 28, 57, 19, 35, 45}, increase of 18

27. {4, 11, 5, 2, 9, 3, 2, 4, 14}, multiplied by 4

28. {72, 152, 104, 80, 120, 96, 104, 176}, multiplied by $\frac{1}{8}$

Identify any outliers in the data set. Explain.

29. {1, 9, 9, 11, 16, 19, 25}

30. {13, 21, 25, 25, 28, 31, 40}

31. {4.6, 5.2, 16.24, 18.03, 22.1, 22.1}

32. {0.54, 0.6, 2.41, 3.87, 3.9, 4.3, 12.34}

33. Create a set of data with 6 values that has a mean of 8, a median of 12, and a range of 16.

34. Create a set of data with 9 values that has a mean of 25, a median of 15, and a range of 40.

11.5 Histograms (pp. 933–946)

Learning Target: Display and interpret data in histograms.

Display the data in a histogram. Identify any gaps, clusters, and outliers.

35.

Heights of Gymnasts	
Heights (inches)	**Frequency**
50–54	6
55–59	8
60–64	0
65–69	3

36.

Minutes Studied	
Minutes	**Frequency**
0–19	5
20–39	9
40–59	12
60–79	3

37. The histogram shows the number of crafts each member of a crafts club made for a fundraiser.

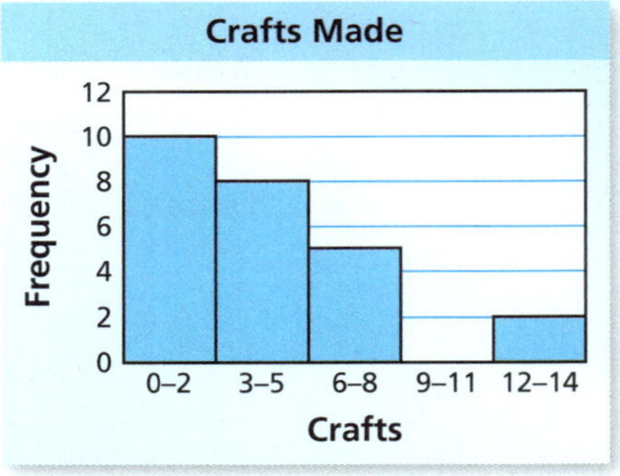

a. Which interval contains the most data values?

b. How many members made at least 6 crafts?

c. Can you use the histogram to determine the total number of crafts made? Explain.

11.6 Shapes of Distributions *(pp. 947–956)*

Learning Target: Describe and compare shapes of distributions.

38. Describe the shape of the distribution.

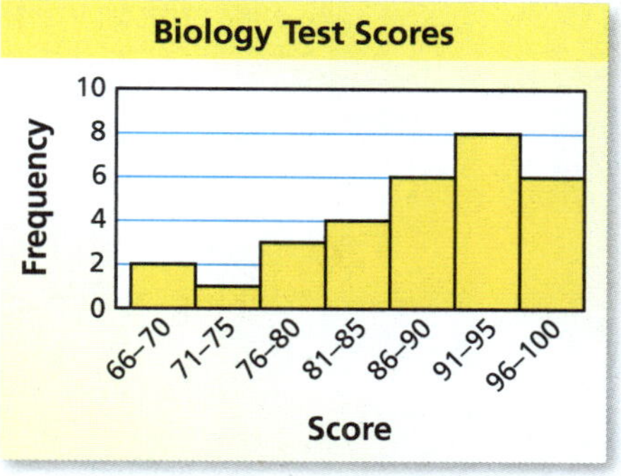

39. The frequency table shows the math test scores for the same class of students as Exercise 38. Display the data in a histogram. Which test has higher scores?

Score	66–70	71–75	76–80	81–85	86–90	91–95	96–100
Frequency	1	5	10	8	2	2	2

40. The table shows the numbers of neutrons for several elements in the nonmetal group of the periodic table. Make a histogram of the data, starting with the interval 0–9. Describe the shape of the distribution.

Number of Neutrons			
0	45	16	8
7	6	16	

11.7 Box-and-Whisker Plots *(pp. 957–970)*

Learning Target: Display and interpret data in box-and-whisker plots.

Make a box-and-whisker plot for the data.

41. Ages of volunteers at a hospital:
14, 17, 20, 16, 17, 14, 21, 18

42. Masses (in kilograms) of lions:
120, 200, 180, 150, 200, 200, 230, 160

Identify the five-number summary.

43.

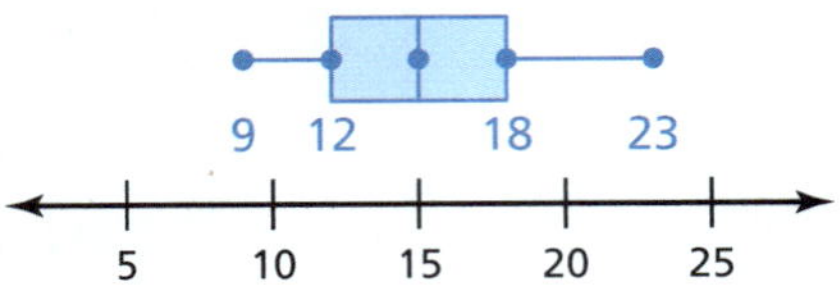

44.

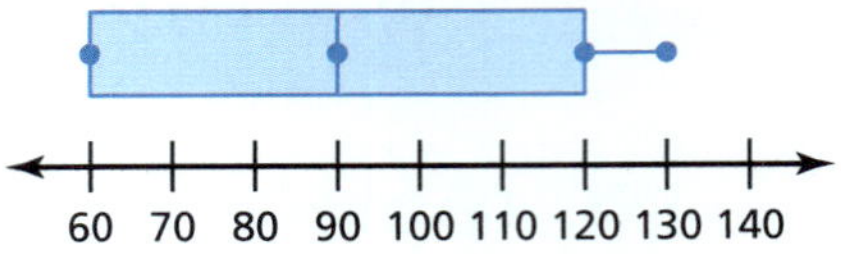

45. The box-and-whisker plot represents the lengths (in minutes) of movies being shown at a theater.

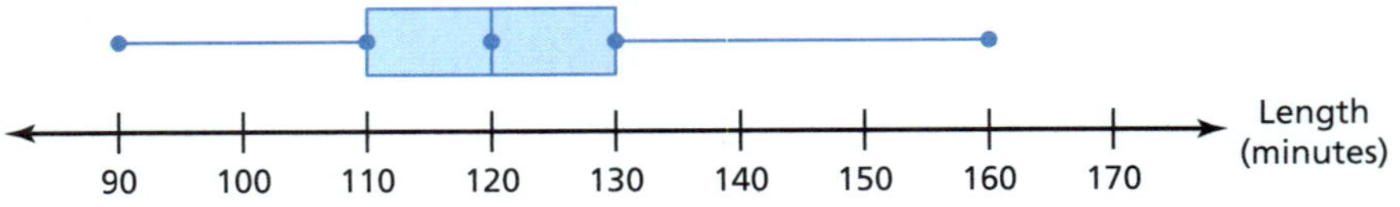

a. What percent of the movies are no longer than 120 minutes?

b. Is there more variability in the movie lengths longer than 130 minutes or shorter than 110 minutes? Explain.

c. Find and interpret the interquartile range of the data.

11.8 Choosing Appropriate Measures *(pp. 971–980)*

Learning Target: Determine which measures of center and variation are appropriate to describe a data set.

Choose appropriate measures to describe the center and the variation.

46.

Stem	Leaf
1	1 3 5 7 7 7 9 9
2	0 0 3 4 4 6 6 8 8
3	2 2 5 5 7

Key: 2 | 4 = 24 meters per minute

47.

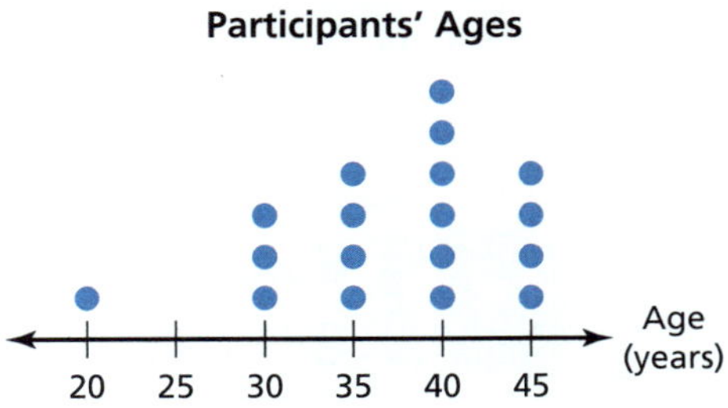

48. Describe the center and the variation of the data set in Exercise 47.

11 Practice Test WITH CalcChat®

Find the mean, median, mode, range, and interquartile range of the data.

1. 5, 6, 4, 24, 10, 6, 9, 8

2. 32, 58, 19, 36, 44, 57, 11, 26, 74

Choose appropriate measures to describe the center and the variation.

3.

Stem	Leaf
6	0 2 3 3 3 7 7
7	
8	1 1 4 6
9	0 0 3 5 5 5 8 9 9

Key: 8 | 1 = 81 ounces

4.

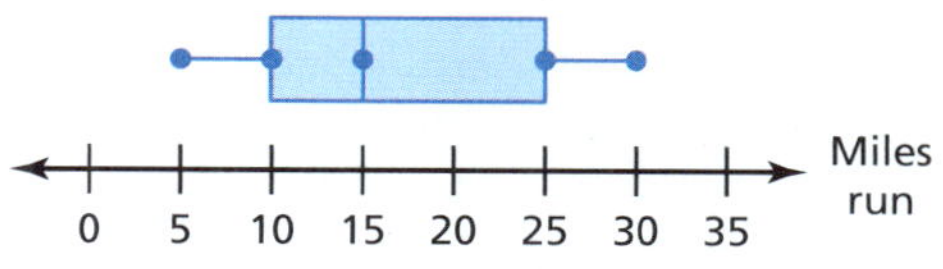

5. The table shows the lengths of several bearded dragons captured for a study. Find the mean, median, and mode of the data in centimeters and in millimeters. How does converting to millimeters affect the mean, median, and mode?

Lengths (centimeters)					
36	58	42	43	55	57
52	46	41	52	56	50

GO DIGITAL

6. The frequency table shows the number of hours per week people watched television.

Television Watched per Week	
Hours	**Frequency**
0–9	14
10–19	26
20–29	13
30–39	8

a. Display the data in a histogram. Identify any gaps, clusters, and outliers.

b. What is the shape of the distribution?

c. How many people watched less than 20 hours of television per week?

7. Make a box-and-whisker plot for the lengths (in inches) of fish in a pond: {12, 13, 7, 8, 14, 6, 13, 10}.

8. The double box-and-whisker plot compares the battery lives (in hours) of two brands of cell phones.

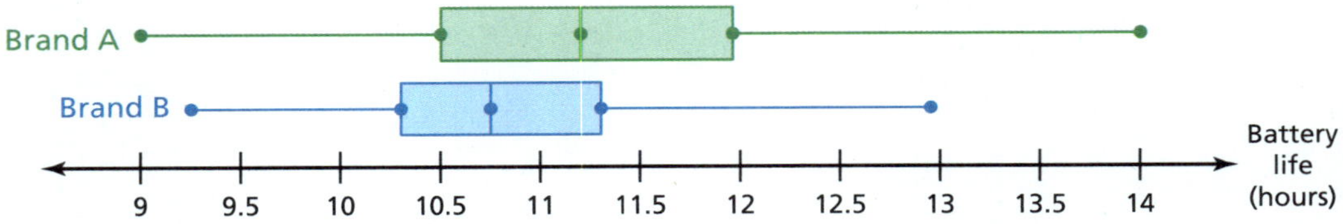

a. What is the range of the upper 75% of battery life for each brand of cell phone?

b. Which brand of cell phone typically has a longer battery life? Explain.

c. In the box-and-whisker plot, there are 190 cell phones of Brand A that have at most 10.5 hours of battery life. About how many cell phones are represented in the box-and-whisker plot for Brand A?

11 Review & Refresh WITH CalcChat®
while Building Fluency

Give the coordinates of the point shown in the coordinate plane. In what quadrant is the point?

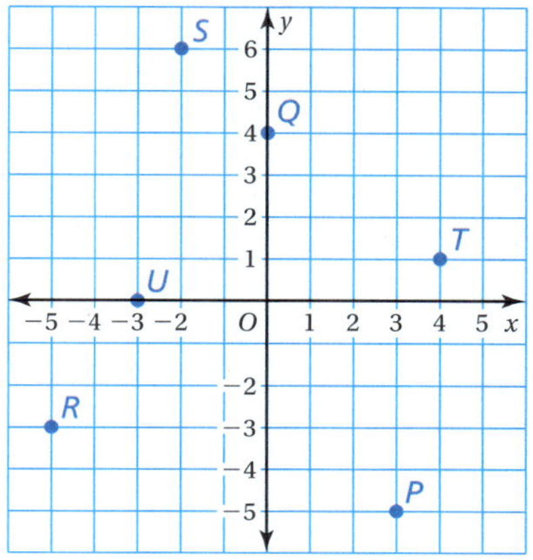

1. P

2. Q

3. R

4. S

5. T

6. U

Use the Distributive Property to evaluate the expression. Explain each step.

7. $11(2 + 9)$

8. $5(-12 + 7)$

9. $(15 - 5) \times 4$

Compare the numbers using <, >, or =.

10. $56.32 \ \square \ 56.30$

11. $0.032 \ \square \ 0.040$

12. $-9.065 \ \square \ -9.066$

13. What is the sum of the values of points K and L? Explain.

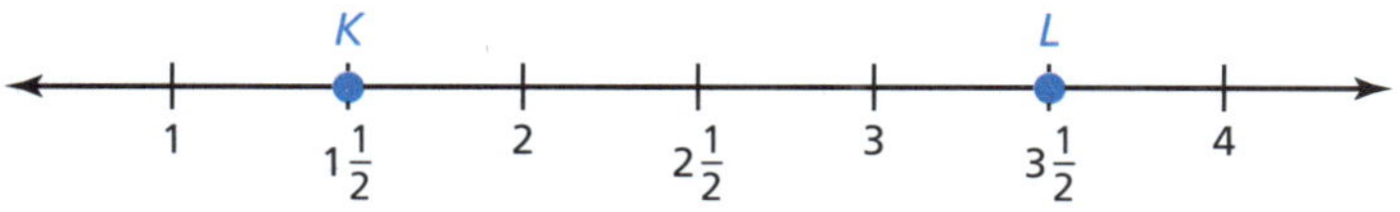

GO DIGITAL

Find the area of the triangle.

14.

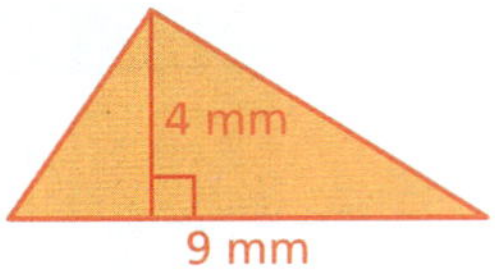

15.

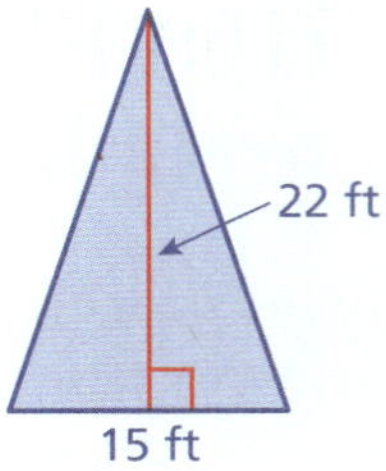

Divide.

16. $192 \div 3$

17. $504 \div 42$

18. $1323 \div 63$

19. You are at the mall and want to see a movie starting in 50 minutes. Do you arrive at the movie on time when you spend 20 minutes getting dinner, 15 minutes playing arcade games, and 8 minutes standing in line? Explain.

20. A beach has the sign shown. If the waves at the beach are 75 inches tall, can you swim at the beach? Explain.

Find the GCF of the numbers.

21. 8, 16, 32

22. 14, 35, 42

23. 64, 144, 304

11 B.E.S.T. Test Prep WITH CalcChat®
Cumulative Practice

1. Which statement can be represented by a negative integer?

   Ⓐ The temperature rises 15 degrees.

   Ⓑ A hot-air balloon ascends 450 yards.

   Ⓒ You earn $50 completing chores.

   Ⓓ A submarine submerges 260 feet.

Test-Taking Strategy
Use Intelligent Guessing

What is the mean length of these hyena fangs: 4 in., 3 in., 3 in., 4 in., 5 in., 5 in.?

Ⓐ 6 in. Ⓑ $\frac{1}{3}$ ft Ⓒ 2 in. Ⓓ 5 in.

2. What is the height h (in inches) of the prism?

GRIDDED RESPONSE

Volume = 5880 in.^{3}

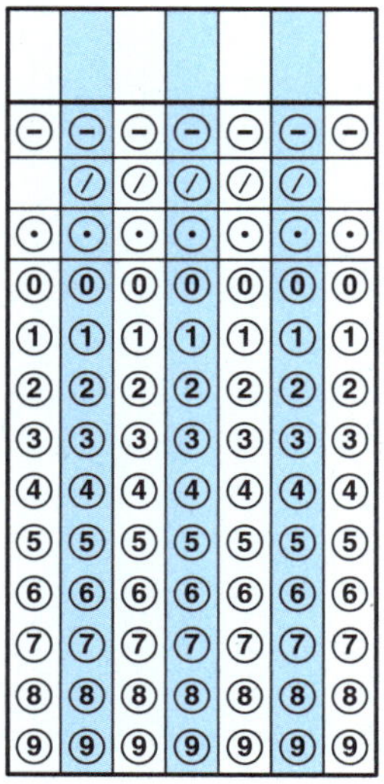

3. Which of the following are solutions to the inequality $4x < 48$?

   Ⓐ $x = 4.98$

   Ⓑ $x = \frac{11}{2}$

   Ⓒ $x = 6\frac{2}{3}$

   Ⓓ $x = 9$

   Ⓔ $x = 12.01$

   Ⓕ $x = 12\frac{5}{6}$

GO DIGITAL

4. GRIDDED RESPONSE

At the end of the school year, your teacher counted the number of absences for each student. The results are shown in the histogram. How many students had fewer than 10 absences?

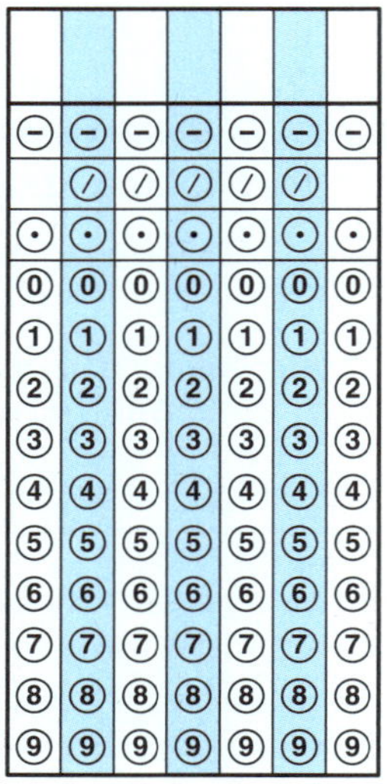

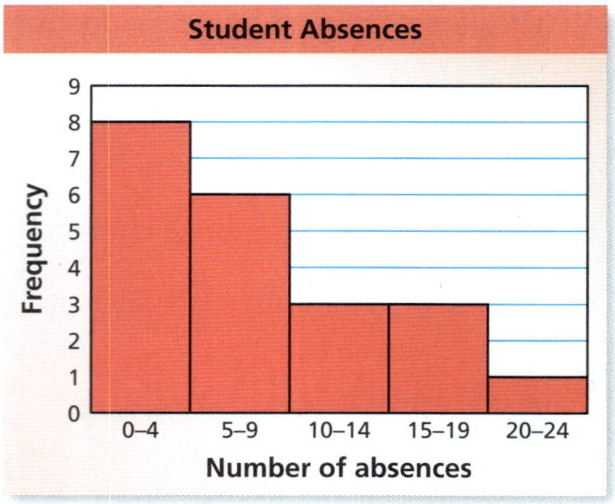

5. On Wednesday, a town received 17 millimeters of rain. This was x millimeters more rain than the town received on Tuesday. Which expression represents the amount of rain, in millimeters, the town received on Tuesday?

Ⓐ $17x$

Ⓑ $17 - x$

Ⓒ $x + 17$

Ⓓ $x - 17$

6. GRIDDED RESPONSE

What is the value of the expression?

$$58.06 \times 12.22$$

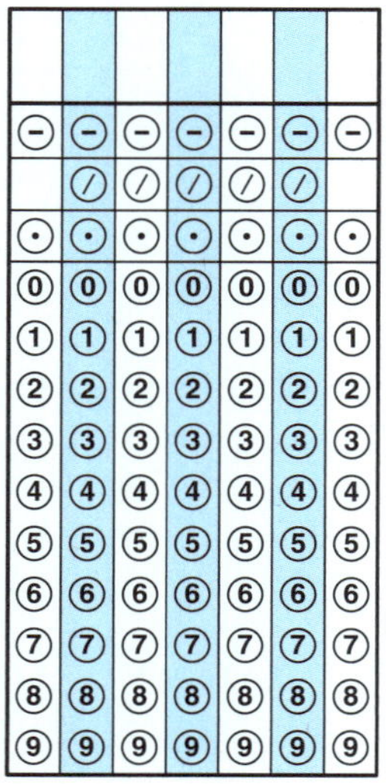

7. The number of hours that each of six students spent reading last week is shown in the bar graph.

For the data in the bar graph, which measure is the *least*?

Ⓐ mean

Ⓑ median

Ⓒ mode

Ⓓ range

8. A family wants to buy tickets to a theme park. There are separate ticket prices for adults and children.

How much does it cost to buy 2 adult tickets and 3 child tickets?

Ⓐ $50

Ⓑ $120

Ⓒ $130

Ⓓ $600

12

Probability and Populations

Chapter Learning Target:
Understand probability and representing data.

Chapter Success Criteria:
- ◆ I can find the sample space of an experiment.
- ◆ I can find experimental and theoretical probabilities.
- ■ I can use an unbiased sample to make a conclusion about a population.
- ■ I can compare populations.

◆ Surface ■ Deep

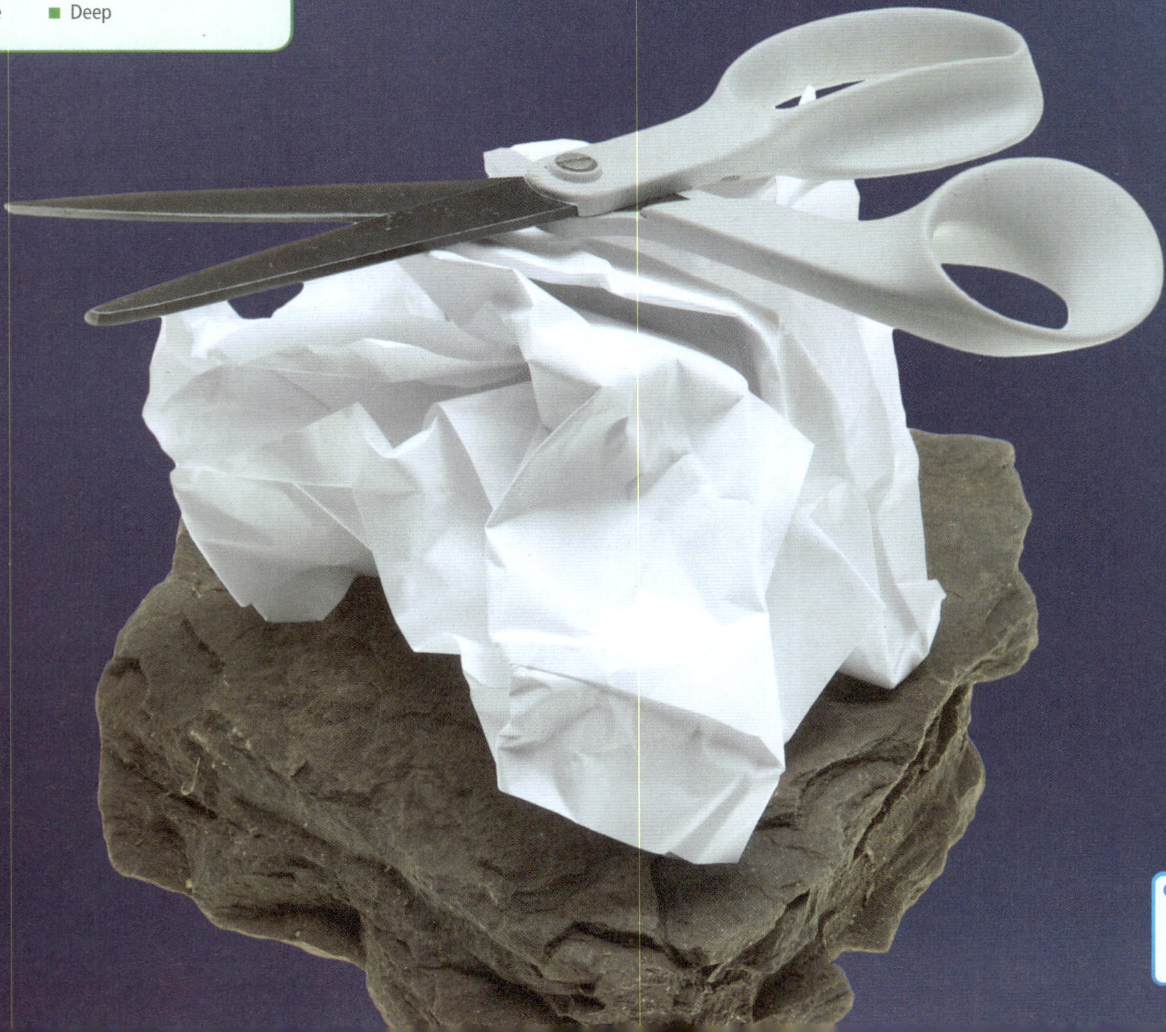

GO DIGITAL

MTR 7 *Mathematicians who apply mathematics to real-world contexts connect mathematical concepts to everyday experiences.*

STEAM Video Massively Multiplayer Rock Paper Scissors

1. The table shows the ways that you can win, lose, or tie in Rock Paper Scissors. You and your opponent throw the signs for rock, paper, or scissors at random. What portion of the time do you expect to win? lose? tie?

		Your Throw		
		Rock	**Paper**	**Scissors**
Opponent's Throw	**Rock**	tie	win	lose
	Paper	lose	tie	win
	Scissors	win	lose	tie

In the video, Tory and Enid play a game of Rock Paper Scissors with multiple people.

MTR 4 **2. DISCUSS MATHEMATICAL THINKING** You play Rock Paper Scissors 15 times. About how many times do you expect to win? Explain your reasoning.

STEAM Performance Task Fair and Unfair Carnival Games

After completing this chapter, you will be able to use the concepts you learned to answer the questions in the *STEAM Performance Task.*

You will be given information about a version of Rock Paper Scissors used at a carnival. Then you will be asked to design your own "unfair" carnival game using a spinner or a number cube, and test your game with a classmate.

GO DIGITAL

Getting Ready for Chapter 12 WITH CalcChat®

Chapter Exploration

1 MTR **Actively Participate in Effortful Learning Collectively**
Work with a partner to prepare for concepts in this chapter.

1. Play Rock Paper Scissors 30 times. Tally your results in the table.

GAME RULES
Rock **breaks** scissors.
Paper **covers** rock.
Scissors **cut** paper.

		Player A		
		 Rock	 Paper	 Scissors
Player B	 Rock			
	 Paper			
	 Scissors			

2. How many possible results are there?

3. Of the possible results, in how many ways can Player A win? In how many ways can Player B win? In how many ways can there be a tie?

2 MTR **4. MAKE A CONNECTION** Is one of the players more likely to win than the other player? Explain your reasoning.

Vocabulary

The following terms are defined in this chapter. Think about what each might mean and record your thoughts.

sample space	experimental probability	sample
probability	theoretical probability	unbiased sample
relative frequency	population	biased sample

GO DIGITAL

12.1 Probability

Learning Target: Understand how the probability of an event indicates its likelihood.

Success Criteria:
- I can find the sample space of an experiment.
- I can use probability and relative frequency to describe the likelihood of an event.
- I can use relative frequency to make predictions.

Data Analysis and Probability
MA.7.DP.2.1 Determine the sample space for a simple experiment.
MA.7.DP.2.2 Given the probability of a chance event, interpret the likelihood of it occurring. Compare the probabilities of chance events.

Exploration 1 Determining Likelihood

Work with a partner. Use the spinners shown.

Spinner 1

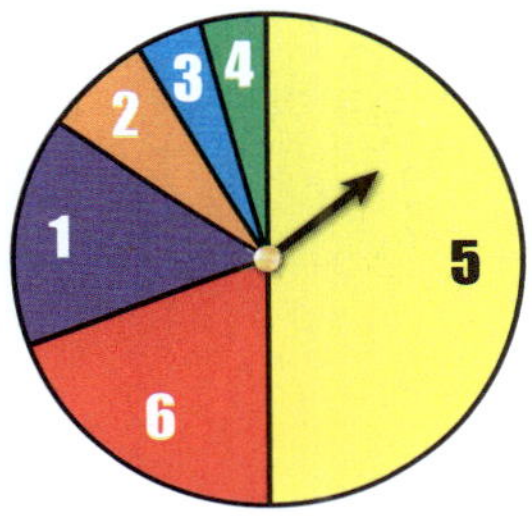

Spinner 2

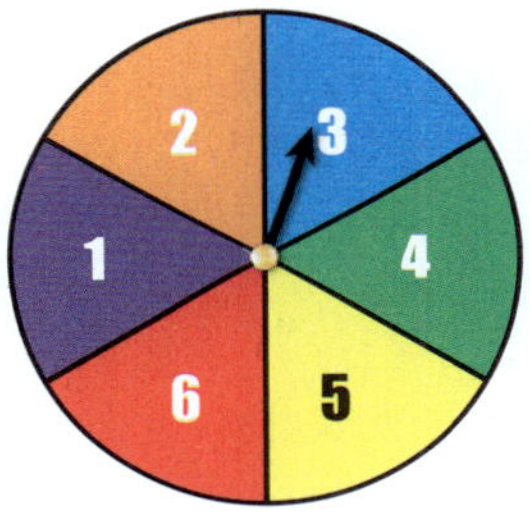

a. For each spinner, determine which numbers you are more likely to spin and which numbers you are less likely to spin. Explain your reasoning.

GO DIGITAL

b. Spin each spinner 20 times and record your results in two tables. Do the data support your answers in part (a)? Explain why or why not.

Spinner 1	
Number	**Frequency**
1	
2	
3	
4	
5	
6	

Spinner 2	
Number	**Frequency**
1	
2	
3	
4	
5	
6	

c. How can you use percents to describe the likelihood of spinning each number? Explain.

7 MTR **APPLY MATHEMATICS**

You are playing a board game and can choose to spin either spinner. Which spinner should you choose when you want to spin an even number? Explain your reasoning.

12.1 Lesson

Key Vocabulary

experiment, *p. 1005*
sample space, *p. 1005*
outcomes, *p. 1005*
event, *p. 1005*
favorable outcomes, *p. 1005*
probability, *p. 1006*
relative frequency, *p. 1007*

Key Idea

Outcomes and Events

An **experiment** is an investigation or a procedure that has varying results. The **sample space** of an experiment is the set of all possible results, or **outcomes**, of the experiment. A collection of one or more outcomes is an **event**. The outcomes of a specific event are called **favorable outcomes**.

For example, randomly choosing a marble from a group of marbles is an experiment. Each marble in the group is an outcome. Selecting a green marble from the group is an event.

Sample Space

Event: Choosing a green marble
Number of favorable outcomes: 2

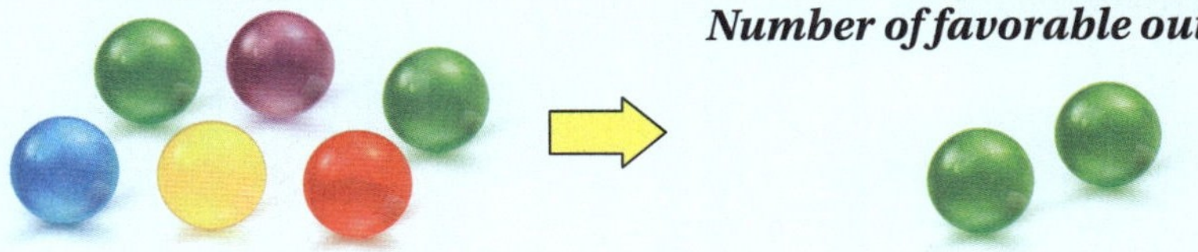

Example 1 Identifying Outcomes

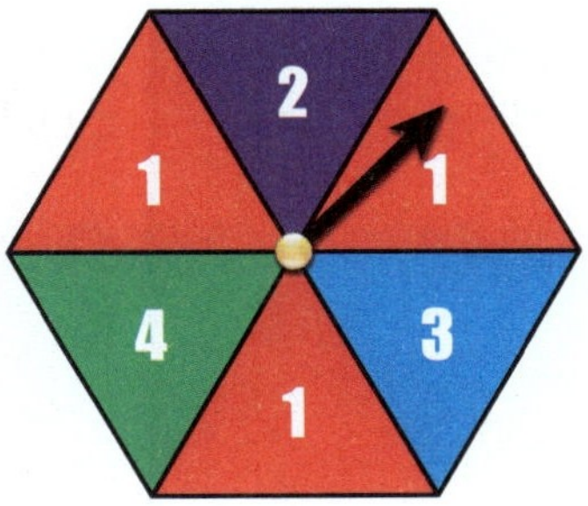

You spin the spinner.

a. Find the sample space. How many possible outcomes are there?

The sample space is the set of all possible outcomes of spinning the spinner, {1, 2, 1, 3, 1, 4}. So, there are six possible outcomes.

b. What are the favorable outcomes of spinning an even number?

even	*not* even
2, 4	1, 1, 3, 1

The favorable outcomes of spinning an even number are 2 and 4.

c. In how many ways can spinning a number less than 2 occur?

less than 2	*not* less than 2
1, 1, 1	2, 3, 4

The possible outcomes of spinning a number less than 2 are 1, 1, and 1. So, spinning a number less than 2 can occur in 3 ways.

GO DIGITAL

Try It

1. You randomly choose one of the tiles shown from a hat.

   a. Find the sample space. How many possible outcomes are there?

   b. What are the favorable outcomes of choosing a vowel?

   c. In how many ways can choosing a consonant occur?

Key Idea

Probability

The **probability** of an event is a number that represents the likelihood that the event will occur. Probabilities are between 0 and 1, including 0 and 1. The diagram relates likelihoods (above the diagram) and probabilities (below the diagram).

Probabilities can be written as fractions, decimals, or percents.

Example 2 Describing Likelihood

There is an 80% chance of rain, a 50% chance of thunderstorms, a 15% chance of hail, and a 20% chance of high winds tomorrow.

a. Describe the likelihood that there is rain tomorrow.

The probability of rain tomorrow is 80%.

▶ Because 80% is close to 75%, it is *likely* that there will be rain tomorrow.

b. Describe the likelihood that there are thunderstorms tomorrow.

The probability of thunderstorms tomorrow is 50%.

▶ Because the probability is 50%, thunderstorms are *equally likely to happen or not happen.*

c. Which is more likely tomorrow, hail or high winds?

The probability of hail tomorrow is 15%, and the probability of high winds tomorrow is 20%.

▶ Both probabilities are between 0% and 25%, so each event is *unlikely*. However, because 20% > 15%, high winds are slightly more likely than hail tomorrow.

Try It **Describe the likelihood of the event given its probability.**

2. The probability that you land a jump on a snowboard is $\frac{1}{10}$.

3. There is a 100% chance that the temperature will be less than 120°F tomorrow.

Relative frequency is a measure of probability.

When you conduct an experiment, the **relative frequency** of an event is the fraction or percent of the time that the event occurs.

$$\text{relative frequency} = \frac{\text{number of times the event occurs}}{\text{total number of times you conduct the experiment}}$$

GO DIGITAL

Example 3 Using Relative Frequencies

You flip a bottle and record the number of times it lands upright and the number of times it lands on its side. Describe the likelihood that the bottle lands upright on your next flip.

Upright	II
Side	𝍸 𝍸 𝍸 𝍸 III

The bottle landed upright 2 times in a total of 25 flips.

$$\text{relative frequency} = \frac{\text{number of times the event occurs}}{\text{total number of times you conduct the experiment}}$$

$$= \frac{2}{25}$$

The bottle landed upright 2 times.

There was a total of 25 flips.

▶ The relative frequency is $\frac{2}{25}$, or 8%. So, it is unlikely that the bottle lands upright.

Try It

Shots Made	𝍸 IIII
Shots Missed	𝍸 I

4. You attempt three-point shots on a basketball court and record the number of made and missed shots. Describe the likelihood of each event.

a. You make your next shot.

b. You miss your next shot.

In-Class Practice

1 I don't understand yet. 2 I can do it with help. 3 I can do it on my own. 4 I can teach someone else.

5. IDENTIFYING OUTCOMES You roll a number cube. Find the sample space.

GO DIGITAL

In-Class Practice

1 I don't understand yet. 2 I can do it with help. 3 I can do it on my own. 4 I can teach someone else.

Red	卌 卌 卌 卌 |
Blue	卌 卌 卌 卌 |

6. USING RELATIVE FREQUENCIES A bag contains only red marbles and blue marbles. You randomly draw a marble from the bag and replace it. The table shows the results of repeating this experiment. Find the likelihood of each event.

a. The next marble you choose is red.

b. The next marble you choose is neither red nor blue.

Example 4 Modeling Real Life

7 MTR

Each turn in a game, you randomly draw a token from a bag and replace it. The table shows the number of times you draw each type of token. How many times can you expect to draw a positive point value in 35 turns?

Token	Frequency
+3 points	卌 卌
+1 point	卌 ||
−2 points	|||

Understand the problem.

You are given the number of times that you draw each type of token from a bag. You are asked to determine the number of times you can expect to draw a positive point value in 35 turns.

Make a plan.

Find the relative frequency of drawing a positive point value. Then use the relative frequency and the percent equation to answer the question.

Solve and check.

The favorable outcomes of drawing a positive point value are drawing a +3 token or a +1 token. So, the relative frequency of drawing a positive point value is $\frac{10+7}{20} = \frac{17}{20}$, or 85%.

To determine the number of times you can expect to draw a positive point value, answer the question "What is 85% of 35?"

$a = p\% \cdot w$	Write percent equation.
$= 0.85 \cdot 35$	Substitute 0.85 for $p\%$ and 35 for w.
$= 29.75$	Multiply.

▶ You can expect to draw a positive point value about 30 times.

Check Reasonableness
The table shows that positive point values are drawn 17 of 20 times. So, in 35 turns, you can expect to draw positive point values less than $17 \times 2 = 34$ times. ✓

GO DIGITAL

In-Class Practice

1 I don't understand yet. **2** I can do it with help. **3** I can do it on my own. **4** I can teach someone else.

Pop Quiz	No Pop Quiz
III	~~IIII~~ ~~IIII~~ II

7. The table shows the number of days you have a pop quiz and the number of days you do not have a pop quiz in three weeks of school. How many days can you expect to have a pop quiz during a 180-day school year? Explain.

8. In a football game, the teams pass the ball on 40% of the plays. Of the passes thrown, greater than 75% are completed. You watch the film of a randomly chosen play. Describe the likelihood that the play results in a complete pass. Explain your reasoning.

12.1 Practice WITH CalcChat® AND CalcView®

Review & Refresh

Find and interpret the interquartile range of the data.

1. 3, 8, 11, 6, 5, 9, 6

2. 9, 12, 15, 12, 19, 10

3. 25, 21, 26, 25, 26, 24, 35, 31

4. 1.9, 2.6, 2.4, 1.7, 2.3, 2.4, 1.8

Write the indicated ratio. Then find and interpret the value of the ratio.

5. rolled oats : chopped peanuts

6. sunflower seeds to pumpkin seeds

7. pumpkin seeds : rolled oats

Granola
(dry ingredients)

rolled oats	2 cups
chopped peanuts	1/2 cup
sunflower seeds	1/3 cup
pumpkin seeds	1/4 cup

Solve the inequality. Graph the solution.

8. $x + 5 < 9$

9. $b - 2 \geq -7$

10. $1 > -\frac{w}{3}$

11. $6 \leq -2g$

GO DIGITAL

Concepts, Skills, & Problem Solving

DETERMINING LIKELIHOOD Determine which numbers you are more likely to spin and which numbers you are less likely to spin. Explain your reasoning. (See Exploration 1.)

12.

13.

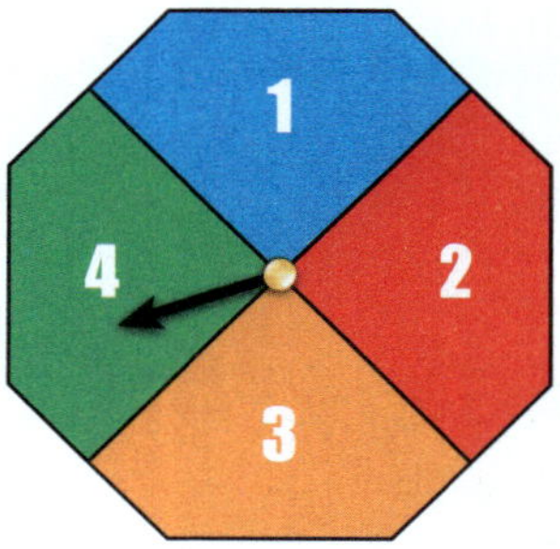

IDENTIFYING OUTCOMES You spin the spinner shown. (See Example 1.)

14. Find the sample space. How many possible outcomes are there?

15. What are the favorable outcomes of spinning a number no greater than 3?

16. In how many ways can spinning an even number occur?

17. In how many ways can spinning a prime number occur?

IDENTIFYING OUTCOMES You randomly choose one marble from the bag. (a) Find the number of ways the event can occur. (b) Find the favorable outcomes of the event.

18. Choosing blue

19. Choosing green

20. Choosing purple

21. Choosing yellow

22. Choosing *not* red

23. Choosing *not* blue

4 MTR **24.** **YOU BE THE TEACHER** Your friend finds the number of ways that choosing *not* purple can occur. Is your friend correct? Explain your reasoning.

purple	*not* purple
purple	red, blue, green, yellow

Choosing not purple can occur in 4 ways.

4 MTR **DISCUSS MATHEMATICAL THINKING** **Tell whether the statement is *true* or *false*. If it is false, change the italicized word to make the statement true.**

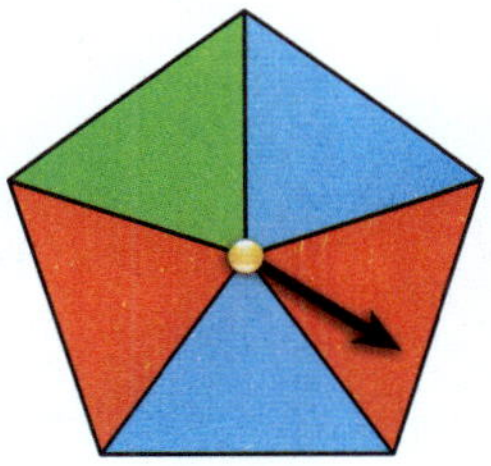

25. Spinning blue and spinning *green* have the same number of favorable outcomes on Spinner A.

26. There are *three* possible outcomes of spinning Spinner A.

27. Spinning *red* can occur in four ways on Spinner B.

28. Spinning not green can occur in *three* ways on Spinner B.

DESCRIBING LIKELIHOOD **Describe the likelihood of the event given its probability.** (See Example 2.)

Spinner B

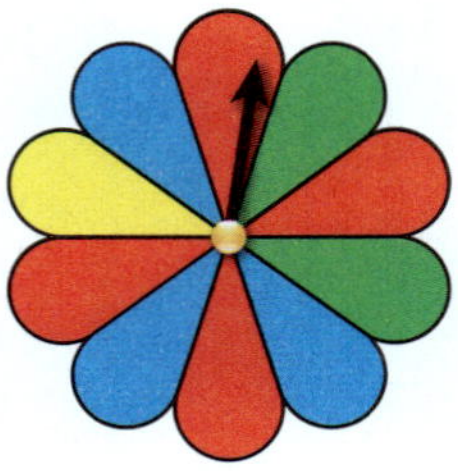

29. Your soccer team wins $\frac{3}{4}$ of the time.

30. There is a 0% chance that you will grow 12 feet.

31. The probability that the sun rises tomorrow is 1.

32. It rains on $\frac{1}{5}$ of the days in June.

6 MTR

33. **ASSESS REASONABLENESS** You have a 50% chance of being chosen to explain a math problem and a 32% chance of being chosen to explain a science problem in front of the class.

a. Describe the likelihood that you are chosen to explain the math problem and the likelihood you are chosen to explain the science problem.

b. Your friend says you are likely to be chosen to explain the math problem because you have a greater chance of being chosen to explain the math problem than the science problem. Is your friend's claim reasonable? Explain.

7 MTR

34. **MODELING REAL LIFE** You roll a number cube and record the number of times you roll an even number and the number of times you roll an odd number. Find the sample space, and then describe the likelihood of each event.

Even	𝍸 𝍸 𝍸 𝍸 𝍸 I
Odd	𝍸 𝍸 𝍸 𝍸 IIII

a. You roll an even number on your next roll.

b. You roll an odd number on your next roll.

7 MTR ▶ **35. MODELING REAL LIFE** You want to determine whether a coin is *fair*. You flip the coin and record the number of times you flip heads and the number of times you flip tails. (See Example 3.)

Heads	~~IIII~~ ~~IIII~~ ~~IIII~~ ~~IIII~~ II
Tails	III

a. Describe the likelihood that you flip heads on your next flip.

b. Describe the likelihood that you flip tails on your next flip.

c. Do you think the coin is a *fair* coin? Explain.

7 MTR **36. MODELING REAL LIFE** At a carnival, each guest randomly chooses 1 of 50 rubber ducks and then replaces it. The table shows the numbers of each type of duck that have been drawn so far. Out of 150 draws, how many can you expect to *not* be a losing duck? Justify your answer. (See Example 4.)

Win	~~IIII~~ I
Lose	~~IIII~~ ~~IIII~~ ~~IIII~~
Free Turn	IIII

37. REASONING A dodecahedron has twelve sides numbered 1 through 12. Describe the likelihood that each event will occur when you roll the dodecahedron. Then identify which event is most likely to occur and which event is least likely to occur. Explain your reasoning.

a. rolling an odd number

b. rolling a multiple of 4

c. rolling a number less than 10

38. **Dig Deeper** You play Rock Paper Scissors 18 times with a friend. For you, choosing rock has the same number of favorable outcomes as choosing either paper or scissors. How many times did you use rock?

39. **REASONING** You randomly choose one of the cards and set it aside. Then you randomly choose a second card. Describe how the number of possible outcomes changes after the first card is chosen.

5 MTR **STRUCTURE** **A Punnett square is a grid used to show possible gene combinations for the offspring of two parents. In the Punnett square shown, a boy is represented by *XY*. A girl is represented by *XX*.**

40. Complete the Punnett square. Explain why the events "having a boy" and "having a girl" are equally likely.

Mother's Genes

Father's Genes	X	X
X	XX	
Y		

41. Two parents each have the gene combination *Cs*. The gene *C* is for curly hair. The gene *s* is for straight hair. Any gene combination that includes a *C* results in curly hair. When all outcomes are equally likely, what is the probability of a child having curly hair?

12.2 Experimental and Theoretical Probability

Learning Target: Develop probability models using experimental and theoretical probability.

Success Criteria:
- I can explain the meanings of experimental and theoretical probabilities.
- I can find experimental and theoretical probabilities.
- I can compare experimental and theoretical probabilities.
- I can use simulations to find experimental probabilities.

Data Analysis and Probability
MA.7.DP.2.3 Find the theoretical probability of an event related to a simple experiment.
MA.7.DP.2.4 Use a simulation of a simple experiment to find experimental probabilities and compare them to theoretical probabilities.
Also MA.7.DP.2.1, MA.7.DP.2.2

Exploration 1 Conducting Experiments

Work with a partner. Conduct the following experiments and find the relative frequencies.

Experiment 1

- Flip a quarter 25 times and record whether each flip lands heads up or tails up.

Experiment 2

- Toss a thumbtack onto a table 25 times and record whether each toss lands point up or on its side.

a. Combine your results with those of your classmates. Do the relative frequencies change? What do you notice?

GO DIGITAL

b. Everyone in your school conducts each experiment, and you combine the results. How do you expect the relative frequencies to change? Explain.

c. How many times in 1000 flips do you expect a quarter to land heads up? How many times in 1000 tosses do you expect a thumbtack to land point up? Explain your reasoning.

d. In a *uniform probability model,* each outcome is equally likely to occur. Can you use a uniform probability model to describe either experiment? Explain.

1 MTR ANALYZE A PROBLEM

You know the number of possible outcomes in a uniform probability model. Can you find the probability of each outcome? Explain your reasoning.

12.2 Lesson

Key Vocabulary
experimental probability, *p. 1019*
theoretical probability, *p. 1019*
simulation, *p. 1021*

Key Idea

Experimental Probability

Probability that is based on repeated trials of an experiment is called **experimental probability**.

$$P(\text{event}) = \frac{\text{number of times the event occurs}}{\text{total number of trials}}$$

Example 1 Finding an Experimental Probability

Heads	Tails
6	19

The table shows the results of spinning a penny 25 times. What is the experimental probability of spinning heads?

Heads was spun 6 times in a total of $6 + 19 = 25$ spins.

Experimental probabilities are found the same way as relative frequencies.

$$P(\text{event}) = \frac{\text{number of times the event occurs}}{\text{total number of trials}}$$

$$P(\text{heads}) = \frac{6}{25}$$

Heads was spun 6 times.

There was a total of 25 spins.

▶ The experimental probability is $\frac{6}{25}$, 0.24, or 24%.

Try It **The table shows the results of rolling a number cube 50 times. Find the experimental probability of the event.**

Number Rolled	1	2	3	4	5	6
Frequency	10	4	8	11	11	6

1. rolling a 3

2. rolling an odd number

Key Idea

Theoretical Probability

When all possible outcomes are equally likely, the **theoretical probability** of an event is the quotient of the number of favorable outcomes and the number of possible outcomes.

$$P(\text{event}) = \frac{\text{number of favorable outcomes}}{\text{number of possible outcomes}}$$

GO DIGITAL

Example 2 Finding a Theoretical Probability

You randomly choose one of the letters shown. What is the theoretical probability of choosing a vowel?

$$P(\text{vowel}) = \frac{\text{number of favorable outcomes}}{\text{number of possible outcomes}} = \frac{3}{8}$$

There are 3 vowels.

There is a total of 8 letters.

▶ The probability of choosing a vowel is $\frac{3}{8}$, or 37.5%.

Try It

3. What is the theoretical probability of randomly choosing an X?

Example 3 Comparing Probabilities

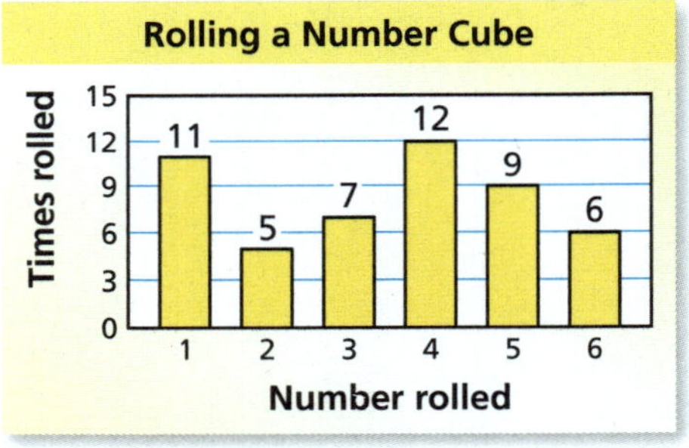

The bar graph shows the results of rolling a number cube 50 times. How does the experimental probability of rolling an odd number compare with the theoretical probability?

Step 1: Find the experimental probability of rolling an odd number.

The bar graph shows 11 ones, 7 threes, and 9 fives. So, an odd number was rolled $11 + 7 + 9 = 27$ times in a total of 50 rolls.

$$P(\text{odd}) = \frac{\text{number of times an odd number was rolled}}{\text{total number of rolls}}$$

$$= \frac{27}{50}, \text{ or } 54\%$$

Step 2: Find the theoretical probability of rolling an odd number.

$$P(\text{odd}) = \frac{\text{number of favorable outcomes}}{\text{number of possible outcomes}} = \frac{3}{6} = \frac{1}{2}, \text{ or } 50\%$$

▶ The experimental probability of rolling an odd number is 54%, which is close to the theoretical probability of 50%.

GO DIGITAL

Try It

4. In Example 3, how does the experimental probability of rolling a number greater than 1 compare with the theoretical probability?

A **simulation** is an experiment that is designed to reproduce the conditions of a situation or process. Simulations allow you to study situations that are impractical to create in real life.

Example 4 Using a Simulation

Use technology to simulate rolling a number cube 500 times. How does the experimental probability of rolling an odd number compare with the theoretical probability?

You can simulate the rolling of a number cube 500 times by randomly generating a set of 500 numbers from one through six and organizing the results in a bar graph.

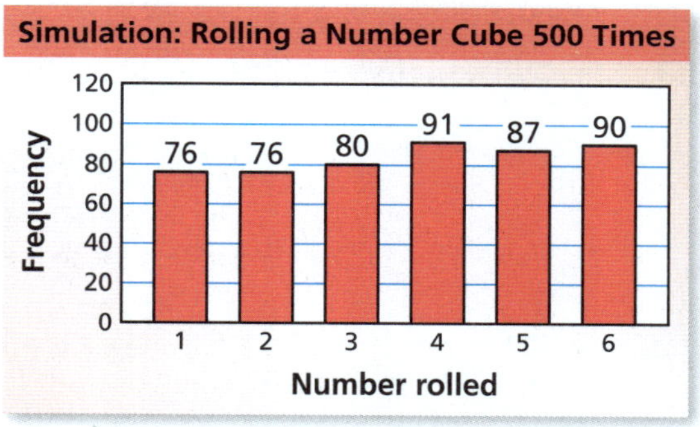

In general, as the number of trials increases, the experimental probability gets closer to the theoretical probability.

$$P(\text{odd}) = \frac{\text{number of times an odd number was rolled}}{\text{total number of rolls}} = \frac{76 + 80 + 87}{500} = \frac{243}{500}, \text{ or } 48.6\%$$

▶ The experimental probability of rolling an odd number is 48.6%, which is close to the theoretical probability of 50%.

Try It

5. In Example 4, how does the experimental probability of rolling a number greater than 1 compare with the theoretical probability?

GO DIGITAL

In-Class Practice

1 I don't understand yet. | 2 I can do it with help. | 3 I can do it on my own. | 4 I can teach someone else.

6. **USING A SIMULATION** Design and perform simulations for flipping a coin 50 times and 500 times. Compare the experimental probability of flipping heads to the theoretical probability in each simulation.

4 MTR

7. **DIFFERENT WORDS, SAME QUESTION** You flip a coin and record the results in the table. Which is different? Find "both" answers.

Heads	Tails
36	24

What is the experimental probability of flipping heads?

What fraction of the flips can you expect to result in heads?

What percent of the flips result in heads?

What is the relative frequency of flipping heads?

Example 5 B.E.S.T. Test Prep: Modeling Real Life 7 MTR

Prize	Probability
Bobblehead	$\frac{1}{6}$
Shirt	$\frac{1}{2}$
Other	$\frac{1}{3}$

The table shows the theoretical probabilities of winning different items when spinning a prize wheel. The wheel has 18 sections. There are

Ⓐ 3 Ⓑ 6 Ⓒ 9 Ⓓ 18	**bobblehead sections and**	Ⓐ 3 Ⓑ 6 Ⓒ 9 Ⓓ 18	**shirt sections.**

Use the equation for theoretical probability.

$$P(\text{bobblehead}) = \frac{\text{number of bobblehead sections}}{\text{total number of sections}}$$

$$\frac{1}{6} = \frac{b}{18}$$ Substitute. Let b be the number of bobblehead sections.

$$3 = b$$ Multiply each side by 18.

$$P(\text{shirt}) = \frac{\text{number of shirt sections}}{\text{total number of sections}}$$

$$\frac{1}{2} = \frac{s}{18}$$ Substitute. Let s be the number of shirt sections.

$$9 = s$$ Multiply each side by 18.

▶ So, the first correct answer is Ⓐ and the second correct answer is Ⓒ.

GO DIGITAL

In-Class Practice

1 I don't understand yet. 2 I can do it with help. 3 I can do it on my own. 4 I can teach someone else.

8. The table shows the theoretical probabilities of each result when randomly drawing tickets from a hat. There are 7 winning tickets in the hat. How many tickets are in the hat? Explain your reasoning.

Ticket	Probability
Win	14%
Lose	60%
Draw again	26%

9. **Dig Deeper** You randomly choose two different songs on a music playlist that has 25 songs. The probability that the first song is a hip-hop song is 28%. The first song you choose is a hip-hop song. What is the probability that the second song is also a hip-hop song? Explain your reasoning.

12.2 Practice WITH CalcChat® AND CalcView®

Review & Refresh

Describe the likelihood of the event given its probability.

1. You randomly guess the correct answer of a multiple choice question $\frac{1}{4}$ of the time.

2. There is a 95% chance that school will *not* be canceled tomorrow.

Find the annual interest rate.

3. $I = \$16$, $P = \$200$, $t = 2$ years

4. $I = \$26.25$, $P = \$500$, $t = 18$ months

Find the missing values in the ratio table. Then write the equivalent ratios.

5.

Dogs	3		9
Cats	1	2	

6.

Water	2	8	6
Flour	3		

GO DIGITAL

Concepts, Skills, & Problem Solving

CONDUCTING AN EXPERIMENT Use the bar graph below to find the relative frequency of the event. (See Exploration 1.)

7. spinning a 6

8. spinning an even number

FINDING AN EXPERIMENTAL PROBABILITY Use the bar graph to find the experimental probability of the event. (See Example 1.)

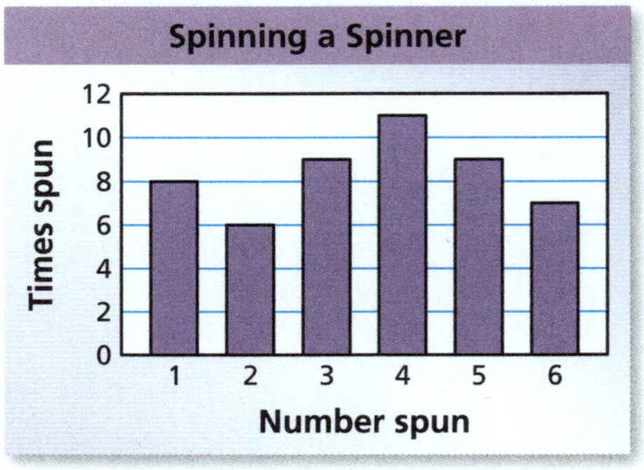

9. spinning a number less than 3

10. *not* spinning a 1

11. spinning a 1 or a 3

12. spinning a 7

4 MTR **13.** **YOU BE THE TEACHER** Your friend uses the bar graph above to find the experimental probability of spinning a 4. Is your friend correct? Explain your reasoning.

> There are 6 possible outcomes. So, the experimental probability of spinning a 4 is $\frac{1}{6}$.

7 MTR **14.** **MODELING REAL LIFE** You check 20 laser pointers at random. Three of the laser pointers are defective. What is the experimental probability that a laser pointer is defective?

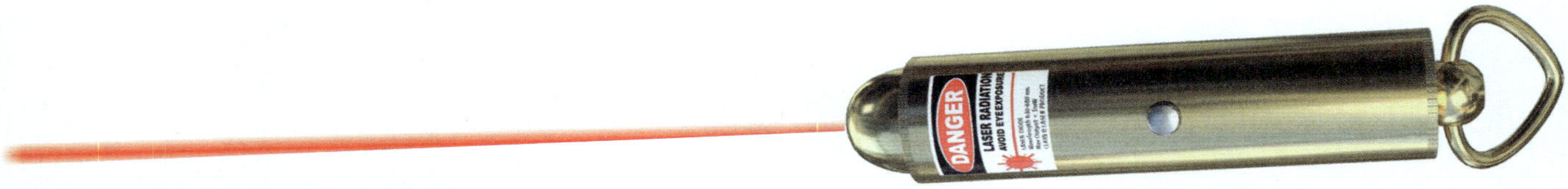

FINDING A THEORETICAL PROBABILITY **Use the spinner to find the theoretical probability of the event.** (See Example 2.)

▶ **15.** spinning red

16. spinning a 1

17. spinning an odd number

18. spinning a multiple of 2

19. spinning a number less than 6

20. spinning a 9

21. **REASONING** Each letter of the alphabet is printed on an index card. What is the theoretical probability of randomly choosing any letter except Z?

COMPARING PROBABILITIES **The bar graph shows the results of spinning the spinner below 20 times. Compare the theoretical and experimental probabilities of the event.** (See Example 3.)

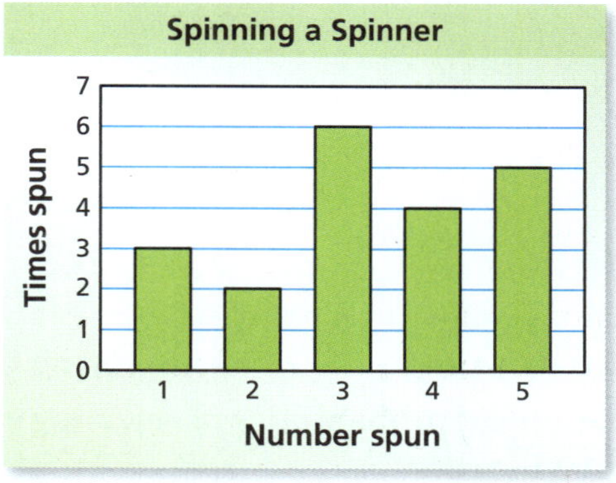

22. spinning a 4

▶ **23.** spinning a 3

24. spinning a number greater than 4

25. spinning an odd number

2 MTR **26.** **MODEL A PROBLEM** Use technology to simulate 200 spins of the spinner from Exercises 22–25. How does the experimental probability of spinning a number greater than 2 compare with the theoretical probability? (See Example 4.)

3 MTR **27.** **CHOOSE A METHOD** Should you use *theoretical* or *experimental* probability to predict the number of times you will spin a 3 in 10,000 spins of the spinner for Exercises 22–25? Explain.

7 MTR **28.** **MODELING REAL LIFE** In a charity raffle, each donor randomly draws a marble from a bag and replaces it. A blue marble represents a *win*, and a red marble represents a *loss*. The theoretical probability of drawing a winning marble is $\frac{3}{10}$. The bag contains 9 winning marbles. (See Example 5.)

a. How many marbles are in the bag?

b. Out of 20 donors, how many do you expect to draw a winning marble?

COMPARING THEORETICAL PROBABILITIES **Use <, >, and = to compare the theoretical probabilities of selecting cards from a standard deck of 52 playing cards.**

29. $P(\text{red})$ ☐ $P(\text{club})$

30. $P(\text{red})$ ☐ $P(\text{black})$

31. $P(\text{heart})$ ☐ $P(\text{diamond})$

32. $P(\text{spade})$ ☐ $P(\text{black})$

Hearts	Diamonds	Spades	Clubs
A♥	A♦	A♠	A♣
K♥	K♦	K♠	K♣
Q♥	Q♦	Q♠	Q♣
J♥	J♦	J♠	J♣
10♥	10♦	10♠	10♣
9♥	9♦	9♠	9♣
8♥	8♦	8♠	8♣
7♥	7♦	7♠	7♣
6♥	6♦	6♠	6♣
5♥	5♦	5♠	5♣
4♥	4♦	4♠	4♣
3♥	3♦	3♠	3♣
2♥	2♦	2♠	2♣

33. PROBLEM SOLVING There are 6 females and 10 males in a class.

a. What is the theoretical probability that a randomly chosen student is female?

b. One week later, there are 24 students in the class. The theoretical probability that a randomly chosen student is a female is the same as last week. How many males joined the class?

34. NUMBER SENSE The table at the right shows the results of flipping two coins 12 times each.

HH	HT	TH	TT
1	3	2	6

a. What is the experimental probability of flipping two tails? Using this probability, how many times can you expect to flip two tails in 600 trials?

HH	HT	TH	TT
23	29	26	22

b. The table at the left shows the results of flipping the same two coins 100 times each. What is the experimental probability of flipping two tails? Using this probability, how many times can you expect to flip two tails in 600 trials?

c. Why is it important to use a large number of trials when using experimental probability to predict results?

35. OPEN-ENDED Describe a real-life situation where you would use theoretical probability to find the likelihood of an event. Then describe a real-life situation where you would use experimental probability to find the likelihood of an event. Explain your reasoning.

36. DISCUSS MATHEMATICAL THINKING The table shows the possible outcomes of rolling a pair of number cubes. You simulate rolling a pair of number cubes 60 times and record your results in the bar graph shown.

	1	2	3	4	5	6
1	1 1	2 1	3 1	4 1	5 1	6 1
2	1 2	2 2	3 2	4 2	5 2	6 2
3	1 3	2 3	3 3	4 3	5 3	6 3
4	1 4	2 4	3 4	4 4	5 4	6 4
5	1 5	2 5	3 5	4 5	5 5	6 5
6	1 6	2 6	3 6	4 6	5 6	6 6

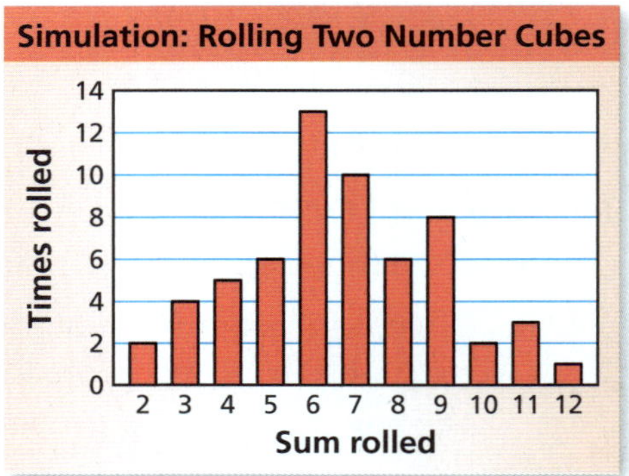

a. Use technology to compare the theoretical and experimental probabilities of rolling each sum.

b. Which sum do you expect to be most likely after 500 trials? 1000 trials? Explain your reasoning.

c. Predict the experimental probability of rolling each sum after 10,000 trials. Explain your reasoning.

37. MAKE A PLAN Choose an event and design a simulation.

a. Use your simulation to find the experimental probability of the event you chose. How does the experimental probability change when you adjust the length of the simulation?

b. Find the theoretical probability of the event. What do you notice?

12.3 Samples and Populations

Learning Target: Understand how to use random samples to make conclusions about a population.

Success Criteria:
- I can explain why a sample is biased or unbiased.
- I can explain why conclusions made from a biased sample may not be valid.
- I can use an unbiased sample to make a conclusion about a population.

Data Analysis and Probability
MA.7.DP.1.3 Given categorical data from a random sample, use proportional relationships to make predictions about a population.

A **population** is an entire group of people or objects. A **sample** is a part of a population. You can gain information about a population by examining samples of the population.

Exploration 1 Using Samples of Populations

Work with a partner. You want to make conclusions about the favorite extracurricular activities of students at your school.

a. Identify the population. Then identify five samples of the population.

b. When a sample is selected *at random*, each member of the population is equally likely to be selected. Are any of the samples in part (a) selected at random? Explain your reasoning.

GO DIGITAL

c. How are the samples below different? Is each conclusion valid? Explain your reasoning.

You ask 20 members of the school band about their favorite activity. The table shows the results. You conclude that band is the favorite activity of 70% of the students in your school.

Favorite Activity	
Band	14
Other	6

You ask every eighth student who enters the school about their favorite activity. One student says glee club for every nine who name a different activity. You conclude that glee club is the favorite activity of 10% of the students in your school.

d. Write a survey question about a topic that interests you. How can you choose people to survey so that you can use the results to make a valid conclusion?

4 MTR **COMMUNICATE CLEARLY**

Can the size of a sample affect the validity of a conclusion about a population? Explain.

12.3 Lesson

Key Vocabulary
population, *p. 1031*
sample, *p. 1031*
unbiased sample, *p. 1033*
biased sample, *p. 1033*

An **unbiased sample** is representative of a population. It is selected at random and is large enough to provide accurate data.

A **biased sample** is not representative of a population. One or more parts of the population are favored over others.

Example 1 B.E.S.T. Test Prep: Identifying an Unbiased Sample

You want to estimate the number of students in a high school who listen to reggaeton, a popular style of music in Puerto Rico and many other Caribbean countries. Which sample is unbiased?

Ⓐ 4 students in the hallway

Ⓑ all students on the soccer team

Ⓒ 50 Puerto Rican students at random

Ⓓ 100 students at random during lunch

Choice A is not large enough to provide accurate data.

Choice B is not selected at random.

Choice C is not representative of the population, because Puerto Rican students are favored over other students.

Conclusions made from the sample in Choice C may be inaccurate because Puerto Rican students may be more likely to listen to reggaeton than other students.

Choice D is representative of the population because it is selected at random and is large enough to provide accurate data.

▶ So, the correct answer is Ⓓ.

GO DIGITAL

Try It

1. **WHAT IF?** You want to estimate the number of Puerto Rican students in a high school who listen to reggaeton. Which sample is unbiased? Explain.

2. You want to estimate the number of eighth-grade students in your school who find it relaxing to listen to music. You consider two samples.
   - fifteen randomly selected members of the band
   - every fifth student whose name appears on an alphabetical list of eighth-grade students

   Which sample is unbiased? Explain.

The results of an unbiased sample are proportional to the results of the population. So, you can use unbiased samples to make conclusions about a population. Biased samples are not representative of a population. So, you should not use them to make conclusions about a population.

Example 2 Determining Whether Conclusions Are Valid

You want to know how the residents of your town feel about adding a new landfill. Determine whether each conclusion is valid.

a. You survey the 100 residents who live closest to the new landfill. The table shows the results. You conclude that 10% of the residents of your town support the new landfill.

New Landfill	
Support	10
Do Not Support	90

The sample is not representative of the population, because residents who live close to the landfill may be less likely to support it.

▶ So, the sample is biased, and the conclusion is not valid.

b. You survey 100 residents at random. The table shows the results. You conclude that 60% of the residents of your town do not support the new landfill.

New Landfill	
Support	40
Do Not Support	60

The sample is representative of the population, because it is selected at random and is large enough to provide accurate data.

▶ So, the sample is unbiased, and the conclusion is valid.

GO DIGITAL

Try It

3. Four out of five randomly chosen teenagers support the new landfill. So, you conclude that 80% of the residents of your town support the new landfill. Is the conclusion valid? Explain.

In-Class Practice

1 I don't understand yet. | 2 I can do it with help. | 3 I can do it on my own. | 4 I can teach someone else.

4 MTR **4.** **WHICH ONE DOESN'T BELONG?** You want to estimate how many people in your city read comic books. Which sample does *not* belong with the other three? Explain your reasoning.

your 5 closest friends

25 students in your school at random

75 people at random from a list of city residents

the first 50 people to arrive at a comic book convention

5. **ANALYZING A CONCLUSION** At a football game, you ask 50 randomly chosen students from your school whether they play a school sport. The table shows the results. You conclude that 80% of students in your school play a school sport. Is your conclusion valid? Explain.

Play a School Sport	
Yes	40
No	10

Example 3 Modeling Real Life

7 MTR

You ask 75 randomly chosen students at a school how many movies they watch each week. There are 1200 students in the school. Estimate the number of students in the school who watch one movie each week.

Movies per Week	
Zero	30
One	21
Two or More	24

Understand the problem.

You are given the numbers of movies watched each week by a sample of 75 students. You are asked to make an estimate about the population, all students in the school.

Make a plan.

The sample is representative of the population because it is selected at random and is large enough to provide accurate data. So, the sample is unbiased, and its results are proportional to the results of the population. Use a ratio table to estimate the number of students in the school who watch one movie each week.

Solve and check.

	×4	×4	
Students (one movie)	21	84	336
Total Students	75	300	1200
	×4	×4	

Another Method Use a proportion.

$$\frac{21}{75} = \frac{n}{1200}$$

$336 = n$ ✓

▶ So, about 336 students in the school watch one movie each week.

In-Class Practice

1 I don't understand yet. 2 I can do it with help. 3 I can do it on my own. 4 I can teach someone else.

6. You want to estimate the mean photo size on your cell phone. You choose 30 photos at random from your phone. The total size of the sample is 186 megabytes. Explain whether you can use the sample to estimate the mean size of photos on your cell phone. If so, what is your estimate?

Bicycle-Sharing System	
Zero Times	24
One Time	9
Two Times	7
Three or More Times	10

7. Dig Deeper You ask 50 randomly chosen employees of a company how many times they used a bicycle-sharing system last month. The table shows the results. There are 600 people employed by the company. Estimate the number of employees who used a bicycle-sharing system at least one time last month.

GO DIGITAL

12.3 Practice WITH CalcChat® AND CalcView®

Review & Refresh

Choose the most appropriate measures to describe the center and the variation.

1.

2.

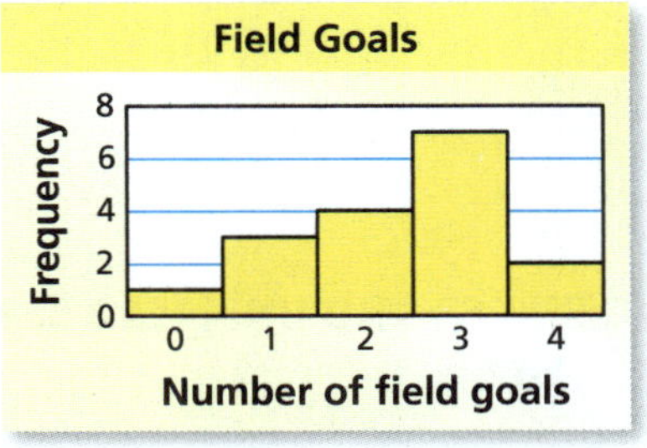

Solve the inequality.

3. $x - 5 < 9$

4. $q + 2 \geq -13$

5. $3 < -3 + r$

Concepts, Skills, & Problem Solving

USING SAMPLES OF POPULATIONS **You ask 50 randomly chosen artists in your town about their favorite art form. Determine whether your conclusion is valid. Justify your answer.** (See Exploration 1.)

Favorite Art Form	
Drawing	20
Photography	12
Painting	6
Ceramics	5
Other	7

6. You conclude that drawing is the favorite art form of 60% of artists in your town.

7. You conclude that ceramics is the favorite art form of 10% of people in your town.

GO DIGITAL

IDENTIFYING POPULATIONS AND SAMPLES **Identify the population and the sample.**

8.

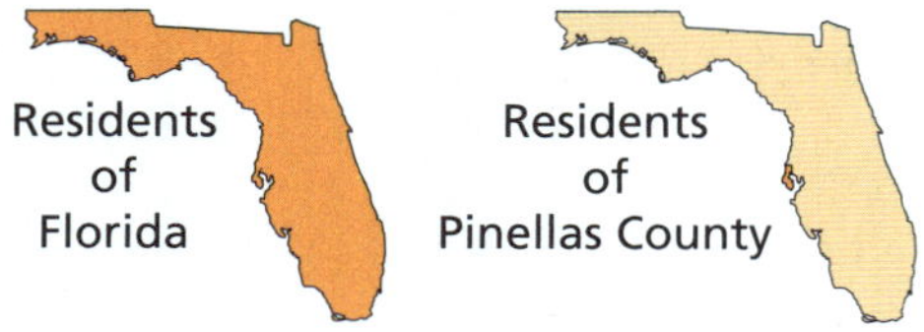

9.

4 cards

All cards in a deck

IDENTIFYING BIASED AND UNBIASED SAMPLES **Determine whether the sample is *biased* or *unbiased*. Explain.** (See Example 1.)

10. You want to estimate the number of books students in your school read over the summer. You survey every fourth student who enters the school.

11. You want to estimate the number of people in a town who think that a park needs to be remodeled. You survey every 10th person who enters the park.

7 MTR

12. MODELING REAL LIFE You want to determine the number of students in your school who have visited a science museum. You survey 50 students at random. Twenty have visited a science museum, and thirty have not. So, you conclude that 40% of the students in your school have visited a science museum. Is your conclusion valid? Explain. (See Example 2.)

South Florida Science Center and Aquarium

13. USING A SAMPLE Which sample is better for making an estimate? Explain.

Estimate the number of defective pencils produced per day.	
Sample A	A random sample of 500 pencils from 20 machines
Sample B	A random sample of 500 pencils from 1 machine

CONDUCTING SURVEYS Determine whether you should survey the population or a sample. Explain.

14. You want to know the average height of seventh graders in the United States.

15. You want to know the favorite types of music of students in your homeroom.

4 MTR

16. DISCUSS MATHEMATICAL THINKING Does increasing the size of a sample necessarily make the sample more representative of a population? Give an example to support your explanation.

17. REASONING A person surveys residents of a town to determine whether a skateboarding ban should be overturned. Describe how the person can conduct the survey so that the sample is biased toward overturning the ban.

7 MTR

18. MODELING REAL LIFE You ask 20 randomly chosen environmental scientists from your state to name their favorite way to eliminate waste. There are 200 environmental scientists in your state. Estimate the number of environmental scientists in your state whose favorite way to eliminate waste is recycling. (See Example 3.)

Favorite Way to Eliminate Waste	
Reducing	14
Reusing	4
Recycling	2

7 MTR

19. MODELING REAL LIFE To predict the result of a mayoral election, you survey 50 likely voters at random. The table shows the results. Describe whether the sample can be used to predict the outcome of the election. If so, what is your prediction for the number of votes received by the winner, assuming that 500 people vote?

Candidate Preference	
Candidate A	33
Candidate B	12
Other	5

20. Dig Deeper You ask 100 randomly chosen dog owners in your town how many dogs they own. The results are shown in the table. There are 500 dog owners in your town.

Number of Dogs	Frequency
1	54
2	38
3	3
4	1
5	4

a. Estimate the median number of dogs per dog owner in your town. Justify your answer.

b. Estimate the mean number of dogs per dog owner in your town. Justify your answer.

12.4 Comparing Populations

Learning Target: Compare populations using measures of center, measures of variation, and random samples.

Success Criteria:
- I can compare populations using measures of center and variation.
- I can compare random samples using measures of center and variation.
- I can compare populations using multiple random samples.

Data Analysis and Probability

MA.7.DP.1.2 Given two numerical or graphical representations of data, use the measure(s) of center and measure(s) of variability to make comparisons, interpret results and draw conclusions about the two populations.

Exploration 1 Using Random Samples

Work with a partner. You want to compare the numbers of hours spent on homework each week by male and female students in your state. You take a random sample of 15 male students and 15 female students throughout the state.

Male Students				
1.5	3	0	2.5	1
8	2.5	1	3	0
6.5	1	5	0	5

Female Students				
4	0	3	1	1
5	1	3	5.5	10
2	0	6	3.5	2

a. Compare the data in each sample.

b. Are the samples likely to be representative of all male and female students in your state? Explain.

GO DIGITAL

c. You take 100 random samples of 15 male students in your state and 100 random samples of 15 female students in your state and record the median of each sample. The double box-and-whisker plot shows the distributions of the sample medians. Compare the distributions in the double box-and-whisker plot with the distributions of the data in the tables.

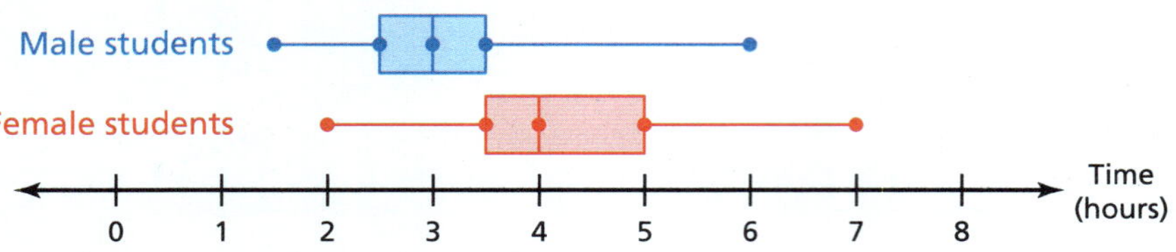

d. What can you conclude from the double box-and-whisker plot? Explain.

e. How can you use random samples to make accurate comparisons of two populations?

3 MTR **REFLECT ON YOUR METHOD**

How does taking multiple random samples allow you to make conclusions about two populations?

12.4 Lesson

You can use measures of center and variation to compare two populations.

Example 1 Comparing Populations

Two data sets contain an equal number of values. The double box-and-whisker plot represents the values in the data sets.

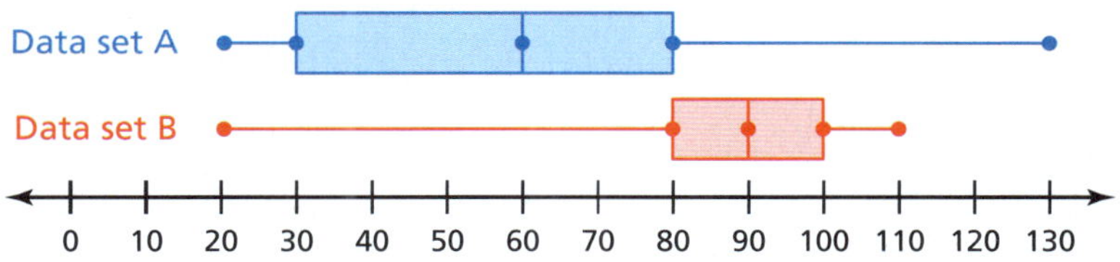

a. Compare the data sets using measures of center and variation.

Because you are given a double box-and-whisker plot without the original data, use the median and the IQR.

Data set A	**Data set B**
Median = 60	Median = 90
IQR = 80 − 30 = 50	IQR = 100 − 80 = 20

▶ So, Data set B has a greater measure of center, and Data set A has a greater measure of variation.

b. Which data set is more likely to contain a value of 95?

About 25% of the data values in Data set A are between 80 and 130. About 25% of the data values in Data set B are between 90 and 100.

▶ So, Data set B is more likely to contain a value of 95.

c. Which data set is more likely to contain a value that differs from the center by at least 30?

The IQR of Data set A is 50 and the IQR of Data set B is 20. This means it is more common for a value to differ from the center by 30 in Data set A than in Data set B.

▶ So, Data set A is more likely to contain a value that differs from the center by at least 30.

5 MTR USE STRUCTURE

Explain how you know that about 25% of the data values in Data set B are between 90 and 100.

____

____

____

____

____

____

____

____

____

____

____

GO DIGITAL

Try It

1. Which data set is more likely to contain a value of 70?

2. Which data set is more likely to contain a value that differs from the center by no more than 3?

You do not need to have all the data from two populations to make comparisons. You can use random samples to make comparisons.

Example 2 Comparing Random Samples

Two bags each contain 1000 numbered tiles. The double box-and-whisker plot represents a random sample of 12 numbers from each bag. Compare the samples using measures of center and variation. Can you determine which bag contains tiles with greater numbers?

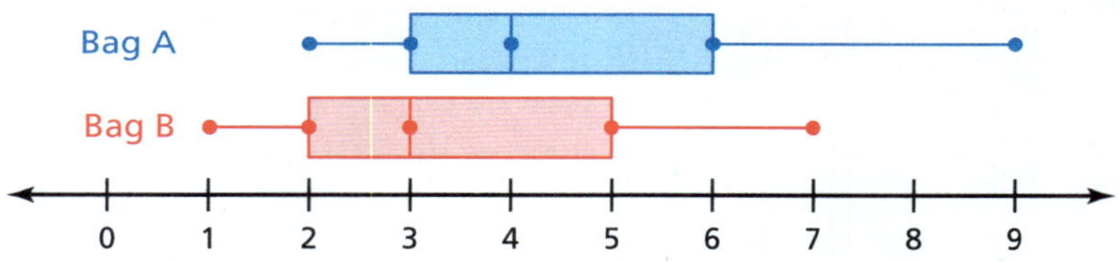

Use the median and the IQR.

Bag A	**Bag B**
Median = 4	Median = 3
IQR = 6 − 3 = 3	IQR = 5 − 2 = 3

▶ The variation in the samples is about the same, but the sample from Bag A has a greater median. The sample size is too small, however, to conclude that tiles in Bag A generally have greater numbers than tiles in Bag B.

You are more likely to make valid comparisons when the sample size is large and there is little variability in the data.

GO DIGITAL

Try It

3. The double dot plot shows the weekly reading habits of a random sample of 10 students in each of two schools. Compare the samples using measures of center and variation. Can you determine which school's students read less? Explain.

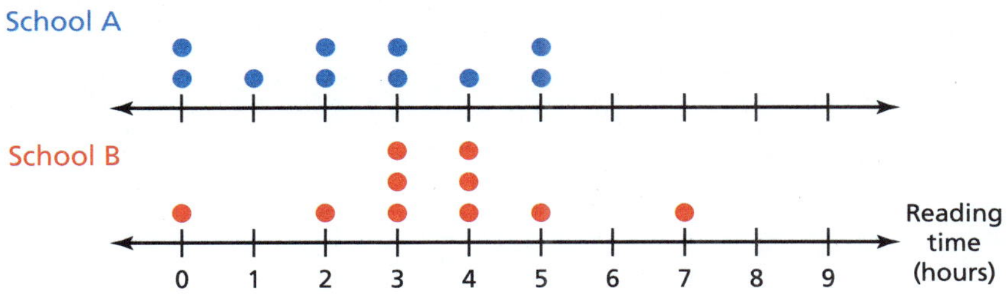

Example 3 Using Multiple Random Samples

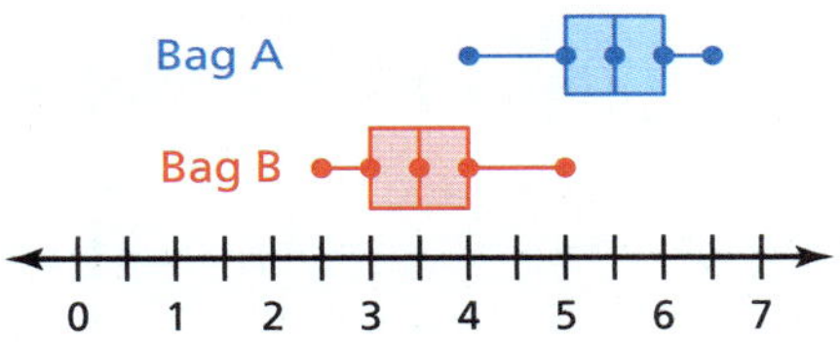

The double box-and-whisker plot represents the medians of 50 random samples of 12 numbers from each bag in Example 2. Compare the variability of the sample medians to the variability of the samples in Example 2. Can you determine which bag contains tiles with greater numbers?

Use the median and the IQR.

Bag A	**Bag B**
Median $= 5.5$	Median $= 3.5$
$\text{IQR} = 6 - 5 = 1$	$\text{IQR} = 4 - 3 = 1$

▶ The IQR of the sample medians for each bag is 1, which is less than the IQR of the samples in Example 2. Most of the sample medians for Bag A are greater than the sample medians for Bag B. So, tiles in Bag A generally have greater numbers than tiles in Bag B.

Try It

4. **WHAT IF?** Each value in the box-and-whisker plot of the sample medians for Bag A decreases by 2. Does this change your answer?

GO DIGITAL

In-Class Practice

1 I don't understand yet. | 2 I can do it with help. | 3 I can do it on my own. | 4 I can teach someone else.

5. **COMPARING POPULATIONS** Compare the data sets using measures of center and variation.

Data set A: {3, 4, 1, 3, 2, 5, 2, 3, 3, 4, 2, 4}

Data set B: {1, 3, 3, 2, 1, 1, 2, 1, 3, 2, 2, 3}

6. **COMPARING RANDOM SAMPLES** Two boxes each contain 600 numbered tiles. The back-to-back stem-and-leaf plot shows a random sample of 8 numbers from each box. Compare the samples using measures of center and variation. Can you determine which box contains tiles with greater numbers? Explain.

Box A		Box B
	1	0 0 0 5
5	2	0 5 5 5
5 5 0	3	
5 0 0	4	
0	5	

Key: 5 | 2 | 0 = 25 and 20

7. **USING MULTIPLE RANDOM SAMPLES** Two crates each contain 750 objects. The double box-and-whisker plot shows the median weights of 50 random samples of 10 objects from each crate. Can you determine which crate weighs more? Explain.

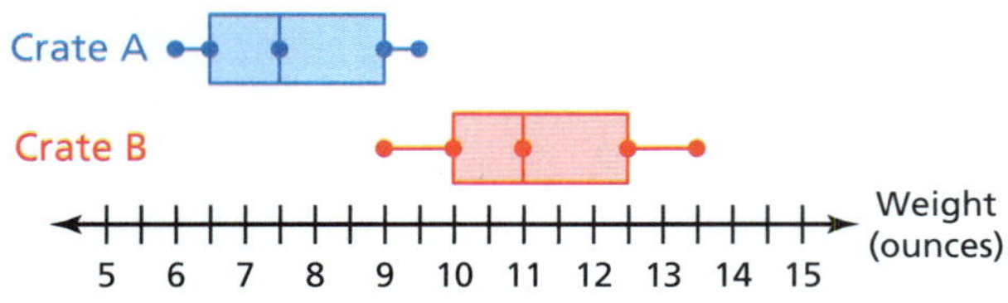

Example 4 Modeling Real Life 7 MTR

The double box-and-whisker plot represents the medians of 50 random samples of 10 speeding tickets issued in two states. Compare the costs of speeding tickets in the two states.

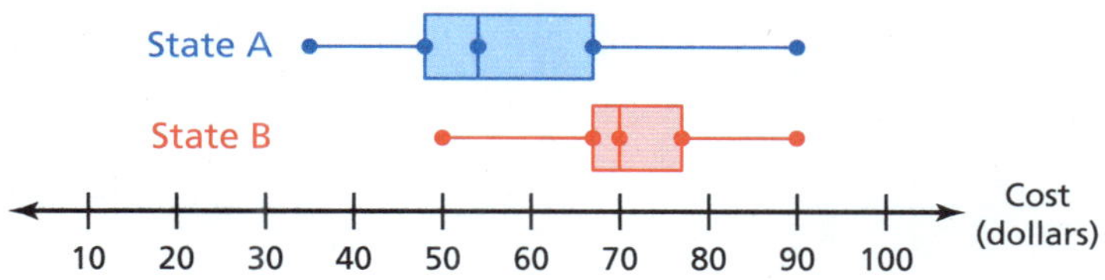

There is enough data to draw conclusions about the costs of speeding tickets in the two states. Find the measures of center and variation for the sample medians from each state. Then compare the data.

Use the median and the IQR.

State A	**State B**
Median $\approx$ 54	Median = 70
IQR $\approx 67 - 48 = 19$	IQR $\approx 77 - 67 = 10$

The variation for State A is greater than the variation for State B, and the measure of center for State B is greater than the measure of center for State A.

▶ So, you can conclude that the cost of speeding tickets varies more in State A, but speeding tickets generally cost more in State B.

GO DIGITAL

In-Class Practice

1 I don't understand yet. **2** I can do it with help. **3** I can do it on my own. **4** I can teach someone else.

8. The double box-and-whisker plot represents the medians of 100 random samples of 20 battery lives for two cell phone brands. Compare the battery lives of the two brands.

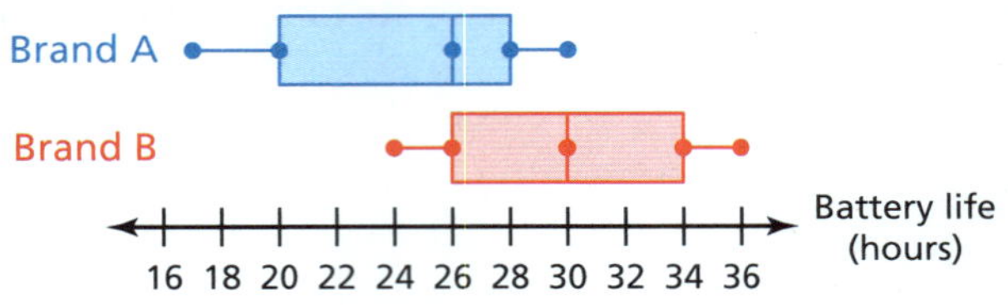

9. The double box-and-whisker plot represents the medians of 50 random samples of 10 wait times at two patient care facilities. Which facility should you choose? Explain your reasoning.

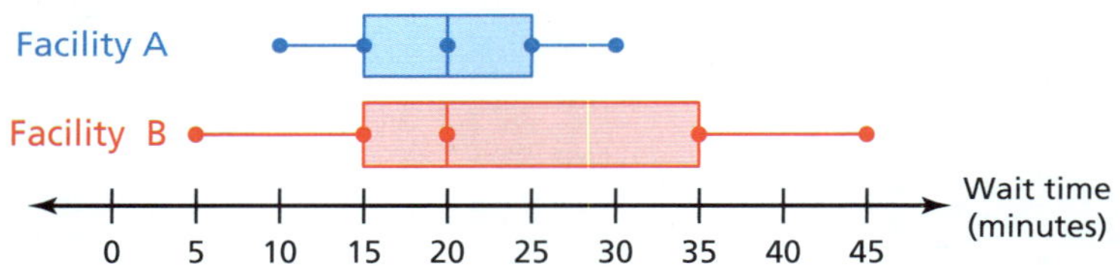

12.4 Practice WITH CalcChat® AND CalcView®

Review & Refresh

You ask 100 randomly chosen high school students whether they support a new college in your town. Determine whether your conclusion is valid.

1. You conclude that 85% of high school students in your town support the new college.

New College	
Support	85
Do not support	15

2. You conclude that 15% of residents in your town do not support the new college.

Solve the equation. Check your solution.

3. $b - 3 = 22$

4. $1.5d = -4.5$

5. $\frac{t}{-3} = -4$

GO DIGITAL

Concepts, Skills, & Problem Solving

USING RANDOM SAMPLES You want to compare the numbers of hours spent on recreation each week by teachers and non-teachers in your state. You take 100 random samples of 15 teachers and 100 random samples of 15 non-teachers throughout the state and record the median value of each sample. The double box-and-whisker plot shows the distributions of sample medians. (See Exploration 1.)

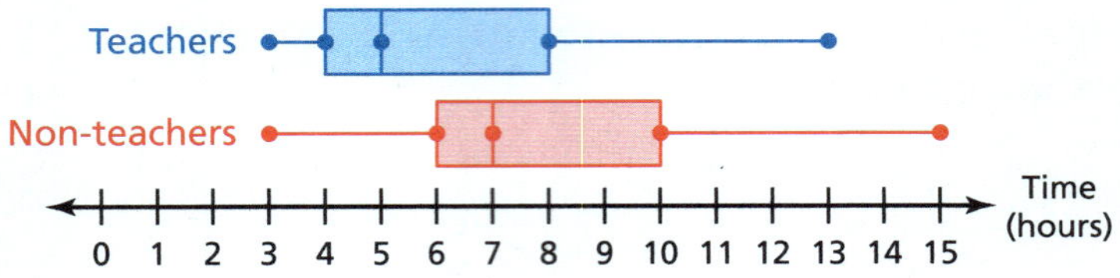

6. Are the samples likely to be representative of all teachers and non-teachers in your state?

7. What can you conclude from the double box-and-whisker plot? Explain.

COMPARING POPULATIONS **Two data sets contain an equal number of values. The double box-and-whisker plot represents the values in the data sets.** (See Example 1.)

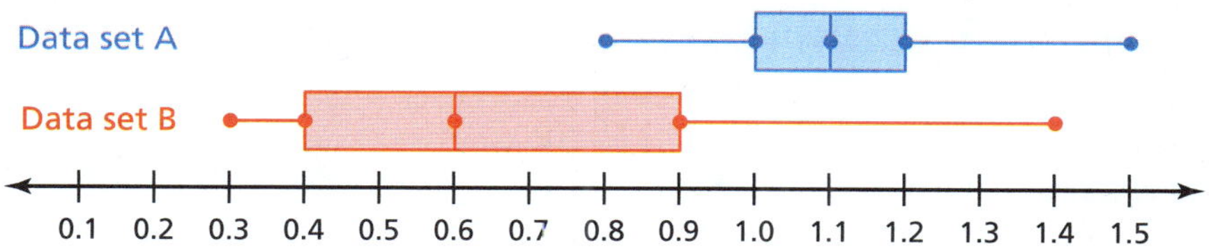

8. Compare the data sets using measures of center and variation.

9. Which data set is more likely to contain a value of 1.1?

10. Which data set is more likely to contain a value that differs from the center by 0.3?

11. **COMPARING RANDOM SAMPLES** The double dot plot shows the weekly running habits of athletes at two colleges. Compare the samples using measures of center and variation. Can you determine which college's athletes spend more time running? Explain. (See Example 2.)

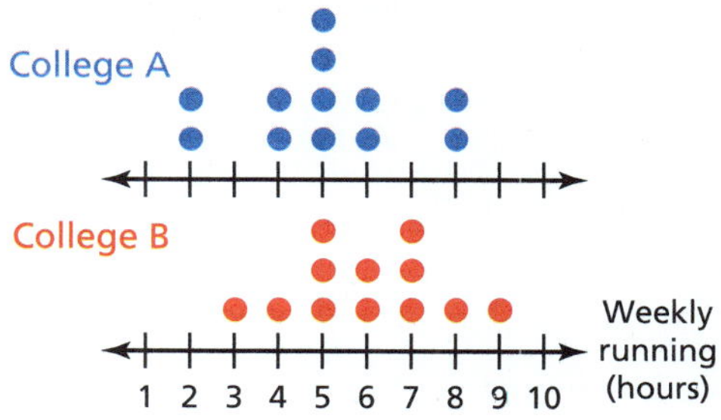

12. MODELING REAL LIFE The back-to-back stem-and-leaf plot shows the test scores for two classes taught by the same teacher. Compare the samples using measures of center and variation. Are the test scores significantly greater for one class than the other? Explain.

Class A		Class B
0 0 5 5	7	0 0 0 5 5 5 5 5
0 0 5 5 5 5	8	0 0 0 0 0 5 5 5 5
0 0 0 5 5 5 5 5 5	9	0 0 0 5 5
0 0 0	10	0

Key: 5|7|0 = 75 and 70

13. USING MULTIPLE RANDOM SAMPLES Two lakes each contain about 2000 fish. The double box-and-whisker plot shows the medians of 50 random samples of 14 fish lengths from each lake. Can you determine which lake contains longer fish? Explain. (See Example 3.)

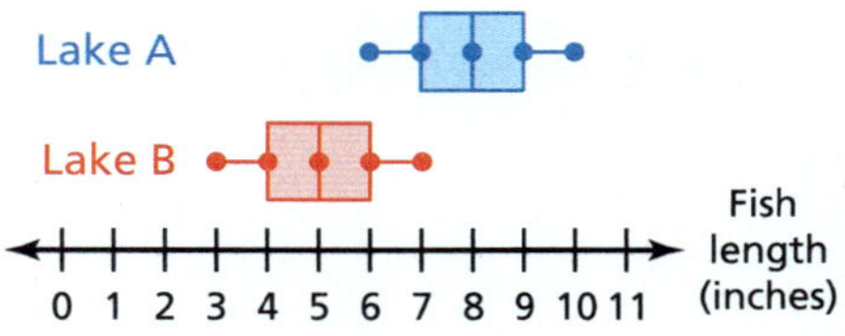

14. MODELING REAL LIFE Two laboratories each produce 800 chemicals. A chemist takes 10 samples of 15 chemicals from each lab, and records the number that pass an inspection. Are the samples likely to be representative of all the chemicals for each lab? If so, which lab has more chemicals that will pass the inspection? Justify your answer.

Research Lab A				
14	13	15	15	14
14	15	15	13	12

Research Lab B				
12	10	12	14	11
9	14	11	11	15

7 MTR **15. MODELING REAL LIFE** A farmer grows two types of corn seedlings. There are 1000 seedlings of each type. The double box-and-whisker plot represents the median growths of 50 random samples of 7 corn seedlings of each type. Compare the growths of each type of corn seedling. Justify your result. (See Example 4.)

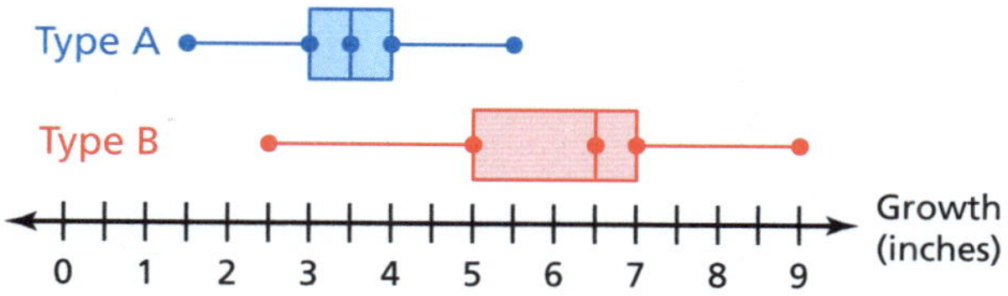

4 MTR **16. MAKING AN ARGUMENT** The histograms show the populations of large cities in Florida and New York. Your friend says that you cannot compare the populations in the histograms because you cannot calculate the measures of center and variation. Is your friend correct? Explain your reasoning.

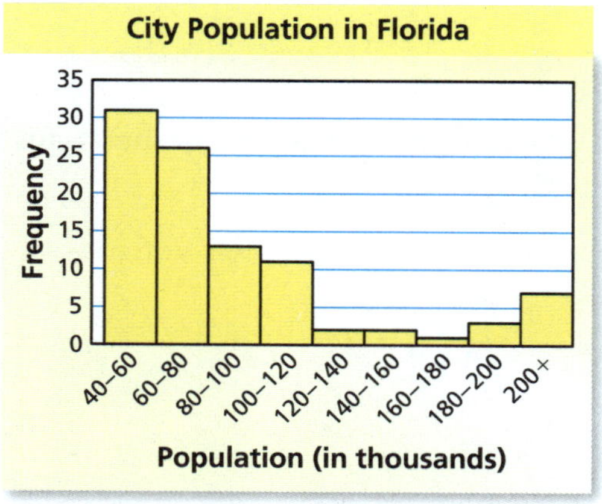

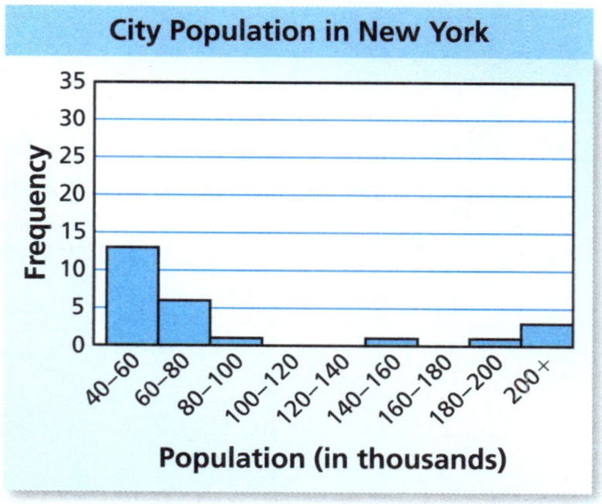

17. **Dig Deeper** You want to compare the number of words per sentence in a sports magazine to the number of words per sentence in a political magazine.

a. The data represent random samples of the number of words in 10 sentences from each magazine. Compare the samples using measures of center and variation. Can you use the data to make a valid comparison about the magazines? Explain.

Sports magazine: 9, 21, 15, 14, 25, 26, 9, 19, 22, 30

Political magazine: 31, 22, 17, 5, 23, 15, 10, 20, 20, 17

b. The double box-and-whisker plot represents the means of 200 random samples of 20 sentences from each magazine. Compare the variability of the sample means to the variability of the sample numbers of words in part (a).

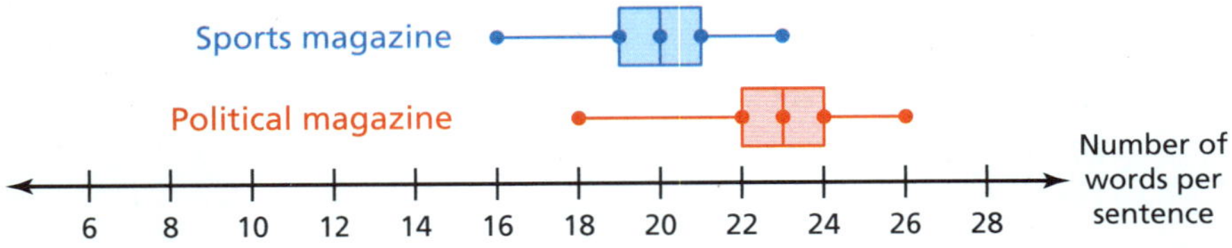

c. Make a conclusion about the numbers of words per sentence in each magazine.

7 MTR

18. **INVESTIGATE** You want to compare the average amounts of time students in sixth, seventh, and eighth grade spend on homework each week.

a. Design an experiment involving random sampling that can help you make a comparison.

b. Perform the experiment. Can you make a conclusion about which grade spends the most time on homework? Explain your reasoning.

12 Connecting Concepts

5 MTR *Mathematicians who use patterns and structure to help understand and connect mathematical concepts relate previously learned concepts to new concepts.*

1 MTR **1. ANALYZE A PROBLEM** A game at a carnival uses the number cube shown. You win the game if you roll an even number. You watch 20 people play the game and calculate that the experimental probability of winning is 10% less than the theoretical probability. How many people won the game?

A number cube has numbers 1–6 on each face.

Using the Problem-Solving Plan

Understand the problem. You have the information needed to calculate the theoretical probability of winning the game, and you know how the experimental probability relates to the theoretical probability. You want to find the number of people who won the game.

Make a plan. Find the theoretical probability of winning the game, and use it to find the experimental probability. Then set up a proportion to find the number of people who won the game.

Solve and check. Use the plan to solve the problem. Then check your solution.

2. The dot plots show the values in two data sets. Compare the data sets using the measure of center and variation.

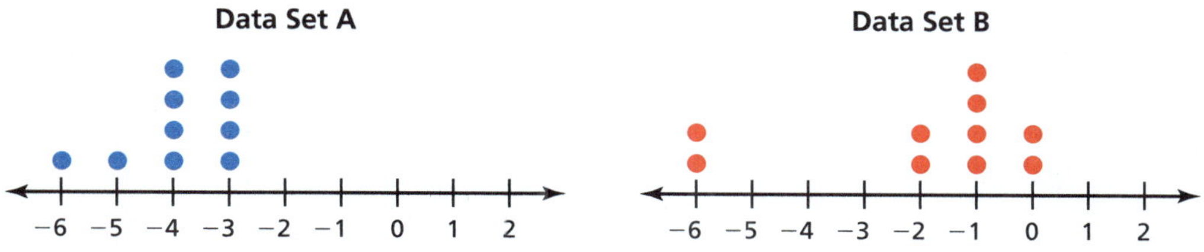

STEAM Performance Task — Fair and Unfair Carnival Games

GO DIGITAL

At the beginning of this chapter, you watched a STEAM video called "Massively Multiplayer Rock Paper Scissors." You are now ready to complete the STEAM Performance Task, where you will design your own "unfair" carnival game using a spinner and a number cube. Be sure to use mathematical thinking and reasoning, and the problem-solving plan as you work through the performance task.

12 Chapter Review WITH CalcChat®

Review Vocabulary

Write the definition and give an example of each vocabulary term.

experiment, *p. 1005*

probability, *p. 1006*

population, *p. 1031*

sample space, *p. 1005*

relative frequency, *p. 1007*

sample, *p. 1031*

outcomes, *p. 1005*

experimental probability, *p. 1019*

unbiased sample, *p. 1033*

event, *p. 1005*

theoretical probability, *p. 1019*

biased sample, *p. 1033*

favorable outcomes, *p. 1005*

simulation, *p. 1021*

GO DIGITAL

Graphic Organizers

You can use a **Four Square** to organize information about a concept. Each of the four squares can be a category, such as definition, vocabulary, example, non-example, words, algebra, table, numbers, visual, graph, or equation. Here is an example of a Four Square for ***probability***.

Probability

Definition
The likelihood that an event will occur

Words
Probabilities are between 0 and 1, including 0 and 1. They can be written as fractions, decimals, or percents.

Visual

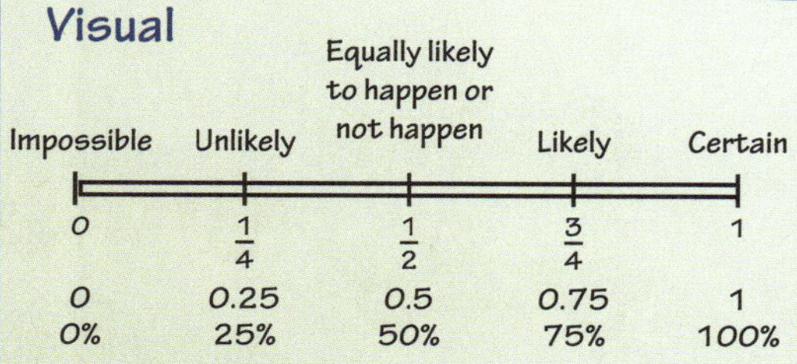

Example
There is a 15% chance of snow tomorrow. Describe the likelihood of snow tomorrow.
The probability of snow tomorrow is 15%. Because 15% is close to 25%, it is unlikely that it will snow tomorrow.

Choose and complete a graphic organizer to help you study the concept.

1. sample space
2. event
3. favorable outcomes
4. relative frequency
5. experimental probability
6. theoretical probability
7. simulation
8. population

Chapter Learning Target: Understand probability and representing data.

Chapter Success Criteria:
- ◆ I can find the sample space of an experiment.
- ◆ I can find experimental and theoretical probabilities.
- ■ I can use an unbiased sample to make a conclusion about a population.
- ■ I can compare populations.

◆ Surface
■ Deep

Rate your understanding after each section.

1	2	3	4
I don't understand yet.	I can do it with help.	I can do it on my own.	I can teach someone else.

12.1 Probability *(pp. 1003–1016)*

Learning Target: Understand how the probability of an event indicates its likelihood.

You randomly choose one toy race car.

1. Find the sample space. How many possible outcomes are there?

2. What are the favorable outcomes of choosing a car that is *not* green?

3. In how many ways can choosing a green car occur?

You spin the spinner. (a) Find the number of ways the event can occur. (b) Find the favorable outcomes of the event.

4. spinning a 1

5. spinning a 3

6. spinning an odd number

7. spinning an even number

GO DIGITAL

Describe the likelihood of the event given its probability.

8. There is a 0% chance of snow in July for Florida.

9. The probability that you are called on to answer a question in class is $\frac{1}{25}$.

10. There is an 85% chance the bus will be on time.

11. Tomorrow there is a 70% chance the pollen count will be high, a 65% chance the dust count will be high, and a 15% chance the grass count will be high.

a. Describe the likelihood that the grass count is high tomorrow.

b. Which is more likely tomorrow, a high pollen count or a high dust count?

12. During a basketball game, you record the number of rebounds from missed shots for each team. (a) Describe the likelihood that your team rebounds the next missed shot. (b) How many rebounds should your team expect to have in 15 missed shots?

Your Team	𝍸 II
Opposing Team	III

12.2 Experimental and Theoretical Probability (pp. 1017–1030)

Learning Target: Develop probability models using experimental and theoretical probability.

The bar graph shows the results of spinning a spinner 100 times. Use the bar graph to find the experimental probability of the event.

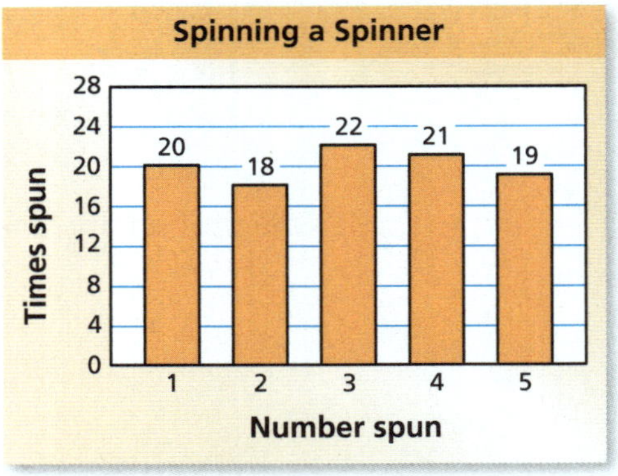

13. spinning a 2

14. spinning an even number

15. *not* spinning a 5

Use the spinner to find the theoretical probability of the event.

16. spinning blue

17. spinning a 1

18. spinning an even number

19. spinning a 4

20. You simulate selecting a numbered ball out of a bag 150 times by using a spreadsheet to randomly generate a list of 150 numbers from 1 through 6. The table shows the results.

Number Selected	1	2	3	4	5	6
Frequency	34	28	16	22	19	31

How does the experimental probability of choosing a number *greater* than 3 compare with the theoretical probability?

12.3 Samples and Populations (pp. 1031–1040)

Learning Target: Understand how to use random samples to make conclusions about a population.

21. You want to estimate the number of students in your school whose favorite subject is biology. You survey the first 10 students who arrive at biology club. Determine whether the sample is *biased* or *unbiased*. Explain.

22. You want to know how the residents of your town feel about building a new baseball stadium. You randomly survey 100 people who enter the current stadium. Eighty support building a new stadium, and twenty do not. So, you conclude that 80% of the residents of your town support building a new baseball stadium. Is your conclusion valid? Explain.

Favorite Subject	
English	22
Art	10
Science	4
Other	4

23. The favorite subject of 40 randomly chosen seventh-grade students at a school are shown. There are 460 seventh-grade students at the school.

a. Estimate the number of seventh-grade students at the school who prefer English.

b. Estimate the number of seventh-grade students at the school who prefer art.

12.4 Comparing Populations (pp. 1041–1054)

Learning Target: Compare populations using measures of center, measures of variation, and random samples.

24. The double box-and-whisker plot represents the points scored per game by two football teams during the regular season.

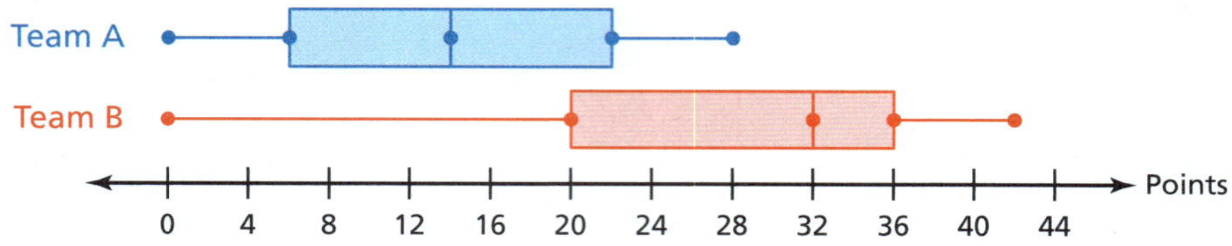

a. Compare the data sets using measures of center and variation.

b. Which team is more likely to score 18 points in a game?

25. The double dot plot shows the median gas mileages of 10 random samples of 50 vehicles for two car models. Compare the samples using measures of center and variation. Can you determine which car model gets better gas mileage? Explain.

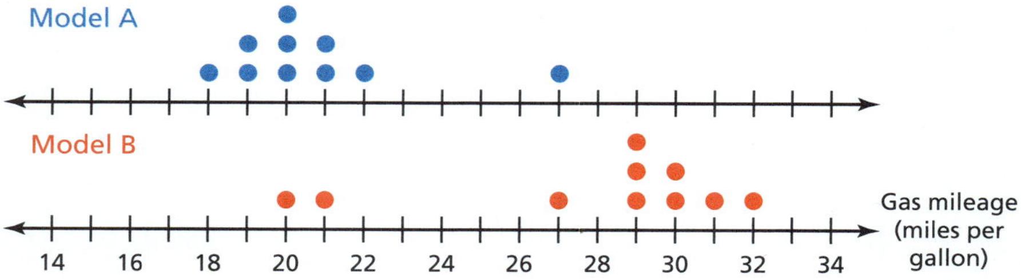

26. You compare the average amounts of time people in their twenties and thirties spend driving each week. The double box-and-whisker plot represents the medians of 100 random samples of 8 people from each age group. Can you determine whether one age group drives more than the other? Explain.

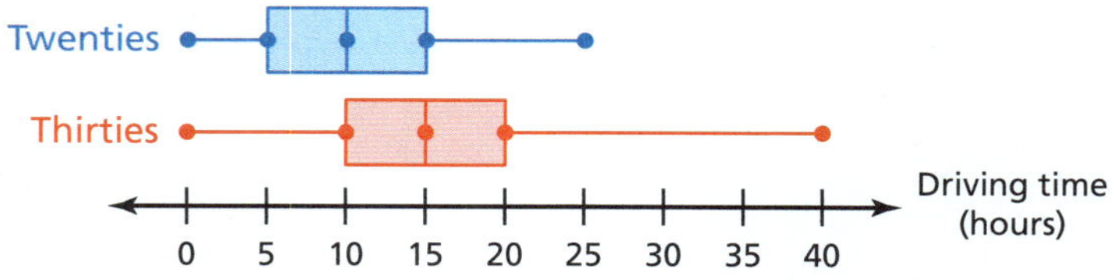

12 Practice Test WITH CalcChat®

You randomly choose one game piece.

1. Find the sample space. How many possible outcomes are there?

2. What are the favorable outcomes of choosing a piece that is *not* yellow?

3. How many ways can choosing a blue piece occur?

4. Describe the likelihood of choosing a red piece.

Use the bar graph to find the experimental probability of the event.

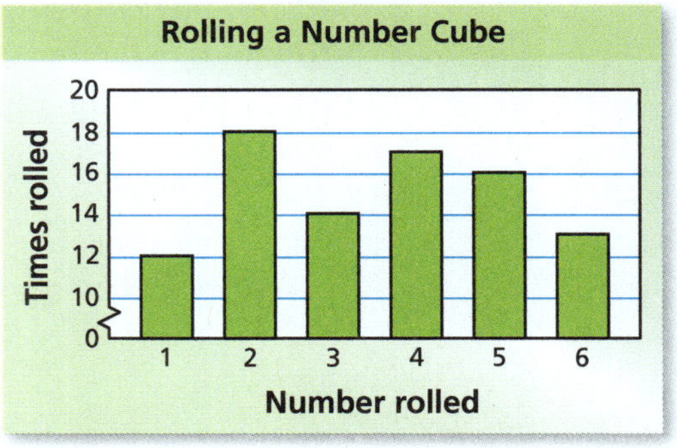

5. rolling a 1 or a 2

6. rolling an odd number

7. *not* rolling a 5

8. rolling a number less than 7

GO DIGITAL

9. You simulate rolling a number cube 220 times by using a spreadsheet to randomly generate a list of 220 numbers from 1 through 6. The table shows the results.

Number Rolled	1	2	3	4	5	6
Frequency	42	28	36	40	43	31

How does the experimental probability of rolling an even number compare with the theoretical probability?

10. You want to estimate the number of students in your school who prefer to bring a lunch from home rather than buy one at school. You survey five students who are standing in the lunch line. Determine whether the sample is *biased* or *unbiased*. Explain.

11. Of 60 randomly chosen students from a school surveyed, 15 chose the aquarium as their favorite field trip. There are 720 students in the school. Predict the number of students in the school who choose the aquarium as their favorite field trip.

12. Two airplanes each hold about 400 pieces of luggage. The double dot plot shows a random sample of 8 pieces of luggage from each plane. Compare the samples using measures of center and variation. Can you determine which plane has heavier luggage? Explain.

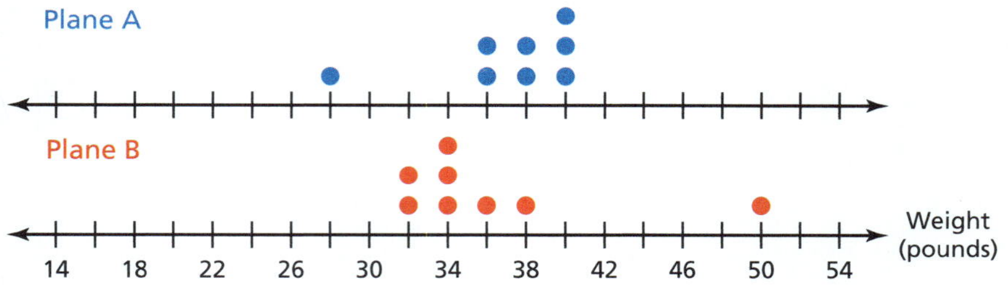

12 Review & Refresh WITH CalcChat®
while Building Fluency

Divide. Write fractions in simplest form.

1. $\frac{8}{15} \div \frac{3}{5}$

2. $-\frac{2}{3} \div \frac{4}{9}$

3. $4\frac{1}{6} \div 2\frac{1}{3}$

Use the Distributive Property to simplify the expression.

4. $4(3t - 5)$

5. $-7(2 - n)$

6. $\frac{1}{2}(-2h + 1)$

Find the mean, median, mode, and range of the data set.

7. $\{2.2, 8.3, 5.4, 12.7, 5.4\}$

8. $\{30.5, 14.6, 29.3, 30.5, 19.2, 24.1\}$

9. In Exercise 7, which measures of center and variation change when each number in the data set is doubled?

GO DIGITAL

10. The minimum height limit on a roller coaster is 5 feet. is 63 inches tall. Can your friend ride the roller coaster?

11. A car wash charges a 5% service fee for every transaction. Your total is \$14 plus the service fee. How much do you spend at the car wash?

Find the selling price.

12. Cost to store: \$40
Markup: 5%

13. Cost to store: \$96
Markup: 64.5%

14. Cost to store: \$182.50
Markup: 30%

Factor the algebraic expression.

15. $4b - 12$

16. $-8 + 24x$

17. $-10p + 60$

18. A subway car has a maximum capacity of 110 people. There are currently 84 people in the subway car. Write and graph an inequality that represents the number of people who can enter the subway car.

The Coral Castle in Homestead, Florida, is said to have been built by a single man in 1920, defying the laws of physics.

19. A coral rock at the Coral Castle is 3 meters tall. The wall of the castle is 244 centimeters tall. What is the difference of the heights (in centimeters) of the coral rock and the wall of the castle?

12 B.E.S.T. Test Prep WITH CalcChat®

Cumulative Practice

1. A school athletic director asked each athletic team member to name his or her favorite professional sports team. The results are below:

- D.C. United: 3
- Florida Panthers: 8
- Jacksonville Jaguars: 26
- Jacksonville Sharks: 7
- Miami Dolphins: 22
- Miami Heat: 15
- Miami Marlins: 20
- Minnesota Lynx: 4
- New York Knicks: 5
- Orlando Magic: 18
- Tampa Bay Buccaneers: 17
- Tampa Bay Lightning: 12
- Tampa Bay Rays: 28
- Other: 15

One athletic team member is picked at random. What is the likelihood that this team member's favorite professional sports team is *not* located in Florida?

Ⓐ certain

Ⓑ likely, but not certain

Ⓒ unlikely, but not impossible

Ⓓ impossible

2. Find the value of x.

GRIDDED RESPONSE

$$-\frac{1}{4} + \frac{3}{5} = \frac{x}{20}$$

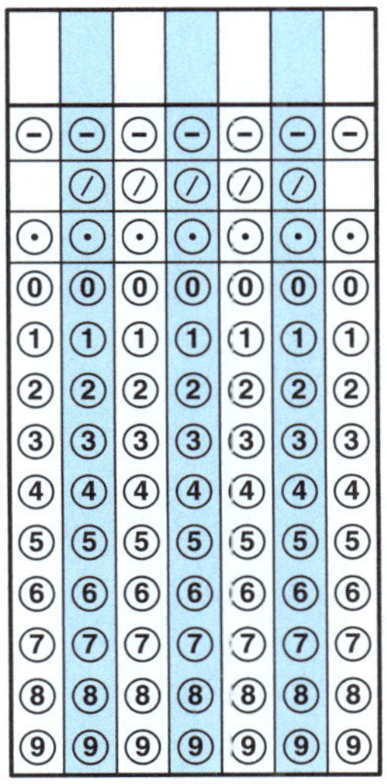

GO DIGITAL

3. A pair of running shoes is on sale for 25% off the original price. Which price is closest to the sale price of the running shoes?

Ⓐ \$93

Ⓑ \$99

Ⓒ \$124

Ⓓ \$149

4. GRIDDED RESPONSE

Each student in your class voted for his or her favorite day of the week. Their votes are shown in the table.

Favorite Day of the Week	
Friday	8
Saturday	10
Sunday	6
Other	6

A student from your class is picked at random. What is the probability that this student's favorite day of the week is Sunday?

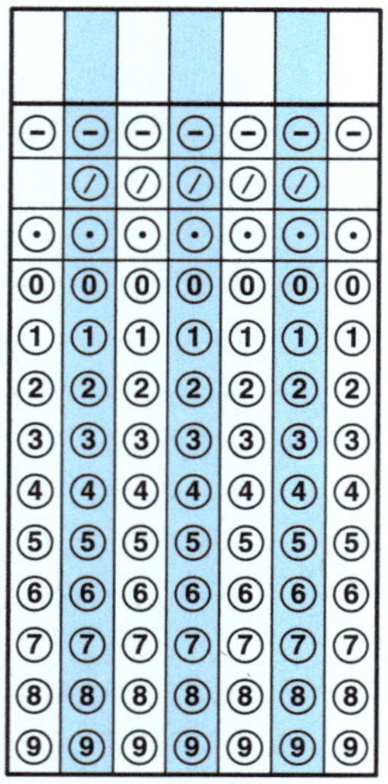

5. Which value is a solution of the inequality?

$$3 - 2y < 7$$

Ⓐ -6

Ⓑ -3

Ⓒ -2

Ⓓ -1

6. Your friend solved the proportion in the box below.

$$\frac{16}{40} = \frac{p}{27}$$

$$16 \cdot p = 40 \cdot 27$$

$$16p = 1080$$

$$\frac{16p}{16} = \frac{1080}{16}$$

$$p = 67.5$$

What should your friend do to correct the error that he made?

Ⓐ Add 40 to 16 and 27 to p.

Ⓑ Subtract 16 from 40 and 27 from p.

Ⓒ Multiply 16 by 27 and p by 40.

Ⓓ Divide 16 by 27 and p by 40.

Selected Answers

Chapter 7

Chapter Exploration

1. What number can be added to 3 to get 7?; 4
2. What number can be subtracted from 5 to get 2?; 3
3. What number can be multiplied by 3 to get 12?; 4
4. What number can be divided by 5 to get 3?; 15
5. What number can 20 be divided by to get 4?; 5
6. $x + 7 = 12$; 5
7. $11 - x = 3$; 8
8. $10x = 30$; 3
9. $x \div 7 = 3$; 21
10. $x + 7 = 21$; 14 years old

Section 7.1

7.1 Review & Refresh

1. $3(2 + 9)$
2. $9(w + 8)$
3. $6(7 + 4n)$
4. $6(3h + 5k)$
5. $2^5 \cdot 3$
6. 13
7. 3
8. 28
9. 5
10. 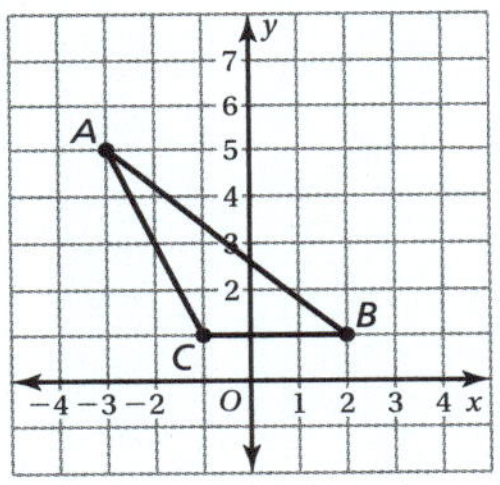

11. 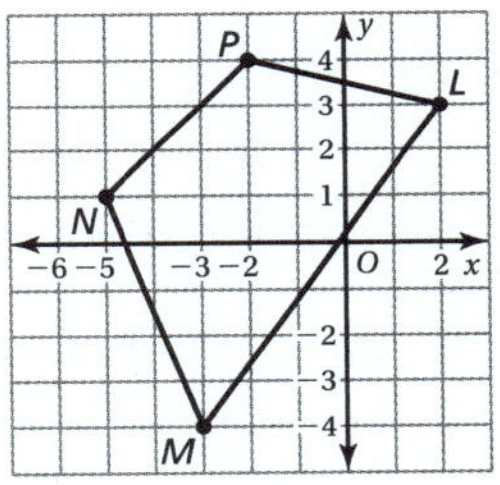

12. 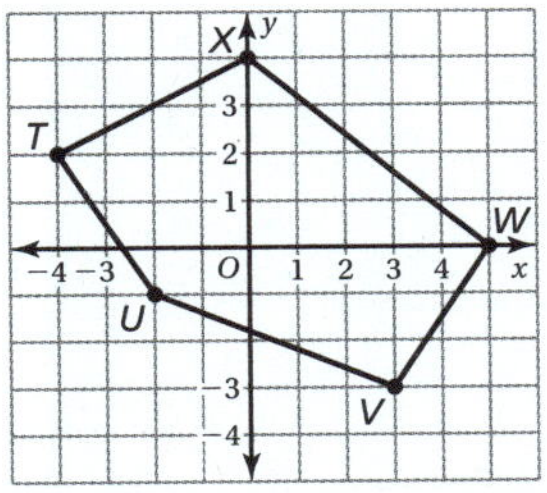

7.1 Concepts, Skills, & Problem Solving

13. $4x = 8$; 2
15. $y - 9 = 8$
17. $-9b = 36$
19. $54 = t + 9$
21. $n - 9 = 27$
23. yes; "5 less than a number n" means "$n - 5$" and "is 12" means "$= 12$".
25. $40 = \frac{1}{3}d$
27. $30 = 4x$
29. **a.** 76 hours
    b. *Sample answer:* Apollo 11 was pulling away from Earth and against Earth's gravity on the journey to lunar orbit, but was moving towards Earth and with Earth's gravity on the journey from lunar orbit.
31. $14w - \frac{w^2}{4} = 75$

Section 7.2

7.2 Review & Refresh

1. $x - 3 = 12$
2. $y + 11 = 19$
3. $4 = w \div 6$
4. $7z = 35$
5. $(-4, 3)$
6. $(-2, -4)$
7. $(4, 2)$
8. $(1, -2)$
9. -64
10. 0
11. 360

7.2 Concepts, Skills, & Problem Solving

13. $x + \frac{2}{3} = \frac{8}{9}; x = \frac{2}{9}$
15. yes
17. no
19. yes
21. yes
23. 6; *Sample answer:* use mental math and think "How many more than 7 is 13?"
25. -14; *Sample answer:* use mental math and think "What number added to 19 is 5?"
27. 4.7; *Sample answer:* use mental math and think "What number is 4.3 greater than 0.4?"
29. $\frac{3}{7}$; *Sample answer:* draw a model and think "$\frac{5}{7}$ is how much greater than $\frac{2}{7}$?"
31. $\frac{8}{9}$; *Sample answer:* draw a model and think "$1\frac{1}{3}$ is how much less than $2\frac{2}{9}$?"
33. 3; *Sample answer:* use mental math and think "18 divided by what number is 6?"
35. 9; *Sample answer:* use mental math and think "4 times what number is 36?"
37. 15; *Sample answer:* draw a model and think "3 is $\frac{1}{5}$ of what number?"
39. 8; *Sample answer:* use mental math and think "7.2 divided by what number is 9?"
41. yes; *Sample answer:* By thinking "what number is 0.6 greater than 3.4?", 4 is 0.6 greater than 3.4. So, $x = 4$.
43. 4.1 miles
45. 8 groups
47. no
49. *Sample answer:* The smaller bracket points to 1 of the 8 pieces in the tape diagram, so label that $\frac{1}{8}$. The second bracket points to 3 of the 8 pieces, so label that $\frac{3}{8}$.
51. Answers will vary. *Sample answer:* You are saving up to buy a video game priced at \$60. You currently have \$29. How much do you need to save?; You buy 3 bottles of water and spend a total of \$4.80. What is the price of one bottle of water?; \$31; \$1.60

53. 6; *Sample answer:* Try setting $x > 3$, so both sides of the equation are positive. Test the positive factors of 18 that are greater than 3, such as 6, 9, and 18.

Section 7.3

7.3 Review & Refresh

1. yes **2.** no **3.** yes **4.** 2

5. 32 **6.** 10 **7.** $\frac{14}{27}$ **8.** 14

9. 9 **10.** 25

11. 36 **12.** 63.8 sec

7.3 Concepts, Skills, & Problem Solving

13. $x + 4 = -8; x = -12$

15. $y = 10$ **17.** $r = 22$ **19.** $k = 12$ **21.** $f = 46$

23. $x = -20$ **25.** $w = -30$

27. no; Subtract 7 to get $x = 6$.

29. $366 = x - 30$; 396 m **31.** $w - 13 = 15; w = 28$

33. $93 = g + 58; g = 35$

35. $a = 22; b = 31; c = 13$; The constant sum is 75, and $a + 53 = 75$, $b + 44 = 75$, and $c + 62 = 75$.

37. **a.** $5

b. no; It costs $20 to ride each ride twice.

Section 7.4

7.4 Review & Refresh

1. $y = 11$ **2.** $m = 1$

3. $p = -3$ **4.** 64

5. $\frac{2}{45}$ **6.** $\frac{5}{21}$ **7.** $\frac{7}{10}$ **8.** $4\frac{2}{3}$

9. 0.36 **10.** 0.39

11. 11.3353 **12.** 3.6572

7.4 Concepts, Skills, & Problem Solving

13. $x = 1$ **15.** $s = 70$ **17.** $a = 4$ **19.** $y = 10$

21. $x = 15$ **23.** $v = -45$

25. $s = -40$

27. no; Multiply by 4 to get $x = 112$.

29. $60 = 12x$; *Sample answer:* $6x$ is half of $12x$, so $6x = 30$.

31. 9 **33.** $3x = 45$; 15

35. *Sample answer:* $2 = f \div 30$ represents dividing f flowers into flower arrangements containing 30 flowers each, resulting in 2 flower arrangements; $f \div 20 = 3$ represents dividing f flowers into flower arrangements containing 20 flowers each, resulting in 3 flower arrangements; They are equivalent equations because they have the same solution, $f = 60$.

37. length: 20 in.; width: 5 in.

Section 7.5

7.5 Review & Refresh

1. $x = -10$ **2.** $y = 54$

3. $t = 48$ **4.** $x = 9$

5. 4 **6.** 12 **7.** 8 **8.** 64

9. 3 **10.** 32.18 **11.** 6.625

7.5 Concepts, Skills, & Problem Solving

13. 5 or any number greater than 5

15. 10 or any number less than 10

17. $a > 6$ **19.** $b \geq -3$

21. $25 \leq 7 + p$ **23.** $\frac{x}{3} \leq 5$

25. no **27.** no **29.** yes **31.** yes

33. A **35.** C

37.

39.
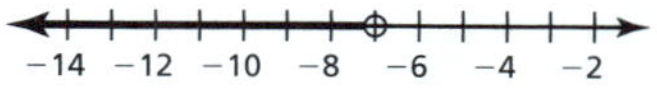

41.
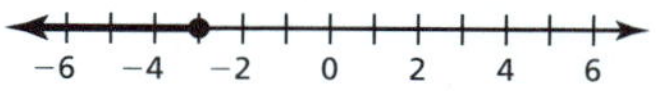

43.

45.

47.

49. $x \leq 1$; A number x is at most 1.

51. $x > 0$; A number x is more than 0.

53. no; The graph shows $x < -1$, not $x > -1$.

55. **a.** $b \leq 3$;

b. $\ell \geq 18$;
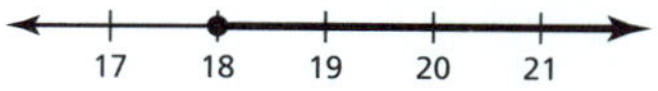

57. **a.** $s \leq 140$;

b. $s > 140$;
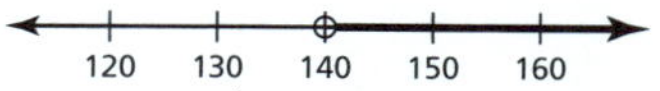

c. no; There are 300 milligrams of sodium in one serving, which is greater than 140.

59. sometimes; The only time this is not true is if $x = 5$.

61. $p + 50 \leq 425$

Section 7.6

7.6 Review & Refresh

1. $p > 5$ **2.** $3z \leq -4.8$

3. $n + \frac{2}{3} \geq 5\frac{1}{3}$ **4.** $x = 9$

5. $w = -27$ **6.** $b = -22$ **7.** $h = 80$

8. a.

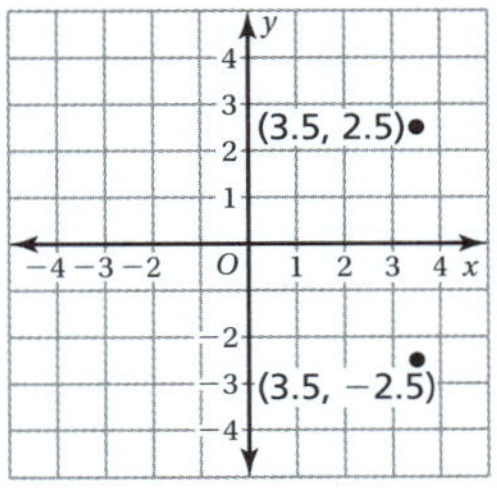

b.

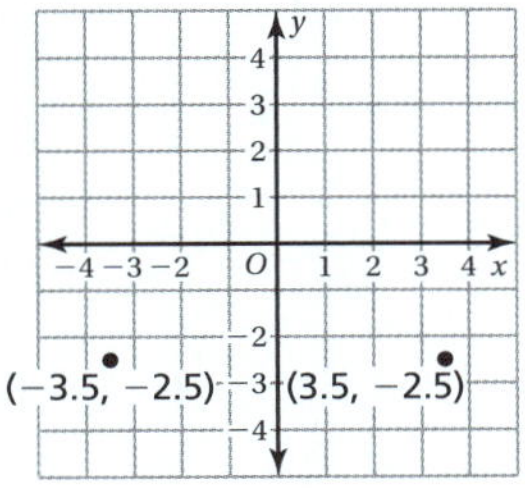

7.6 Concepts, Skills, & Problem Solving

9. $-1 < 4$; yes; $1 < 6$

11. $-4 < -1$; yes; $-2 < 1$

13. $a > 6$

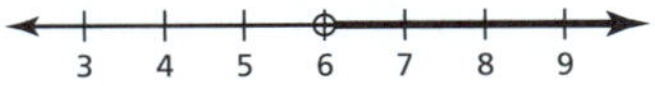

15. $k \leq -11$

17. $n < 9$

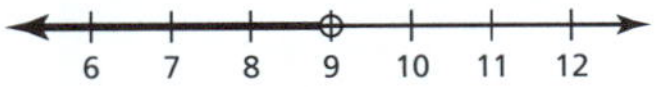

19. $p \geq 1\frac{3}{4}$

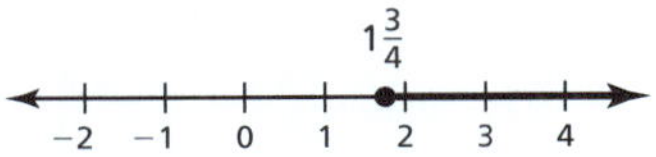

21. $z \geq 3.1$

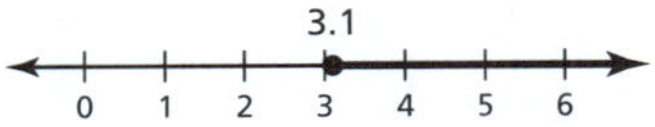

23. $-\frac{4}{5} > s$

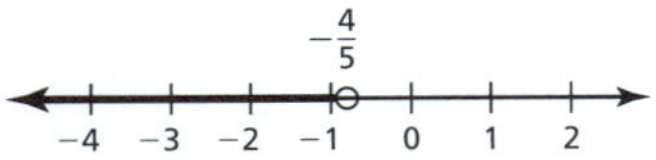

25. $r < -0.9$

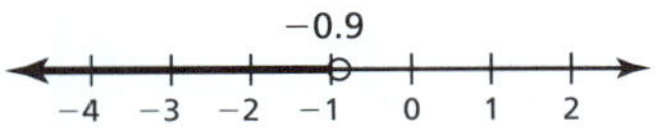

27. no; The graph should have an open circle at 5.

29. a. $x + 17 \leq 30$; $x \leq 13$

b. no; *Sample answer:* No more than 13 passengers can board the airboat.

31. $x > 5$ in. **33.** 4

35. a. $x \leq 1350$ watts

b. *Sample answer:* aquarium and television; yes; $200 + 150 = 350$ watts, $200 + 1100 = 1300$ watts

Section 7.7

7.7 Review & Refresh

1. $h < 2$

2. $c \geq 9$

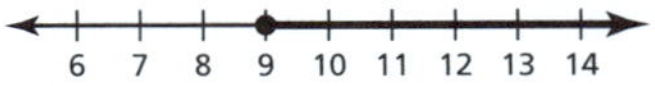

3. $n \geq -\frac{1}{10}$

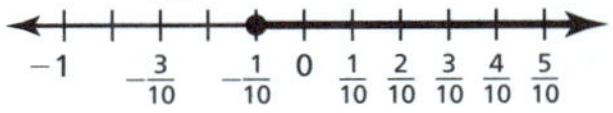

4. $w = -16$ **5.** $v = 15$

6. $x = 6$ **7.** $m = -12$

8. $-\frac{1}{21}$

7.7 Concepts, Skills, & Problem Solving

9. $-2 < 5$; yes; no; $-4 < 10$, $4 \nless -10$

11. $6 > -3$; yes; no; $12 > -6$, $-12 \ngtr 6$

13. $c \leq -36$

15. $x < -32$

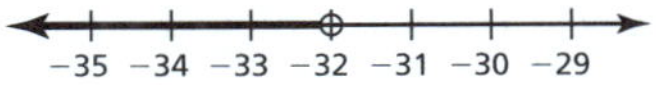

17. $k > 2$

19. $y \leq -3$

21. $9.2x \geq 299$; $x \geq 32.5$ h

23. $n \geq -3$

25. $h \leq -24$

27. $y > \frac{14}{3}$

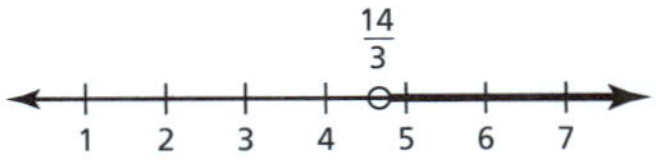

29. $m < -27$

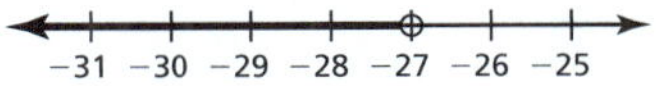

31. $b > 6$

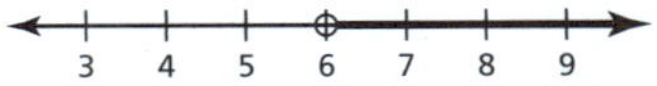

33. yes; The properties of inequalities were all used correctly to find the solution.

35. $\frac{p}{7} < -3$; $p < -21$

37. $-2x > 30$; $x < -15$

39. $-2.5s < -20$; $s > 8$ sec

41. $10x \geq 120$; $x \geq 12$ cm

43. at least 5 days; The solution of $37d \geq 185$ is $d \geq 5$.

45. **a–c.** Answers will vary.

47. $m > -1$ and $m < 5$

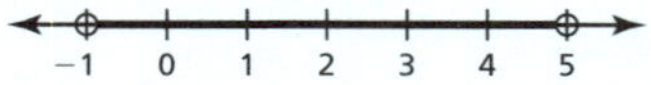

49. $x \geq 3$

Chapter 7 Review

1. $m \cdot (-2) = 8$ **2.** $t - 6 = -7$

3. $m + 5 = 7$ **4.** $8 = g \div 3$ **5.** $\frac{1}{3}h = 6$

6. no **7.** yes **8.** yes **9.** no

10. $y = 5.46$; *Sample answer:* Think "How many more than 5.54 is 11?"

11. $t = 3$; *Sample answer:* Think "How many times -3 is -9?"

12. $n = \frac{12}{7}$; *Sample answer:* Draw a tape diagram and think "What number multiplied by $\frac{7}{27}$ is $\frac{4}{9}$?"

13. $s = 15.2$; *Sample answer:* Think "What number divided by 5 is 3.04?"

14. $5m = 35$; $m = 7$ **15.** $x = 9$

16. $m = 4$ **17.** $p = -9$

18. $x + 5 = -9$; $x = -14$ **19.** $82 = b - 24$; $b = 106$

20. 12 lb **21.** $q = 9$

22. $k = -63$ **23.** $a = 75$

24. $48 **25.** $m < 5$

26. $-14 \geq 2h$ **27.** yes

28. no **29.** no

30.

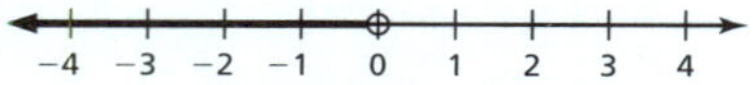

31.

32.

33. $s \leq 35$;

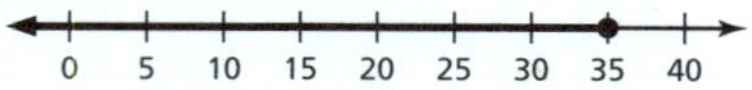

34. $d < 7$

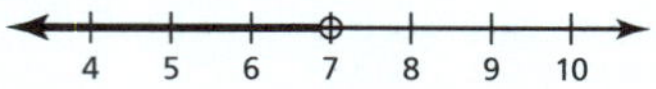

35. $t \leq -10$

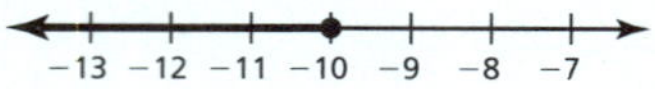

36. $z \geq -14.4$

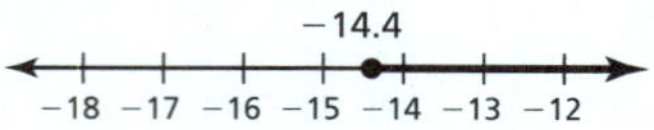

37. **a.** $p + 115 \leq 500$; $p \leq 385$ people

b. yes; *Sample answer:* The cruise ship can hold less than or equal to 385 more people.

38. *Sample answer:* $x + 5 < 2$

39. $q < -3$

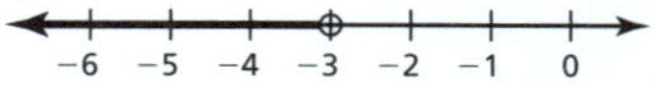

40. $r \geq -18$

41. $3 < s$

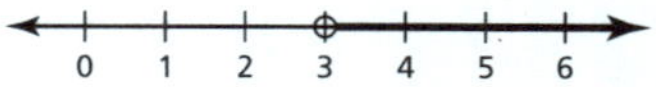

42. $-3p > 21$; $p < -7$ **43.** $k(5) \leq -25$; $k \leq -5$

44. $\frac{3}{7} < \frac{m}{4}$; $\frac{12}{7} < m$ **45.** $\frac{3}{4}b \leq 24$; $b \leq 32$

Chapter 7 Practice Test

1. $7s = 84$ **2.** $b = 22$

3. $v = 22$ **4.** $x = 14$

5. $m = -18$ **6.** $2s = 366$

7. 83 **8.** $t < 7$ **9.** $8 \geq 3p$

10. $x > -2$

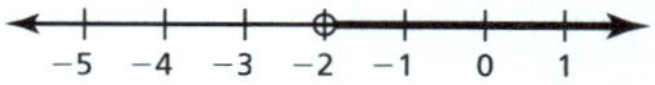

11. $y \leq \frac{7}{9}$

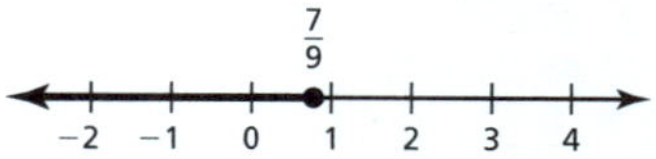

12. $z \leq -6$

13. $p \leq -20.8$

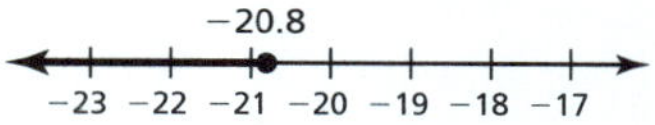

14. 52 shirts **15.** 30 **16.** $w \geq 74$

17. **a.** $8g \leq 100$; $g \leq 12.5$

b. 12; 12 is the greatest whole number that satisfies the inequality.

Chapter 7 Review & Refresh

1. $\frac{1}{12}$ **2.** -3.15 **3.** $-\frac{3}{20}$

4.

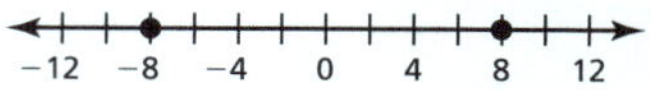

5.

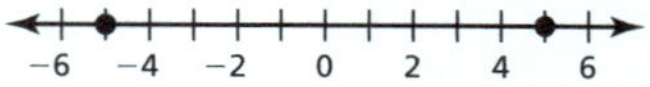

6.

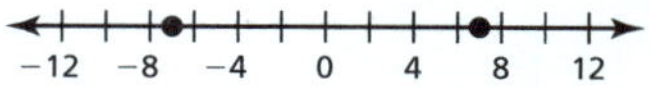

7.

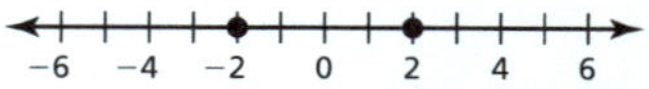

8. 12 in. **9.** -1174 **10.** t^3 **11.** $3y^4$

12. $-6p^5$ **13.** $15(2 + 3)$

14. $9(2 - 7)$ **15.** $3(-2 + 11)$

16. -5 ft **17.** 4 articles per day

18. -17 **19.** 9 **20.** -63

21. a. 18 ft
    b. 16.64 ft^2
    c. 8.32 ft^2

Chapter 7 B.E.S.T. Test Prep

1. C   2. 53   3. D   4. A
5. 300   6. D   7. C   8. A
9. C; C

Chapter 8

Chapter Exploration

1. *Sample answer:* $\frac{4}{12}$; as a fraction
2. *Sample answer:* $\frac{12}{18}$; as a fraction
3. *Sample answer:* $\frac{5}{9}$; as a fraction
4. *Sample answer:* $\frac{4}{9}$; as a fraction
5. *Sample answer:* $\frac{6}{9}$; as a fraction
6. *Sample answer:* $\frac{16}{24}$; as a fraction
7. *Sample answer:* $\frac{4}{12}$; as a fraction
8. *Sample answer:* $\frac{5}{12}$; as a fraction
9. *Sample answer:* $\frac{6}{6}$; as a fraction
10. 1 and 7; 2, 5, and 6; *Sample answer:* $\frac{12}{18} = \frac{6}{9} = \frac{16}{24}$
11. *Sample answers:*

R	B	B	B	B	B

R	B	B	B
B	R	B	B
B	B	B	B

12. *Sample answers:*

Y	Y	Y	G

G	G	G	Y
Y	Y	Y	Y
Y	Y	Y	Y

13. *Sample answers:*

G	G	B	G	G

G	B	G	G	G
G	G	G	B	G

14. *Sample answers:*

Y	Y	R	R	R	R	R

R	R	R	Y	R	R	Y
Y	R	R	Y	R	R	R

15. no; $\frac{8}{14} > \frac{9}{17}$; 2

Section 8.1

8.1 Review & Refresh

1. $x \leq -6$

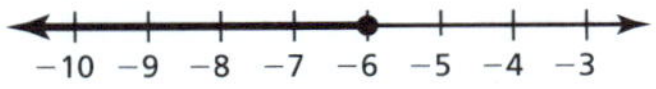

2. $d > 12$

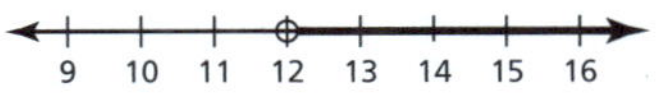

3. $g \geq -8$

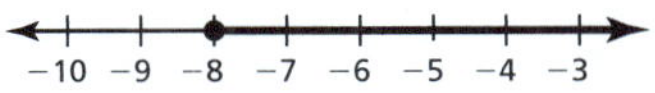

4. 56   5. 0   6. 47   7. 50
8. $(-3, -4)$ 9. $(2, -1)$ 10. $(1, 3)$

8.1 Concepts, Skills, & Problem Solving

11. Add 2 parts of iced tea for every 1 part of lemonade added.
13. 2 to 5, or 2 : 5
15. 2 to 6, or 2 : 6
17. 12 : 16
19. *Sample answer:* 1 out of every 7 contestants wins a prize.
21. *Sample answer:* 2 days in a weekend per 5 weekdays
23. a. 4; The number of sunny days is 4 times the number of rainy days.
    b. 5 : 1
25. equivalent
27. not equivalent
29. not equivalent
31. equivalent
33. 18
35. 21
37. *Sample answer:* 2 cups of water and 4 cups of cornstarch, 3 cups of water and 6 cups of cornstarch; $\frac{1}{2} = \frac{2}{4} = \frac{3}{6}$
39. 21
41. yes
43. no; The ratio of boys to people in the competition is 3 : 5, and there is no ratio of a whole number to 9 that is equivalent to the ratio 3 : 5.
45. 67.5 in.; *Sample answer:* 2 : 3 : 4 is equivalent to 15 : 22.5 : 30, and 15 + 22.5 + 30 = 67.5.
47. 5 girls; *Sample answer:* $\frac{12}{10} = \frac{18}{15}$

Section 8.2

8.2 Review & Refresh

1. equivalent
2. equivalent
3. not equivalent
4. not equivalent
5. $\frac{1}{2}$
6. $1\frac{3}{4}$
7. $13\frac{7}{16}$
8. \$105.45

8.2 Concepts, Skills, & Problem Solving

9. 800 m
11. beginner trail: 400 m, expert trail: 1600 m
13. 6 h   15. 2   17. 9   19. 27
21. 12 blueberries, 4 strawberries
23. 4 blueberries, 8 strawberries
25. 36 blueberries, 12 strawberries

27. 288

29. 20 boys, 28 girls

31. $200

Section 8.3

8.3 Review & Refresh

1. 2
2. 21
3. 15
4. 48
5. 8
6. 16
7. 45
8. 3
9. 1
10. 4
11. 28 yd
12. 46 mm

8.3 Concepts, Skills, & Problem Solving

13. 1440; *Sample answer:* $2 \times 8 = 16$ and $180 \times 8 = 1440$

15. 495; *Sample answer:* $(2 \div 2) \times 5.5 = 5.5$ and $(180 \div 2) \times 5.5 = 495$

17.

Burgers	3	6	9
Hot dogs	5	10	15

3 : 5, 6 : 10, 9 : 15

19.

Adults	2	1	3	18
Children	14	7	21	126

2 : 14, 1 : 7, 3 : 21, 18 : 126

21.

Plums	14	42	6	48
Grapes	7	21	3	24

14 : 7, 42 : 21, 6 : 3, 48 : 24

23. 16
25. $60
27. 1840

29. **a.** 392; no; *Sample answer:* 392 < 400

b. $16,400; no; *Sample answer:* $16,400 < $16,800

31. yours; $\frac{2}{22} > \frac{3}{37}$

33. $A = 12, B = 18$

35. **a.** yes; *Sample answer:* Find a ratio equivalent to 161 : 28 that has 12 on the right.

b. 21

37. 96

39. $270

Section 8.4

8.4 Review & Refresh

1.

Chickens	8	16	24
Eggs	6	12	18

8 : 6, 16 : 12, 24 : 18

2.

Fish	6	3	12	15
Snails	2	1	4	5

6 : 2, 3 : 1, 12 : 4, 15 : 5

3.

Miles	3	1	5	7
Minutes	21	7	35	49

3 : 21, 1 : 7, 5 : 35, 7 : 49

4.

Books	6	10	2	8
Movies	15	25	5	20

6 : 15, 10 : 25, 2 : 5, 8 : 20

5. $\frac{2}{3}$
6. $\frac{1}{16}$
7. $1\frac{7}{12}$
8. 2
9. 19.241
10. 1.043
11. 7.919
12. 180

8.4 Concepts, Skills, & Problem Solving

13. Answers will vary. Multiply the hand clap rate in Exploration 2(a) by 2.5.

15. Answers will vary. Multiply the hand clap rate in Exploration 2(a) by 11.25.

17. $20/guest

19. 6 necklaces/h

21. 19 students/class

23. 110 calories/serving

25. $2.5/oz

27. $\frac{1}{6}, \frac{17}{6}$

29. **a.** 330 L of blood

b. 24 seconds

31. equivalent

33. equivalent

35. your friend; 2.5 min

37. **a.** about 0.12 mi/min

b. about 8.0 min/mi

c. b; "10-minute miles" is 10 minutes per mile.

Section 8.5

8.5 Review & Refresh

1. 51 beats/min
2. 12 shirts/clothing rack
3. $20/volunteer
4. 7.5 mL/h
5. A
6. C
7. B
8. D

8.5 Concepts, Skills, & Problem Solving

9. *Sample answer:* $\frac{1}{3} = \frac{4}{12}, \frac{6}{18} = \frac{4}{12}, \frac{12}{4} = \frac{18}{6}$; Use the columns to form equivalent ratios.

11. $x = 6$

13. $p = 2$

15. $k = 2$

17. $m = 12$

19. $w = 3$

21. no; $\frac{16}{8} \neq \frac{1}{2}$, so the solution is incorrect. Your friend should multiply each side by 2 to find the solution.

23. *Sample answer:* $\frac{1}{3} = \frac{x}{18}$

25. yes

27. Answers will vary. Multiply the rate of 10 feet per number of seconds measured by $\frac{3600}{5280}$, or $\frac{15}{22}$.

Section 8.6

8.6 Review & Refresh

1. $b = 4$ **2.** $x = 15$

3. $k = 9$ **4.** $w = 7$

5. 36

6. $2 \times 2 \times 2 \times 7$, or $2^3 \times 7$

7. 2×37 **8.** $3 \times 3 \times 7$, or $3^2 \times 7$

9. $2 \times 2 \times 3 \times 11$, or $2^2 \times 3 \times 11$

10. 6^2 **11.** 18^4 **12.** 12^5

8.6 Concepts, Skills, & Problem Solving

13.

Cups	16	32	48	64
Gallons	1	2	3	4

Sample answer: Since there are 16 cups in 1 gallon, the number of gallons can be multiplied by 16 to get the number of cups.

15. 6 **17.** 2.5

19. 0.00008 **21.** 60

23. 480 **25.** 200

27. 5 **29.** 4

31. 69 **33.** 11

35. 6 **37.** 3

39. 180 **41.** 6 **43.** 2

45. yes; The student converted the rate correctly.

47. $<$ **49.** $=$ **51.** $<$ **53.** $>$

55. a. 300 qt/h

b. $\frac{1}{3}$ c/sec

57. your friend.; *Sample answer:* 2.5 miles per hour is the same rate as 220 ft/min. Since 220 ft/min is a slower rate than 227 ft/min, your friend arrives at school first.

59. 240; 2.4

61. a. Orca; Pacific Whitesided Dolphin and Bottlenose Dolphin

b. no; The Orca swims at 50.6 ft/sec, which is 34.5 mi/hr. This is a faster rate than 24.7 mi/hr.

63. a. Your friend's car

b. *Sample answer:* Since your car moves at about 6.7 meters per second, and your friend's car moves 8 meters per second, you could place your car about $\frac{7}{8}$ the distance from the finish line to make the race more competitive.

Chapter 8 Review

1. 3 : 2, or 3 to 2 **2.** 6 : 3, or 6 to 3

3. a. $\frac{2}{3}$; The number of hydrogen atoms is $\frac{2}{3}$ times the number of nitrogen atoms.

b. 3 : 1

4. equivalent **5.** not equivalent

6. not equivalent **7.** equivalent

8. *Sample answer:* 10 : 14, 15 : 21

9. 8 **10.** no; your friend

11. 4 h **12.** 32 h

13. 15 **14.** 27

15. 35 sit-ups, 10 pull-ups

16.

Levers	6	12	18
Pulleys	3	6	9

6 : 3, 12 : 6, 18 : 9

17.

Nails	5	10	15
Screws	2	4	6

5 : 2, 10 : 4, 15 : 6

18. *Sample answer:*

×2 ÷8 +4 −2

Pumpkins Grown	4	8	1	5	3
Seeds Planted	12	24	3	15	9

×2 ÷8 +12 −6

19. 60 sec **20.** Surfshop A

21. 3 stunts/movie **22.** 1200 stitches/min

23. 40 mi/h, 0.025 h/mi

24. a. 5

b. 704

25. $d = 36$ **26.** $m = 27$

27. $k = 1$ **28.** $y = 11$

29. no; *Sample answer:*

$$\frac{4}{16} = \frac{x}{100}$$

$$25 = x$$

At this pace, it will take the runner 25 hours to complete the race.

30. 20 **31.** 4

32. 3.5 **33.** 120

34. 52 **35.** 1.6

36. 0.36 **37.** 23.9 ft/sec

Chapter 8 Practice Test

1. 3 : 3, or 3 to 3 **2.** not equivalent

3.

Lemons	4	12	36
Limes	2	6	18

4 : 2, 12 : 6, 36 : 18

4.

Rabbits	2	12	10
Hamsters	9	54	45

2 : 9, 12 : 54, 10 : 45

5. 56 mi/h **6.** $h = 40$

7. 400 **8.** 7.5

9. 1680 **10.** 48

11. Team A scores 810 runs, Team B scores 630 runs

12. yours

13. Windsurfer B; 720 ft/min is 12 ft/sec. 12 ft/sec is a slower rate than 15 ft/sec.

14. 64.75 ft; 213.90625 ft^2

Chapter 8 Review & Refresh

1. 17.25 **2.** 26.754

3. 978 **4.** 82

5–6.

Number of Cats	Number of Dogs	Ratio of Cats to Dogs
12	4	3 : 1
6	21	2 : 7

7. 1 : 3 **8.** B

9. no; *Sample answer:* You are catching sailfish at a rate of $1\frac{1}{3}$ sailfish per hour. Your friend is catching sailfish at a rate of $1\frac{1}{2}$ per hour. Your friend is catching the sailfish at a faster rate.

10.

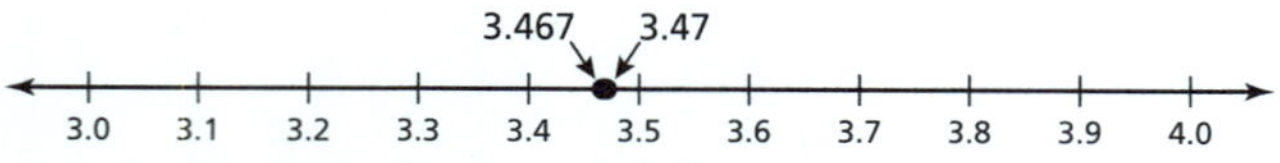

3.47 is greater because it has a 7 in the hundredths place, which is greater than the 6 in the hundredths place in 3.467.

11. $3(4 + 6) = 3(4) + 3(6)$ Distributive Property

$= 12 + 18$ Multiply.

$= 30$ Add 12 and 18.

12. $(5 - 1) \times 4 = 4(5) - 4(1)$ Distributive Property

$= 20 - 4$ Multiply.

$= 16$ Subtract.

Chapter 8 B.E.S.T. Test Prep

1. 1.25 **2.** C

3. D **4.** B

5 41 **6.** C

7. D **8.** B; B

9. A

Chapter 9

Chapter Exploration

1. **a.** M
b. O
c. J
d. N
e. L
f. P
g. Q
h. K

2. *Sample answer:* miles per hour, used as a rate of speed; *Percent* means rate per hundred.

3. **a.** $\frac{4}{100}$
b. $\frac{1}{25}$
c. 0.04
d. 4%

4. **a.** $\frac{15}{100}$
b. $\frac{3}{20}$
c. 0.15
d. 15%

5. **a.** $\frac{30}{100}$
b. $\frac{3}{10}$
c. 0.3
d. 30%

6. a. $\frac{50}{100}$

b. $\frac{1}{2}$

c. 0.5

d. 50%

7. a. $\frac{100}{100}$

b. 1

c. 1

d. 100%

8. a. $\frac{0}{100}$

b. 0

c. 0

d. 0%

Section 9.1

9.1 Review & Refresh

1.	3000	2.	112
3.	1.5	4.	21,120
5.	6500	6.	720
7.	44	8.	13
9.	170	10.	21
11.	43	12.	144
13.	3 : 1	14.	8 : 4
15.	4 : 1	16.	3 : 8

9.1 Concepts, Skills, & Problem Solving

17. 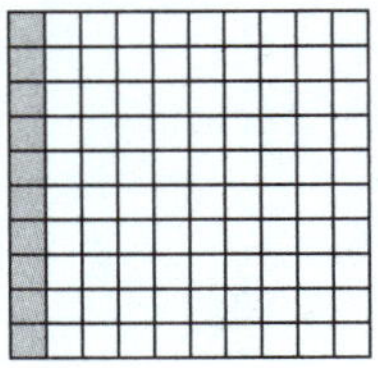

19. 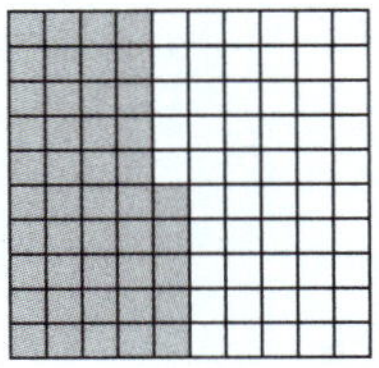

21. $\frac{9}{10}$ 23. $\frac{7}{100}$ 25. $\frac{79}{100}$ 27. $1\frac{22}{25}$

29. $2\frac{6}{25}$ 31. $\frac{1}{250}$ 33. 10% 35. 55%

37. 8% 39. 1.2%

41. 185% 43. 300.5%

45. $\frac{56}{100} = 56\%$ 47. 25%

49. $1\frac{1}{4}$; No, you have more than you need.

51. 81.25% 53. 82.5%

55. a. $\frac{7}{40}, \frac{17}{1000}, \frac{9}{3125}, \frac{3}{200}$

b. no; Alaska's area is more than 10 times as great as Florida's.

c. AK, TX, CA, MT, NM, AZ, NV, CO, OR, WY, MI, MN, UT, ID, KS, NE, SD, WA, ND, OK, MO, FL, WI, GA

57. Shade half of one of the squares.

Section 9.2

9.2 Review & Refresh

1. 14% 2. 40%

3. 0.4% 4. 508%

5. 25 red marbles and 15 green marbles

6. 48 red marbles and 16 green marbles

7. $\frac{2}{21}$ 8. $\frac{1}{12}$ 9. $4\frac{1}{5}$ 10. $7\frac{7}{12}$

9.2 Concepts, Skills, & Problem Solving

11. 51%, 0.51 13. 0.78

15. 0.185 17. 0.33 19. 0.476 21. 1.66

23. 0.006 25. 74% 27. 89% 29. 99%

31. 48.7% 33. 368% 35. 0.7%

37. no; 0.86 = 0.86 = 86%

39. A

41. *Sample answer:* 0.11, 0.13, 0.19

43. $\frac{47}{200}$, 0.235

45. a. car: 0.2, school bus: 0.48, bicycle: 0.08

b. car: $\frac{1}{5}$, school bus: $\frac{12}{25}$, bicycle: $\frac{2}{25}$

c. 24%

d. *Answer should include, but is not limited to:* A bar graph showing either the *number* of students or the *portion* of students in the class that get to school in various ways.

47. a. 70%

b. 6.5

c. 16%; *Sample answer:* subtracting percents; It gives whole number calculations.

Section 9.3

9.3 Review & Refresh

1. 0.12 2. 0.983 3. 0.046

4.

Cashews	5	45
Almonds	4	36

5 : 4, 45 : 36

5.

Cost (dollars)	2	3	13.5
Time (minutes)	50	75	337.5

2 : 50, 3 : 75, 13.5 : 337.5

6. 31.8
7. 162.9
8. 18.7693
9. 5.7084
10. 3
11. 17
12. 176

9.3 Concepts, Skills, & Problem Solving

13. $\frac{3}{10}$, 40%, 0.65, 80%
15. $\frac{1}{8}$, 15%, 0.25, $\frac{27}{100}$
17. 20% > 0.02
19. 50% < $\frac{13}{25}$
21. 76% > 0.67
23. 0.12 > 1.2%
25. 140% > 0.14
27. 70% = $\frac{7}{10}$
29. 0.63, $\frac{13}{20}$, 68%
31. 0.015, 1.6%, $\frac{3}{20}$
33. $\frac{21}{50}$, $\frac{87}{200}$, 43.7%, 0.44
35. Russia, Brazil, United States, India
37. 21%, 0.211, $\frac{11}{50}$, $\frac{111}{500}$
39. *D*
41. *C*
43. **a.** dolphin, rabbit, lion, squirrel, tiger
    b. *Sample answer:* 33%
    c. first

Section 9.4

9.4 Review & Refresh

1. 0.42, 42%
2. 0.007, 0.7%
3. 0.45, 45%
4. 0.8, 80%
5. 3
6. −0.6
7. −2.5
8. $x = -0.2$

9.4 Concepts, Skills, & Problem Solving

9. 16
11. 50
13. 40%
15. $\frac{12}{25} = \frac{p}{100}$; $p = 48$
17. $\frac{5}{w} = \frac{25}{100}$; $w = 20$
19. $\frac{a}{124} = \frac{75}{100}$; $a = 93$
21. $\frac{a}{40} = \frac{0.4}{100}$; $a = 0.16$
23. yes; Your friend wrote and solved the correct percent proportion.
25. $6000
27. $\frac{14.2}{w} = \frac{28.4}{100}$; $w = 50$
29. $\frac{a}{\frac{7}{8}} = \frac{25}{100}$; $a = \frac{7}{32}$
31. 62.5%
33. **a.** a scale along the vertical axis
    b. 6.25%; *Sample answer:* Although you do not know the actual number of votes, you can visualize each bar as a model with the horizontal lines breaking the data into equal parts. The sum of all the parts is 16. Raven has the least parts with 1, which is 100% ÷ 16 = 6.25%.
    c. 31 votes
35. $5x$

Section 9.5

9.5 Review & Refresh

1. $\frac{30}{100} = \frac{6}{w}$; $w = 20$
2. $\frac{42}{80} = \frac{p}{100}$; $p = 52.5$
3. $\frac{p}{100} = \frac{27}{36}$; $p = 75$
4. $\frac{a}{80} = \frac{120}{100}$; $a = 96$
5. 14
6. 2
7. $4\frac{1}{10}$
8. 11.8
9. $\frac{3}{8} = \frac{x}{160}$

9.5 Concepts, Skills, & Problem Solving

11. 40%
13. $a = 0.2 \cdot 150$; 30
15. $35 = 0.35 \cdot w$; 100
17. $29 = p\% \cdot 20$; 145%
19. $51 = p\% \cdot 300$; 17%
21. yes; The percent was converted to a decimal and multiplied by the "whole".
23. $200
25. 8%
27. 26 years old
29. 56 signers
31. **a.** 80 students
    b. 30 students
33. about 28.9%; *Sample answer:* The total with tax is $19.64. $25–$19.64 is a tip of $5.36, which is about 28.9% of the original amount.

Section 9.6

9.6 Review & Refresh

1. $a = 0.25 \cdot 64$; 16
2. $39.2 = p \cdot 112$; 35%
3. $5 = 0.05 \cdot w$; 100
4. $18 = 0.32 \cdot w$; 56.25
5. $-\frac{2}{7}$
6. −0.696
7. $\frac{1}{3}$

9.6 Concepts, Skills, & Problem Solving

9. 20%
11. 6%
13. decrease; 68.8%
15. increase; 225%
17. decrease; 12.5%
19. decrease; 37.5%
21. 12.5% decrease

23. **a.** 20%

b. 4800 years old; *Sample answer:* The amount of the error and the original amount are the same, giving the same percent of error.

25. decrease; 25%

27. decrease; 70%

29. **a.** 38% increase

b. yes; *sample answer:* The percent change from 138,000 to 190,440 is 38%, which is equal to the last 6-year percent change.

31. Box A is acceptable, Box B is unacceptable

33. less than; *Sample answer:* Let x represent the number. A 10% increase is equal to $x + 0.1x$, or $1.1x$. A 10% decrease of this new number is equal to $1.1x - 0.1(1.1x)$, or $0.99x$. Because $0.99x < x$, the result is less than the original number.

35. 10 girls

Section 9.7

9.7 Review & Refresh

1. increase; 25%
2. decrease; 60%
3. decrease; 80%
4. increase; 120%
5. $-\frac{2}{21}$
6. -9.5116
7. $6\frac{3}{8}$

9.7 Concepts, Skills, & Problem Solving

9. 85% off $90;

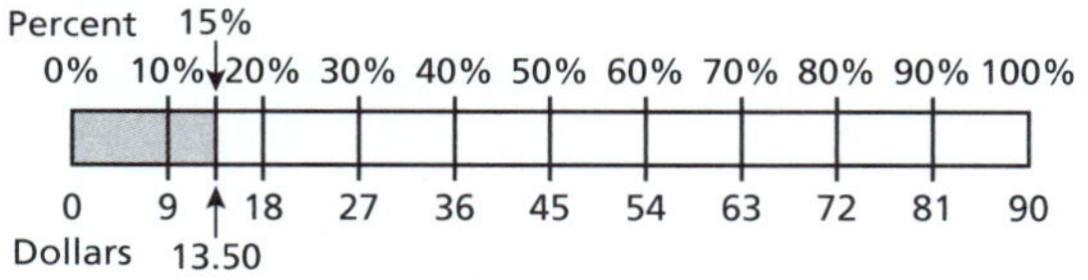

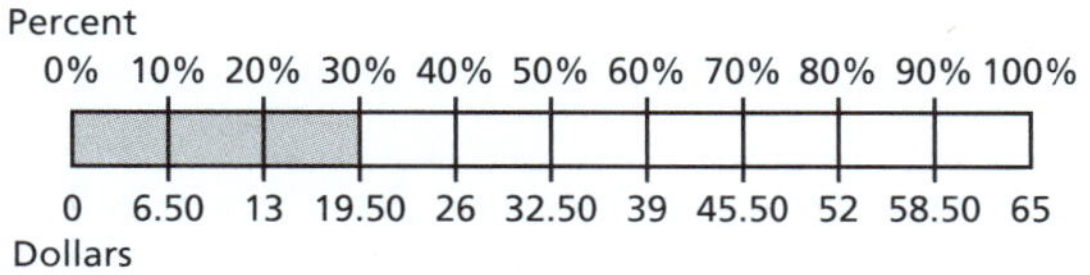

11. $35.70
13. $76.16
15. $120
17. $450
19. $172.40
21. 20%
23. $55
25. $175
27. no; The selling price is 120% of the cost to the store, or 1.2 • $60.
29. yes; $p \cdot 1.18 = 1.18p$; $p + 0.18p = 1.18p$
31. $54 for chlorine tablets and $6.60 for pH test kit
33. $30

Section 9.8

9.8 Review & Refresh

1. $9.60
2. $3.15
3. $x < -3$

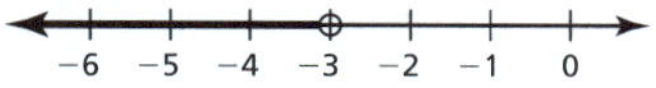

4. $b \geq 1$

5. $w \leq -9$

9.8 Concepts, Skills, & Problem Solving

7. 8%; $315

9. **a.** $300

b. $1800

11. **a.** $292.50

b. $2092.50

13. **a.** $1722.24

b. $6922.24

15. 3%
17. 4%
19. 2 yr
21. 1.5 yr
23. $1440
25. 2 yr
27. $2720
29. $6700.80
31. $8500
33. 5.25%
35. 4 yr
37. 12.5 yr; Substitute $2000 for P, $2000 for I, 0.08 for r, and solve for t.
39. Year 1 = $520
Year 2 = $540.80
Year 3 = $562.43

Chapter 9 Review

1. $\frac{3}{25}$
2. $\frac{22}{25}$
3. $\frac{1}{125}$
4. $1\frac{27}{100}$
5. $\frac{1}{40}$
6. $\frac{9}{50}$
7. 60%
8. 172%
9. 142%
10. 40%
11. 528%
12. 1.75%
13. *Sample answer:* $\frac{11}{25}$
14. *Sample answer:* 378%
15. $\frac{23}{50}$
16. $\frac{2}{5}$
17. 35%
18. 0.76
19. 0.06
20. 0.17
21. 0.008
22. 0.016
23. 3.34
24. 15%
25. 77%
26. 56%
27. 106%
28. 124%
29. 9.7%
30. *Sample answer:* 0.063
31. *Sample answer:* 2.7%
32. 0.38
33. $\frac{1}{2} < 52\%$
34. $\frac{12}{5} < 245\%$
35. $0.46 > 43\%$
36. $0.023 < 22\%$
37. 2.15, 218%, 220%, $\frac{9}{4}$
38. $\frac{7}{8}$, 0.88, 0.89, $\frac{9}{10}$, 92%

39. *Sample answer:* 53%

40. debate, theater, gymnastics, chorus, band

41. 60%
42. 78%

43. 18
44. 18

45. 50
46. 25

47. 30%
48. 49

49. 39
50. 37.5%

51. 2
52. 147.9 million km^2

53. **a.** 25; 4
b. 19

54. $a = 0.24 \cdot 25$; 6
55. $9 = p\% \cdot 20$; 45%

56. $60.8 = p\% \cdot 32$; 190%
57. $91 = 1.3 \cdot w$; 70

58. $10.2 = 0.85 \cdot w$; 12
59. $a = 0.83 \cdot 20$; 16.6

60. 120 parking spaces
61. 64%

62. increase; 500%
63. decrease; 48%

64. 6.25%
65. 25% decrease

66. \$42.50
67. \$93.75

68. \$30

69. **a.** \$62.50
b. \$66.25

70. **a.** \$36
b. \$336

71. **a.** \$280
b. \$2280

72. 1.7%
73. 7.1%

74. 3 years
75. 6 years
76. 4%

77. **a.** Bank A: \$261; Bank B: \$179.25
b. *Sample answer:* A person would choose Bank A if they wanted a short term loan. A person would choose Bank B if they wanted lower monthly payments.

Chapter 9 Practice Test

1. 0.0096

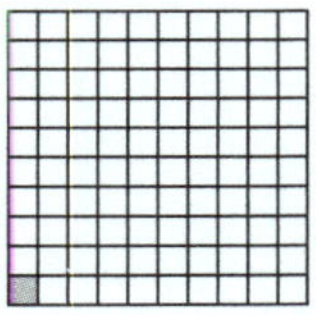

2. 0.03

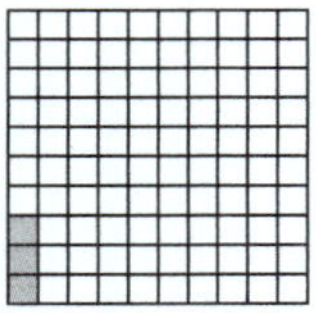

3. 0.65

4. 42%

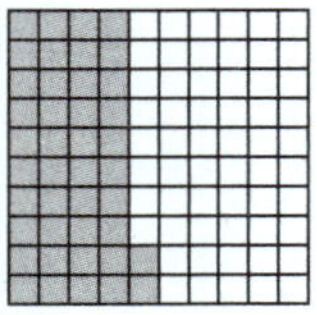

5. 788%

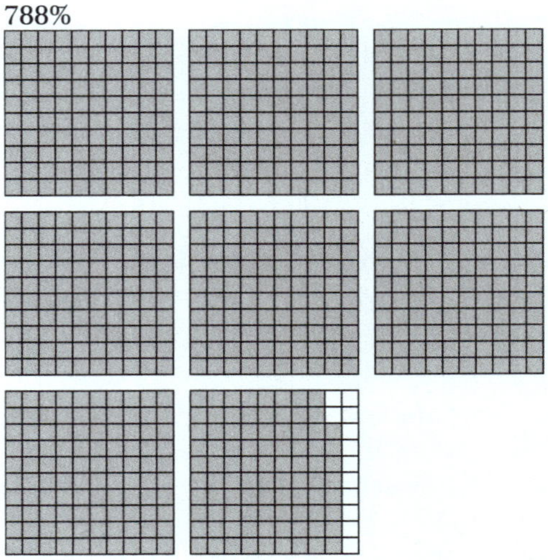

6. 58%

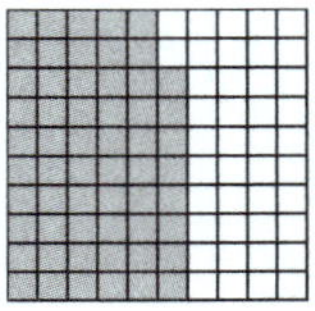

7. $\frac{7}{10}$, 0.74, $\frac{4}{5}$, 85%
8. $\frac{6}{5}$, $\frac{5}{4}$, 1.28, 130%, 1.32

9. 75%
10. 160%

11. 36
12. 750

13. increase; 150%
14. decrease; 25%

15. \$14.25
16. \$9.63

17. \$216
18. \$900

19. **a.** $\frac{1}{4}$; 0.25, 25%
b. 14.3%

20. 1,440,000 ft^3

21. **a.** 12%
b. 84 cars; To get the same percent error, the amount of error needs to be the same. Because your estimate was 9 cars below the actual number, an estimate of 9 cars above the actual number will give the same percent error.

Chapter 9 Review & Refresh

1. 25
2. 13

3. $\frac{2}{5}$
4. 0.125, 12.5%

5. 0.15; 15%
6. 2.25; 225%

7. 21 hours
8. 51.5 kilometers

9. Point M
10. $(-5, -3)$

11. Point J
12. Point N

13. 23.4
14. -2.52

15. -34.87
16. $\frac{7}{20}$

17. $-\frac{1}{6}$

18. $-\frac{5}{32}$

19. $12t > 400$

20.

Meters	20	60	**160**	220
Minutes	0.6	**1.8**	4.8	**6.6**

20 : 0.6; 60 : 1.8; 160 : 4.8; 220 : 6.6

21.

Sugar (cups)	2	**6**	8	12
Water (cups)	4.5	13.5	**18**	27

2 : 4.5; 6 : 13.5; 8 : 18; 12 : 27

Chapter 9 B.E.S.T. Test Prep

1. C
2. B
3. 28
4. D
5. B; D
6. 27
7. A
8. 220
9. B
10. A

Chapter 10

Chapter Exploration

1. a. rectangle: length = 8 units, width = 4 units; triangle: base = 8 units, height = 4 units
   b. rectangle: 32 units^2; triangle: 16 units^2

   The area of the triangle is one-half the area of the rectangle.
2. a. rectangle: length = 8 units, width = 4 units; parallelogram: base = 8 units, height = 4 units
   b. rectangle: 32 units^2; parallelogram: 32 units^2

   The areas are the same.
3. a. rectangle: length = 8 units, width = 4 units; trapezoid: bases are 3 units and 5 units, height = 4 units
   b. rectangle: 32 units^2; trapezoid: 16 units^2

   The area of the trapezoid is one-half the area of the rectangle.

Section 10.1

10.1 Review & Refresh

1. \$84
2. \$78
3. \$36
4. \$84
5. Mult. Ident. Prop. of One
6. Comm. Prop. of Mult.
7. Assoc. Prop. of Add.

10.1 Concepts, Skills, & Problem Solving

9. 14 units^2
11. 6 cm^2
13. 1620 in.^{2}
15. 1125 cm^2
17. yes; $A = \frac{1}{2}bh$
19. $b = 7$ m
21. $h = 15$ in.
23. 4
25. Check students' work.

Section 10.2

10.2 Review & Refresh

1. 63 in.^{2}
2. 26 km^2
3. 25 ft^2
4. $27(1 + 3)$
5. $14(3 - 1)$
6. $17(3 + 5)$
7. 0.02, $\frac{1}{5}$, $\frac{11}{50}$, 22.2%, 0.23
8. 0.7%, 7.7%, $\frac{7}{10}$, 0.77, 0.777
9. $\frac{21}{25}$, 84.4%, $\frac{17}{20}$, 0.856, 0.86
10. 0.009, 9%, 9.9%, $\frac{99}{100}$, 0.999

10.2 Concepts, Skills, & Problem Solving

11. 12 units^2; *Sample answer:* Decompose into two isosceles triangles; Decompose into two right triangles and a rectangle.
13. 25 cm^2
15. 125 m^2
17. 55 mi^2
19. 95 m^2
21. 20 km
23. 120 ft^2
25. 36 units^2
27. a. 462 m^2
    b. *Sample answer*: The restaurant can add triangular portions to make the resulting patio a 15 meter by 34 meter rectangle; 510 m^2.

Section 10.3

10.3 Review & Refresh

1. 15 ft^2
2. 108 m^2
3.

Length (inches)	4	**8**	12
Width (inches)	3	6	**9**

4 : 3; 8 : 6; 12 : 9

4.

Red Blood Cells	600	1200	**1800**
White Blood Cells	1	**2**	3

600 : 1; 1200 : 2; 1800 : 3

10.3 Concepts, Skills, & Problem Solving

5. 12 units^2
7. 24 units^2
9. 840 mm^2
11. 3750 cm^2

Selected Answers

13. 894 mi^2

15. 6 $in.^2$

17. 5 ft

19. 16 in.

21. 22

23. n^2bh or n^2A

Section 10.4

10.4 Review & Refresh

1. 72 $in.^2$
2. 10.5 km^2
3. 255 mi^2
4. 2.7
5. $-\frac{1}{4}$
6. $-\frac{3}{4}$
7. *Sample answer:* 48 airplanes each hour

10.4 Concepts, Skills, & Problem Solving

9. 16 $units^2$

11. 192 m^2

13. 180 cm^2

15. 14.25 mi^2

17. 28 $in.^2$

19. 105 ft^2

21. 16 ft^2

23. 17 km

25. 4 mm

27. about 21 people per square mile of land; Check students' work.

29. 4

31. x must be greater than 0 and less than 15; no; x would be 0.

Section 10.5

10.5 Review & Refresh

1. 15 ft^2
2. 18 km^2
3. 39 $in.^2$
4. 24
5. 75
6. 168
7. 210
8. 36
9. 40
10. 70

10.5 Concepts, Skills, & Problem Solving

11. front: 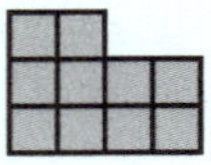 side: 

top:  10

13. front: 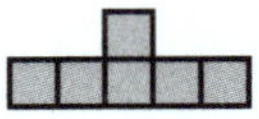

side: 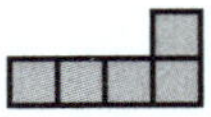

top: 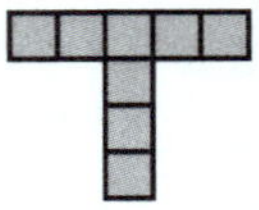

15. 10 faces, 24 edges, and 16 vertices

17. 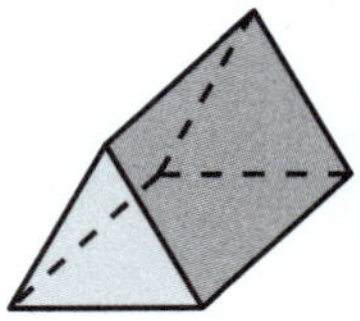

19. 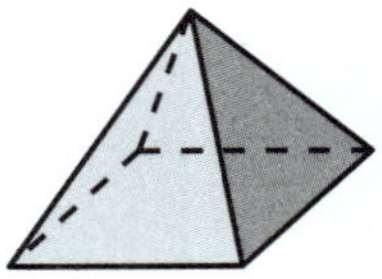

21. 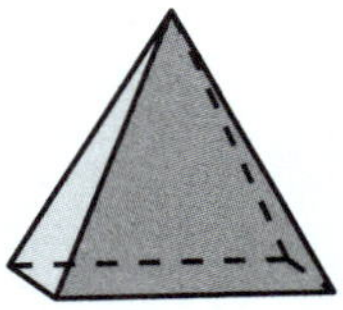

23. front:  side:  top: 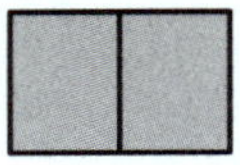

25. front: 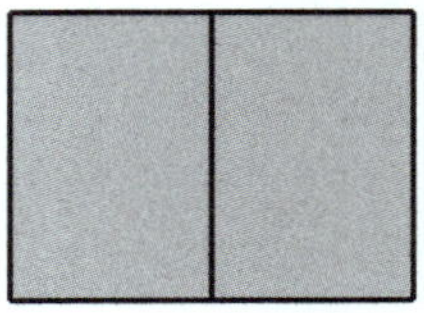 side: 

top: 

27. front:  side: 

top: 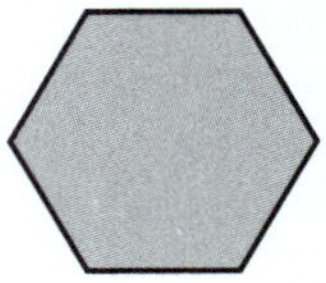

29. 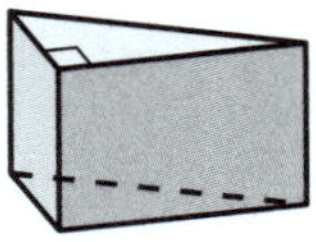

31. *Answer should include, but is not limited to:* an original drawing of a house;
a description of any solids in the drawing.

33. a.

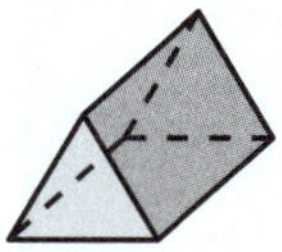

6 vertices
9 edges

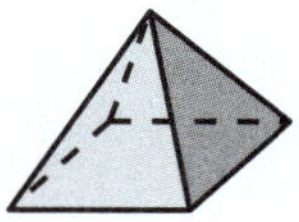

5 vertices
8 edges

b. More than one solid can have the same number of faces, so knowing the number of edges and vertices can help you to draw the intended solid.

Section 10.6

10.6 Review & Refresh

1. front:

side:

top:

2. front:

side:

top:

3. front:

side:

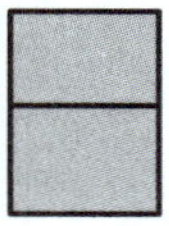

top:

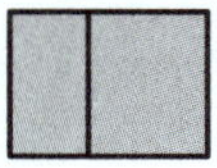

4. 18 **5.** 22
6. 3 **7.** 12
8. $s = 17$ **9.** $x = 11$
10. $r = 8$ **11.** $m = 65$

10.6 Concepts, Skills, & Problem Solving

13. 126 units^2; $2(5)(6) + 2(5)(3) + 2(6)(3) = 126$
15. 180 ft^2 **17.** 130 ft^2
19. 156.25 m^2 **21.** $\frac{2}{3}$ ft^2
23. 448 in.^{2}; It is the surface area of the box.
25. a. 384 in.^{2}
b. 94 in.^{2}
27. B, D **29.** 364 ft^2

Section 10.7

10.7 Review & Refresh

1. 82 ft^2 **2.** 664 cm^2
3. 384 in.^{2} **4.** C
5. B **6.** D
7. A

10.7 Concepts, Skills, & Problem Solving

9. 133 units^2;
$7 \cdot 7 + 4 \cdot \left(\frac{1}{2}\right) \cdot 7 \cdot 6 = 133$
11. 72 m^2 **13.** 119 in.^{2}
15. 1944 cm^2 **17.** 21,274.4 ft^2
19. 81 micrometers^2 **21.** 4
23. no; You can place the four triangles on top of the square and it covers the entire square. But when you lift up the triangles, they do not touch. So, they do not form a pyramid.

Section 10.8

10.8 Review & Refresh

1. 51 ft^2 **2.** 74.25 m^2
3. 864 yd^2 **4.** $x + 8$; 6
5. $y - (-9)$; 21

10.8 Concepts, Skills, & Problem Solving

7. $\frac{1}{18}$ unit^3 **9.** $\frac{3}{10}$ in.^{3}
11. $\frac{8}{125}$ ft^3 **13.** $3\frac{1}{8}$ cm^3
15. $h = 20$ cm **17.** $w = 14$ in
19. a. *Sample answer:* 297 in.^{3}
b. no; The container only holds 196 cubic inches.
21. 1000; Multiply by the conversion factor $\frac{1 \text{ cm}^3}{1000 \text{ mm}^3}$;
Multiply by the conversion factor $\frac{1000 \text{ mm}^3}{1 \text{ cm}^3}$.
23. 1152 cm^3
25. *Answers should include, but not limited to:*
a. a sketch of a tree house that has a surface area of at most 1400 square feet and a volume of at least 250 cubic feet
b. For the dimensions to be reasonable, the tree house should be able to fit people and fit in a tree.

Chapter 10 Review

1. 80 km^2 **2.** 175 cm^2
3. $b = 10$ mi **4.** $h = 3$ cm
5. 105 m^2 **6.** 108 mm^2

7. *Sample answer:*

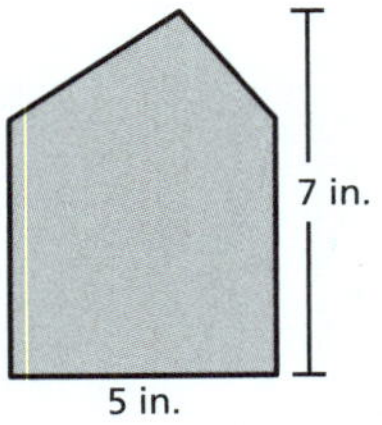

8. 500 yd^2

9. 88 mm^2

10. 45 cm^2

11. 24 ft^2

12. 864 $in.^2$

13. 14 ft

14. 1.5 m

15. *Sample answer:*

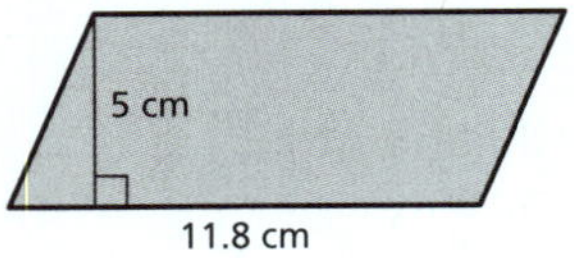

16. 600 m^2

17. 6 $in.^2$

18. 49 mi^2

19. 76 ft^2

20. 369 cm^2

21. 90 $in.^2$

22. 4 mi

23. 7 mm

24. 28 ft

25. 636 $in.^2$

26. 80,000 cm^2

27. 6 faces, 12 edges, 8 vertices

28. 6 faces, 10 edges, 6 vertices

29. 9 faces, 21 edges, 14 vertices

30.

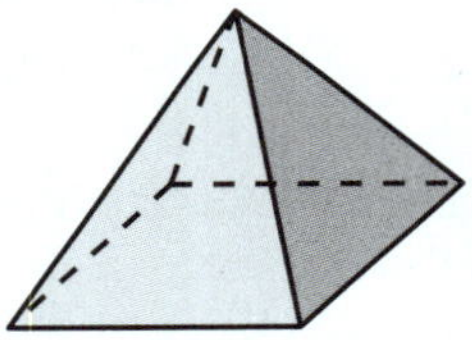

31.

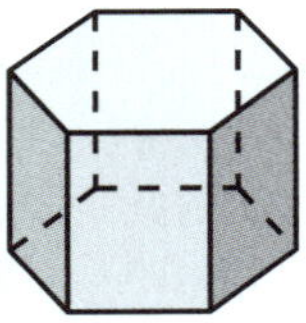

32. front:

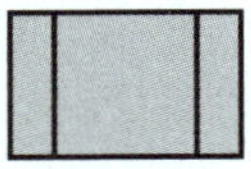

side:

top:

33. front:

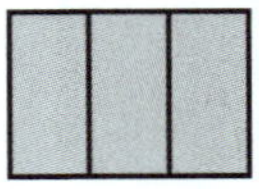

side:

top:

34. front:

side:

top:

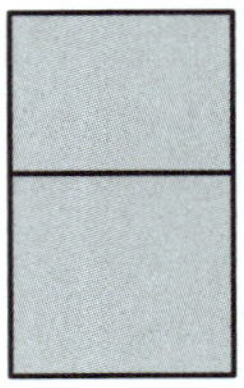

35. 100 $in.^2$

36. 132 cm^2

37. 243 m^2

38. 258.25 ft^2

39. 294 yd^2

40. $433\frac{1}{2}$ mi^2

41. 14

42. 1144 ft^2

43. 180.6 cm^2

44. 286.2 m^2

45. 80 $in.^2$

46. 5 ft^3

47. $\frac{11}{18}$ cm^3

48. $\frac{27}{512}$ $in.^3$

49. 7.5 ft

50. $162\frac{1}{2}$

51. *Sample answer:*

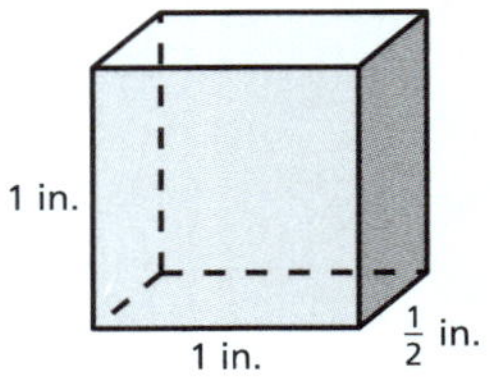

Chapter 10 Practice Test

1. 19,800 cm^2

2. 154 $in.^2$

3. 4 cm^2

4. 478 ft^2

5. 354 m^2

6. $\frac{35}{8}$ cm^3, or $4\frac{3}{8}$ cm^3

7. $\frac{64}{125}$ yd^3

8. *Sample answer:*

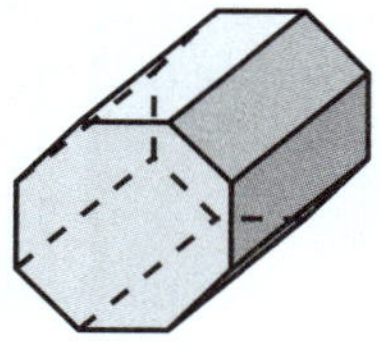

9. 8 faces, 18 edges, 12 vertices

10. 14 m **11.** 44.8 in.^{2}

12. 138 in.^{2} **13.** 8

14. 217,800 ft^2; 1,241,263 ft^2

Chapter 10 Review & Refresh

1. $t = -24$ **2.** $h = 4$

3. $p = \frac{2}{9}$ **4.** 5

5. 2 **6.** 1

7. 255 **8.** 15 angelfish

9. 17 contestants

10. **a.** 3600 times

b. 5 minutes

11. $\frac{5}{2}$ **12.** $\frac{53}{7}$

13. $\frac{-18}{5}$ **14.** 72

15. 40 **16.** 48

17. 6.076 ft^2 **18.** $3\frac{1}{4}$ ft

Chapter 10 B.E.S.T. Test Prep

1. B **2.** 51

3. C **4.** C

5. $m = 7$ **6.** C

7. D **8.** 20

9. C

Chapter 11

Chapter Exploration

1. Check students' work.

2. Answers will vary. Listen for students to mention ideas such as least and greatest numbers, numbers that occur most often, and numbers that look to be in the middle of the data set.

3. Check students' work.

4. (a)-(c). Check students' work.

Section 11.1

11.1 Review & Refresh

1. $\frac{1}{15}$ ft^3 **2.** $\frac{8}{7}$ cm^3

3. $2\frac{2}{5}$ in.^{3} **4.** (4, 3)

5. 0.2, 24%, $\frac{1}{4}$, 0.32, $\frac{7}{20}$ **6.** $\frac{3}{4}$, 78%, 85%, $\frac{7}{8}$, 0.88

11.1 Concepts, Skills, & Problem Solving

7. 12; yes

9. *Sample answer:* 9th; no

11. yes; The answers will vary.

13. yes; The answers will vary.

15. **a.** 18; 18 players are on the team

b. *Sample answer:* Use a tape measure; inches

c. *Sample answer:* "What are the heights of players on an NBA championship team?"; The heights are spread out, but most of the heights (in inches) are in the mid-to-low 80s.

17.

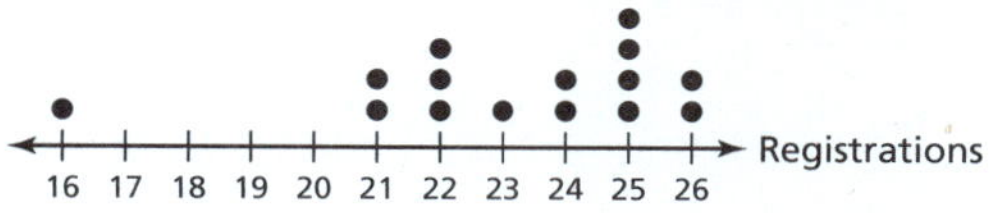

Most of the registrations are in a cluster from 21 to 26. The peak is 25. There is a gap between 16 and 21.

19. *Sample answer:* 45 mi/h; Most of the data cluster around 45 and 45 miles per hour is a common speed limit.

21. **a.** yes; It is a statistical question because you would anticipate variability in the hours spent on homework each night by students.

b. Most of the hours cluster around 2. The peak is 2. There is no gap.

c. Most students spend between 1 and 3 hours on homework during a school night.

23. *Sample answer:* rain gauge; inches

25. *Sample answer:* "How many letters are there in the English alphabet?"; "How many letters are there in a word?"; The number of letters in the English alphabet is fixed but different words will have different numbers of letters.

Section 11.2

11.2 Review & Refresh

1. yes; The answers will vary.

2. no; There is only one answer.

3. no; There is only one answer.

4. yes; The answers will vary.

5. $\frac{21}{25}$ **6.** $\frac{71}{100}$

7. $3\frac{53}{100}$ **8.** $\frac{1}{500}$

9. 1.3 **10.** 14.56

11. 4.2 **12.** 24.8

11.2 Concepts, Skills, & Problem Solving

13. 12 **15.** 15

17. 2 **19.** 103

21. B

23. *Sample answer:* 20, 21, 21, 21, 21, 22; 20, 20.5, 20.5, 21.5, 21.5, 22

25. The Tigers' mean height is greater; no; neither data set has outliers.

27. **a.** \$3.50; Divide \$84 by 24 people.

b. *Sample answer:* The class mean will not change since the new student's weekly allowance is the same as the mean.

Section 11.3

11.3 Review & Refresh

1. 4.7 **2.** 9

3. 36.25 **4.** 39

5. $x \leq 18$

6. 11.00, 16.50; 1 : 5.5, 2 : 11, 6 : 16.5

7. 600, 120; 5 : 100, 30 : 600, 6 : 120

8. 236 m^2 **9.** 63.5 ft^2

10. 540.48 yd^2

11.3 Concepts, Skills, & Problem Solving

11. 5.5

13. median = 7; mode = 3

15. median = 92.5; mode = 94

17. median = 17; mode = 12

19. no; The data were not ordered from least to greatest.

21. singing

23. mean = 7.61; median = 7.42; no mode

25. mean = 4.9; median = 5.0; no mode

27. With outlier: mean = 103, median = 85, mode = 85;
Without outlier: mean = 85, median = 85, mode = 85;
mean

29. With outlier: mean = 101, median = 102, mode = 110;
Without outlier: mean = 105.625, median = 106, mode = 110;
mean

31. mean = 35.875; median = 44; mode = 48;
Sample answer: The median is best because the mean is less than most of the data and the mode is the greatest value.

33. mean = 12; median = 8; mode = 2;
Sample answer: The median is best because the mean is greater than most of the data and the mode is the least value.

35. 91°F

37. Answers will vary. The mean, median, and mode should be calculated for comparison to answer the question along with the data.

39. *Sample answer:* A mode can be an unreliable measure of a data set because there can be no mode or more than one mode.

41. **a.** mean = $8x$; median = $5x$; mode = $3x$

b. yes; $23x$

Section 11.4

11.4 Review & Refresh

1. mean = 7.1; median = 7; mode = 4

2. mean = 77; median = 78; no mode

3. mean = 16; median = 16; modes = 15, 17

4. mean = 23; median = 22; no mode

5. > **6.** <

7. > **8.** >

9. 480 mm^2 **10.** 110 $in.^2$

11. 304 ft^2

11.4 Concepts, Skills, & Problem Solving

13. *Sample answer:*

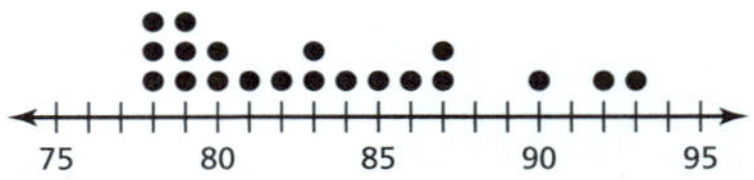

15. 33; The data vary by no more than 33 units.

17. 23; The data vary by no more than 23 units.

19. 7.3; The data vary by no more than 7.3 units.

21. 5; The middle half of the data vary by no more than 5 units.

23. 7; The middle half of the data vary by no more than 7 units.

25. 11; The middle half of the data vary by no more than 11 units.

27. The range and IQR remain the same.

29. The range and IQR are multiplied by 2.

31. **a.** range: \$32; IQR: \$19.50

b. The range and IQR are decreased by 5%.

33. no outliers **35.** 55 and 125

37. The range and IQR are halved.

39. *Sample answer:* An outlier increases the range of a data set because there is a wider spread between the greatest and least values.

41. Your collection: mean = 1929, median = 1930, no mode, range = 54, IQR = 48;
Your friend's collection: mean = 1929, median = 1929.5, no mode, range = 15, IQR = 6
Sample answer: The measures of center for the data sets are almost identical. But the measures of variation for your friend's coin collection are much less than the measures for your coin collection. This means that the years of the coins in your friend's collection are closer together than the years of the coins in your collection.

43. **a.** Show A: mean = 20, median = 19.5, range = 13, IQR = 5;
Show B: mean = 21, median = 20.5, range = 23, IQR = 6;

The mean and the median ages for the shows, and the interquartile ranges of ages for the shows, are about the same. The range of the ages for Show A is less than the range for Show B, so the ages for Show B are more spread out.

b. Show A: median, range; Show B: median, range

c. Show A: The measures do not change by a large amount because 21 is towards the middle of the data set.
Show B: The mean, median, and IQR all change by small amounts but the range changes by a large amount because 36 is an outlier of the data set.

Section 11.5

11.5 Review & Refresh

1. 6.5; the middle half of the data vary by no more than 6.5 units

2. 19.5; the middle half of the data vary by no more than 19.5 units

3. 45 **4.** 27 **5.** 22.4 **6.** 51.2

7. 84

11.5 Concepts, Skills, & Problem Solving

9. *Sample answer:*

Interval	Tally	Total
20–29	I	1
30–39	III	3
40–49	~~IIII~~ II	7
50–59	IIII	4

11.

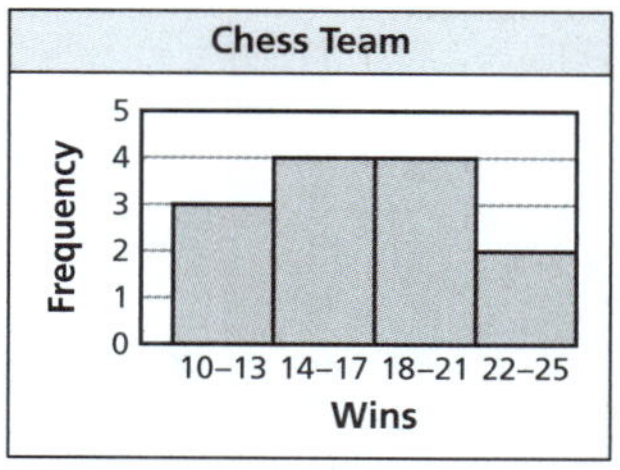

no gaps; no clusters; no outliers

13.

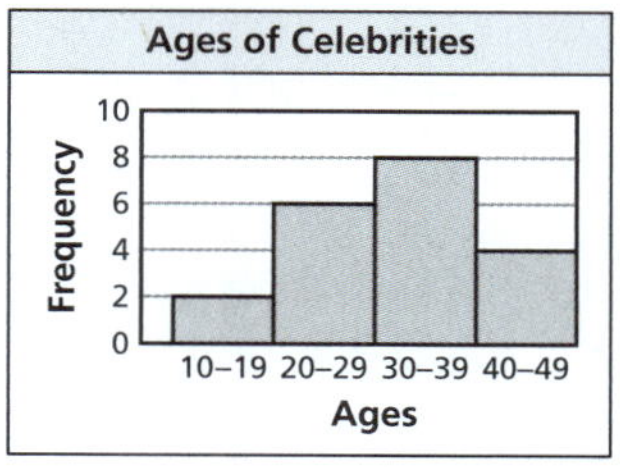

no gaps; cluster from 20–39; no outliers

15.

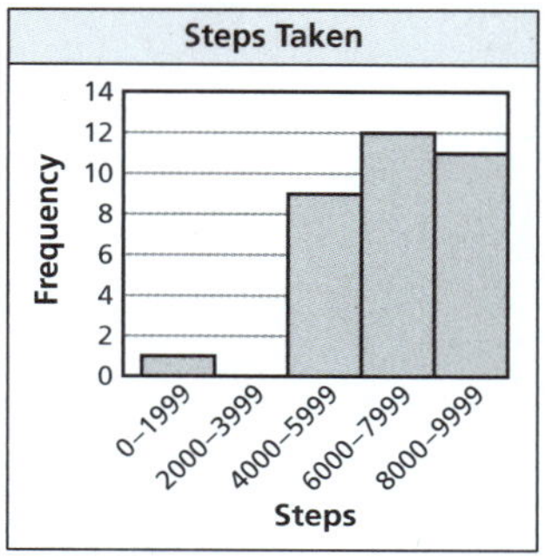

gap from 2000–3999; cluster from 4000–9999; outlier from 0–1999

17. **a.** 4–5 magazines read

b. 20 students

c. 85%

19. **a.** no; The histogram shows that only one state fell in the interval of 40%–44.9%. This state did not necessarily have 40% of possible voters vote.

b. yes; 36 states are between 50% and 64.9%.

c. no; The 55%–59.9% interval has the highest frequency, but does not necessarily contain the mode of the data.

21. stem-and-leaf plot; You need to know the specific data values, the intervals in the histogram do not give enough information.

23. **a.**

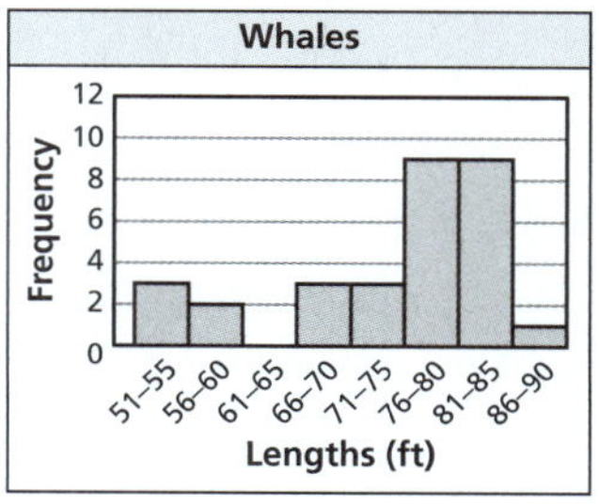

b.

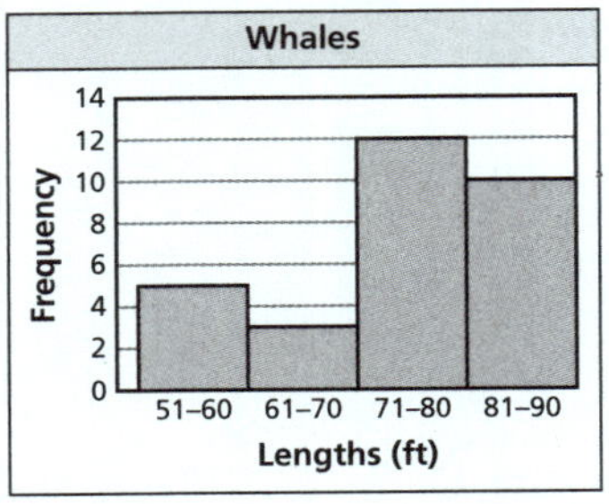

c. *Sample answer:* The second histogram has four intervals and it does not have a gap as in the first histogram.

25. no; You only know what interval each of the data values falls into, not the specific values.

Section 11.6

11.6 Review & Refresh

1.

2.

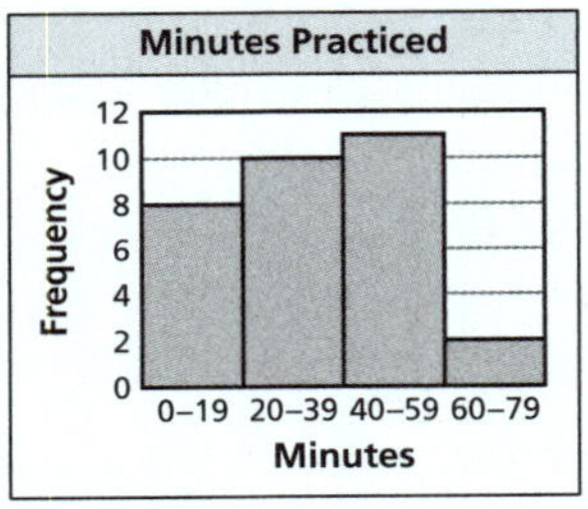

3.

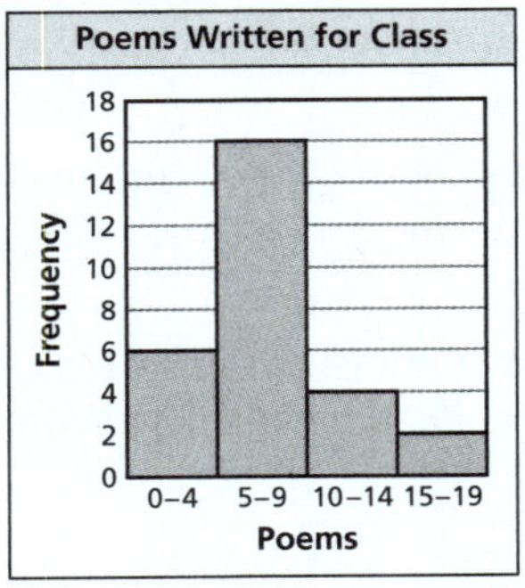

4. $25 per day

5. 40 km/h

11.6 Concepts, Skills, & Problem Solving

7. flat

9. symmetric

11. skewed left

13. symmetric; The data on the left are a mirror image of the data on the right.

15. Check students' work.

17. a. skewed right; A: Mode, B: Median, C: Mean; *Sample answer:* A is the value with the highest frequency, about half of the data are to the left of B and about half are to the right of B

b. skewed left; A: Mean, B: Median, C: Mode; *Sample answer:* C is the value with the highest frequency, about half of the data are to the left of B and about half are to the right of B

Section 11.7

11.7 Review & Refresh

1. skewed left

2. skewed right

3. >

4. <

5. >

6. $14(3 + 1)$

7. $6(2x - 3)$

8. $4(7n + 5)$

9. $5(12g - 5h)$

11.7 Concepts, Skills, & Problem Solving

11. *Sample answer:* The top plot has a distribution that is skewed left while the bottom plot has a distribution that is skewed right. The top plot has less variability in its scores within its quartiles compared to the bottom plot. The bottom plot has a lower median and quartile values in comparison to the top plot.

13.

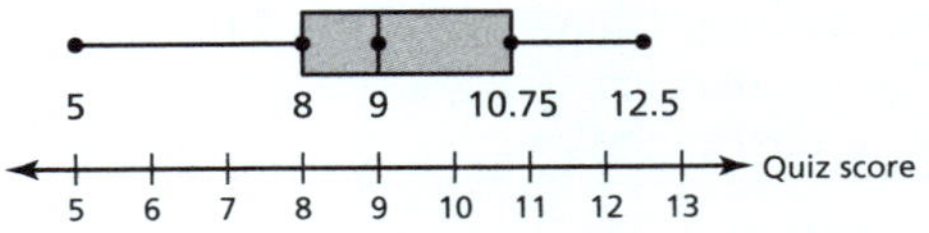

15.

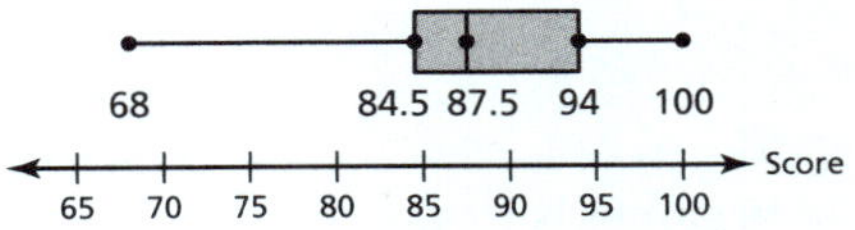

17.

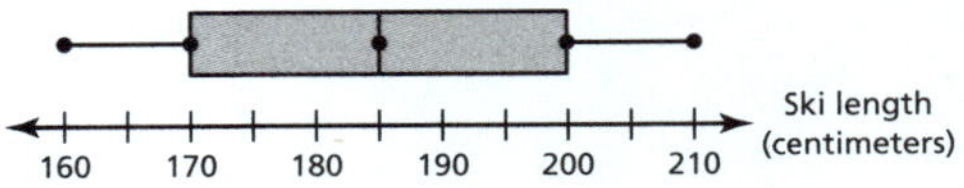

19.

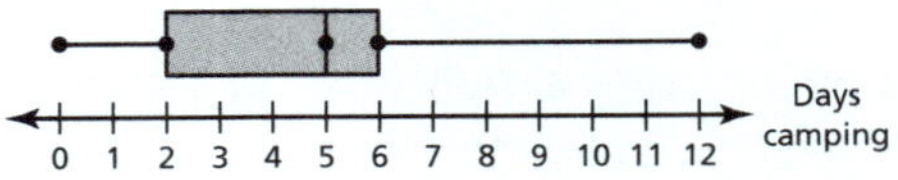

12 days camping

21. least value: 12; Q_1: 14; median: 16; Q_3: 24; greatest value: 30

23. least value: 54; Q_1: 56; median: 58; Q_3: 60; greatest value: 60

25. a.

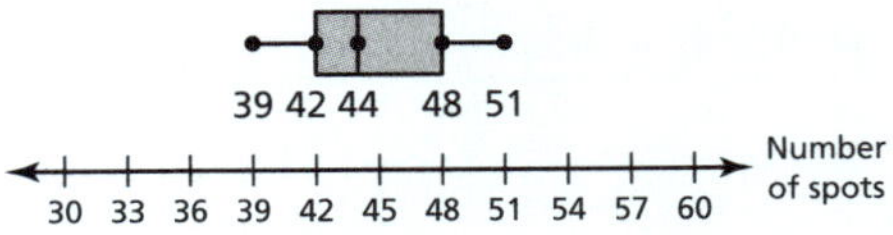

b. *Sample answer:* dot plot: advantages: gives the overall shape of the distribution; disadvantages: a lot of work to plot a very large set of numbers; box-and-whisker plot: advantages: handles large data sets easily and details of the distribution can be found; disadvantages: exact values cannot be obtained.

27. skewed left; The left whisker is longer than the right whisker, and most of the data are on the right.

29. skewed right; The right whisker is longer than the left whisker, and most of the data are on the left.

31.

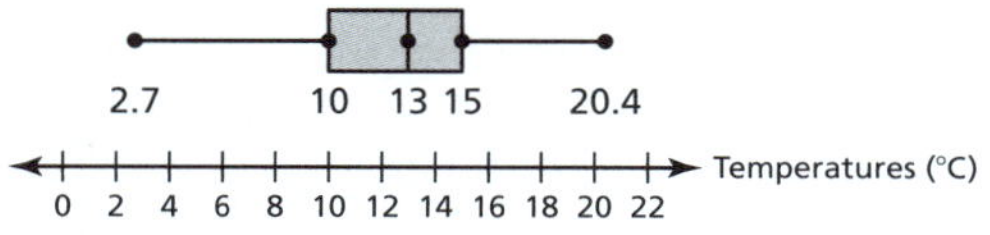

33. The left whisker is shorter, the length of the box is slightly greater, the least value increased, the median increased, the first quartile increased, and the third quartile increased. In this case, removing the outlier had no effect on the greatest value.

35. When the least value and the first quartile are equal, there is no whisker on the left. When the greatest value and the third quartile are equal, there is no whisker on the right.

37. **a.** Team 1; The range and IQR are both smaller for Team 1 than for Team 2.

b. Team 1; In 75% of the games, Team 1 scored 10 goals or more. However, Team 2 scored 10 goals or less in 75% of the games.

c. yes; Team 2 played 16 games and they scored 10 or fewer goals in 75% of those games. So, 75% of 16 games is 12 games.

Section 11.8

11.8 Review & Refresh

1.

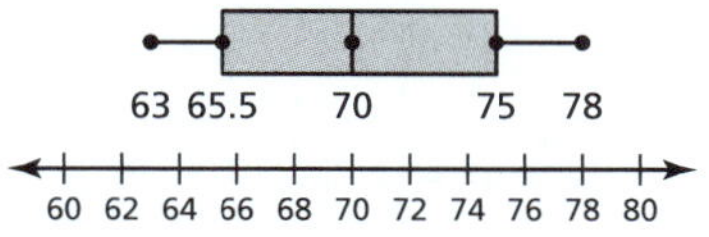

2.

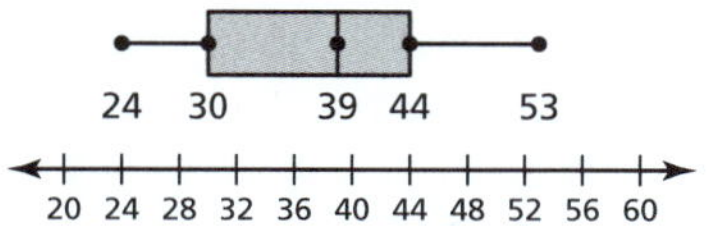

3. 28 qt

4. 1,532,000 m

5.

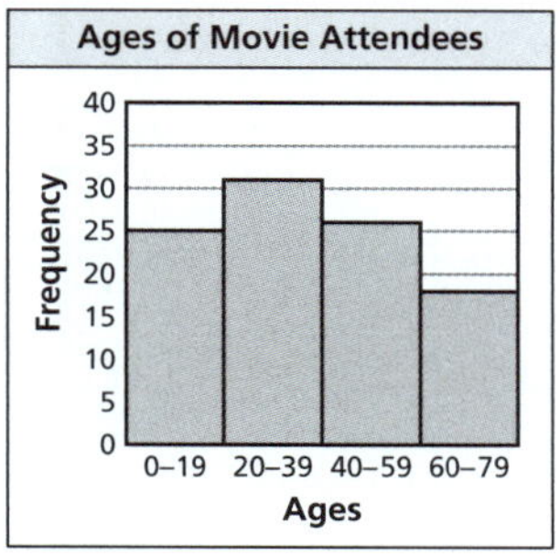

6.

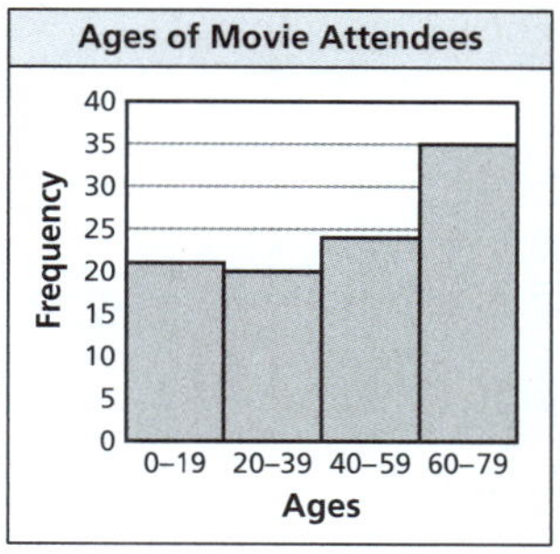

11.8 Concepts, Skills, & Problem Solving

7. mean = 7; median = 7.5; range = 7; IQR = 2.5; median and IQR; There is an outlier.

9. mean or median, range or IQR

11. mean or median, range or IQR

13. **a.** The time spent volunteering centered around approximately 5.78 hours and the middle half of the data varies by no more than 2 hours.

b. The test scores centered around 82 and the middle half of the data varies by no more than 18 points.

15. *Sample answer:* The distribution of household income in the United States. The data for this distribution are likely to contain outliers, indicating that median and IQR will be the most appropriate.

17. *Sample answer:* Pile A; Pile A has less variability than Pile B.

19. no; The greatest possible value for the IQR is $9 - 1 = 8$. So, any value greater than $9 + 1.5(8) = 21$ is an outlier. So, there are no outliers.

Chapter 11 Review

1. no; There is only one answer for the question.

2. yes; The answers will vary.

3. **a.** 20; There are 20 families who live in the city block.

b. *Sample answer:* How many televisions does a family on the city block have?; Most families on the block have between 1 and 4 televisions.

4.

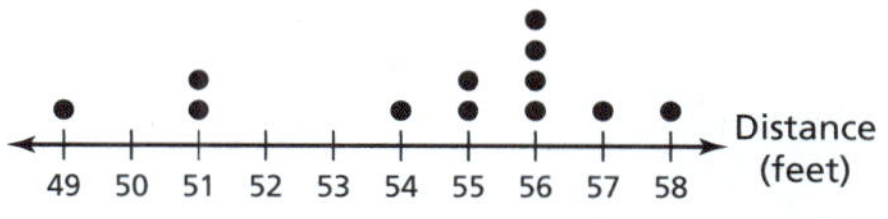

Most of the data are clustered around 56. There is a peak at 56. There are gaps between 49 and 51 and between 51 and 54.

5.

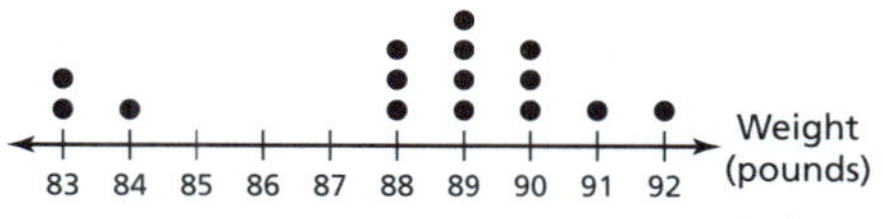

Most of the data are clustered around 89. There is a peak at 89. There is a gap between 84 and 88.

6. *Sample answer:* 68 beats per minute

7. 1126°F

8. Company A averaged more profit than Company B.

9. The outlier lowers the mean by 4 points; *Sample answer:* The typical score for a student in the class is an 80.

10. median = 6; mode = 8

11. median = 46; no mode

12. median = 7.27; no mode

13. median = $\frac{5}{7}$; mode = $\frac{3}{4}$

14. With outlier: mean = 63, median = 57, mode = 50;
Without outlier: mean = 57, median = 54.5, mode = 50; mean

15. median; The mode is less than most of the data, the mean is greater than most of the data.

16. *Sample answer:* the eye colors of the students in the classroom; The median can only be calculated for numerical data sets.

17. 77

18. 87

19. 82

20. 78

21. 36.5

22. 66.5

23. range = 12.5 kg; IQR = 7 kg; The weights vary by no more than 12.5 kilograms and the middle half of the weights vary by no more than 7 kilograms; The data set has no outliers.

24. no; *Sample answer:* Set 1: 1, 2, 3, 4, 5, 6;
Set 2: 1, 2, 3, 4, 5, 8

25. The range and IQR remain the same.

26. The range and IQR remain the same.

27. The range and IQR are multiplied by 4.

28. The range and IQR are multiplied by $\frac{1}{8}$.

29. no outliers

30. no outliers

31. no outliers

32. 12.34

33. *Sample answer:* {−2, 0, 12, 12, 12, 14}

34. *Sample answer:* {12, 13, 14, 14, 15, 22, 37, 46, 52}

35. 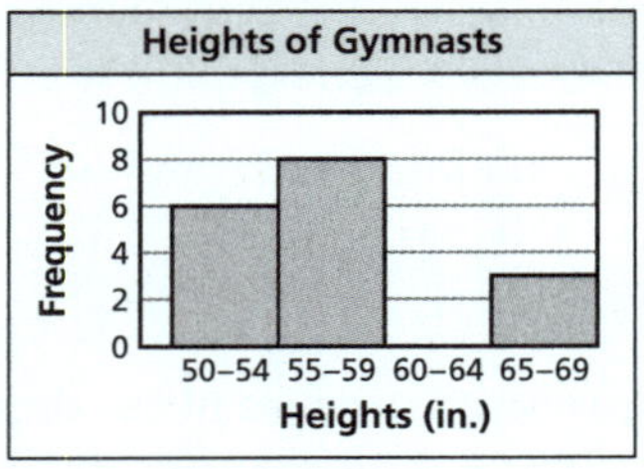

gap from 60–64; cluster from 50–59; no outliers

36. 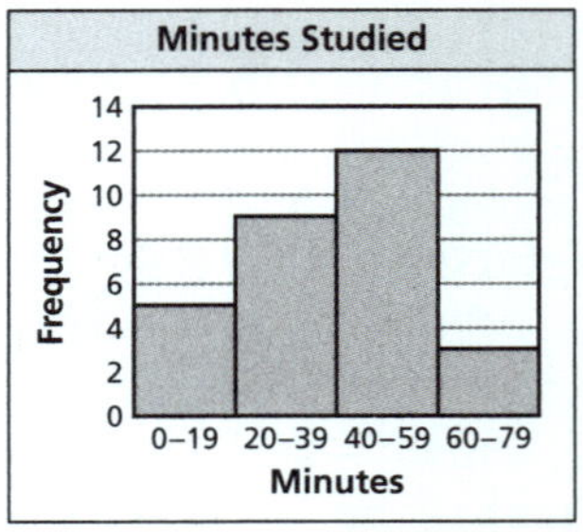

no gaps; cluster from 20–59; no outliers

37. a. 0–2 crafts

b. 7

c. no; *Sample answer:* In a histogram, you can only tell the intervals that the data values are contained in, not the exact values.

38. skewed left

39. 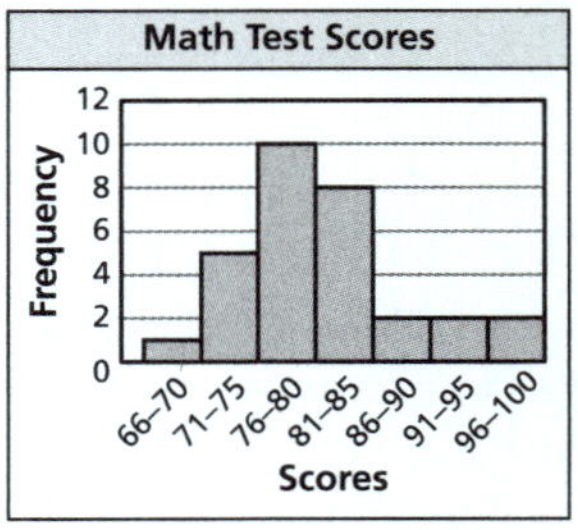

biology

40. 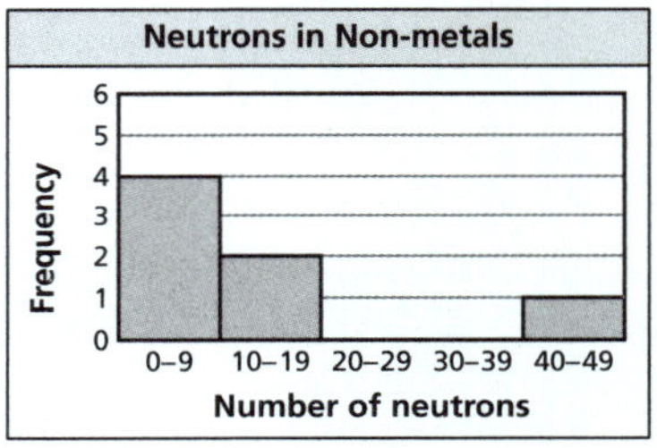

skewed right

41. 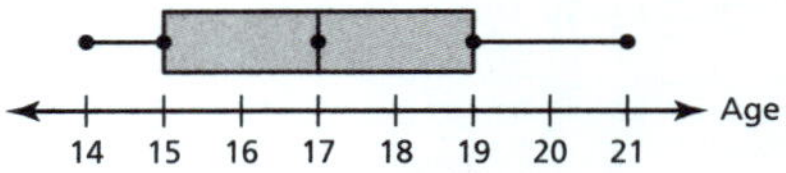

42. 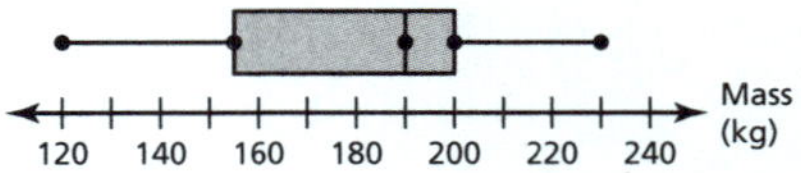

43. least value: 9; Q_1: 12; median: 15; Q_3: 18; greatest value: 23

44. least value: 60; Q_1: 60; median: 90; Q_3: 120; greatest value: 130

45. a. about 50%

b. longer than 130 minutes; the right whisker is longer

c. 20 minutes; The middle half of the data vary by no more than 20 minutes.

46. mean or median, range or IQR

47. median, IQR

48. median = 40, IQR = 5

Chapter 11 Practice Test

1. mean = 9; median = 7; mode = 6; range = 20; IQR = 4

2. mean ≈ 39.7; median = 36; no mode; range = 63; IQR = 35

3. mean or median, range or IQR

4. mean or median, range or IQR

5. centimeters: mean = 49 cm, median = 51 cm, mode = 52 cm; millimeters: mean = 490 mm, median = 510 mm, mode = 520 mm;
Sample answer: The mean, median, and mode are all multiplied by 10.

6. a.

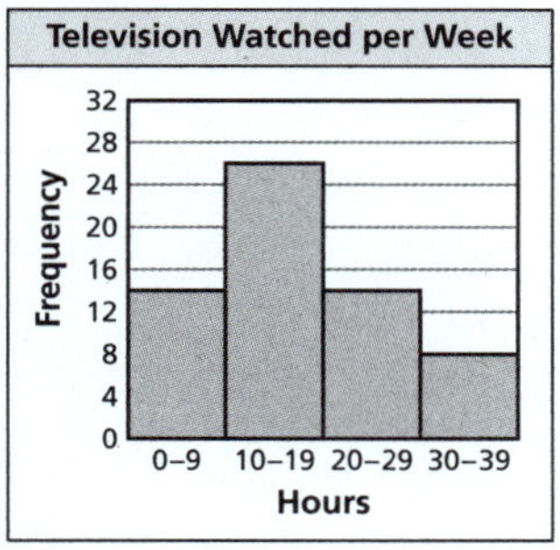

no gaps; cluster from 10–19; no outliers

b. skewed right

c. 40

7.

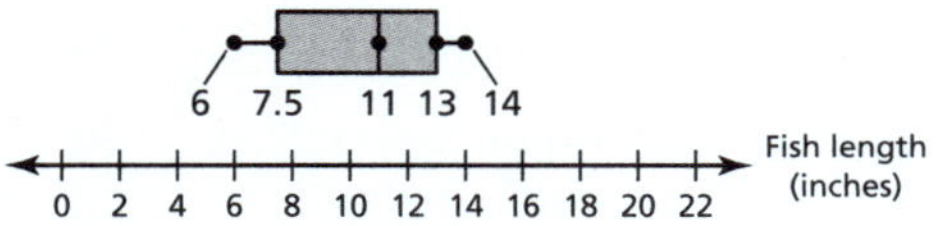

8. a. Brand A: 3.5 h; Brand B: 2.7 h

b. Brand A; Brand A has a higher mean, greatest value, and first and third quartile.

c. 760

Chapter 11 Review & Refresh

1. (3, −5)

2. (0, 4)

3. (−5, −3)

4. (−2, 6)

5. (4, 1)

6. (−3, 0)

7. $11(2 + 9) = 11(2) + 11(9)$ Distributive Property
$= 22 + 99$ Multiply.
$= 121$ Add.

8. $5(-12 + 7) = 5(-12) + 5(7)$ Distributive Property
$= -60 + 35$ Multiply.
$= -25$ Add.

9. $(15 - 5) \times 4 = 15(4) - 5(4)$ Distributive Property
$= 60 - 20$ Multiply.
$= 40$ Subtract.

10. >

11. <

12. >

13. 5; $1\frac{1}{2} + 3\frac{1}{2} = 5$

14. 18 mm^2

15. 165 ft^2

16. 64

17. 12

18. 21

19. yes; You spent only 43 minutes on the three activities.

20. no; 75 inches is greater than 6 feet, or 72 inches.

21. 8

22. 7

23. 16

Chapter 11 B.E.S.T. Test Prep

1. D

2. 14 in.

3. A, B, C, D

4. 14

5. B

6. 709.4932

7. A

8. B

Chapter 12

Chapter Exploration

1. Answers will vary. Students should play Rock Paper Scissors 30 times and record their results in the table.

2. 9 results

3. 3 ways; 3 ways; 3 ways

4. no; Both players have 3 out of 9 possible ways of winning.

Section 12.1

12.1 Review & Refresh

1. 4; The middle half of the data vary by no more than 4.

2. 5; The middle half of the data vary by no more than 5.

3. 4; The middle half of the data vary by no more than 4.

4. 0.6; The middle half of the data vary by no more than 0.6.

5. $2 : \frac{1}{2}$; The amount of rolled oats in the recipe is 4 times the amount of chopped peanuts.

6. $\frac{1}{3} : \frac{1}{4}$; The amount of sunflower seeds in the recipe is $\frac{4}{3}$ the amount of pumpkin seeds.

7. $\frac{1}{4} : 2$; The amount of pumpkin seeds in the recipe is $\frac{1}{8}$ the amount of rolled oats.

8. $x < 4$

9. $b \geq -5$

Selected Answers

10. $w > -3$

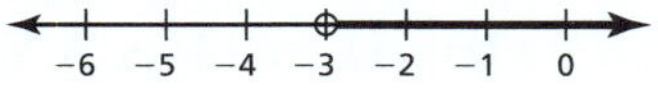

11. $g \leq -3$

12.1 Concepts, Skills, & Problem Solving

13. equal chance; *Sample answer:* All numbers have the same area, so you are equally likely to spin each number.

15. 1, 2, 3

17. 4 ways

19. **a.** 1 way
b. green

21. **a.** 1 way
b. yellow

23. **a.** 7 ways
b. red, red, red, purple, purple, green, yellow

25. false; red

27. true

29. likely

31. certain

33. **a.** equally likely to be chosen or not chosen; unlikely to be chosen
b. no; *Sample answer:* Being chosen to explain the math problem and being chosen to explain the science problem are two separate events.

35. **a.** likely
b. unlikely
c. no; *Sample answer:* A fair coin would result in an equal number of heads and tails for the relative frequency.

37. **a.** equally likely to happen or not happen; Rolling an odd number has a probability of $\frac{1}{2}$, or 50%, which is equally likely to happen or not happen.
b. unlikely; Rolling a multiple of 4 has a probability of $\frac{1}{4}$, or 25%, which is unlikely.
c. likely; Rolling a number less than 10 has a probability of $\frac{3}{4}$, or 75%, which is likely.

Rolling a number less than 10 is the most likely and rolling a multiple of 4 is the least likely.

39. With all five cards available, the number of possible outcomes is 5. With only four cards left, the number of possible outcomes is reduced to 4.

41. $\frac{3}{4}$, or 75%

Section 12.2

12.2 Review & Refresh

1. unlikely

2. likely

3. 4%

4. 3.5%

5.

Dogs	3	**6**	9
Cats	1	2	**3**

3 : 1; 6 : 2; 9 : 3

6.

Water	2	8	6
Flour	3	**12**	**9**

2 : 3; 8 : 12; 6 : 9

12.2 Concepts, Skills, & Problem Solving

7. $\frac{7}{50}$, or 14%

9. $\frac{7}{25}$, or 28%

11. $\frac{17}{50}$, or 34%

13. no; Your friend found the theoretical probability.

15. $\frac{1}{4}$, or 25%

17. $\frac{1}{2}$, or 50%

19. $\frac{5}{8}$, or 62.5%

21. $\frac{25}{26}$, or about 96.2%

23. theoretical: $\frac{1}{5}$, or 20%; experimental: $\frac{3}{10}$, or 30%. The experimental probability is close to the theoretical probability.

25. theoretical: $\frac{3}{5}$, or 60%; experimental: $\frac{7}{10}$, or 70%. The experimental probability is close to the theoretical probability.

27. theoretical; *Sample answer:* Using experimental probability would be very time consuming to spin the spinner 10,000 times.

29. $>$

31. $=$

33. **a.** $\frac{3}{8}$, or 37.5%
b. 5 males

35. *Sample answer:* games of chance; You can use theoretical probability to predict the chances of certain cards drawn or numbers picked; batting average in baseball; The batter's historical batting average is used to predict the likelihood of a hit.

37. **a.** Answers will vary. As the length of the simulations increase, the experimental probabilities will get closer to the same percentage.
b. Answers will vary. The theoretical probability will be close to the experimental probabilities, especially those with longer simulations.

Section 12.3

12.3 Review & Refresh

1. mean or median, range or IQR
2. mean or median, range or IQR
3. $x < 14$
4. $q \geq -15$
5. $6 < r$

12.3 Concepts, Skills, & Problem Solving

7. no; A sample of only artists is not likely to be representative of the whole town.
9. Population: All cards in a deck
   Sample: 4 cards
11. biased; The sample is not representative of the population because people who go to a park are more likely to think that the park needs to be remodeled.
13. Sample A; It is representative of the population.
15. a population; There are few enough students in your homeroom to not make the surveying difficult.
17. *Sample answer:* The person could ask, "Do you agree with the town's unfair ban on skateboarding on public property?"
19. yes; The sample is representative of the population, selected at random, and large enough to provide accurate data; 330 votes

Section 12.4

12.4 Review & Refresh

1. valid
2. invalid
3. $b = 25$
4. $d = -3$
5. $t = 12$

12.4 Concepts, Skills, & Problem Solving

7. non-teachers spend more time on recreation each week than teachers; *Sample answer:* 75% of non-teachers spend at least 6 hours on recreation, 50% of teachers spend less than 5 hours on recreation.
9. Data set A
11. yes; *Sample answer:* The variation for College A and College B are the same, but the measure of center for College B is greater than the measure of center for College A. You can conclude that College B's athletes generally spend more time running than College A's athletes.
13. yes; *Sample answer:* The variation for Lake A and Lake B are the same, but the measure of center for Lake A is greater than the measure of center for Lake B. You can conclude that Lake A generally contains larger fish than Lake B.
15. *Sample answer:* The measures of center and variation for Type B are greater than Type A. You can conclude that growths of the Type B corn seedlings are larger and vary by more than the Type A corn seedlings.
17. **a.** The mean, median, and IQR for the sports magazine, 19, 20, and 11, are close to the mean, median, and IQR for the political magazine, 18, 18.5, and 7. However, the sample size is small, and the variability is too great to conclude that the number of words per sentence is about the same.
    b. The sample means vary much less than the sample numbers of words per sentence.
    c. The number of words per sentence is generally greater in the political magazine than in the sports magazine.

Chapter 12 Review

1. {1, 2, 7, 11, 13, 15, 21}; 7 possible outcomes
2. blue, red
3. 5 ways
4. **a.** 2
   b. green 1, purple 1
5. **a.** 3
   b. orange 3, blue 3, purple 3
6. **a.** 5
   b. green 1, purple 1, orange 3, blue 3, purple 3
7. **a.** 3
   b. blue 2, orange 2, green 2
8. impossible
9. unlikely
10. likely
11. **a.** unlikely
    b. high pollen count
12. **a.** likely
    b. 10 or 11 rebounds
13. $\frac{9}{50}$, 0.18, or 18%
14. $\frac{39}{100}$, 0.39, or 39%
15. $\frac{81}{100}$, 0.81, or 81%
16. $\frac{1}{4}$, or 25%
17. $\frac{3}{8}$, or 37.5%
18. $\frac{5}{8}$, or 62.5%
19. $\frac{1}{8}$, or 12.5%
20. The theoretical probability of choosing a number greater than 3 is 50%. The experimental probability is 48%, which is close to the theoretical probability.
21. biased; The sample is not selected at random and is not representative of the population because students in the biology club like biology.

22. no; The sample is not representative of the population because people going to the baseball stadium are more likely to support building a new baseball stadium. So, the sample is biased and the conclusion is not valid.

23. a. 253

    b. 115

24. a. The IQRs are the same and Team B has a greater median.

    b. Team A

25. The median and IQR are greater for Model B; yes; *Sample answer:* The variation for Model A and Model B are close. Most of the sample medians for Model B are greater than the sample medians for Model A. So, in general, Model B has a better gas mileage than Model A.

26. yes; *Sample answer:* The variation for each age group is the same, but the measure of center for people in their thirties is greater than the measure of center for people in their twenties. So, in general, people in their thirties drive more than people in their twenties.

Chapter 12 Practice Test

1. {yellow, red, blue, yellow, red, yellow, green, yellow, blue, yellow}; 10

2. red, blue, red, green, blue

3. 2

4. unlikely

5. $\frac{1}{3}$, or about 33.3%

6. $\frac{7}{15}$, or about 46.7%

7. $\frac{37}{45}$, or about 82.2%

8. 1, or 100%

9. The theoretical probability of rolling an even number is 50%. The experimental probability is 45%, which is close to the theoretical probability.

10. biased; The sample size is too small and students standing in line are more likely to say they prefer to buy their lunches at school.

11. 180 students

12. The IQRs are the same and Plane A has a greater median; no; The sample size is not large enough to provide accurate data.

Chapter 12 Review & Refresh

1. $\frac{8}{9}$

2. $-\frac{3}{2}$ or $-1\frac{1}{2}$

3. $1\frac{11}{14}$

4. $12t - 20$

5. $-14 + 7n$

6. $-h + \frac{1}{2}$

7. mean 6.8, median 5.4, mode 5.4, range 10.5

8. mean 24.7, median 26.7, mode 30.5, range 15.9

9. mean, median, mode, and range

10. yes

11. $14.70

12. $42

13. $157.92

14. $237.25

15. $4(b - 3)$

16. $-8(1 - 3x)$

17. $-10(p - 6)$

18. $p \leq 26$;

19. 56 cm

Chapter 12 B.E.S.T. Test Prep

1. C

2. 7

3. A

4. $\frac{1}{5}$ or 20%

5. D

6. C

English-Spanish Glossary

English	Spanish
A	
absolute value *(p. 223)* The distance between a number and 0 on a number line; The absolute value of a number a is written as $\lvert a \rvert$.	**valor absoluto** *(p. 223)* La distancia entre un número y 0 en una recta numérica; El valor absoluto de un número a es escrito como $\lvert a \rvert$.
additive inverse *(p. 276)* The opposite of a number	**inverso aditivo** *(p. 276)* El opuesto de un número
algebraic expression *(p. 422)* An expression that contains numbers, operations, and one or more variables	**expresión algebraica** *(p. 422)* Una expresión que contiene números, operaciones, y uno o más variables
B	
base (of a power) *(p. 5)* The base of a power is the repeated factor.	**base (de una potencia)** *(p. 5)* La base de una potencia es el factor repetido.
biased sample *(p. 1033)* A sample that is not representative of a population	**muestra sesgada** *(p. 1033)* Una muestra que no es representiva de una población
box-and-whisker plot *(p. 959)* A data display that shows the variability of a data set along a number line using the least value, the greatest value, and the quartiles of the data	**diagrama de cajas y bigotes** *(p. 959)* Una presentación de datos que muestra la variabilidad de un conjunto de datos a lo largo de una línea de números usando el valor menor, el valor mayor y los cuartiles de los datos
C	
coefficient *(p. 422)* The numerical factor of a term that contains a variable	**coeficiente** *(p. 422)* El factor numérico de un término que contiene una variable
common factors *(p. 49)* Factors that are shared by two or more numbers	**factores comunes** *(p. 49)* Factores compartidos por dos o más números
common multiples *(p. 59)* Multiples that are shared by two or more numbers	**comunes múltiplos** *(p. 59)* Múltiplos compartidos por dos o más números
complex fraction *(p. 385)* A fraction that has at least one fraction in the numerator, the denominator, or both	**fracción compleja** *(p. 385)* Una fracción que tiene al menos una fracción en el numerador, el denominador, o ambos
composite figure *(p. 795)* A figure made up of triangles, squares, rectangles, and other two-dimensional figures	**figura compuesta** *(p. 795)* Una figura hecha de triángulos, cuadros, rectángulos, y otras figuras bidimensionales
constant *(p. 422)* A term without a variable	**constante** *(p. 422)* Un término que no tiene una variable

conversion factor *(p. 650)* A rate that equals 1

factor de conversión *(p. 650)* Una tasa que es igual a 1

coordinate plane *(p. 233)* A plane formed by the intersection of a horizontal number line and a vertical number line

plano de coordenadas *(p. 233)* Un plano formado por la intersección de una recta numérica horizontal y una recta numérica vertical

D

discount *(p. 741)* A decrease in the original price of an item

descuento *(p. 741)* Una disminución en el precio original de un artículo

E

edge *(p. 827)* A line segment or curve where two faces of a solid intersect

arista *(p. 827)* Un segmento de recta o curva donde se intersecan dos caras de un sólido

equation *(p. 494)* A mathematical sentence that uses an equal sign, =, to show that two expressions are equal

ecuación *(p. 494)* Una oración matemática que usa un signo de igualdad, =, para mostrar que dos expresiones son iguales

equivalent expressions *(p. 25)* Expressions with the same value

expresiones equivalentes *(p. 25)* Expresiones con el mismo valor

equivalent rates *(p. 629)* Rates that have the same unit rate

tasas equivalentes *(p. 629)* Tasas que tienen la misma tasa de unidad

equivalent ratios *(p. 594)* Two ratios that describe the same relationship

razones equivalentes *(p. 594)* Dos razones que describen la misma relación

evaluate *(p. 15)* To use the order of operations to find the value of a numerical expression

evaluar *(p. 15)* Usar el orden de operaciones para hallar el valor de una expresión numérica

event *(p. 1005)* A collection of one or more outcomes

evento *(p. 1005)* Un grupo de una o más resultadas

experiment *(p. 1005)* An investigation or a procedure that has varying results

experimento *(p. 1005)* Una investigación o un método que tiene resultados variados

experimental probability *(p. 1019)* A probability based on repeated trials of an experiment

probabilidad experimenta *(p. 1019)* Una probabilidad basada en ensayos repetidos de un experimento

exponent *(p. 5)* The exponent of a power indicates the number of times the base is used as a factor.

exponente *(p. 5)* El exponente de una potencia indique cuantas veces el base sirve como un factor.

F

face *(p. 827)* A flat or curved surface of a solid

cara *(p. 827)* Una superficie plana o curva de un sólido

factor pair *(p. 39)* A set of two nonzero factors whose product results in a definite number

par de factor *(p. 39)* Un conjunto de dos factores distintos de cero cuyo producto resulta en un número definido

factor tree *(p. 40)* A diagram used to find the factors of a composite number

árbol de factores *(p. 40)* Un diagrama utilizado para hallar los factores de un número compuesto

factoring an expression *(p. 468)* Writing a numerical expression or algebraic expression as a product of factors

factorizando una expresión *(p. 468)* Escribiendo una expresión numérica o algebraica como un producto de factores

favorable outcomes *(p. 1005)* The outcomes of a specific event

resultados favorables *(p. 1005)* Los resultados de un evento especifico

first quartile *(p. 921)* The median of the lower half of a data set; represented by Q_1

primer cuartil *(p. 921)* La mediana de la mitad inferior de un conjunto de datos; representado por Q_1

five-number summary *(p. 959)* The five numbers that make up a box-and-whisker plot

resumen de cinco números *(p. 959)* Los cinco números que componen un diagrama de cajas y bigotes

frequency *(p. 934)* The number of data values in an interval

frecuencia *(p. 934)* El número de valores de datos en un intervalo

frequency table *(p. 934)* A chart that groups data values into intervals

table de frecuencia *(p. 934)* Un gráfico que agrupa valores de datos en intervalos

G

graph of an inequality *(p. 540)* A graph that shows all the solutions of an inequality on a number line

gráfica de una desigualdad *(p. 540)* Una gráfica que muestra todas las soluciones de una desigualdad en una recta numérica

greatest common factor (GCF) *(p. 49)* The greatest of the common factors shared by two or more numbers

máximo factor común (MFC) *(p. 49)* El mayor de los factores comunes compartido por dos o más números

H

histogram *(p. 935)* A bar graph that shows the frequency of data values in intervals of the same size

histograma *(p. 935)* Un gráfico de barras que muestra la frecuencia de valores de datos en intervalos del mismo tamaño

I

inequality *(p. 538)* A mathematical sentence that compares expressions; contains the symbols $<$, $>$, $\leq$, or $\geq$

desigualdad *(p. 538)* Una oración matemática que compara las expresiones; contiene los símbolos $<$, $>$, $\leq$, o $\geq$

integers *(p. 193)* The set of whole numbers and their opposites

números enteros *(p. 193)* El conjunto de números naturales y sus opuestos

interest *(p. 750)* Money paid or earned for the use of money

interés *(p. 750)* Dinero pagado o ganado por el uso del dinero

interquartile range *(p. 921)* A measure of variation for a data set, which is the difference of the third quartile and the first quartile

rango de intercuartiles *(p. 921)* Una medida de variación para un conjunto de datos, el cual es la diferencia del tercer cuartil y el primer cuartil

inverse operations *(p. 519)* Operations that "undo" each other, such as addition and subtraction, or multiplication and division

operaciones inversas *(p. 519)* Operaciones que deshacer unos de otros, tales como suma y resta, o multiplicación y división

L

least common multiple (LCM) *(p. 59)* The least of the common multiples shared by two or more numbers

mínimo común múltiplo (MCM) *(p. 59)* El menor de los factores comunes compartido por dos o más números

like terms *(p. 445)* Terms of an algebraic expression that have the same variables raised to the same exponents

términos semejantes *(p. 445)* Términos de una expresión algebraica que tienen las mismas variables elevadas a los mismos exponentes

linear expression *(p. 446)* An algebraic expression in which the exponent of each variable is 1

expresión lineal *(p. 446)* Una expresión algebraica en la que el exponente de cada variable es 1

M

markup *(p. 741)* The increase from what a store pays to the selling price

sobreprecio *(p. 741)* El aumento de lo que una tienda paga al precio de venta

mean *(p. 897)* The sum of the data divided by the number of data values

media *(p. 897)* La suma de los datos dividido por el número de valores de datos

measure of center *(p. 907)* A measure that describes the typical value of a data set

medida de centro *(p. 907)* Una medida que describe el valor típico de un conjunto de datos

measure of variation *(p. 921)* A measure that describes the spread, or distribution, of a data set

medida de variación *(p. 921)* Una medida que describe la extensión, o distribución, de un conjunto de datos

median *(p. 907)* For a data set with an odd number of ordered values, the median is the middle value; For a data set with an even number of ordered values, the median is the mean of the two middle values.

mediana *(p. 907)* Para un conjunto de datos con un número impar de valores ordenados, la mediana es el valor del medio; Para un conjunto de datos con un número par de valores ordenados, la mediana es la media de los dos valores del medio.

metric system *(p. 649)* A decimal system of measurement, based on powers of 10, that contains units for length, capacity, and mass

sistema métrico decimal *(p. 649)* Un sistema de medición, basado en potencias de 10, que incluye unidades para longitud, capacidad, y masa

mode *(p. 907)* The data value or values that occur most often; Data can have one mode, more than one mode, or no mode.

moda *(p. 907)* El valor o valores de datos que ocurre(n) con más frecuencia; Datos pueden tener una moda, más que una moda, o ninguna moda.

multiplicative inverses *(p. 111)* Two numbers whose product is 1

inversos multiplicativos *(p. 111)* Dos números cuyo producto es 1

N

negative numbers *(p. 193)* Numbers that are less than 0

números negativos *(p. 193)* Números que son menos de 0

net *(p. 837)* A two-dimensional representation of a solid

red *(p. 837)* Una representación bidimensional de un sólido

numerical expression *(p. 15)* An expression that contains only numbers and operations

expresión numérica *(p. 15)* Una expresión que contiene solamente números y operaciones

O

opposites *(p. 193)* Two numbers that are the same distance from 0 on a number line, but on opposite sides of 0

opuestos *(p. 193)* Dos números que están a la misma distancia de 0 en una recta numérica, pero en lados opuestos de 0

order of operations *(p. 15)* The order in which to perform operations when evaluating expressions with more than one operation

orden de operaciones *(p. 15)* El orden en el que para realizar operaciones al evaluar expresiones con más de una operación

origin *(p. 233)* The point, represented by the ordered pair (0, 0), where the horizontal and vertical number lines intersect in a coordinate plane

origen *(p. 233)* El punto, representado por el par ordenado (0, 0), donde las rectas numéricas horizontales y verticales se intersecan en un plano de coordenadas

outcomes *(p. 1005)* The possible results of an experiment

resultados *(p. 1005)* Los resultados posibles de un experimento

outlier *(p. 899)* A data value that is much greater than or much less than the other values in a data set.

valor atípico *(p. 899)* Una valor de datos que es mucho mayor o mucho menor que los otros valores en un conjunto de datos

P

percent *(p. 680)* The value of a part-to-whole ratio where the whole is 100

porcentaje *(p. 680)* El valor de una razón de número de partes por un entero en donde el entero es 100

percent of change *(p. 731)* The percent that a quantity changes from the original amount;

$$\text{percent of change} = \frac{\text{amount of change}}{\text{original amount}}$$

porcentaje de cambio *(p. 731)* El porcentaje que cambia una cantidad de la cantidad original

$$\text{porcentaje de cambio} = \frac{\text{cambio}}{\text{cantidad original}}$$

percent of decrease *(p. 731)* The percent of change when the original amount decreases;

$$\text{percent of decrease} = \frac{\text{original amount} - \text{new amount}}{\text{original amount}}$$

porcentaje de disminución *(p. 731)* El porcentaje de cambio cuando la cantidad original disminuye;

$$\text{porcentaje de disminución} = \frac{\text{cantidad original} - \text{nueva cantidad}}{\text{cantidad original}}$$

percent error *(p. 733)* The percent that an estimated amount differs from the actual amount;

$$\text{percent error} = \frac{\text{amount of error}}{\text{actual amount}}$$

error porcentual *(p. 733)* El porcentaje en que una cantidad estimada difiere de la cantidad real;

$$\text{error porcentual} = \frac{\text{error}}{\text{cantidad correcta}}$$

percent of increase *(p. 731)* The percent of change when the original amount increases;

$$\text{percent of increase} = \frac{\text{new amount} - \text{original amount}}{\text{original amount}}$$

porcentaje de aumento *(p. 731)* El porcentaje de cambio cuando la cantidad original aumenta;

$$\text{Porcentaje de aumento} = \frac{\text{nueva cantidad} - \text{cantidad original}}{\text{cantidad original}}$$

perfect square *(p. 6)* The square of a whole number

cuadrado perfecto *(p. 6)* El cuadrado de un número natural

polygon *(p. 245)* A closed figure in a plane that is made up of three or more line segments that intersect only at their endpoints; for example, parallelograms and triangles

polígono *(p. 245)* Una figura cerrada en un plano, hecha de tres o más segmentos de líneas que intersectan solamente à sus puntos finales; por ejemplo, paralelogramos y triángulos

polyhedron *(p. 827)* A solid whose faces are all polygons

poliedro *(p. 827)* Un sólído tridimensional cuyas caras son todas polígonos

population *(p. 1031)* A population is an entire group of people or objects.

población *(p. 1031)* Una población es un grupo entero de personas u objetos.

positive numbers *(p. 193)* Numbers that are greater than 0

números positivos *(p. 193)* Números que son mayores que 0

power *(p. 5)* A product of repeated factors

potencia *(p. 5)* Un producto de factores repetidos

prime factorization *(p. 40)* Writing a composite number as a product of its prime factors

factorización prima *(p. 40)* Escribir un número compuesto como el producto de sus factores primeros

principal *(p. 750)* An amount of money borrowed or deposited

principal *(p. 750)* Una cantidad de dinero prestada o depositada

prism *(p. 827)* A polyhedron that has two parallel, identical bases; The lateral faces are parallelograms.

prisma *(p. 827)* Un poliedro que tiene dos bases idénticas y paralelas; Las caras laterales son paralelogramos.

probability *(p. 1006)* A measure of the likelihood, or chance, that an event will occur

probabilidad *(p. 1006)* Una medida de la probabilidad o posibilidad de que ocurrirá un evento

proportion *(p. 638)* An equation stating that the values of two ratios are equivalent

proporción *(p. 638)* Una ecuación que se indica que los valores de dos razones son equivalentes

pyramid *(p. 827)* A polyhedron that has one base; The lateral faces are triangles.

pirámide *(p. 827)* Un poliedro que tiene una base; Las caras laterales son triángulos.

Q

quadrants *(p. 233)* The four regions created by the intersection of the horizontal and vertical number lines in a coordinate plane

cuadrantes *(p. 233)* Las cuatro regiones creadas por la intersección de las rectas numéricas horizontales y verticales en un plano de coordenadas

quartiles *(p. 921)* Values that divide a data set into four equal parts

cuartiles *(p. 921)* Valores que dividen un conjunto de datos en cuatro partes iguales

R

range *(p. 921)* The difference of the greatest value and the least value of a data set

rango *(p. 921)* La diferencia del valor mayor y el valor menor de un conjunto de datos

rate *(p. 629)* A ratio of two quantities using different units

tasa *(p. 629)* Una razón de dos cantidades usando unidades diferentes

ratio *(p. 593)* A comparison of two quantities; The ratio of a to b can be written as $a : b$.

razón *(p. 593)* Una comparación de dos cantidades; La razón de a a b puede escribirse como $a : b$.

ratio table *(p. 617)* A table used to find and organize equivalent ratios

tabla de razones *(p. 617)* Una tabla usada para encontrar y organizar las razones equivalentes

rational number *(p. 212)* A number that can be written as $\frac{a}{b}$ where a and b are integers and $b \neq 0$

número racional *(p. 212)* Un número que puede ser escrito como $\frac{a}{b}$ donde a y b son enteros y $b \neq 0$

reciprocals *(p. 111)* Two numbers whose product is 1

recíprocos *(p. 111)* Dos números cuyo producto es 1

relative frequency *(p. 1007)* The fraction or percent of the time that an event occurs in an experiment

frecuencia relativa *(p. 1007)* La fracción o porcentaje de tiempo en que un evento ocurre en un experimento

S

sample *(p. 1031)* A part of a population

muestra *(p. 1031)* Una parte de una población

sample space *(p. 1005)* The set of all possible outcomes of an experiment

espacio de muestra *(p. 1005)* El conjunto de todos los resultados posibles de un experimento

simple interest *(p. 750)* Money paid or earned only on the principal

interés simple *(p. 750)* Dinero pagado o ganado solamente en el principal

simplest form (of an algebraic expression) *(p. 445)* An algebraic expression is in simplest form when it has no like terms and no parentheses.

mínima expresión (de una expresión algebraica) *(p. 445)* Una expresión algebraica está en su mínima expresión cuando tiene ningunos términos semejantes y ningunos paréntesis.

simplest form (of a fraction) *(p. 50)* A fraction where the numerator and denominator have no common factors other than 1

forma más simple (de una fracción) *(p. 50)* Una fracción donde el numerador y el denominador no tienen otros factores comunes más que 1

simulation *(p. 1021)* An experiment that is designed to reproduce the conditions of a situation or process

simulación *(p. 1021)* Un experimento que está diseñado para reproducir las condiciones de una situatción o proceso

solid *(p. 827)* A three-dimensional figure that encloses a space

sólido *(p. 827)* Una figura tridimensional que encierra un espacio

solution *(p. 505)* A value that makes an equation true

solución *(p. 505)* Un valor que hace una ecuación verdadera

solution of an inequality *(p. 539)* A value that makes an inequality true

solución de una desigualdad *(p. 539)* Un valor que hace una desigualdad verdadera

solution set *(p. 539)* The set of all solutions of an inequality

conjunto de solución *(p. 539)* El conjunto de todas las soluciones de una desigualdad

statistical question *(p. 887)* A question for which a variety of answers is expected; The interest is in the distribution and tendency of those answers.

pregenta estadística *(p. 887)* Una pregunta para la cual se espera una variedad de repuestas; El interés es en la distribución y la tendencia de aquellas respuestas.

statistics *(p. 887)* The science of collecting, organizing, analyzing, and interpreting data

estadísticas *(p. 887)* La ciencia de recolectar, organizar, analizar e interpretar datos

surface area *(p. 837)* The sum of the areas of all the faces of a polyhedron

área de la superficie *(p. 837)* La suma de las áreas de todas las caras de un poliedro

T

term *(p. 422)* A part of an algebraic expression; a number or variable by itself, or the product of numbers and variables

término *(p. 422)* Un component de una expresión algebraica; un número o variable por sí mismo, o el producto de números y variables

theoretical probability *(p. 1019)* The quotient of the number of favorable outcomes and the number of possible outcomes when all possible outcomes are equally likely

probabilidad teórica *(p. 1019)* El cociente del número de resultados favorables y el número de posibles resultados cuando todos los resultados posibles son igualmente probables

third quartile *(p. 921)* The median of the upper half of a data set; represented by Q_3

tercer cuartil *(p. 921)* La mediana de la mitad superior de un conjunto de datos; representado por Q_3

U

U.S. customary system *(p. 649)* A system of measurement that contains units for length, capacity, and weight

sistema estadounidense *(p. 649)* Un sistema de medición que incluye unidades para longitud, capacidad, y peso

unbiased sample *(p. 1033)* A sample that is representative of a population

muestra no sesgada *(p. 1033)* Una muestra que es representativa de una población

unit analysis *(p. 650)* A process used to decide which conversion factor will produce the appropriate units

análisis de unidades *(p. 650)* Un proceso utilizado para decidir qué factor de conversión producirá las unidades apropiadas

unit rate *(p. 629)* A rate that compares a quantity to one unit of another quantity

tasa unitaria *(p. 629)* Una tasa que compara una cantidad a una unidad de otra cantidad

V

value of a ratio *(p. 594)* The number $\frac{a}{b}$ associated with the ratio $a : b$

valor de una razón *(p. 594)* El número $\frac{a}{b}$ asociadas con la razón $a : b$

variable *(p. 422)* A symbol that represents one or more numbers

variable *(p. 422)* Un símbolo que representa a uno o más números

Venn diagram *(p. 47)* A diagram of overlapping circles used to show the relationships between two or more sets

Diagrama de Venn *(p. 47)* Un diagrama de círculos solapados usado para mostrar las relaciones entre dos o más conjuntos

vertex (of a polyhedron) *(p. 827)* A point where three or more edges of a polyhedron intersect

vértice (de un poliedro) *(p. 827)* Un punto donde se intersecan tres o más aristas de un poliedro

volume *(p. 855)* A measure of the amount of space that a three-dimensional figure occupies; Volume is measured in cubic units such as cubic feet (ft^3) or cubic meters (m^3).

volumen *(p. 855)* Una medida de la cantidad de espacio que una figura tridimensional ocupa; Volumen es medido en unidades cúbicas como pies cúbicos ($pies^3$) o metros cúbicos (m^3).

Index

A

D

Index

E

F

L

M

N

O

P

Q

R

T

U

V

Index

W

X

Y

Z

Credits

Chapter 7

490 Vitalii Nesterchuk/Shutterstock.com; **495** Ivo Petkov/Shutterstock.com; **497** Mariakray/iStock/Getty Images Plus; **498** Albert Russ/Shutterstock.com, jonnysek/iStock/Getty Images Plus; **501** *Exercise 28* JoopS/iStock/Getty Images Plus, Su Nitram/iStock/Getty Images Plus, MiguelMalo/iStock/Getty Images Plus; **502** spaxiax/iStock/Getty Images Plus; **505** dolgachov/iStock/Getty Images Plus; **509** Naviya/Shutterstock.com; **512** IlexImage/E+/getty Images; **514** Philippe Clement/iStock/Getty Images Plus; **516** Tatiana Terekhina/iStock/Getty Images Plus; **521** belinski/iStock/Getty Images Plus; **522** fivespots/Shutterstock.com; **525** hsun337i/iStock/Getty Images Plus; **526** drpnncpptak/Shutterstock.com; **531** *top* WAYHOME Studio/Shutterstock.com; *bottom* fototrav/iStock Unreleased; **532** *left* Sergiy1975/iStock/Getty Images Plus; *right* youngID/DigitalVision Vectors; **535** Harvepino/Shutterstock.com; **537** MJTH/Shutterstock.com; **539** Yobro10/iStock/Getty Images Plus; **542** NASA/Johns Hopkins University Applied Physics Laboratory; **546** Olga Lyubkin/Shutterstock.com; **547** Val Krasn/Shutterstock.com; **549** Cookie Studio/Shutterstock.com; **553** VaLiza/Shutterstock.com; **554** M.Aurelius/Shutterstock.com; **558** AnastasiaOsipova/Shutterstock.com; **559** Tuzemka/Shutterstock.com; **566** Johnny Bravoo/Shutterstock.com; **568** Floortje/iStock/Getty Images Plus; **570** jacoblund/iStock/Getty Images Plus; **572** Imo/iStock/Getty Images Plus; **573** Macrovector/Shutterstock.com; **579** Rawpixel/iStock/Getty Images Plus; **581** ntzolov/E+/Getty Images; **582** WinterStorm/iStock/Getty Images Plus; **583** Kenishirotie/iStock/Getty Images Plus; **584** thekopmylife/iStock/Getty Images Plus; **587** monkeybusinessimages/iStock/Getty Images Plus

Chapter 8

588 Rasi Bhadramani/iStock/Getty Images Plus; **591** ka2shka/iStock/Getty Images Plus; **593** *top* GlobalP/iStock/Getty Images Plus, spxChrome/E+/Getty Images; *bottom* maogg/iStock/Getty Images Plus, asafta/iStock/Getty Images Plus, TokenPhoto/E+/Getty Images, Meral Hydaverdi/Shutterstock.com; **594** Ewa Studio/Shutterstock.com; **596** Nerthuz/iStock/Getty Images Plus, bbevren/iStock/Getty Images Plus; **598** mlorenz/Shutterstock.com; **600** *Exercise 13* amwu/iStock/Getty Images Plus, GlobalP/iStock/Getty Images Plus; *Exercise 14* Lightspring/Shutterstock.com; *Exercise 15* hh5800/iStock/Getty Images Plus, borisyankov/E+/Getty Images; *Exercise 16* HONG VO/iStock/Getty Images Plus, Gonzalo Calle Asprilla/iStock/Getty Images Plus; **601** SpiffyJ/iStock/Getty Images Plus; **603** szefei/iStock/Getty Images Plus; **605** Natalia Lisovskaya/Shutterstock.com; **609** Twoellis/iStock/Getty Images Plus; **610** OSTILL/ iStock/Getty Images Plus; **611** mountainpix/Shutterstock.com; **613** *left* Inna Tarasenko/iStock/Getty Images Plus; *right* desert_fox99/iStock/Getty Images Plus; **615** 4kodiak/iStock/Getty Images Plus; **620** RuthBlack/iStock/Getty Images Plus; **624** GeorgeManga/DigitalVision Vectors/Getty Images; **626** Salparadis/Shutterstock.com; **629** Joe_Potato/iStock/Getty Images Plus; **635** Maria Spb/Shutterstock.com; **637** *left* ericb007/iStock/Getty Images Plus; *right* Rumi Fujishima/DigitalVision Vectors; **639** CK Foto/Shutterstock.com; **641** OSTILL/iStock/Getty Images Plus; **645** jirkaejc/iStock/Getty Images Plus; **646** Nataliia_Makarova/iStock/Getty Images Plus; **647** Rob Marmion/Shutterstock.com; **653** Tom Merton/OJO Images/Getty Images; **654** Nerthuz/Shutterstock.com; **656** Sean Pavone/iStock/Getty Images Plus; **658** Yobro10/iStock/Getty Images Plus; **659** bbevren/iStock/Getty Images Plus; **660** gcafotografia/Shutterstock.com; **661** Holiday.Photo.Top/Shutterstock.com; **664** *left* thomaslenne/iStock/Getty Images Plus, ONYXprj/iStock/Getty Images Plus; *right* Floortje/E+/Getty Images, phanasitti/iStock/Getty Images Plus; *bottom right* scanrail/ iStock/Getty Images Plus; **666** Big Cheese Photo/Big Cheese Photo/Getty Images Plus; **669** ValuaVitaly/iStock/Getty Images Plus, Grafner/iStock/Getty Images Plus; **670** simonkr/E+/Getty Images; **671** GeoffBlack/iStock/Getty Images Plus; **672** bbevren/iStock/Getty Images Plus; **674** *middle* ithinksky/iStock/Getty Images Plus, *left* GlobalP/iStock/Getty Images Plus, top *right* GlobalP/iStock/Getty Images Plus, *right* GlobalP/iStock/Getty Images Plus, *bottom right* GlobalP/iStock/Getty Images Plus

Chapter 9

676 Iaroslav Neliubov/Shutterstock.com; **679** *left* MicroStockHub/iStock/Getty Images Plus; *center* asafta/iStock/Getty Images Plus; *right* Andrey_KZ/iStock/Getty Images Plus; **684** Mark Herreid/Shutterstock.com; **691** Studio_G/Shutterstock.com; **697** *Exercise 45 left* Vladimiroquai/iStock/Getty Images Plus, *Exercise 45 center* jondpatton/E+/Getty Images, *Exercise 45 right* andegro4ka/iStock/Getty Images Plus; **699** VaLiza/Shutterstock.com; **702** Gino Santa Maria/Shutterstock.com; **708** GlobalP/iStock/Getty Images Plus; **709** ValuaVitaly/iStock/Getty Images Plus; **714** Courtesy Kristy Doyle; **716** President's Council on Fitness, Sports & Nutrition; **718** Wavebreakmedia/iStock/Getty Images Plus; **724** alblec/iStock/Getty Images Plus; **727** P.S_2/Shutterstock.com; **729** mocker_bat/iStock/Getty Images Plus; **731** mocker_bat/iStock/Getty Images Plus; **734** Kseniia_Designer/iStock/Getty Images Plus, Creativ/iStock/Getty Images Plus; *bottom* wragg/E+/Getty Images; **736** Tsekhmister/iStock/Getty Images Plus; **742** Roman Samokhin/iStock/Getty Images Plus; **743** studio BM/Shutterstock.com; **746** comzeal/iStock/Getty Images Plus; **748** Fototocam/iStock/Getty Images Plus; **749** LightFieldStudios/iStock/Getty Images Plus; **753** turk_stock_photographer/iStock/Getty Images Plus; **757** Harmony Video Production/Shutterstock.com; **758** Talaj/iStock/Getty Images Plus; **759** artisteer/iStock/Getty Images Plus; **763** FedBul/Shutterstock.com; **766** *top* ksena32/iStock/Getty Images Plus; **767** Photographer: C. J. Walker; **768** DeborahMaxemow/E+/Getty Images; **769** C-You/iStock/Getty Images Plus; **770** Dimitris Leonidas/Shutterstock.com; **772** posteriori/iStock/Getty Images Plus; **773** Roman Valiev/iStock/Getty Images Plus; **777** Monkey Business Images/Shutterstock.com

Chapter 10

778 local_doctor/Shutterstock.com; **781** Siphotography/iStock/Getty Images Plus; **786** bamse009/iStock/Getty Images Plus; **788** BirdHunter591/iStock/Getty Images Plus; **790** Edoma/iStock/Getty Images Plus; **791** jpfotograaf/iStock/Getty Images Plus; **801** DragonImages/iStock/Getty Images Plus; **806** fritschk/Shutterstock.com; **810** onairjiw/iStock/Getty Images Plus; **811** izzet ugutmen/Shutterstock.com; **818** jsp/iStock/Getty Images Plus; **823** Krakenimages.com/Shutterstock.com; **828** CHINTHIKUN/Shutterstock.com; **830** *left* KanKankavee/iStock/Getty Images Plus; *right* undrey/iStock/Getty Images Plus; *bottom* archetype/Shutterstock.com; **832** filipe_lopes/iStock/Getty Images Plus; **833** design56/Shutterstock.com; **834** Franck-Boston/iStock/Getty Images Plus; **835** vadimguzhva/iStock/Getty Images Plus; **839** Diane Schuster/Fermi Gamma-ray Space Telescope/NASA; **840** Iaremenko/iStock/Getty Images Plus; **845** Thomas M Perkins/Shutterstock.com; **849** Goran Bogicevic/Shutterstock.com; **850** demerzel21/iStock/Getty Images Plus; **853** muratart/Shutterstock.com; **855** carballo/Shutterstock.com; **860** alarich/iStock/Getty Images Plus; **863** ©iStockphoto.com/Jill Chen; **869** *Exercise 13* hrstkinkr/ iStock/Getty Images Plus; *Exercise 14* Magdevski/iStock/Getty Images Plus; **876** gokturk_06/Shutterstock; **878** *top* CreativeNature_nl/iStock/Getty Images Plus; *bottom* tiler84/iStock/Getty Images Plus; **881** fotosparrow/Shutterstock.com

Chapter 11

882 lem/Shutterstock.com; **883** Vitalii Bondarenko/iStock/Getty Images Plus; **884** aldomurillo/iStock/Getty Images Plus; **885** Toxitz/iStock/Getty Images Plus; **886** Andyworks/E+/Getyy Images; **887** Paleka/Shutterstock.com; **890** GlobalP/iStock/Getty Images Plus; **893** bazilfoto/iStock/Getty Images Plus; **897** *left* Cunaplus_M.Faba/iStock/Getty Images Plushands texting; ZargonDesign/iStock Unreleased/Getty Images; **898** Lisa F Young/Shutterstock.com; **899** GlobalP/iStock/Getty Images Plus; **904** Dafinchi/iStock/Getty Images Plus; **905** Mccallk69/Shutterstock.com; **908** only_fabrizio/iStock/Getty Images Plus; **912** annalisa e marina durante/Shutterstock.com; **915** IrinaK/Shutterstock.com; **920** Syda Productions/Shutterstock.com; **922** Vladimiroquai/iStock/Getty Images Plus; **929** Tassii/E+/Getty Images; **931** Joe_Potato/iStock/Getty Images Plus; **932** 4x6/iStock/Getty Images Plus; **933** pajtica/Shutterstock.com; **937** diane39/iStock/Getty Images Plus; **939** LanaStock/iStock/Getty Images Plus; **947** lucky-photographer/iStock/Getty Images Plus; **951** pagadesign/E+/Getty Images; **954** J-Elgaard/E+/Getty Images; **958** Yeamake/Shutterstock.com; **959** globalP/iStock/Getty Images Plus; **960** klebercordeiro/iStock/Getty Images Plus; **963** THEGIFT777/iStock/Getty Images Plus; **968** sdbower/iStock/Getty Images Plus; **969** Sailom/Shutterstock.com; **971** *right* pictafolio/E+/Getty Images; *left* AaronAmat/iStock/Getty Images Plus; **979** Boris Ryaposov/Shutterstock.com; **981** Andrei310/iStock/Getty Images Plus; **988** KeithSzafranski/E+/Getty Images; **997** Six Dun/iStock/Getty Images

Chapter 12

1000 dlewis33/E+/Getty Images; **1003** Africa Studio/Shutterstock.com; **1008** Michael Burrell/iStock/Getty Images Plus; **1013** Gearstd/iStock/Getty Images Plus; **1017** *left* mimagephotography/Shutterstock.com; *top right* Meral Hydaverdi/Shutterstock.com, maogg/iStock/Getty Images Plus; *bottom right* George3973/Shutterstock.com ; **1020** Andrei_F/iStock/Getty Images Plus; **1024** Khakimullin Aleksandr/Shutterstock.com; **1026** Francesco Milanese/iStock/Getty Images Plus; **1033** Natalya Timofeeva/Shutterstock.com; **1038** South Florida Science Center and Aquarium; **1041** One Stock Photo/Shutterstock.com; **1047** RiverNorthPhotography/iStock/Getty Images Plus; **1053** enter89/E+/Getty Images; **1058** charles taylor/Shutterstock.com; **1062** LUHUANFENG/iStock/Getty Images Plus; **1066** BorisVetshev/Shutterstock.com; **1068** Believe_In_Me/iStock/Getty Images Plus; **1069** Max Topchii/Shutterstock.com

Cartoon illustrations: Grant Larson Productions, LLC.

Mathematics Reference Sheet

Conversions

U.S. Customary

1 foot = 12 inches
1 yard = 3 feet
1 mile = 5280 feet
1 acre = 43,560 square feet
1 cup = 8 fluid ounces
1 pint = 2 cups
1 quart = 2 pints
1 gallon = 4 quarts
1 gallon = 231 cubic inches
1 pound = 16 ounces
1 ton = 2000 pounds
1 cubic foot ≈ 7.5 gallons

Metric

1 centimeter = 10 millimeters
1 meter = 100 centimeters
1 kilometer = 1000 meters
1 liter = 1000 milliliters
1 kiloliter = 1000 liters
1 milliliter = 1 cubic centimeter
1 liter = 1000 cubic centimeters
1 cubic millimeter = 0.001 milliliter
1 gram = 1000 milligrams
1 kilogram = 1000 grams

Time

1 minute = 60 seconds
1 hour = 60 minutes
1 hour = 3600 seconds
1 year = 52 weeks

Number Properties

Commutative Properties of Addition and Multiplication

$a + b = b + a$

$a \cdot b = b \cdot a$

Associative Properties of Addition and Multiplication

$(a + b) + c = a + (b + c)$

$(a \cdot b) \cdot c = a \cdot (b \cdot c)$

Additive Identity Property of Zero

$a + 0 = a$

Multiplicative Property of Zero

$a \cdot 0 = 0$

Multiplicative Identity Property of One

$a \cdot 1 = a$

Multiplicative Inverse Property

$n \cdot \frac{1}{n} = \frac{1}{n} \cdot n = 1, n \neq 0$

Distributive Property:

$a(b + c) = ab + ac$

$a(b - c) = ab - ac$

Properties of Equality

Addition Property of Equality

If $a = b$, then $a + c = b + c$.

Subtraction Property of Equality

If $a = b$, then $a - c = b - c$.

Multiplication Property of Equality

If $a = b$, then $a \cdot c = b \cdot c$.

Division Property of Equality

If $a = b$, then $a \div c = b \div c, c \neq 0$.

Properties of Inequality

Addition Property of Inequality

If $a > b$, then $a + c > b + c$.

Subtraction Property of Inequality

If $a > b$, then $a - c > b - c$.

Multiplication Property of Inequality

If $a > b$ and c is positive, then $a \cdot c > b \cdot c$.

If $a > b$ and c is negative, then $a \cdot c < b \cdot c$.

Division Property of Inequality

If $a > b$ and c is positive, then $a \div c > b \div c$.

If $a > b$ and c is negative, then $a \div c < b \div c$.

Perimeter and Area

Square	Rectangle	Triangle	Parallelogram	Trapezoid
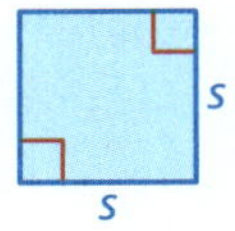	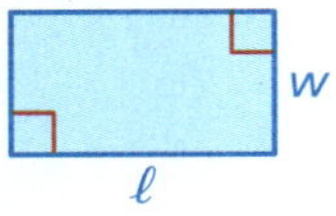	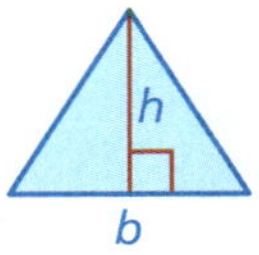	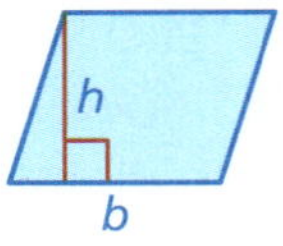	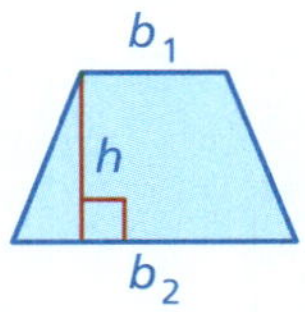
$P = 4s$ $A = s^2$	$P = 2\ell + 2w$ $A = \ell w$	$A = \frac{1}{2}bh$	$A = bh$	$A = \frac{1}{2}h(b_1 + b_2)$

Surface Area

Prism

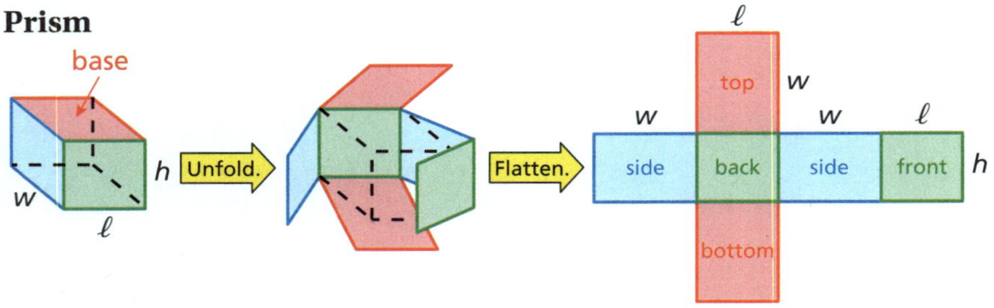

$S =$ areas of bases $+$ areas of lateral faces

Pyramid

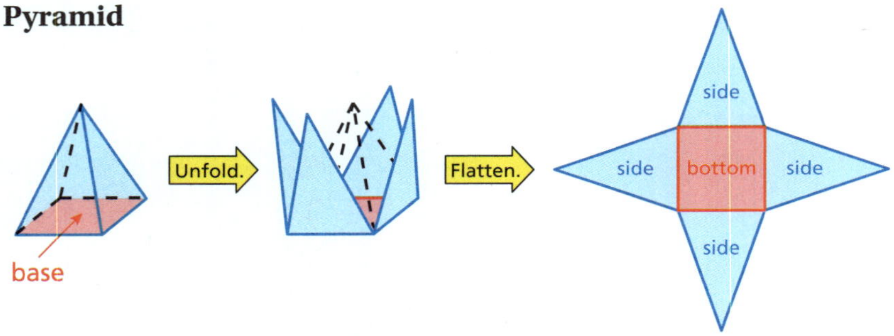

$S =$ area of base $+$ areas of lateral faces

Volume of a Rectangular Prism

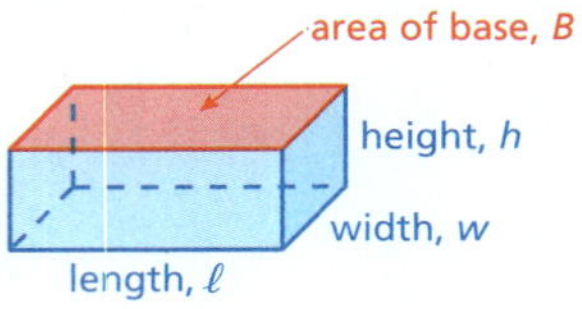

$V = Bh = \ell wh$

Simple Interest

Simple interest formula

$I = Prt$

The Coordinate Plane

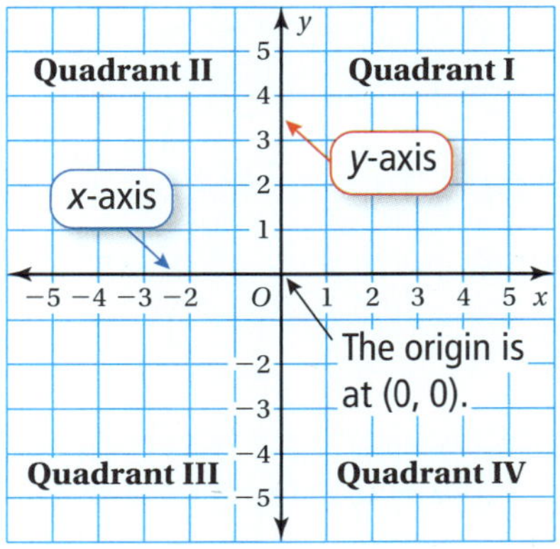